EXPLORATION OF THE DEEP CONTINENTAL CRUST

Crustal Structure of the Bohemian Massif and the West Carpathians

Edited by
V. Bucha and M. Blížkovský

In cooperation with
M. Burda, M. Krs, M. Suk
and J. Šefara

Translated by
G. Plíva

With 159 Figures and 24 Tables

Springer-Verlag
Berlin Heidelberg New York
London Paris Tokyo
Hong Kong Barcelona
Budapest

Prof. Dr. Václav Bucha
Geophysical Institute
Academy of Sciences of the Czech Republic
Boční II
141 31 Prague 4
Czech Republic

Dr. Sc. Milan Blížkovský †
Director of Geofyzika
Ječná 29 a
612 46 Brno
Czech Republic

Scientific reviewers:
RNDr. Václav Hanuš, DrSc.
Ing. Klement Rosa, CSc.
Dr. Jiří Vaněk, DrSc.

Copublished with Academia, Publishing House of the Academy of Sciences of the Czech Republic, Prague, Czech Republic

ISBN-13: 978-3-642-78997-7 e-ISBN-13: 978-3-642-78995-3
DOI: 10.1007/978-3-642-78995-3

CIP data applied for.

Softcover reprint of the hardcover 1st edition 1994

SPIN 10044056 32/3130 – 543210 – Printed on acid-free paper

Preface

The earth is seen as a dynamic body undergoing convection that is driven by its internal heat. The elucidation of the nature and evolution of the earth's lithosphere, that has provided all minerals and energy resources and the space in which we live is one of the principal motives behind the International Program "Dynamics and Evolution of the Lithosphere". A full understanding of the lithosphere requires further research especially on the continents and their margins: while the oceanic lithosphere cools and subsides as it flows away from the zone of upwelling, the continents, because of their different composition, remain buoyant in the mantle and survive over major portions of the earth's history.

In trying to produce full confirmation of the validity of sea floor spreading and lithosphere subduction, Central Europe became more and more a crucial case. The deformation of continental plates is, however, more complex than that of the ocean-floor pattern. No discrete boundary exists between both the African and Eurasian plates. A number of unanswered questions arise e.g. whether African crust has overridden Europe, or European lithosphere collided in the ranges of Atlas of Algeria. The work of German, French and Suiss geologists accomplished much during this century. Central Europe has come to serve as a test site for the refinement of plate tectonics. A new understanding of the crustal dynamics of Central Europe was born with the aim of explaining its structural evolution.

The main purpose of the established continental deep-borehole project (KTB) is to collect extensive geoscientific data on cores, cuttings, drilling fluids and gases and to help in solving the central problem of geosciences, i.e. explain the differences in the thickness of the Earth's crust in Outer-Alpine Europe as well as the causes which led to the mosaic of small tectonic blocks of Europe. The major aspect of the project is to drill through the zone of Erbendorf-Vohenstrauss nappe complex to reach the inferred Saxothuringian/Moldanubian collision zone underneath.

New findings and the knowledge of the deep structure in regions of the Bohemian Massif and the West Carpathians, as contained in this monograph, can contribute to the solution of the International Lithosphere Program and also to the interpretation of the KTB results. This book deals with seismic research, studies of geoelectric inhomogeneities and heat, gravimetric, geomagnetic and palaeomagnetic fields in Central Europe and with discussions of geological and petrological models. Dynamic interpretations of the lithosphere in Central Europe with a special emphasis on the contact zone between the Bohemian Massif and the West Carpathians are also important. That is why the results obtained by Czech and Slovak geoscientists may be useful also in the study and elucidation of processes which were involved in the evolution of the western part of the Bohemian Massif in NE Bavaria; they may provide a good opportunity for comparative research of both territories.

V. Bucha

Contents

1 Review of the Deep Structure Research in the Bohemian Massif and West Carpathians

M. Blížkovský[1], M. Suk[2], M. Burda[3]

The beginning of research into the deep structure in Czechoslovakia is associated with the work of A. Zátopek who linked solutions of the problem with international projects from the very beginning. In 1961 the Czechoslovak National Committee for the Upper Mantle Project (UMP) was formed, and headed by M. Pick. The National UMP Committee had aimed at organizing a complex research into the structure, composition, and dynamics of the Earth's crust. The results of the work were published in two volumes: UMP – Final report – Geophysics edited by M. Pick and UMP–Final report–Geology edited by V. Zoubek (Prague 1971).

The next stage of research into the lithosphere in Czechoslovakia is connected with the participation of Czechoslovak science in the international geodynamic project (IGP) solved in the period 1972 – 1979. The chairman of the Czechoslovak National Committee for Geodynamics was J. Vaněk. The following institutions took part in solving the IGP problems: institutes of the Czechoslovak and Slovak Academies of Sciences, Geological Survey, Prague (ÚÚG), Geofyzika s.e., and universities. Results were published in the Final Report "Geodynamic Investigations in Czechoslovakia" edited by J. Vaněk (Bratislava 1979). A. Zátopek was Editor-in-Chief of the Geophysical Syntheses in Czechoslovakia, the publication in which the research until 1980 is compiled.

Czechoslovak specialists contributed to the research into the lithosphere also in other projects, namely in the projects of the International Program of Geological Correlations, IUGS – UNESCO, e.g. the investigations concerning ophiolites, applications of plate tectonics, and geochemistry of the processes in the Earth's crust and upper mantle.

Since 1966 the Czechoslovak geophysicists have been members of the Committee of Academies of Socialistic Countries for Planetary Geophysical Investigations (CAPG), and since 1980 they have participated in one of its projects "The Structure of the Earth's Crust of Central and Eastern Europe". The results have contributed to the knowledge of the lithosphere structure in Central Europe (Sollogub et al. 1971, 1980; Sollogub and Čekunov 1977, 1988, 1980; Szénás 1972; Giese et al. 1976; Beránek and

[1] Geofyzika Brno, Ječná 29a, 612 46 Brno, Czech Republic
[2] The Masaryk University, College of Natural Sciences, Kotlářská 2, 600 00 Brno, Czech Republic
[3] Geophysical Institute, Boční II, 141 31 Praha 4, Czech Republic

Zounková 1977). Czechoslovak geophysicists have also participated in taking measurements and interpreting the seismic data obtained along the international geotraverse Fennolora (Beránek et al. 1981; Firbas, Ibrmajer, Suk et al. 1989) as well as numerous national refraction and reflection seismic profiles. Since 1983, Czechoslovak specialists have cooperated in the interunion project the "Lithosphere". Results of the work on this project were presented at the International Geological Congress in Moscow, 1984, and in Washington, 1989, at the IUGG Congresses in Hamburg, 1983 and in Vancouver, 1987, at the symposium on Earth's crust modelling in Switzerland in 1985, at the symposiums on Lithosphere programme in Yaroslav (USSR) in 1988, and in Giessen (FRG) in 1989.

The Geophysical Institute of the Czechoslovak Academy of Sciences organized several international meetings from which selected contributions were published. In 1982 a symposium on "Terrestrial Heat Flow Studies and Structure of the Lithosphere" took place, in 1986 the workshop on "Anisotropy and Inhomogeneity of Lithosphere and Asthenosphere", in 1986 a symposium on "Physics of Fracturing and Seismic Energy Release", in 1987 a conference on the "Terrestrial Heat Flow and the Lithosphere Structure", and in 1988 and 1990 "New Trends in Geomagnetism" were held. Symposia on "Seismic Waves in Laterally Inhomogeneous Media" took place in 1978, 1983, 1988 and on "Analyses of Seismicity and Seismic Hazard" held in 1977, 1981, 1985, and 1989 earned an international recognition.

Since 1969, Geofyzika Brno has regularly been organizing seminars devoted to research into the deep geological structure of Czechoslovakia. The Proceedings from these seminars, edited by B. Beránek and, since 1983, by M. Blížkovský represent a comprehensive survey of geophysical results of the research into the structure of the earth's crust in Central Europe.

During 1986 – 89 the project "Endogeneous Processes in Relation to Metallogenesis" was co-ordinated by the Faculty of Science, Charles University, Prague.

Conferences organized in 1976, 1980, and 1982 at Smolenice played a significant role in this research. The following Proceedings were edited by M. Mahel': Czechoslovak Geology and Global Tectonics (Veda, Bratislava 1979) and Main problems of the Geological Evolution and Structure of Czechoslovakia (Bratislava 1980), and the Earth's Crust and its Relation to Mineral Raw Materials (Bratislava 1984).

The establishment of the continental deep-borehole (KTB – Kontinentale Tiefbohrungsprogramm, Emmermann and Giese 1987, Emmermann and Rischmüller 1990) in Germany was stimulated by the demands of investigations into the deep structure of the Bohemian Massif. Since 1980 the authors of this monograph have carried out a lot of new geophysical measurements and modellings of the upper mantle structure. The monograph was directly inspired by needs of the KTB (K. Fuchs, R. Emmermann, H. J. Behr) and contains a lot of geophysical and geological data as well as their complex interpretations, especially in regions with plutonites (granulites, eclogites and granatic peridotites) in Czechoslovakia. The main geological units and their terminology are illustrated in Fig. 1.1.

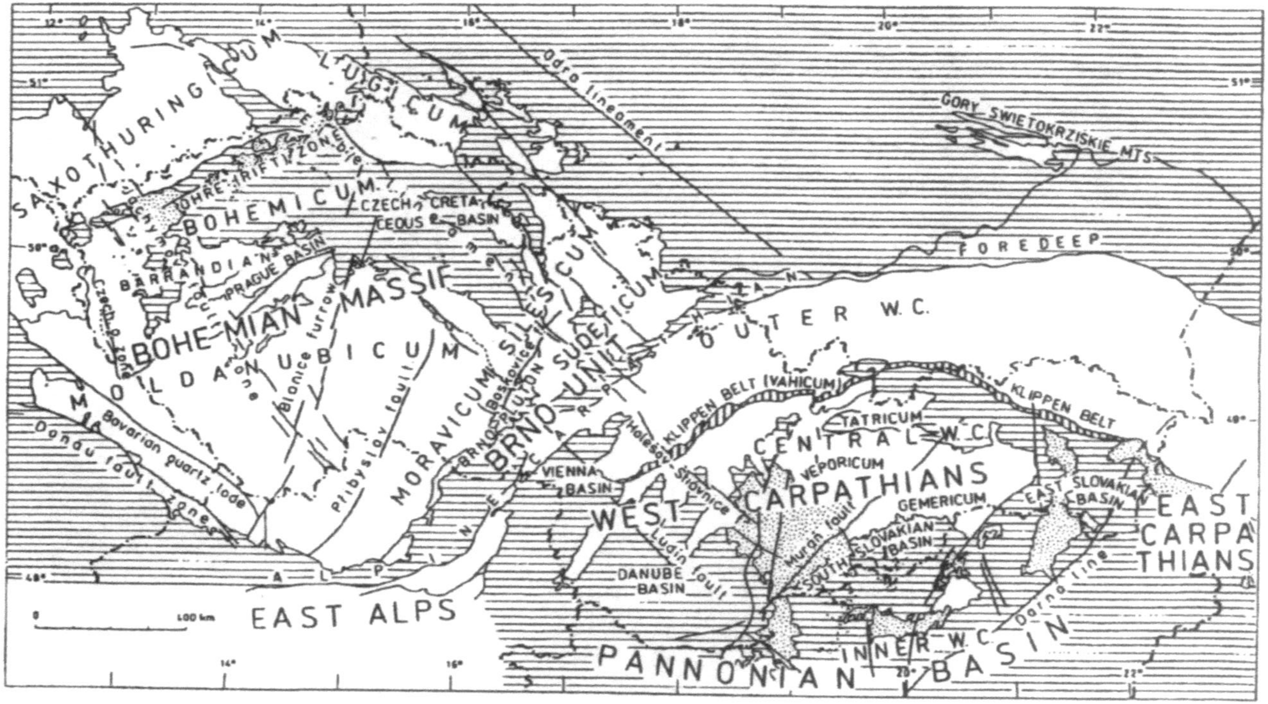

Fig. 1.1 Main geological units in Czechoslovakia and adjacent regions.

2 Seismic Research into the Lithosphere of Central Europe

2.1 Elastic Properties of Rocks and Their Anisotropy

V. Babuška[1], Z. Pros[1]

The anisotropy of physical parameters is one of the fundamental properties of rock-forming minerals as it depends on the configuration of their crystal lattice. Due to the effect of physical fields, the fabric elements of rocks and rock massifs become arranged. Their internal texture is mostly oriented and they also act as an anisotropic medium. As opposed to minerals, this is a statistically predominant or quasi-periodically repeated arrangement of rock fabric elements, depending on shape and texture. Velocities of elastic waves are among the physical parameters which sensitively reflect the regularities of rock texture. Besides the properties and orientation of mineral grains, conditions at grain contacts, as well as the density, quality and orientation of microfractures and other pores, or their saturation with solutions and gases are decisive for the elastic parameters of rocks too. The methods of studying elastic anisotropy can be divided into laboratory and in situ.

Laboratory Methods

Under laboratory conditions *P*–wave anisotropy has been studied most frequently in three mutually perpendicular directions (Birch 1960, 1961; Bajuk et al. 1967). In principle, this measuring system is not satisfactory and thus the measurement orientation should be chosen at least with respect to the fabric elements of rocks, provided they can be detected. In order to define the elastic anisotropy of rocks better, the number of measurement directions was increased. For example Giesel (1963) and Babuška (1966) conducted the measurements on samples of polyhedron shape. On spherical samples, the elastic anisotropy of *P*–waves (Pros and Babuška 1967, 1968; Thill et al. 1969) could be measured in an arbitrary direction with constant accuracy. This method was first applied under laboratory conditions and atmospheric pressure and, consequently, a device was developed for studying elastic anisotropy under hydrostatic pressure of up to 400 MPa,

[1] Geophysical Institute, Boční II, 141 31 Praha 4, Czech Republic

which roughly corresponds to the geostatic pressure at a depth of 15 km (Pros and Podroužková 1974; Pros 1977).

In Situ Methods

For investigating elastic anisotropy of rocks in situ, depending on the size of the studied object, it is possible to choose methods of determining elastic wave velocities ranging from ultrasound tomography (small wavelengths) for detailed studies, via hammer and explosion seismics (medium wavelengths), to teleseismic records of natural earthquakes and nuclear explosions (large wavelengths). With the exception of the first two methods of determining elastic wave velocities on safety pillars in mines and on free blocks in quarries, the in situ work is limited by the accessibility of the investigated object. In order to deploy a suitable observation network only the Earth's surface can be used. Moreover, it is necessary to know the actual distribution of velocities in a particular medium to determine both its structure and its properties. Using the method of intersecting seismic profiles the ocean floor was investigated by Rait et al. (1969) and Shimamura et al. (1983). The upper mantle under continents was studied analogously (Bamford 1977; Drummond 1985). The method of relative time residuals, i.e. the deviations of the actual arrival times of *P*–waves at seismic stations from theoretical arrival times, calculated for the standard Jeffreys – Bullen model of the Earth, was used to investigate the deep structure of the lithosphere of Central Europe (Babuška et al. 1984a). Cidlinský's paper (1980) can be presented as a successful example of reinterpretation of seismic measurements on the site of an underground gas storage on the basis of assumptions of anisotropic properties of the medium and of the knowledge of *P*–wave velocities determined on samples in the laboratory.

The in-situ methods play a principal role in understanding the structures and properties of geological units. The information on the elastic anisotropy phenomenon they yield is, however, incomplete. It is necessary to find ways of improving this situation, e.g. by making better use of the data provided by existing methods. In this respect the results achieved in seismic tomography and holography are promising.

The Effect of the Preferred Orientation of Crystals

Under statistically random orientation of grains crystal aggregates behave like isotropic bodies. Due to the oriented fields of forces in nature, the orientation of crystals becomes preferred, and the physical properties of crystal aggregates anisotropic. The main factors affecting the degree of elastic anisotropy of crystalline rocks are as follows:

1. Anisotropies of rock-forming minerals and their volume in the rock.
2. Type and degree of the preferred orientation of minerals and of other fabric elements.
3. Orientation of extremes of elastic waves velocities in monocrystals with regard to directions significant for the preferred orientation of minerals in force fields.

According to Birch (1961) the coefficient of anisotropy is defined as:

$$k = \frac{v_{max} - v_{min}}{v_{mean}} 100\%$$

where v_{mean} is the mean velocity obtained from all directions of measurements.

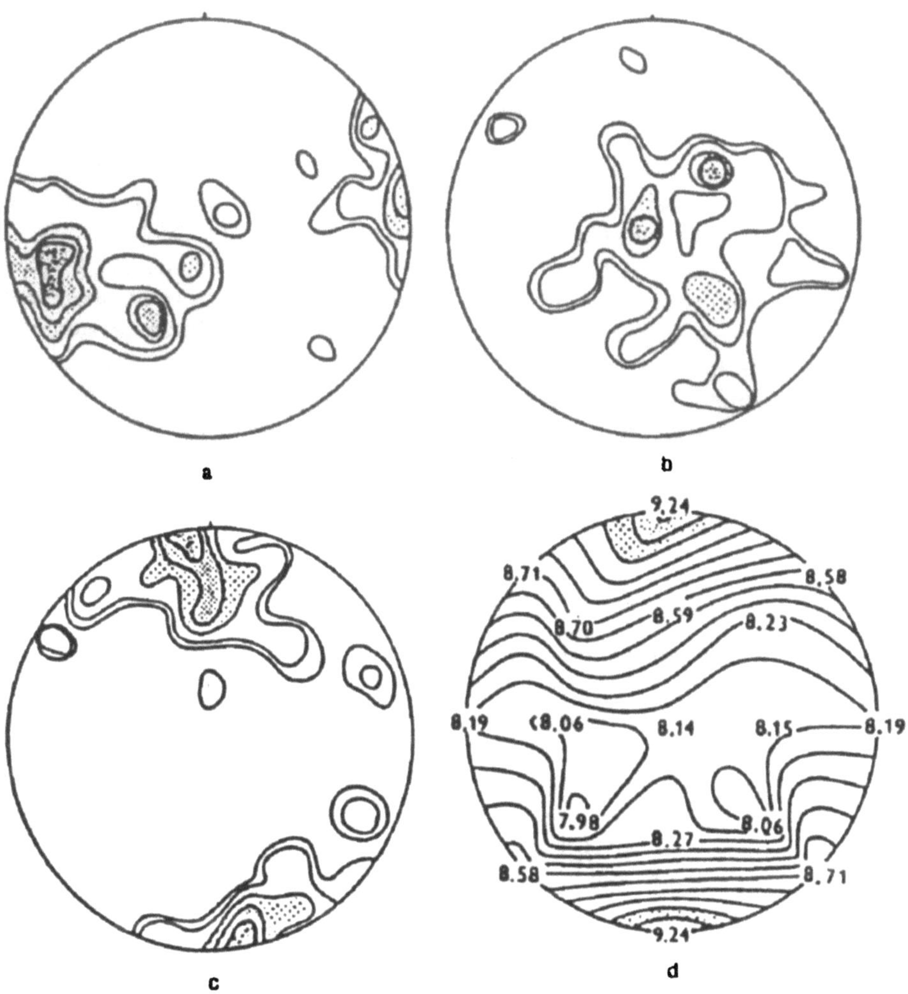

Fig. 2.1 Diagrams of the preferred orientation of 147 grains of olivine in dunite from Twin Sisters (USA) locality and the diagram of elastic wave anisotropy: a) *b*–axes orientation, b) *c*–axes orientation, c) *a*–axes orientation, d) spatial distribution of *P*–wave velocity in dunite, velocity contours range from 8.0 km s^{-1} to 9.20 km s^{-1}.

Apart from garnets, whose elastic anisotropy is negligible, all principal rock-forming minerals can use their own anisotropy to form the elastic anisotropy of rocks. According to the values of coefficients of anisotropy of velocities of longitudinal (k_p) and transverse (k_s) waves rock-forming minerals are arranged in the following sequence: micas ($k_p = 59\%$, $k_s = 100\%$), potassium feldspars (46%, 71%), calcite (33%, 57 %), plagioclases (31%, 55%), hornblende (27%, 31%), quartz (26%, 40%), olivine (25%, 22%), pyroxenes (16%, 21%), garnets (1%, 2%).

Feldspars, quartz, and clinopyroxenes, i.e. minerals with high values of anisotropy coefficients form rock aggregates for which a low elastic anisotropy is characteristic also if a considerable preferred orientation of these minerals is observed. The velocity extremes in the monocrystals of these minerals are, however, mostly not oriented identically with the directions of extremes of tectonic field stress as, for instance, in the case of olivine. That is why olivine ultrabasites have substantially higher coefficients of anisotropy than, e.g. granitoids.

Figure 2.1 shows an elastic anisotropy diagram of dunite together with diagrams of the preferred orientation of olivine, which exhibits maxima of all three crystal axes. Thus, dunite belongs to rocks exhibiting high elastic anisotropy ($k = 15\%$).

As the preferred orientations of orthopyroxene are not as obvious and its elastic anisotropy is lower than that of olivine, the coefficient of anisotropy of ultrabasic rocks decreases with increasing orthopyroxene content. The mean elastic anisotropy of pyroxenites is low (ranging from 3% to 4%). Clinopyroxenes also form aggregates with very low anisotropy (2% to 3% for eclogites). Aggregates of rocks with preferentially oriented micas, calcite, and amphibole generally show marked anisotropy of elastic wave velocity. For micas and amphiboles the external limits of these minerals play an important role in their orientation and in the resulting anisotropy of several metamorphic complexes of the Bohemian Massif (Bartošek et al. 1985).

Effects of Microfractures and Grain Boundaries

In the course of investigating elastic properties of rocks, we have studied in detail the spatial orientation of cleavage planes of rock-forming minerals in granodiorites and in quartzite, as well as the orientation of irregular microcracks and quartz grains boundaries (the shape orientation of crystals) in quartzite, in relation to the *P*–wave velocity anisotropy and its variation in quartzite, in relation to the *P*–wave velocity anisotropy and its changes under hydrostatic pressure. Elastic anisotropy was studied in all directions on samples of spherical shape, 50 mm in diameter, in granodiorite under pressures of up to 300 MPa and in quartzite up to 400 MPa (Babuška et al. 1977, Jech et al. 1985).

Figure 2.2 shows the diagrams of the elastic anisotropy of granodiorite (the Žampach locality in central Bohemia) under atmospheric pressure (a, d) and under increasing hydrostatic pressure (b, c). In the first diagram a significant anisotropy of granodiorite under atmospheric pressure can be observed, the greatest difference in *P*–wave velocities being nearly 1.6 km s^{-1} (coefficient of anisotropy $k = 27.2\%$). But under a hydrostatic pressure of 20 MPa, which corresponds to conditions at a depth of about

700 m, the greatest difference in velocities dropped to only 0.66 km s^{-1}, and in the direction of the velocity minimum, the P–wave velocity increased by 1.6 km s^{-1}. The velocity increase and the anisotropy decrease continue with pressures increasing up to 300 MPa (Fig. 2.2c), when the greatest difference in P–wave velocities is only 0.27 km s^{-1}, the coefficient of anisotropy being $k = 4.4\%$.

From the resulting diagrams of elastic anisotropy, it follows that the sample of granodiorite is markedly anisotropic under atmospheric pressure and only weakly anisotropic under a pressure of 300 MPa. It is obvious from these observations that the

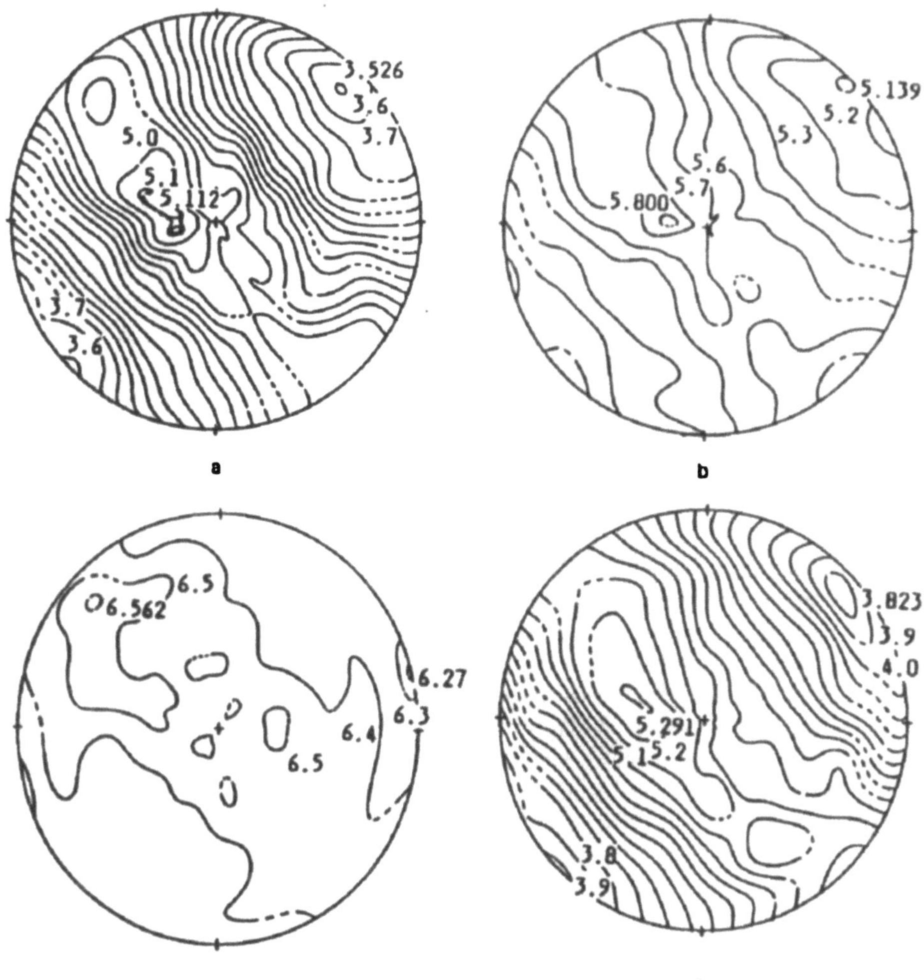

Fig. 2.2 Diagrams of the elastic wave anisotropy and of the P–wave velocity in granodiorite from locality Žampach under atmospheric pressure (a), under hydrostatic pressure of 20 MPa (b) and 300 MPa (c), and under atmospheric pressure at the end of the experiment (d).

elastic anisotropy has probably been caused by an anisotropic distribution of microfractures. In order to be able to confirm this assumption, orientations of cleavage planes of plagioclase, amphibole, and biotite were measured. Apart from this we also measured the orientations of crystal axes of quartz. These minerals form 84% of the rock volume. No dependence of the elastic anisotropy of granodiorite on the preferred orientation of quartz crystals was detected. Fractures in quartz are of an irregular shape and their orientation is random. That is why even they cannot cause elastic anisotropy in a rock, observed under atmospheric pressure. Even though the preferred orientation of cleavage planes of plagioclases is not significant, the maximum of normals to the cleavage (010) approximately corresponds to the direction of the minimum of *P*–wave velocity. The preferred orientation of normals to the cleavage planes of amphibole and biotite is in accordance with the direction of the minimum *P*–wave velocities in granodiorite. Under atmospheric pressure, when the cleavage provides mostly open fractures, the velocities of elastic waves propagating across the fractures are markedly lower. Under high hydrostatic pressure, when the microfractures in rock are tight, the elastic anisotropy is mainly caused by the preferred orientation of minerals.

Another rock studied in detail was quartzite (locality Loket near the town of Vlašim). It is a massive, nearly monomineral rock with an admixture of 3% of fine-grained biotite and chloritized pyroxene. When compared with the sample of granodiorite, the quartzite sample has a lower coefficient of anisotropy under atmospheric pressure ($k = 9.7\%$), and the increase of the velocity in the hydrostatic pressure field is lower than in the case of granodiorite. This indicates that the elastic anisotropy of quartzite is caused prevailingly by the orientation of quartz crystals also under atmospheric pressure. Microscopic measurements of the preferred orientation of microfractures as well as of grain boundaries have shown that the greatest change of *P*–wave velocities under increasing hydrostatic pressure took place in a direction close to the orientation of the maximum of normals to the grain boundaries and microfractures. The position of the maximum of *P*–wave velocities in the anisotropy diagrams, on the other hand, is in accordance with the directions of minimum concentration of normals at microdiscontinuities in a rock. The comparison of results of the measurements indicates that marked cleavage affects the velocities of longitudinal elastic waves as well as their anisotropy under atmospheric pressure and low hydrostatic pressure most.

The velocity anisotropy is strongly dependent on rock fabric. Kern and Wenk (1990) showed that anisotropy increases significantly with the degree of deformation, whereas average velocities and densities do not change. During phyllonitization of granodiorite the fabric, namely the size, shape and orientation of mineral grains, and the anisotropy for both *P* and *S* velocities changed from about 1% in granodiorite to about 12% in strongly deformed phyllonite.

Review of Elastic Anisotropy of Crystalline Rocks

The preferred orientation of minerals, the orientation, number and effectively the quality of fractures represent the main causes of anisotropy of elastic wave velocity. The effect of fractures is most significant below pressures of 200 MPa which corresponds to depths around 7 km. The cleavage planes have the greatest effect upon the velocity anisotropy (e.g. of micas and amphiboles). The elastic anisotropy is thus a result of superposition of two structures, one of which is based on the preferred orientation of minerals and the other on the orientation of cracks and pores in the rock. Investigating the elastic anisotropy and its changes under gradually increasing hydrostatic pressure makes it possible to separate these two structures, to determine their symmetry, and to obtain an idea of the intensity and orientation of processes whose results they represent.

Table 2.1 Mean velocities of longitudinal elastic waves v_p and coefficients of velocity anisotropy k in the main types of crystalline rocks under pressure of 1 GPa (after Babuška 1984).

rock	number of samples	density range [Mgm^{-3}]	v_p range [kms^{-1}]	v_p mean [kms^{-1}]	k range [%]	k mean [%]
granite and granodiorite	14	2.62 – 2.86	5.81 – 6.61	6.40	0.3 – 3.8	1.6
quartz diorite	3	2.76 – 2.91	6.60 – 6.72	6.67	0.1 – 1.1	0.6
gabbro	18	2.77 – 3.21	6.68 – 7.59	7.22	0.7 – 6.2	2.7
basalt and diabase	9	2.79 – 3.01	6.09 – 6.93	6.62	0.1 – 2.2	0.9
pyroxenite	6	3.21 – 3.36	7.83 – 8.29	8.05	0.5 – 6.2	3.7
dunite and peridotite	27	3.20 – 3.34	7.45 – 8.74	8.10	3.1 – 15.0	9.2
serpentinite	4	2.60 – 2.80	6.00 – 6.84	6.49	0.8 – 11.3	4.6
undifferentiated schists	26	2.68 – 3.19	5.83 – 7.54	6.90	0.8 – 21.0	10.1
gneiss	34	2.64 – 3.16	5.73 – 8.54	6.56	0.3 – 21.6	5.9
amphibolite	13	2.96 – 3.26	6.35 – 8.27	7.25	0.7 – 19.8	9.0
granulite	27	2.67 – 3.74	6.48 – 7.48	6.96	0.6 – 5.6	2.4
eclogite	37	3.19 – 3.72	7.55 – 8.61	8.09	0.1 – 11.3	2.6

Some general characteristics of elastic anisotropy of rocks follow from the review of values of the coefficients of elastic anisotropy of *P*–waves calculated from the measurements taken under high hydrostatic pressure (Table 2.1), which are conclusive for the lithosphere as a whole. The mean values of coefficients of anisotropy of the individual groups of igneous rocks are low, only exceptionally reaching several per cent in gabbro samples. On the other hand, numerous metamorphic rocks are characterized by a high anisotropy. They are amphibolites and various kinds of metamorphic schists which are characterized by marked foliation and by high values of the coefficient of anisotropy around 20%. The mean anisotropy of these two groups ranges from 8% to 10%. As the mean anisotropy of granulites is only 2.4%, amphibolites and olivine gabbros are the only rocks which can form an anisotropic medium in the lower crust. In order to use the elastic anisotropy as the parameter for determining the rock fabric, it is also necessary to know the degree of velocity changes due to the composition inhomogeneity of the individual rock types. The data in Table 2.1 confirm the fact that the changes of densities and of mean velocities of crustal rocks are very high even within individual rock types. In view of the very complicated and, even on a regional scale, very inhomogeneous structure of the continental crust, it cannot be assumed that severe effects of anisotropic propagation of seismic waves in the Earth's crust can be observed for natural earthquake spherical waves several kilometres in length. In the course of seismic investigations of geological conditions that form large structures, when seismic waves of short wavelengths are generated, and when the use of anisotropy data can lead to improving the accuracy of measured data, a velocity anisotropy is frequently encountered.

As also follows from Table 2.1, the variations of mean *P*–wave velocities in the rocks of the upper mantle are considerably lower than those of the rocks in the Earth's crust. The highest values of coefficients of elastic anisotropy are observed for olivine – rich ultrabasites (around 10% of olivine) with a maximum of 15% for dunite. In the majority of cases the structures of these ultrabasites can be approximated by orthorhombic symmetry. As already mentioned above, the anisotropies of aggregates with high orthopyroxene and clinopyroxene contents are characterized by low elastic anisotropy only (i.e. 2 – 4%). Under the realistic assumption that stresses exist in the upper part of the mantle, and that olivine is present as a substantional part of mantle rocks, we can presume that in this part of the Earth's body the velocity changes, which are caused by anisotropic structures of large dimensions, are probably equally as significant as the velocity changes caused by inhomogeneities in the rock composition. The marked anisotropy and monoclinal to orthorhombic symmetry of the rock structures of the upper mantle justifies us in distinguishing anisotropic structures and their orientations in the lower part of the lithosphere by observing seismic waves.

2.2 Deep Seismic Sounding in Czechoslovakia

M. Mayerová[1], M. Novotný[1], M. Fejfar[1]

In Czechoslovakia, investigations of the deep parts of the Earth's crust including the Moho discontinuity commenced in 1962 as a part of the project launched by the European Seismological Commission in 1959. Until 1972 measurements were conducted along three international profiles designated by numbers V, VI, and VII. In the period 1975 – 1980 measurements were carried out along regional profiles KIII, KII, 100R, and KI. Positions of all the profiles including the international ones are shown in Fig. 2.3. The aim of the measurements was to provide seismic data for determining the near-surface structure of the Earth's crust and for investigating deep seismic boundaries including the Moho discontinuity. During 1979, seismic measurements were carried out along the trans-Scandinavian profile Fennolora (Beránek et al. 1981). The purpose of these measurements was to investigate the Earth's crustal structure and the upper mantle structure beneath the Scandinavian shield down to depths of several hundreds of kilometres on the basis of detected seismic anomalies in this zone. This profile was extended and branched in two directions, one to Poland and Ukraine, the other over East Germany to Czechoslovakia (Fig. 2.3).

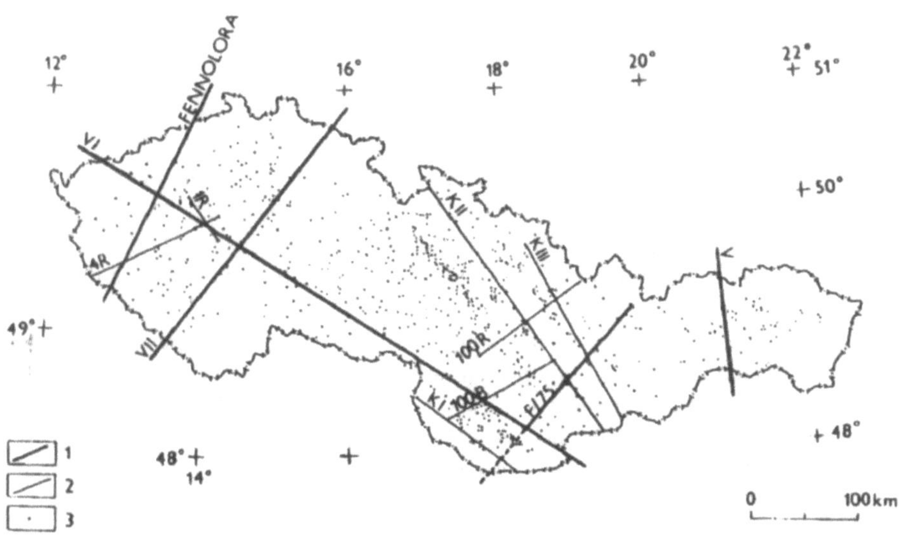

Fig. 2.3 Situation of DSS profiles in Czechoslovakia. 1 – international profiles, 2 – national profiles, 3 – discrete points at which the Moho discontinuity depths were determined from industrial explosions.

---

1 Geofyzika Brno, Ječná 29a, 612 46 Brno, Czech Republic

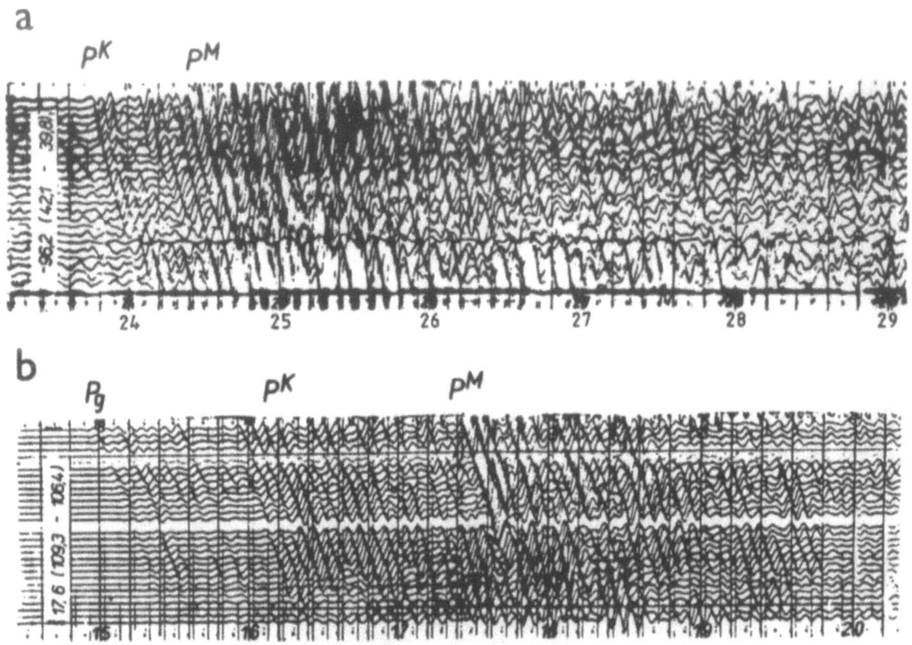

Fig. 2.4 Examples of a seismogram from profile VII.

The system of measurements used along profiles V, VI, VII enabled investigating the seismic waves reflected from the Moho discontinuity at angles close to the critical one or larger. In the Bohemian Massif, P_M waves reflected from the Moho discontinuity are usually most intensive at epicentral distances of 70 – 80 km (the critical distance), which varies with the depth of reflection boundaries and according to the velocity distribution at depth. The reflected wave can, as a rule, be observed up to a distance of 150 – 170 km.

Examples of seismograms obtained along profile VII (Fig. 2.4) show well correlated P_M waves reflected from the Moho discontinuity with a time of 17.4 s at an epicentral distance of 91 km (Fig. 2.4b) and with a time of 24.4 s (Fig. 2.4a) at an epicentral distance of 138 km. Somewhat less striking are P_K wave reflections from internal boundaries within the Earth's crust. The first arrivals (marked as P_g) correspond to refracted waves propagating within the upper part of the consolidated Earth's crust.

Seismic measurements carried out along the regional profiles in later years represented a synthesis of the refraction seismic method and the deep seismic sounding (DSS). The introduction of digital technique in the 1970s allowed geophysicists to develop wave patterns from individual shotpoints. Fig. 2.5 shows an example of these wave patterns along regional profile KII. The developed wave field considerably aids correlation of individual types of waves as well as the entire interpretation of the wave field pattern. The wave field was recorded up to 25 s, so that it contained reflections

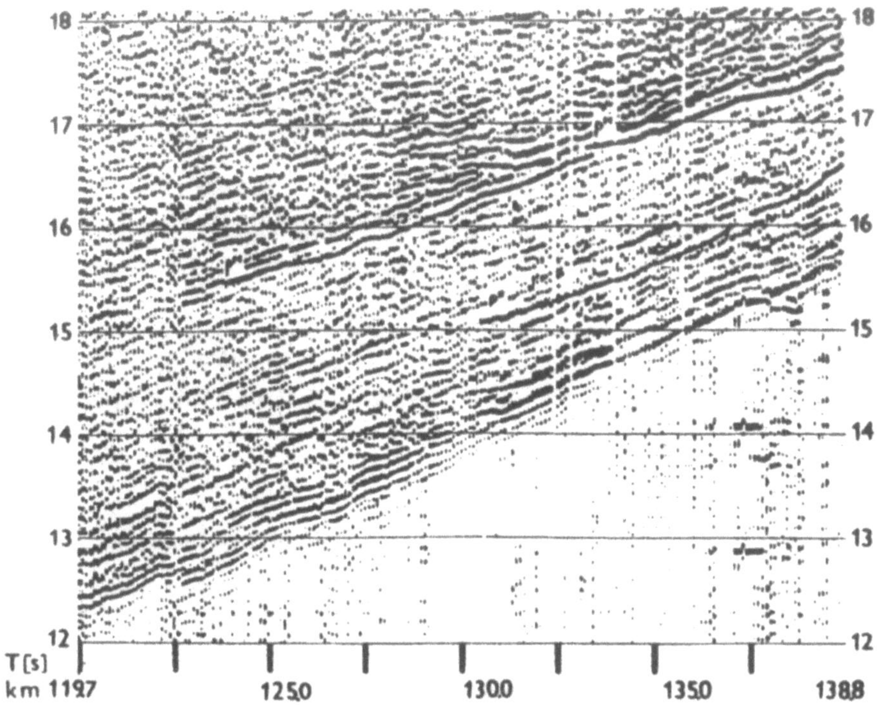

Fig. 2.5 Example of the wave-field pattern obtained along KII profile – the epicentral distance in the range 119.7 – 138.8 km was used.

from deep structures including the Moho discontinuity (see Fig. 2.5). Along the regional profiles, the P_M wave reflected from the Moho discontinuity was followed to epicentral distances of 120 – 130 km.

The measurements provided us with unique sets of time–distance curves of refracted waves thanks to a relatively dense network of shotpoints and by recording the arrivals at small epicentral distances. In order to interpret the set of time–distance curves of refracted waves, the linearization approach (Novotný 1981) was applied, which enables us to derive the 2D velocity distribution in a consequent way. The velocity section obtained by means of this linearization method along profile KII is shown in Fig. 2.6. Even though the velocity mapping by means of refracted waves did not reach the depth of the Moho discontinuity, it was a valuable contribution to the geologic interpretation of the near-surface structure of the Earth's crust. The resolution attained made it possible to specify numerous structural elements in the uppermost part of the crust (Ibrmajer et al. 1983).

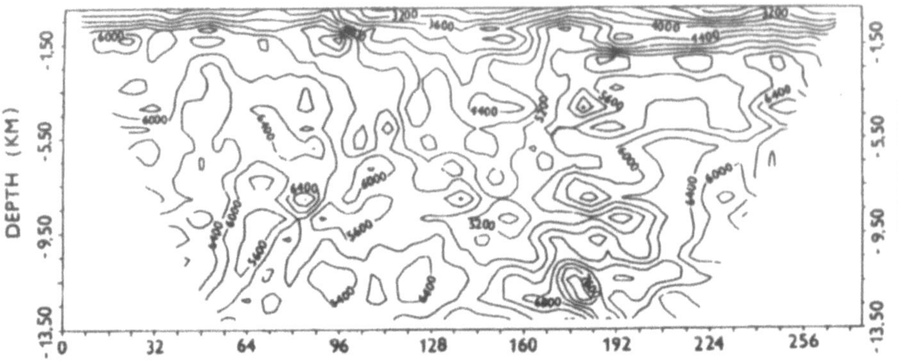

Fig. 2.6 Velocity section along KII profile obtained by an interpretation of time–distance curves of refracted waves using the linearization method.

Main Results of the Seismic Data Interpretation Along Regional Profiles

Regional profiles KIII, KII, 100R, and KI in the West Carpathians (Fig. 2.7), along which the data were obtained in the period from 1975 to 1980, provided us with basic information on the deep structure of the Carpathian allochthon. Results of the first interpretations were published in a number of papers: Beránek and Zounková 1979; Leško et al. 1980; Beránek and Zátopek 1981; Novotný et al. 1981; Ibrmajer J. at al. 1983; Čekunov 1987. As an example of the geophysical interpretation we present some results obtained along profile KIII which intersects, in the NNW – SSE direction, units and nappes of the outer Carpathians, the Klippen Belt, the inner Carpathians and terminates in the region of neovolcanites of central Slovakia. By the Jegorkin transformation (Jegorkin 1966) of the time–distance curves of waves reflected from structural boundaries in the lithosphere the effective velocity section (Fig. 2.8a) was obtained. The integral relation between the mean velocity $v_{mean}(z)$ and the actual velocity of propagation of seismic waves $v(z)$, i.e.

$$\int_0^z \frac{dz}{v(z)} = \frac{z}{v_{mean}(z)}$$

for a constant value of coordinate x along the profile $x = \text{const.}$ was then used for a graphical derivation of the actual velocity section (Fig. 2.8b).

A velocity section reflects, to a certain degree, the geologic conditions along a profile. The reliability of a particular geologic interpretation is given by the density of primary data, i.e. by the average shotpoint step and by the degree of stratification of a medium. In effective and actual velocities, sections along profile KIII (Fig. 2.8) a zone of increased vertical gradient is observable in the upper part down to a depth of around

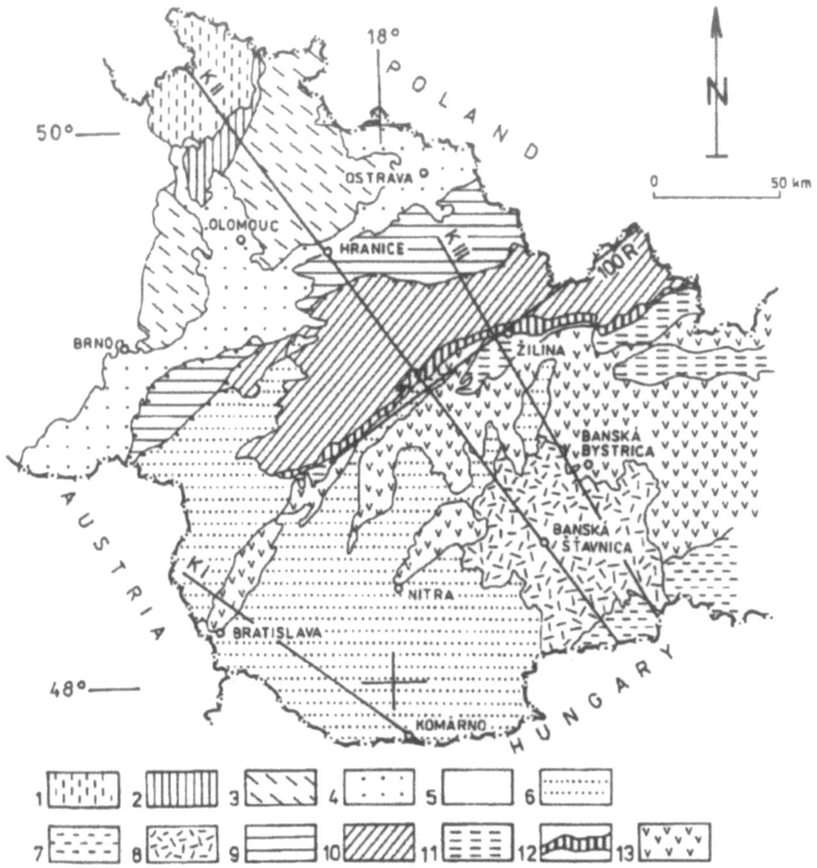

Fig. 2.7 Geologic scheme with positions of regional profiles in the Czechoslovak part of the West Carpathians. 1 – the crystalline unit of the Hrubý Jeseník Mts., 2 – Devonian, 3 – Culm, 4 – Neogene of the Carpathian Foredeep, 5 – Neogene of the Vienna Basin, 6 – Neogene of the Danube Basin, 7 – Neogene of the South–Slovakian Basin, 8 – Neovolcanites, 9 – Outer Flysch, 10 – Magura Flysch, 11 – Paleogene of the Inner West Carpathians, 12 – Klippen Belt, 13 – Mesozoic and crystalline units of the Inner West Carpathians.

10 km. The velocity differentiation is more obvious in the derived section of actual velocities: e.g. a manifestation of the low-velocity channel in the original section (Fig. 2.8a) is markedly enhanced by a transformation to actual velocities (Fig. 2.8b). Beneath the low-velocity channel, which can be observed along the entire length of profile, there is a zone of increased gradient of velocities. The velocity gradient in the lower part of the section is probably associated with the transition zone in the lower part of the Earth's crust (the basaltic layer). It follows from the velocity pattern in the section of profile KIII that rocks of lower velocities have to exist in the upper part of the crust. Granitoids of the Malá Fatra Mts. are not autochthonous but they are a part of the overthrust unit below which sedimentary rocks probably exist (Beránek 1978).

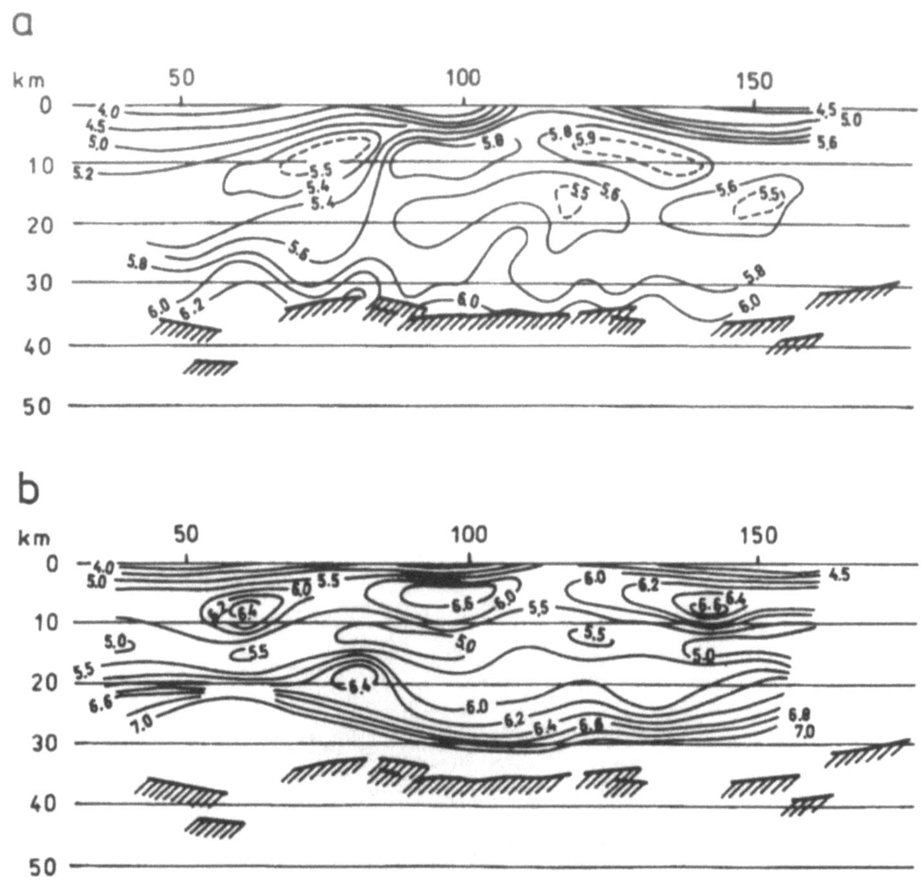

Fig. 2.8 a) Effective velocities pattern in KIII section, b) layer velocities pattern in KIII section.

Among the regional profiles the seismic profile KII is the longest one. Fig. 2.9 illustrates courses of seismic boundaries including the Moho discontinuity along this profile. For constructing the seismic boundaries a simplified procedure was used, placing reflecting elements at the half epicentral distance (assuming that the boundary is dipping only gently). Elements at depths of 10 – 30 km can represent boundaries of rock units or tectonic boundaries. The course of the Moho discontinuity is relatively complex. Records of wavefields show, in numerous places, an interference character with several phases of waves reflected from the Moho discontinuity. Often, later wave groups in the time span of up to 1 s could also have been distinguished. A similar wave pattern was recorded also along the other regional Carpathian profiles. This phenomenon can be caused by a group of layers exhibiting high and low velocities, which form a lamellar character of the Moho discontinuity (Giese et al. 1976; Čekunov 1987).

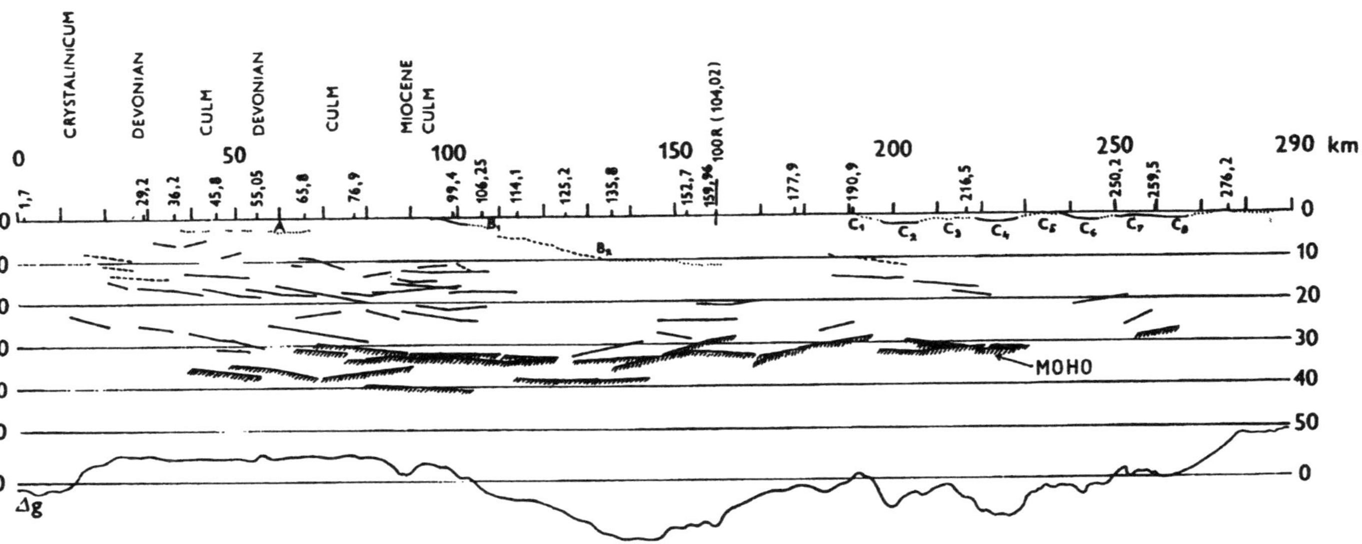

Fig. 2.9 Seismic section of KII profile.

The Moho discontinuity, thus, does not form a sharp boundary but a transient zone up to several kilometres thick. In the seismic section of profile KII the Moho discontinuity was detected at depth (36 km) in its NW part. Approximately beneath the boundary of the Nízký Jeseník Mts. with the Miocene of the Carpathian Foredeep at the ground surface, the Moho discontinuity appears interrupted. This interruption can have the character of a fault. The Moho extends towards SE at depths around 31 – 33 km. In places, where the seismic profile intersects the eastern margin of the Bohemian Massif and the flysch belt of the West Carpathians a duplication of the Moho boundary is observed. The lower Moho boundary ranges here at depths from 38 to 40 km.

2.3 Construction of the Moho Discontinuity Map

M. Mayerová[1] , Z. Nakládalová[1], I. Ibrmajer[1], M. Fejfar[1]

The data on the Moho discontinuity obtained along the international and regional profiles in the West Carpathians in the period 1971 – 1984 were systematically supplemented with results of the DSS obtained by recording seismic waves generated by industrial explosions. In these cases the DSS measurements were performed at discrete points distributed with higher or lower density over the entire territory of Czechoslovakia (see Fig. 2.3). Seismic waves were recorded along short profiles (ca. 3 km) at epicentral distances 80 – 120 km which are particularly suitable for recording the waves reflected from the Moho discontinuity. The interpretation of the discrete DSS data was essentially easier because analogous wavefields from profile measurements were already available for various epicentral distances. The methodology and results of the discrete DSS processing are summarized in (Beránek and Zounková 1977).

The method using industrial explosions could be successfully applied in Czechoslovakia for there is a relatively dense network of uniformly distributed quarries. The character of wave patterns at various epicentral distances as well as the velocity distributions were known for the international and regional profiles on the Czechoslovak territory. In principal, the areal discrete DSS data were tied to the profile measurements during the interpretation.

Corrections for the relief of the Earth's surface were calculated in order to remove the effect of altitudes on calculations of depths of the Moho discontinuity and introduced to all records which were used for further processing. Arrival times of reflection waves were then corrected for the effect of unconsolidated crust both at the recording point and at the shotpoint. The corresponding part of the crust was substituted by a layer with a velocity of 5.6 km s^{-1}. Mean velocities in the sedimentary cover were determined from seismic well-logging measurements and from the compiled maps of mean velocities. The sedimentary formations, necessary for calculating the corrections were taken from maps of thicknesses of sedimentary basins in the Bohemian Massif, from borehole data,

1 Geofyzika Brno, Ječná 29a, 612 46 Brno, Czech Republic

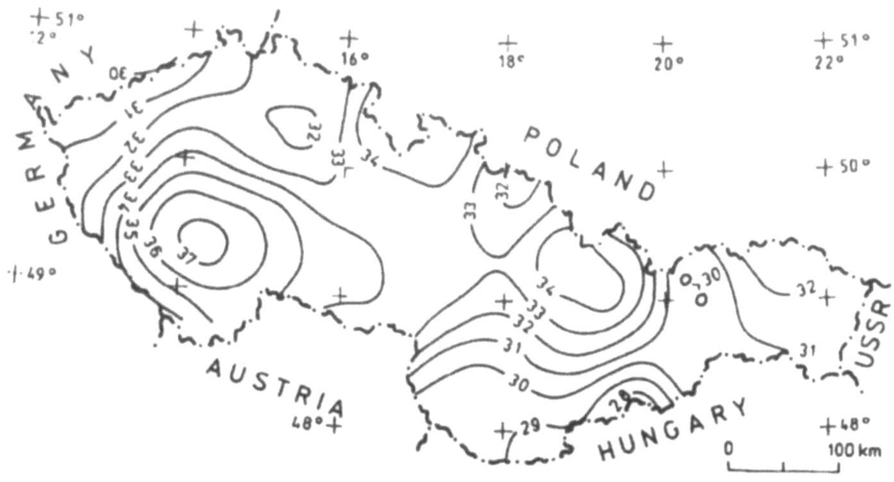

Fig. 2.10 Map of isohypses (in kilometre) of the Moho discontinuity.

and from other available geologic and geophysical materials. The depths to the Moho discontinuity were calculated using the assumption that the Moho is horizontal and that the mean velocities are constant. For both the Bohemian Massif and the West Carpathians the value of 6.3 km s^{-1} was taken as the mean velocity. The depth values were then smoothed and interpolated (Herrmann 1986) to the map of isohypses (Fig. 2.10).

2.4 Deep Reflection Seismic Profiling in Czechoslovakia – the ČESLOKORP (Czechoslovak Crustal Reflection Profiling Project)

I. Ibrmajer[1] , Č. Tomek[1], T. Koráb[2] , L. Dvořáková[1]

Research into the deep structure of the Earth' crust using the traditional oil prospecting method, i.e. the reflection seismic method in the common depth point (CDP) modification, was conducted in numerous countries (projects COCORP, BIRPS, ECORS, DEKORP, LITHOPROBE, ACORP, etc.) at the end of the 1970s. Since the first experiments at the end of the 1960s the research has yielded, in numerous cases, revolutionary results in structural–geologic interpretations of old shields, platforms, Phanerozoic orogenes, continental margins as well as continental rifts. This method greatly contributed to subdividing the Earth's crust in the vertical and horizontal

1 Geofyzika Brno, Ječná 29a, 612 46 Brno, Czech Republic
2 Dionýz Štúr Geological Institute, Mlynská dolina 1, 817 04 Bratislava, Slovakia

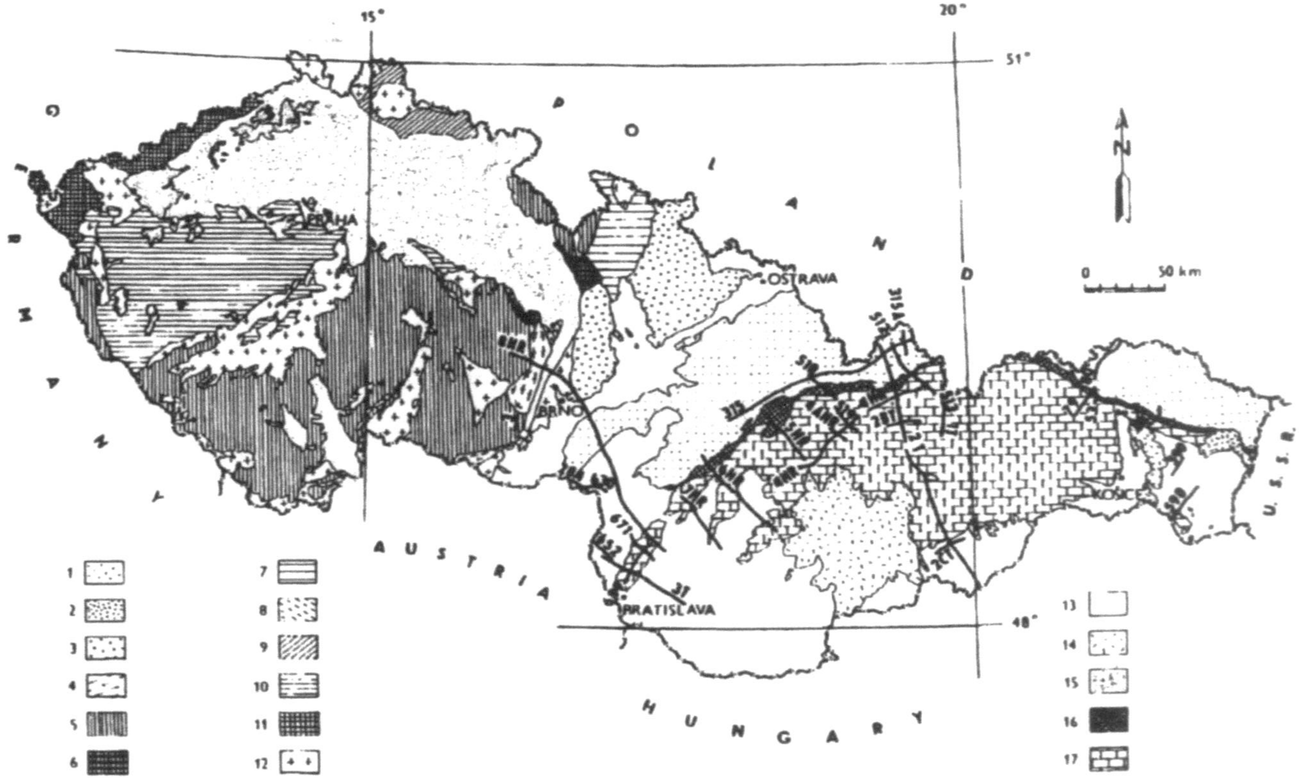

Fig. 2.11 Situation of the ČESLOKORP deep seismic reflection profiles Bohemian Massif: 1 – Tertiary, Cretaceous, Permo–Carboniferous (not distinguished) sediments, 2 – neovolcanites, 3 – Moravo–Silesian Lower Carboniferous and Devonian, 4 – Moravicum, 5 – Moldanubicum, 6 – Zábřeh complex, 7 – Silesicum, 8 – Letovice crystalline complex, 9 – Krkonoše – Jizerské hory crystalline complex, 10 – Teplá – Barrandian region, 11 – Saxothuringicum, 12 – granitoid rocks not distinguished. West Carpathians: 13 – Neogene, 14 – neovolcanites, 15 – Outer and Magura Flysch, 16 – Pieniny Klippen Belt, 17 – Inner West Carpathians.

directions, to characterizing various types of the crust, to studying the lower part of the crust and the transient zones between the lower crust and upper mantle.

The first testing in Czechoslovakia started in 1980. Since then the role of the deep seismic reflection method in investigations of the continental crust has continually increased. Until now, data have been obtained along more than 1 000 km of profiles which include both the short reflection profiles, several tens of kilometres in length, and transects of 100 to 200 km in length. Majority of the profiles of the Czechoslovak project of the deep reflection seismic profiling (ČESLOKORP) has so far been situated in the Carpathian mountain belt and along the contact of the West Carpathians with the Bohemian Massif (Fig. 2.11). Besides Geofyzika Brno the following institutions have taken part in the ČESLOKORP: the Geological Institute of Dionýz Štúr (GÚDŠ), Geological Survey Prague (ÚÚG), Moravian Oil Company (MND) Hodonín. In the course of continental crust investigations in Czechoslovakia, the ČESLOKORP project applies two methodological procedures: measurements along specifically designed deep

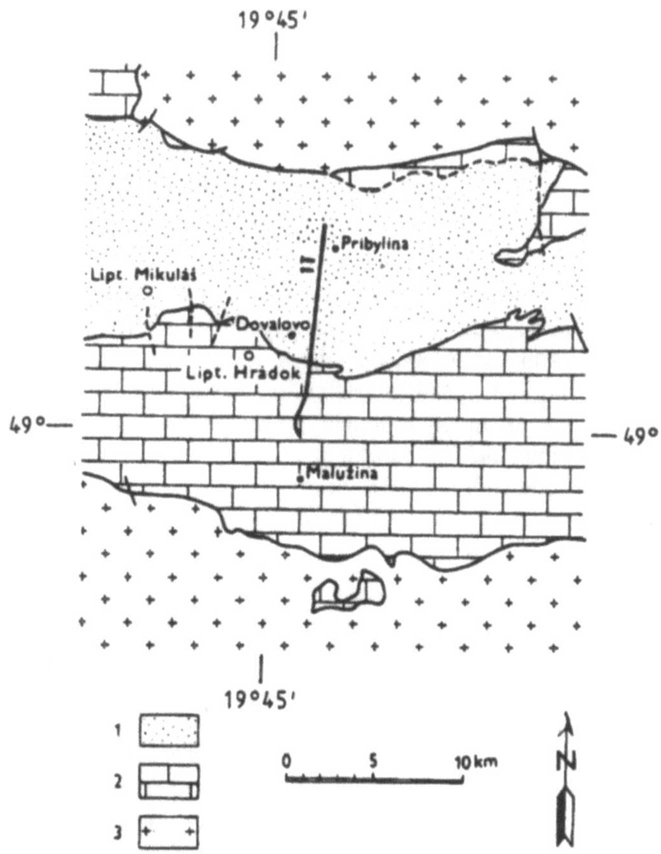

Fig. 2.12 Geologic situation of profile 1T. 1 – Paleogene, 2 – Envelope units of Tatricum, 3 – Tatricum.

seismic reflection profiles, or along oil prospecting profiles carried out in traditional oil-bearing regions with the use of increased times of recording the seismic signal. The deep seismic reflection investigations along the ČESLOKORP profiles are carried out mainly by Geofyzika Brno, which is responsible for preparing the project, conducting the field measurements, performing the numerical data processing, and which has the main share in structural–geologic interpretation of seismic results.

The results presented here include the first testing deep seismic reflection profile 1T situated in central Carpathian Palaeogene, then the specially conducted transect 2T intersecting all basic units of the West Carpathians. They also include examples of oil-prospecting profiles with increased recording times, i.e. profile 3T, stretching from the Vienna Basin over the Malé Karpaty Mts. to the Danube Basin (Tomek et al. 1987a). An example of the deep seismic reflection profiling in the Bohemian Massif is short profile 3 (Tomek et al. 1987b) situated at its eastern margin.

Profile 1T

Measurements along the line were conducted to test abilities of the deep seismic reflection method in central parts of the West Carpathians. The line is oriented in the N – S direction and it stretches in its entire length 14 km in the inner Carpathian Palaeogene of the Liptovská kotlina depression (Fig. 2.12).

The results obtained along the profile were interpreted by Leško and I. Ibrmajer (1983). In the time section (Fig. 2.13) several striking reflections occur. The uppermost reflection at 0.3 – 0.7 s TWT corresponds to the basal layer of the Palaeogene sediments.

Some striking reflections at 1.0 – 2.0 s TWT could be reflected from geologic layers of the Choč and Krížna nappes. There are only sparse real reflections within the time interval to 3.5 s TWT. This could be either due to a tectonically disturbed environment with steeply dipping individual blocks (which are hardly able to be detected by the seismic method) or due to an environment physically more or less homogeneous, corresponding to granitoids (the Tatricum).

Another complex (at 3.5 – 5.0 s TWT) roughly corresponds to the depth interval of 9.5 – 15.0 km. The markedly striking reflection horizon at 4.1 – 4.5 s TWT belongs to a group of subparallel reflections. The deepest complex, approximately at 5.5 – 10.0 s TWT, is characterized by an occurrence of numerous reflection elements with an anomalous dip. At 7.8 s TWT, short subhorizontal reflexes can be observed. A relative change of the wavefield pattern is observable after 7.8 s TWT where number of subhorizontal and horizontal reflections gradually diminishes and steeply dipping reflections start to prevail, which are, however, interpreted as disturbing waves or side reflections. At 10.0 s TWT and higher ones (corresponding depth 32 – 36 km), dynamically striking segments of subhorizontal reflections occur, which through their frequency and position in the time scale probably correspond to the upper boundary of the transient zone between the Earth's crust and upper mantle (i.e. to the Moho).

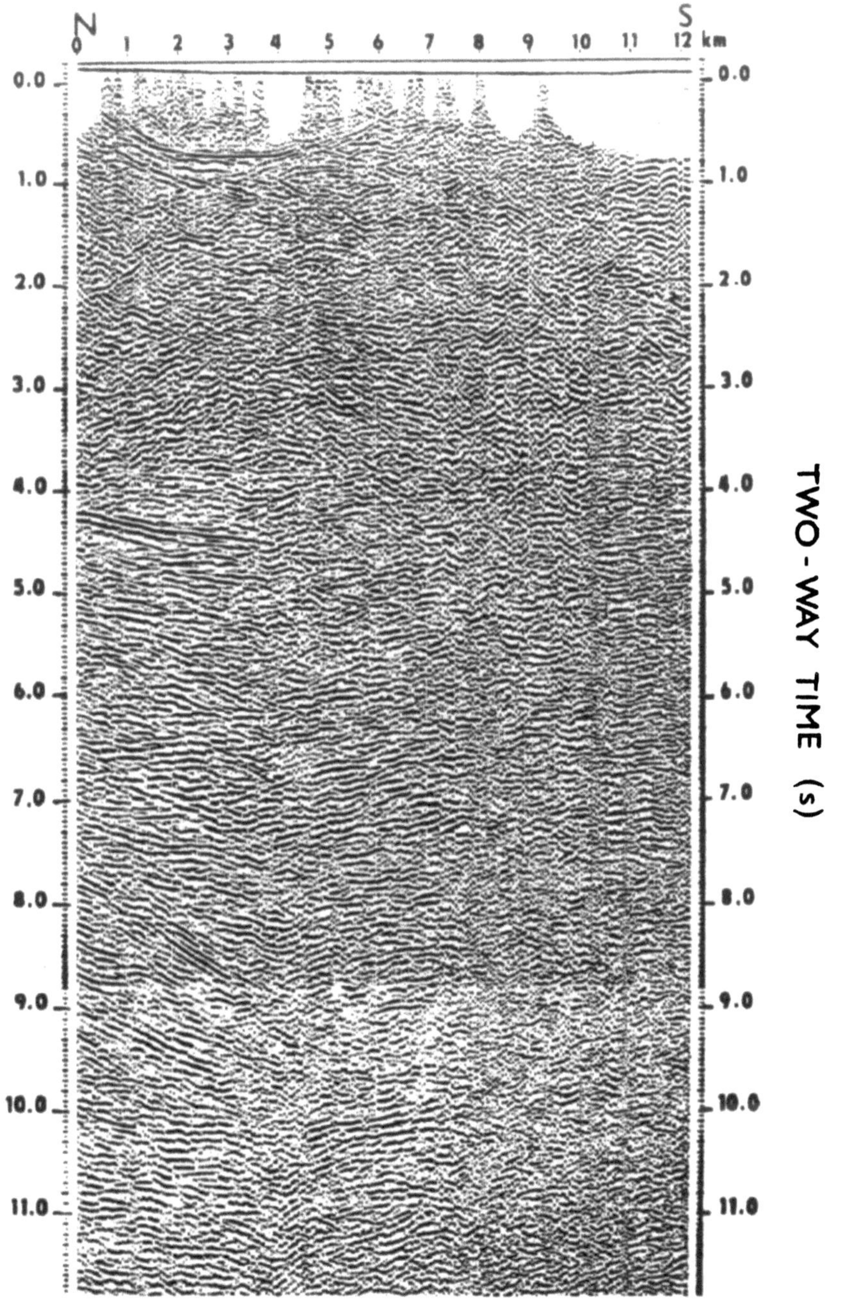

Fig. 2.13 Unmigrated time section of profile 1T.

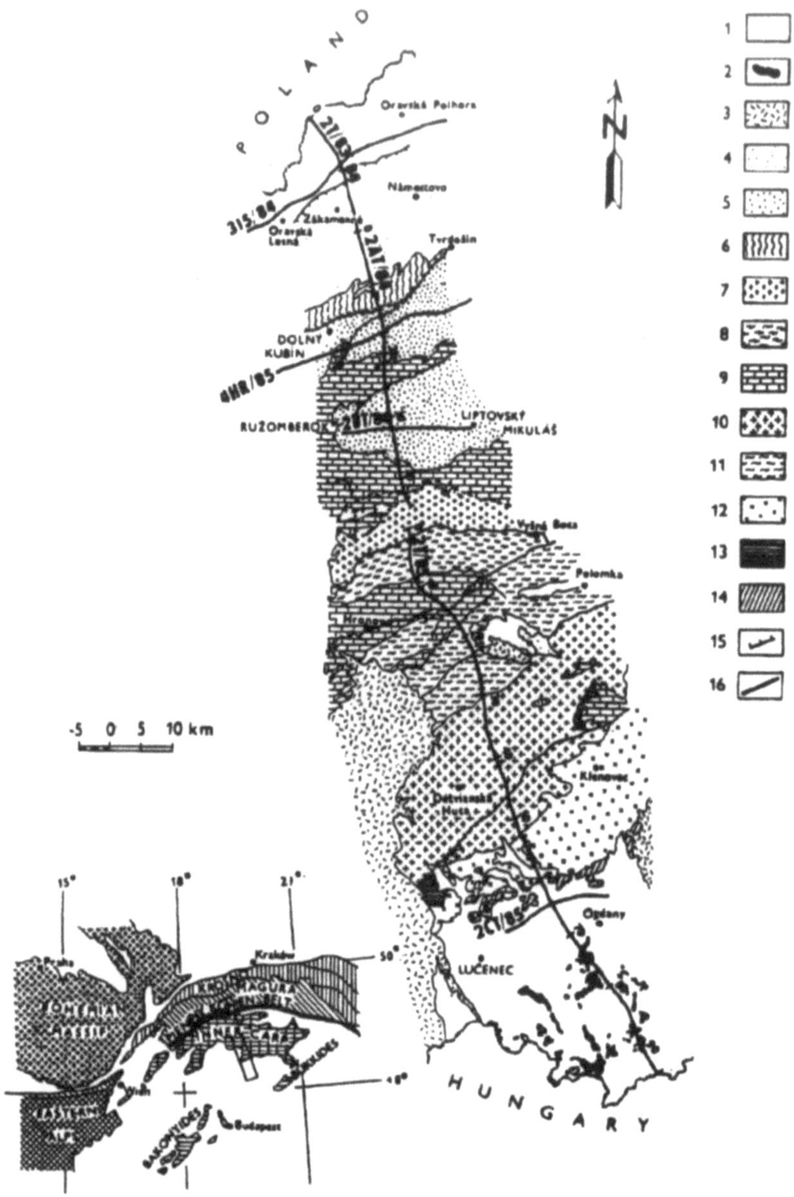

Fig. 2.14 Geologic situation of profile 2T: 1 – Neogene of the Lučenec and Rimava depressions (South-Slovakian basin, 2 – basalts (Pliocene – Quaternary), 3 – neovolcanites, 4 – Magura Flysch, 5 – Paleogene of the Inner West Carpathians, 6 – Pieniny Klippen Belt, 7 – Ďumbier granitoids, 8 – Crystalline schists of Tatricum, 9 – Mesozoic of the sub-Tatricum, 10 – granitoids of Veporicum, 11 – crystalline schists of northern Veporicum, 12 – crystalline schists of southern Veporicum, 13 – Mesozoic Envelope units of Veporicum, 14 – Paleozoic of Gemericum, 15 – tectonic lines, 16 – ČESLOKORP lines.

Profile 2T

Measurements along profile 2T resulted from a joint program of Geofyzika Brno and the Geological Institute of Dionýz Štúr, Bratislava, see e.g. (Tomek et. al. 1989). The aim of the seismic reflection measurements was to tie the profile to the shallow reflection profile NR4–VIII–79K on Polish territory and then to continue it along Pilsko – – Chabenec – Gemerský Jablonec line, more than 170 km long. It was to detect the thicknesses of flysch nappes, the course of the declining basement of the Brunnia, and to specify the type of contact with the Carpatho – Pannonian plate following Royden et al. (1983) and Balla (1984).

The research activity along line 2T has provided principal information also on oil and natural gas deposits. It was revealed that in the northernmost part of the Orava region there is a relatively thick (around 3 km) accretionary complex of the Krosno flysch overlain by the Magura flysch (Tomek et al. 1989).

The whole transect 2T consists of partial profiles 2T/83,84; 2AT/84 and 2T/85 (Fig. 2.14). Seismic results of profiles 2T are presented in a joint section of partial profiles with selected reflections (Fig. 2.15), in which the most significant reflection phenomena are plotted and designated with letters.

Letter A designates the groups of reflections dipping towards SSE. The deepest ones can be observed at 5.0 – 5.5 s TWT. These groups of reflections (Fig. 2.15) are most probably manifestations of crushed zones in the Magura flysch (A_3), then appearances of the tectonic interface between the inner Magura flysch and the outer Krosno flysch in the basement (A_2), and the lowermost one is interpreted as the tectonic interface between the West Carpathian allochthon and autochthonous basement of the Brunnia (A_1) – Fig. 2.16, which could be overlain by autochthonous sediments of probably Devonian up to Karpatian ages. A very flat overthrust of the Carpathians, noticeable in the whole area of the Polish Carpathians, bends in the Orava area sharply along reflections A_1 and A_2. This is interpreted as bending structure of the lower plate (perhaps the passive margin of the Brunnia) in the final stage of the Krosno sea floor subduction.

A series of segmental reflections B at interval 7.5 – 12.0 s TWT bent and dipping southwards probably characterizes the lower part of the crust of the Precambrian Brunnia (Fig. 2.15). Similar intensive reflectivity of the lower crust has been observed in many sections in the Hercynian and older crystalline units on the majority of profiles in Western Europe (BIRPS, ECORS, and DEKORP projects). The bending of lower-crust reflections of the passive margin of the Brunnia caused by the Miocene subduction of the Krosno sea, which terminated here approximately 15 Ma, represent one of the most significant discoveries made on profile 2T. The subhorizontal Moho boundary (letter M_1) is not shown as a set of striking reflections. It is interpreted as the transition zone between the stratified lower crust and the more or less reflectionless uppermost part of the mantle. It seems that the subhorizontal course of the Moho discontinuity was formed during the last 15 Ma.

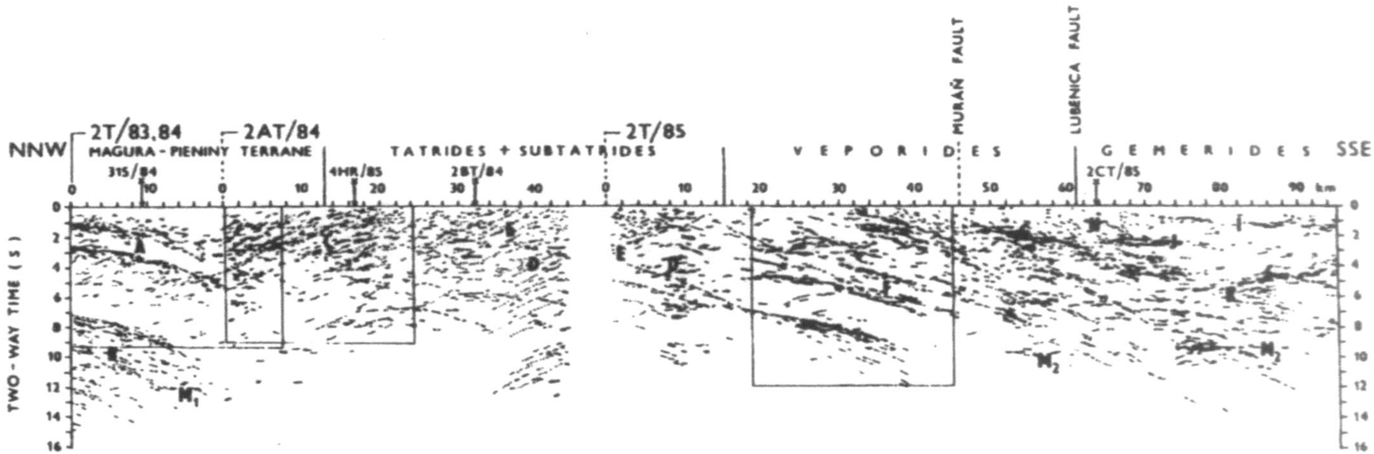

Fig. 2.15 Composite section of selected reflections along 2T profile.

Letter C (Figs. 2.15 and 2.17) shows a complicated reflection pattern consisting of groups of partial reflections of varying dynamics, dipping NNW, that often interfere with shorter subhorizontal reflections. The zone with reflection elements prevailingly dipping NNW includes tectonic topography of the upper plate (the Magura flysch, the Pieniny Klippen Belt, the Tatricum, and the sub-Tatricum). Reflections are most probably due to tectonized rocks at the overthrust faults (see Fountain et al. 1984) with SSE polarity. Individual reflective tectonic zones C_1 to C_2 (Fig. 2.17) form a package 8 to 10 km thick. All the above mentioned terrains, both in outer and Inner Carpathians, are present in the imbricated system. In our opinion a common imbrication took place at the moment of subduction before the Lower Badenian (there is no deformation of the Orava basin).

A less significant system E (Fig. 2.15) in the Nízké Tatry Mts. is similar to the system C. Individual subgroups (E_1, E_2, E_3) of system E intersect the group surface from 39 km (2AT/84) to 12 km (2T/85). These reflexes can be observed, similarly to reflexes C, at 4.0 s TWT. The system of tilted reflections C and E forms a band, approximately 55 km wide, which is probably the zone of reduction of the upper brittle crust of the Carpatho – Pannonian plate. Their "décollement" cannot be observed by seismics, but as there are no tilted reflections, it might exist in the interval 3.5 – 4.0 s TWT (corresponding depth of 10 – 12 km).

In the first part of profile 2T/85, the crustal reflections are of a similar character as in the area N of Nízké Tatry Mts. E reflexes tilted to NNW occur there, too. Beneath them, however, besides the horizontal reflexes of D system, groups of F reflections tilted to SSE occur at higher times, i.e. 4.0 s TWT and more. Further to SSE, very striking reflections of the F system can be observed, which group into bunches and which are generally tilted the SSE at 0.0 to 10 s TWT. Without a major interruption reflections can be observed for at least 10 – 30 km (Fig. 2.18).

With regard to an intensive shear deformation and mylonitization of crystalline schists of the Veporicum we assume that they come from majority of reflections of the F system tectonic zones and that the most significant reflections are those from mylonite zones (Fountain et al. 1984).

Reflection phenomena F_1 can geologically be interpreted as signs of the Lubietová zone of the Veporicum. The base of subsystem F_1 could then correspond to the Čertovica fault. The group of reflections F_2, which are tilted and, where they are near the ground surface, subhorizontal, is striking. It is possible that further NNW less prominent reflections project from the group towards SSE, which could correspond to scales of the Kraklová zone of the Veporicum. A bunch of reflections F_3 ascending to the ground surface at an angle of $25^{\circ} - 30^{\circ}$ is seismically striking. Geologically, it corresponds to a thrust of the Kráľova Hoľa zone of the Vepor crystalline complex over the Kraklová zone at 29 km on the profile. Reflections of the F system in the Veporicum remind us, due to their character, location in the crust, dips, the band-width, and positions in the orogene, of the collision suture of the North and South Appalachian orogenes in the east of the United States (Ando et al. 1984, Nelson et al. 1985) recorded on the seismic reflection profiles of the COCORP project.

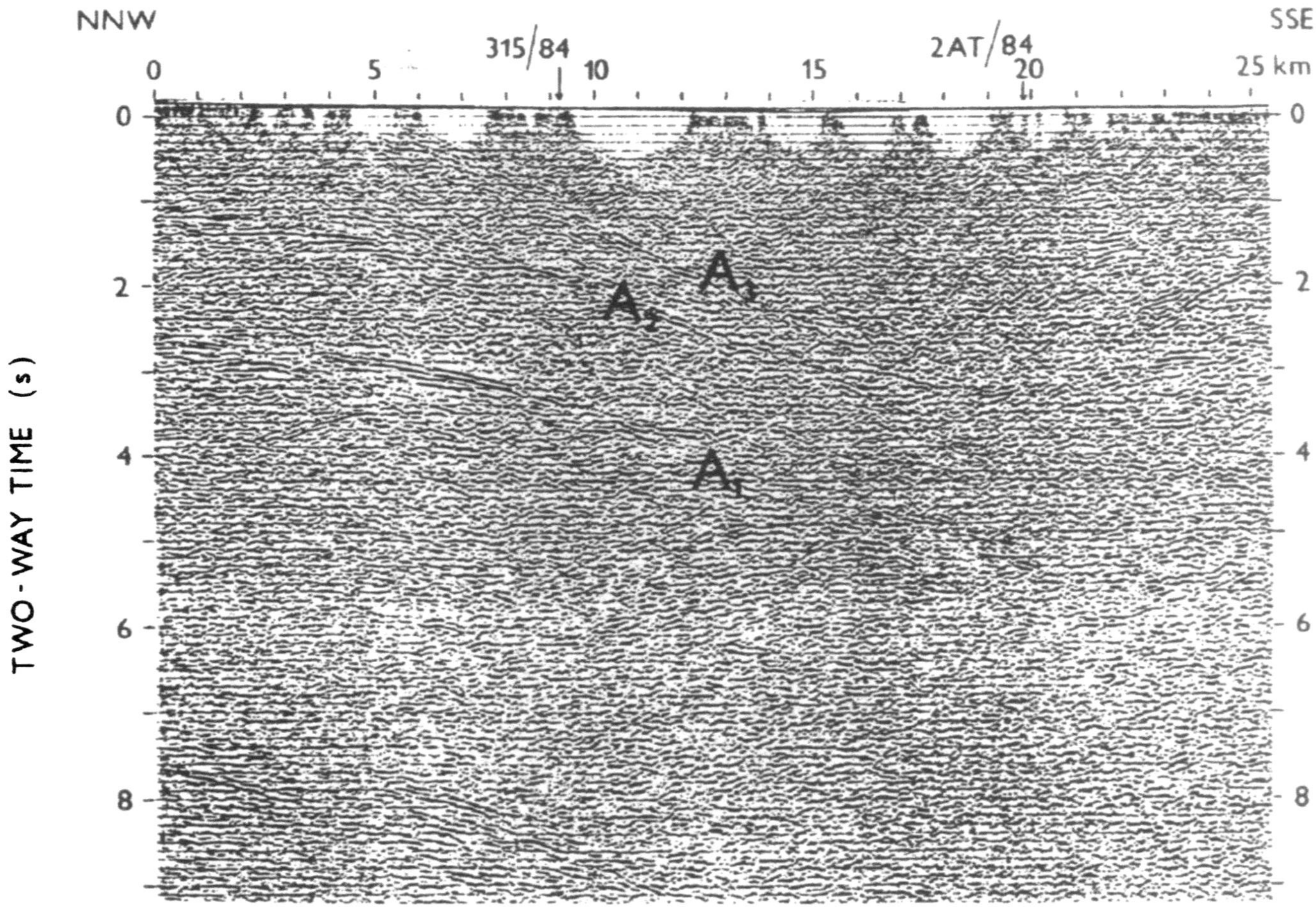

Fig. 2.16 Time section of a part of profile 2T – the continental crust margin bend.

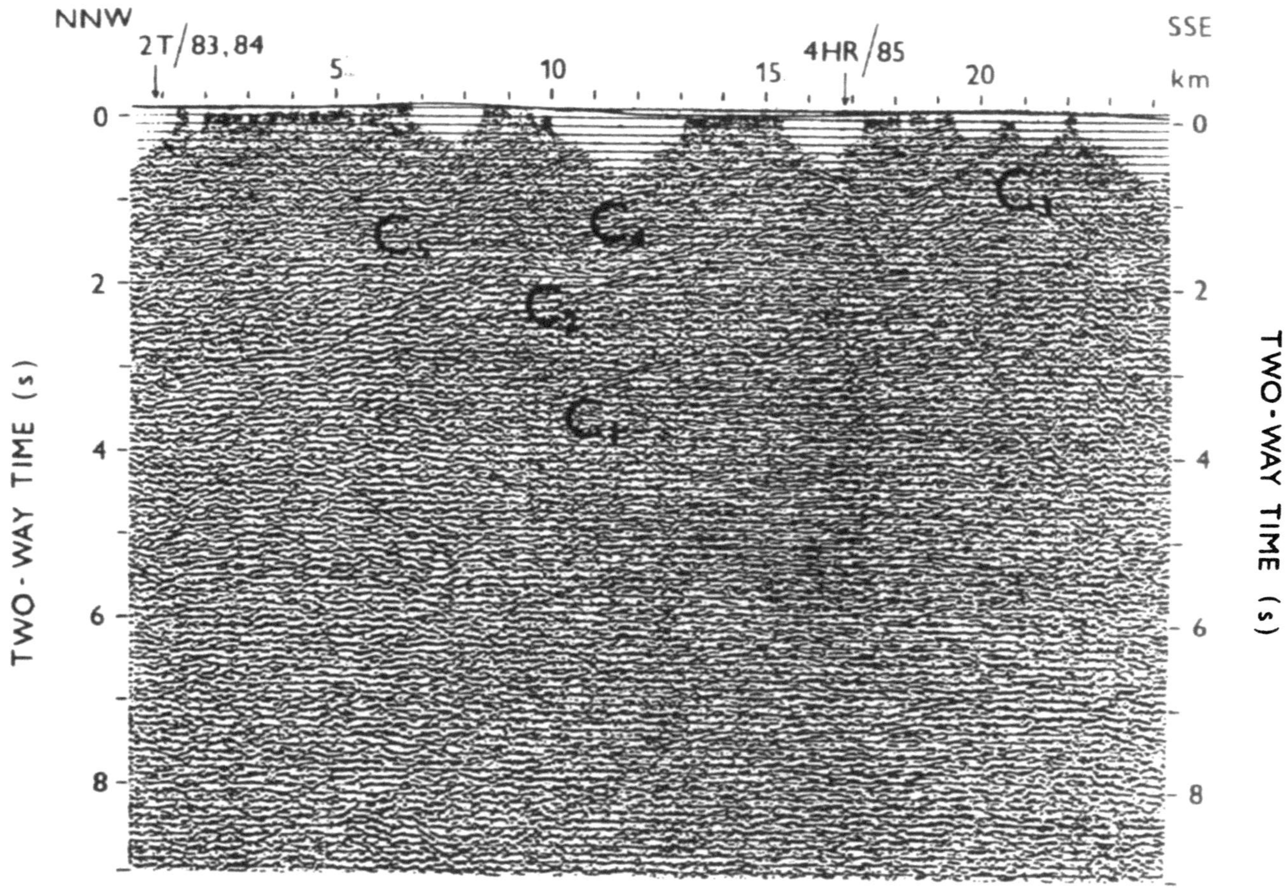

Fig. 2.17 Time section of a part of profile 2T – imbrications in the upper plate.

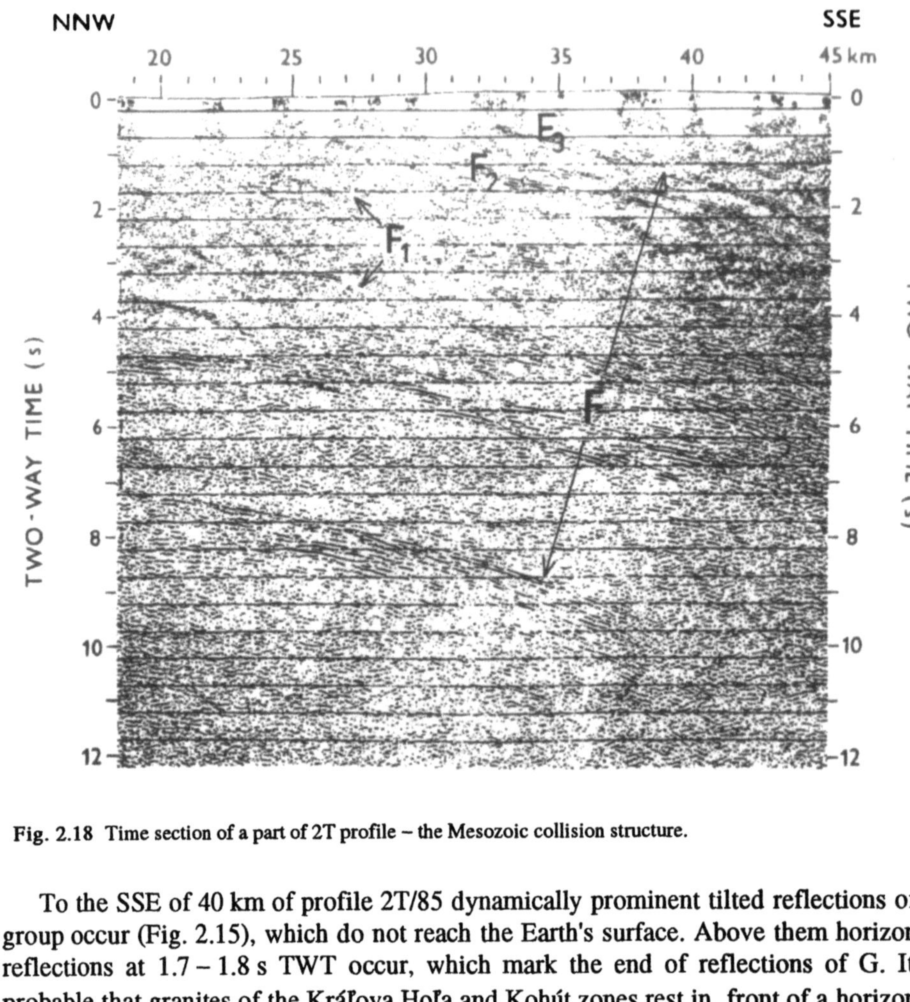

Fig. 2.18 Time section of a part of 2T profile – the Mesozoic collision structure.

To the SSE of 40 km of profile 2T/85 dynamically prominent tilted reflections of G group occur (Fig. 2.15), which do not reach the Earth's surface. Above them horizontal reflections at 1.7 – 1.8 s TWT occur, which mark the end of reflections of G. It is probable that granites of the Kráľova Hoľa and Kohút zones rest in front of a horizontal nappe on the scales of G group which are also included in the collision suture (see also the geologic interpretation by Klinec 1975). At 40 km profile 2T/85 intersects the Muráň fault which is shown in the migrated time-section 2T/85 only indirectly by an interruption of the uppermost F system reflections.

Between 52 – 66 km dynamically less striking subparallel reflections H occur reduced to 3.0 s TWT, steeply tilted to SSE. They probably form a part of the tectonically affected crystalline unit of the south margin of the Veporicum, of the Kohút zone, and of the Gemericum overthrust. The contact of the Veporicum and Gemericum is tectonic and relatively steep. The tilted reflections corresponding to tectonic imbrications extend into both the units. In the Kohút zone, residuals of Gemericum nappes are imbricated together with the Veporicum.

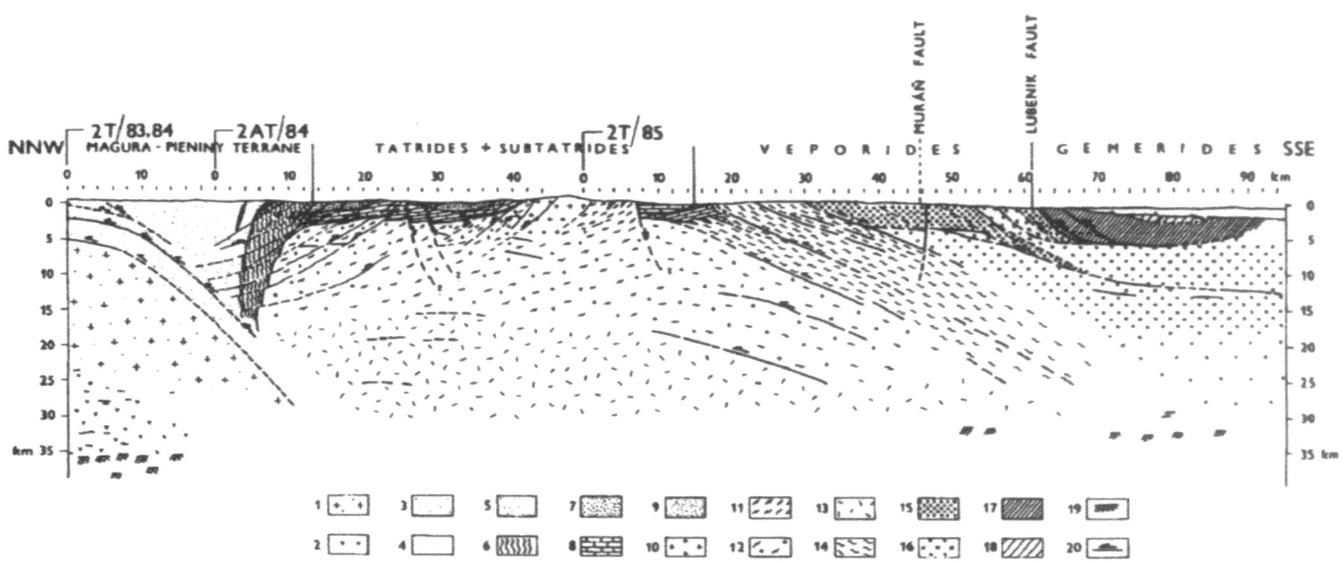

Fig. 2.19 Idealized geologic section of the West Carpathian crust along 2T profile: 1 – upper crust with the sedimentary envelope of the passive margin of the European plate (Brunnia), 2 – lower crust of the passive margin of the European plate (Brunnia), 3 – subduction accretion complex of the Krosno sea, the upper Carpatho–Pannonian plate, 4 – Neogene of Lučenec and Rimava depressions (South-Slovakian basin), 5 – Inner Magura subduction accretion complex, 6 – Pieniny Klippen Belt, 7 – Paleogene of the Inner West Carpathians (Podhale), 8 – sub-Tatricum, 9 – Mesozoic Envelope units of Tatricum, 10 – granitoids of Tatricum, 11 – crystalline schists of Tatricum, 12 – upper crust of Tatricum or of other crystalline complex underlying Tatricum, 13 – lower crust of the same complex, 14 – northern Veporicum, 15 – granitoids of Veporicum, 16 – southern Veporicum, 17 – Gemericum, 18 – Silica nappe, 19 – Moho discontinuity, 20 – overthrust fault.

The Gemericum SSE of this tectonic zone (from 66 km on profile 2T/85) exhibit a different character of reflections as horizontal to subhorizontal reflections of various dynamics and lengths prevail.

The prominent horizontal band of reflections J between 66 – 73 km, which continue as discontinuous reflections towards SSE at 2.0 – 2.5 s TWT, are interpreted as the master décollement of the Gemericum and the south Veporicum.

Similar groups of reflections K between 66 – 95 km at 4.0 – 4.5 s TWT are interpreted as an extension of the overthrust of the Kohút zone and Veporic granites.

On the basis of the seismic data an ideal geologic model was compiled along the whole transect 2T (Fig. 2.19). The model clearly shows two different segments of the Carpathian crust with different periods of final deformation.

In the NNW (10 km on profile 2T/83,84) the palaeosubduction zone occurs with an accretion complex of the Krosno flysch formation tilted, after bending, at an angle of 40°, and with depth the angle probably increases. Then the frontal inner-Magura imbrications and backthrusts at an angle of 30° – 40° on the Orava – Magura nappe, the Pieniny Klippen Belt, the Tatricum with the tectonic envelope of the sub-Tatric nappes, and the sedimentary cover of the central - Carpathian Palaeogene. The northernmost overthrusts at NE Orava are already covered by the Orava basin of a probably Badenian to Sarmatian in age. The time boundaries are thus obvious. That is why the deformation including backthrusts and the upper crust shortening in North Slovakia are classed with the period of final subduction processes (commencing at the Middle Miocene collision).

In the SSE half of the section (from 10 km on partial section 2T/85) reflection segments dipping at angles of 20°°– 30° to SSE predominate. Beneath the Veporicum they form a 20 – 30 km thick collision suture of the Carpathian Cretaceous orogene. This part of the West Carpathians in the Veporicum and Gemericum has not been markedly affected by later deformation movements for the Mesozoic collision suture is well preserved in the time section.

The geologic section also shows the thrust of the Veporicum over the Tatricum along the Čertovica fault and the root of the Krížna nappe in the North Veporicum. The Muráň fault is interpreted as a post-Turonian sinistral strike–slip fault belonging probably to later stages of the collision. The section also shows subhorizontal thrusting of the Vepor granitoids over the crystalline complexes as well as its roots. It also shows thrusting of the imbricated crystalline complex of the Kohút zone together with imbrications of the Gemericum over these granitoids along the Lubeník fault. In our interpretation, the Gemericum is a subhorizontal allochthonous complex only about 7 km thick.

Our interpretation of 2T line revealed its good accord with the perfectly elaborated geologic model of the inner West Carpathians and its lesser accordance with the Tertiary models. The concept of the West Carpathians has so far corresponded to that of Matějka (1927) and of Matějka and Andrusov (1931) being modified to a larger or lesser extent by Biely and Fusán (1967), Andrusov (1968), and Maheľ (1984a, 1986).

Time section 2T confirms the classical model of the thrust of the Veporicum (Matějka and Andrusov 1931). In our geologic section, the Veporicum are presented as an Upper Cretaceous whole-crust collision suture, seismically and tectonically comparable with the sutures of the Palaeozoic collision orogene of the Appalachians and of the recent collisions in the Himalayas. The Gemericum are thrusted over the collision complex of the Veporicum and Tatricum.

The interpretation of time section 2T brings a new insight into the understanding of the history of the Tertiary deformation processes in the West Carpathians. The whole-

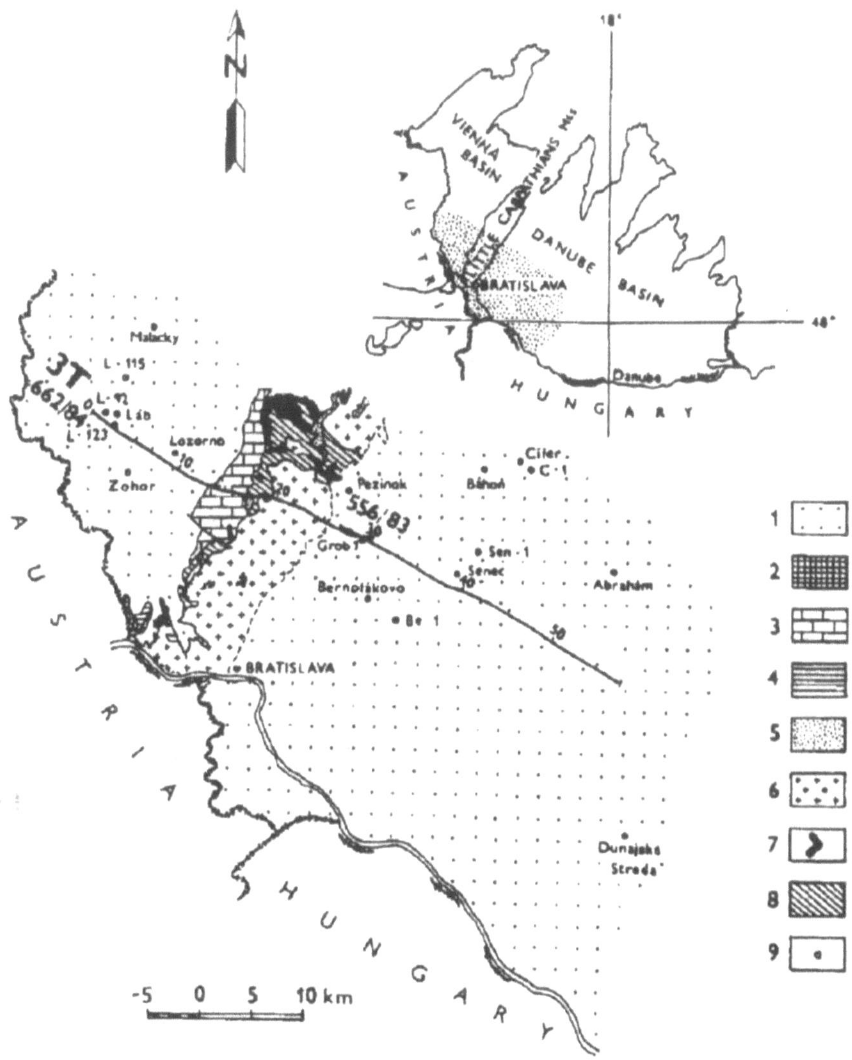

Fig. 2.20 Geologic situation of 3T profile: 1 – Neogene, 2 – Jurassic – Cretaceous (the Malé Karpaty Mts. subunit – limestones and silica rocks), 3 – Jurassic (limestones), 4 – Triassic – Jurassic (the Malé Karpaty Mts. subunit – Devín facies, limestones, dolomites), 5 – Triassic (quartzites, in upper positions spotted schists), 6 – variscan granitoids, 7 – amphibolites, 8 – phyllites, biotite micaschist gneisses, 9 – deep boreholes.

crust flexure of the lower European plate consisting of the Brunnia is indicated by the most significant reflections in the northern part of profile 2T. The flexure can be explained as a result of the termination of subduction movements when also the passive continental margin of the Krosno sea was subducted beneath the Carpathian – Pannonian plate. Beneath the area between Pilsko and Chabenec, the reflection elements in the upper plate consisting of the Magura – Pieniny complex and the Inner Carpathians probably correspond to the phenomena which originated prior to the subduction although the elements in some imbricated backthrusts could have originated during the subduction.

Profile 3T

Deep seismic reflection investigations along profile 3T were to contribute to understanding the structural–tectonic position of the Malé Karpaty Mts. and their wider surroundings and also to provide seismic data on deep parts of the Earth's crust including the Moho discontinuity.

Profile 3T, consisting of two partial profiles 556/83 and 662/84, stretches in total length 58 km in NW–SE direction, from the Vienna Basin over the southern part of the Malé Karpaty Mts. to the Danube Basin (Fig. 2.20). It runs through the Láb horst, the most marked elevation unit in the Vienna Basin, the Zohor – Plavecký Štvrtok graben, through the crystalline core and marginal sediments, then through the marginal block region, through the Tomášovce depression to the Inovec horst-like mega-anticline. Time-section of profile 3T (Fig. 2.21) can be divided into the following three seismically different parts: the Vienna Basin, the Danube Basin, and the Malé Karpaty Mts in between.

In the Vienna Basin, in the NW part of the profile, prominent reflections A_1, A_2, and A_3 occur from its beginning as far as 13 km. These reflections probably originated in the Vienna Basin basement which is markedly discontinued by antithetic normal faults forming the Zohor – Plavecký Štvrtok graben. Reflections within the Vienna Basin are due to various lithostratigraphic boundaries. Beneath 8 km, intensive disturbed waves occur. There is a great energy loss in the Vienna Basin basement, which results in an absence of reflections in the wavefield pattern. The Vienna basin represents tectonically very complicated region with sediments causing a strong attenuation of seismic energy.

Comparing the seismic data, it is obvious that in the Danube Basin (Fig. 2.22) there are far more favourable seismological conditions than in the Vienna Basin. Reflections L in this part of the profile (ca. from 26 km to its end) originated in the basement of the Danube Basin.

Between 18 km and about 22 km (at 0.6 – 0.7 s TWT), horizontal, dynamically prominent reflections B can be observed which approach the ground surface in the WNW direction approximately after 18 km. This probably is the thrust plane of the Bratislava massif thrust over the Envelope unit of the Mesozoic. Above the reflections B, less prominent reflections tilted SE can be observed, which reveal the probably imbricated structure of the Bratislava massif.

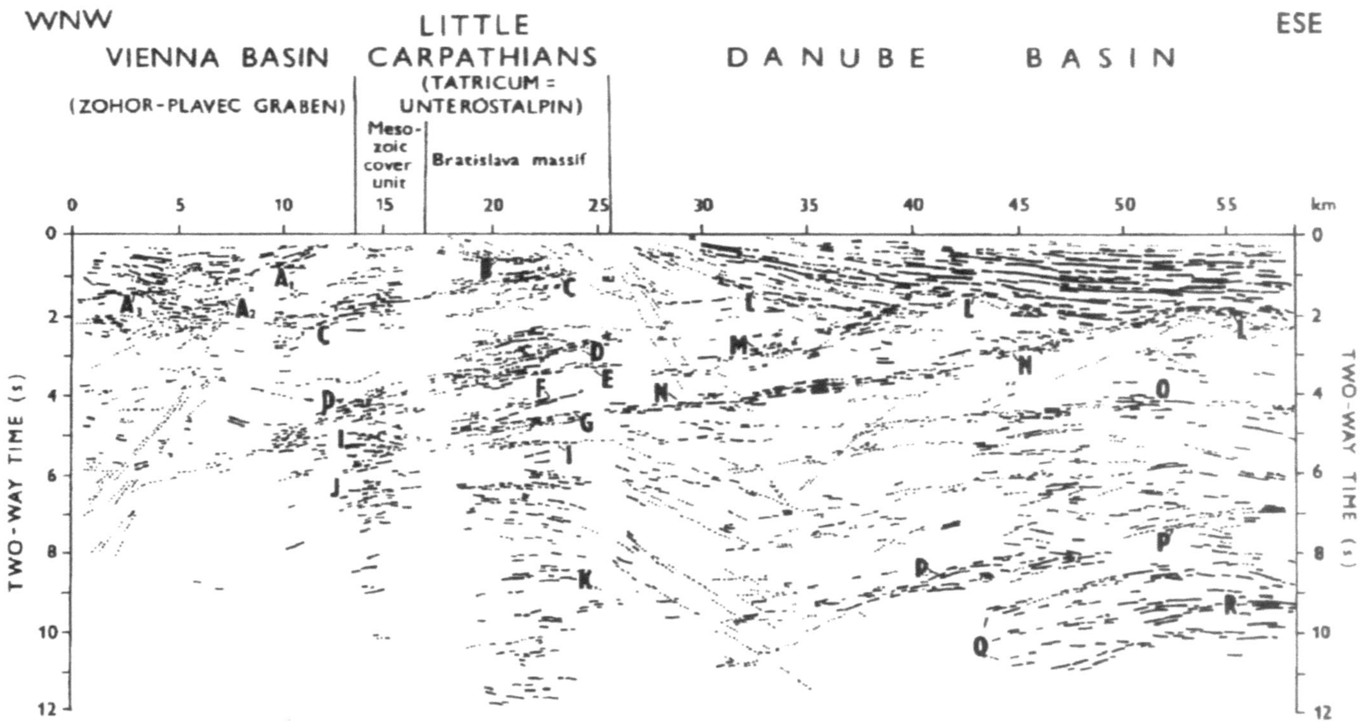

Fig. 2.21 Selected reflections in 3T section.

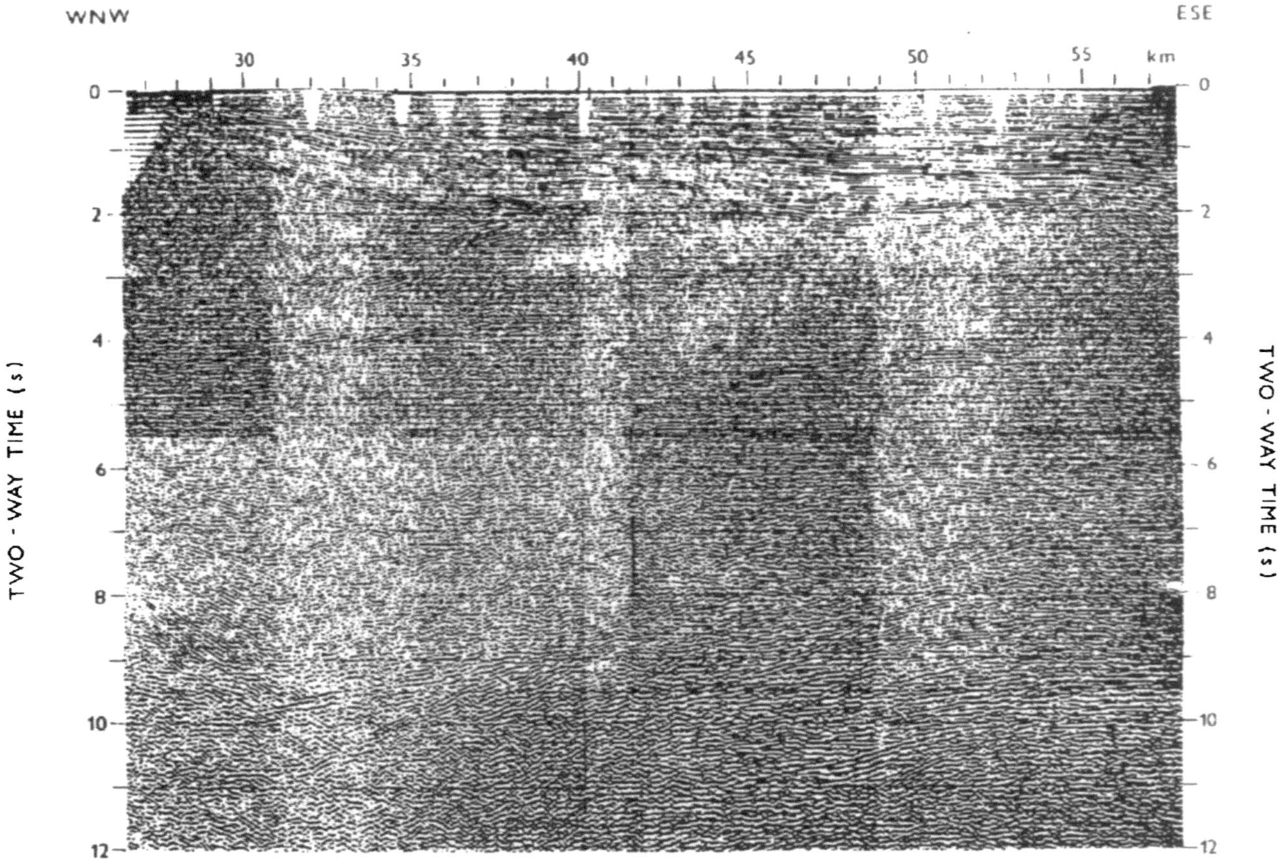

Fig. 2.22 Time section of a part of 3T profile (the Danube Basin).

In the Malé Karpaty Mts. and in the Danube, several prominent NW dipping reflections occur and lie in continuously correlated segments. These are groups of reflections E, G, N, M with small dips and reflections C, D, and F with significant dips. This could be interpreted as thrust planes of great areal extent and depth penetration within the whole upper crust. At 4.0 – 6.0 s TWT groups of horizontal and subhorizontal reflections occur which can be in places easily correlated. These are reflection element complexes I, J, and O. Their wavefield patterns are not clear enough as they are obscured by numerous distorting waves. Their nature may be related to geologic properties of the lower crust.

In the NE part of the profile, the Moho discontinuity can quite reliably be identified through reflexes R at 9.3 – 10.5 s TWT. The reflexes diminish towards NW.

Packages of prominent tilted reflections P and Q (at 7.0 – 11.0 s TWT) could probably originate from thrust planes of the lower crust, of which the lowermost ones extend down to the upper mantle.

From the geologic viewpoint, profile 3T (Fig. 2.23) is an outstanding example of all kinds of crustal movements along faults. The profile shows thrust, normal and strike–slip faults, which were active at different times. While thrust and normal faults can be derived directly from the seismic reflection data, strike–slip faults can be detected only indirectly. The thrust of the Bratislava Massif over the Envelope unit of the Mesozoic is probably the oldest tectonic element observed along the profile. Based on analogy with the thrusting in Tatricum it is ranged with the Upper Cretaceous (Turonian) tectonic movements. Reversed NW dipping thrust faults of various tilts (20° – 35°) are ranged with the collision tectonics phenomena of the Carpatho – Pannonian and Euroasian (the Brunnia) plates.

Low-angle normal faults beneath the Danube Basin were active in the Badenian (the east fault), and in the Sarmatian and Pannonian (the west fault). It cannot be excluded, however, that the origin of listric normal faults was caused by a reactivation of older thrust faults. Listric normal faults attest to a tension in the Pannonian Basin in the Badenian to Pannonian, which resulted in a total crust thinning to a thickness of about 27 km. Normal extension tectonics in the Danube Basin was interrupted for a short time by the Pliocene compression period which caused a high-angle thrust faulting which took place at the beginning of the Pliocene. The thrust fault can be observed beneath 20 km of the profile. This thrust fault may, at the same time, be the sign of a strike–slip fault.

The transtension–sinistral strike–slip fault, running along the western margin of the Malé Karpaty Mts., i.e. along the Záhorský fault, and in Austria along the Mur Mürz fault (Gutdeutsch and Aric 1977), is the youngest tectonic phenomenon in the 3T section.

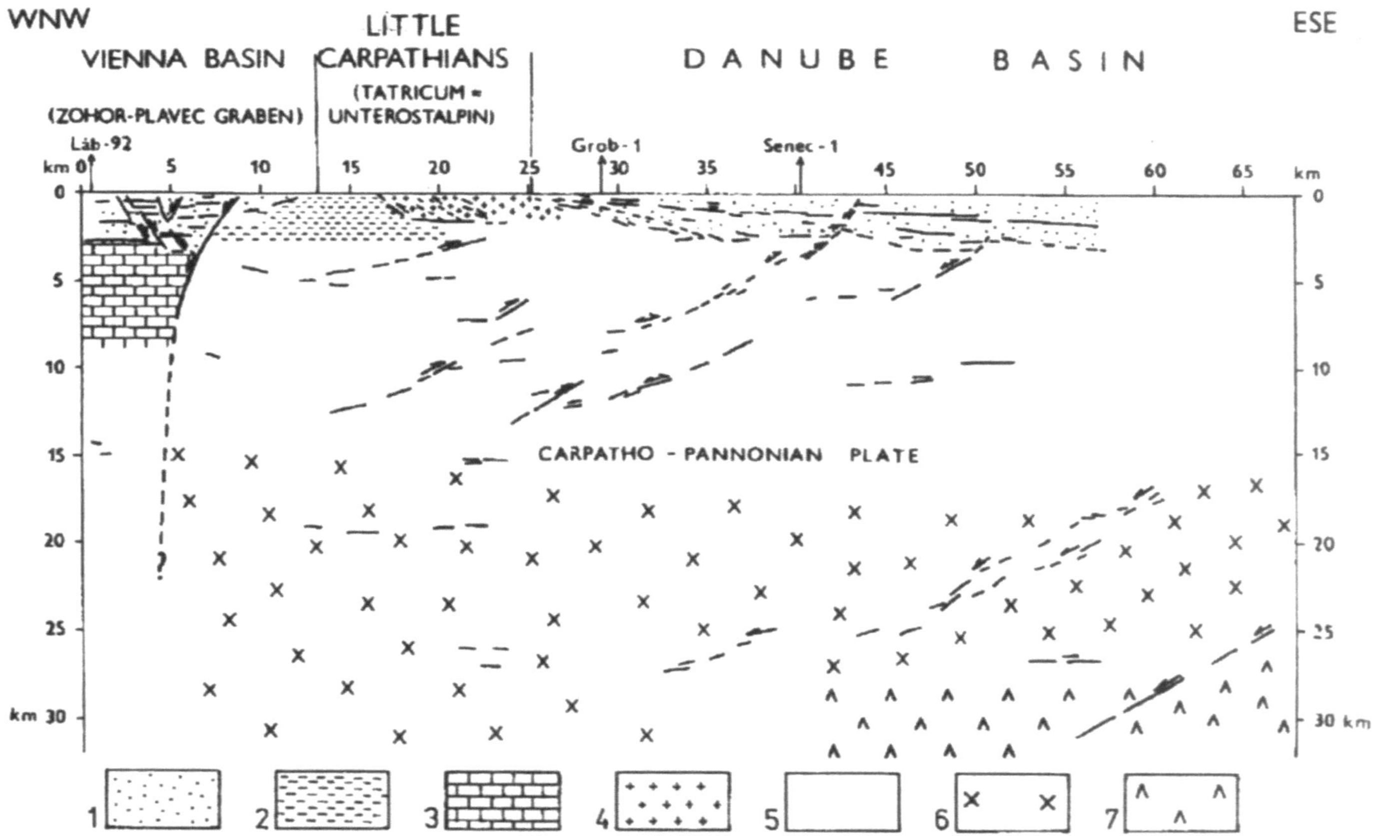

Fig. 2.23 Idealized geologic model section of 3T profile: 1 – Neogene, 2 – Envelope units of Tatricum, 3 – Austro–Alpine nappes, 4 – Tatricum, 5 – upper crust of the Carpatho–Pannonian plate, 6 – lower crust of the Carpatho–Pannonian plate, 7 – upper mantle.

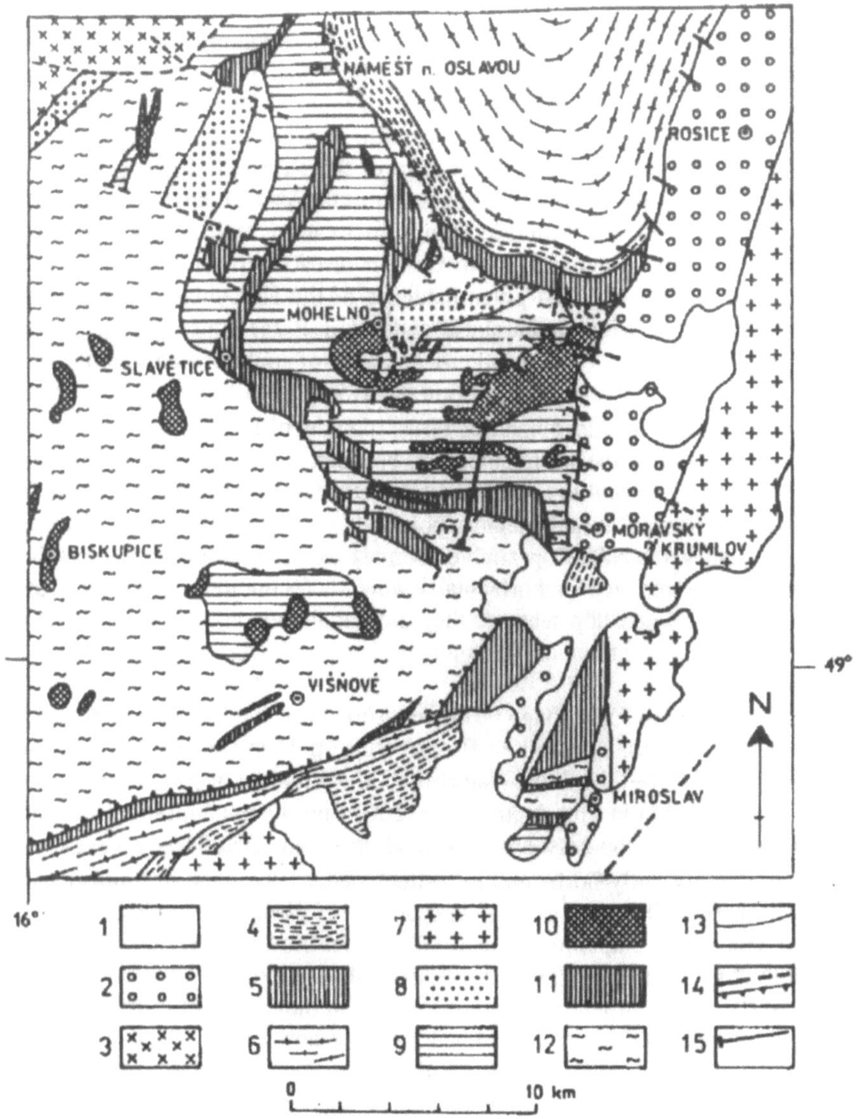

Fig. 2.24 Geologic situation of profile 3: 1 – Neogene sediments, 2 – Permo–Carboniferous of the Boskovice furrow, 3 – Třebíč – Meziříčí massif, Vranov – Olešnice group and Lukov group (Moravicum), 5 – Moravian micaschist belt (Moravicum), 7 – Brno and Dyje massifs, 8 – biotite and sillimanite–biotite paragneisses (Moldanubicum), 9 – 11 – granulitic rocks, serpentinites, amphibolites (the Moldanubian granulite belt), 12 – Gföhl gneisses (Moldanubicum), 13 – boundaries of geologic units, 14 – tectonic boundaries, overthrusts, 15 – localization of the seismic reflection profile.

Profile 3

As part of preparations for deep seismic reflection investigations of the Earth's crust structure in the Bohemian Massif, parametric profile 3 was made in the West Moravian granulite belt at Mohelno (Blížkovský et al. 1986b). The eastern margin of the Bohemian Massif is a region with a very complicated geologic structure (Fig. 2.24), where Suess (1912) assumed a nappe structure of the Hercynides and a long-distance thrusting of the Moldanubicum over the Moravicum (the Moldanubian overthrust). Since then his hypotheses has been criticized many times, and Czech geologists have never accepted it, at least in its extreme form. Current research into the geologic structure carried out in Czechoslovakia has given evidence of the Hercynian nappe structure of the Svratka dome in the Moravicum (Jaroš and Mísař 1974) although their conception somewhat differs from that of Suess. Significant signs of subhorizontal overthrusts were revealed by the seismic reflection data also in the Moravia – Silesian culm. Nevertheless the problem of the large Moldanubian overthrust remains still open.

The eastern margin of the Moldanubicum, between the Svratka and Dyje domes is formed by the Náměšť granulite complex. Its structure is complicated as it was formed by a polyphase deformation and metamorphism. It consists of markedly diverse rock units. In some places, the Náměšť granulite complex forms an antiform where individual rock complexes are deposited subhorizontally which is favourable for the application of seismic methods. The question whether they are autochthonous (forming a part of the layer sequence) or allochthonous (tectonic wedging in overthrusts) is a subject of discussion.

Profile 3 was only 4.5 km long. From south to north it passes through the Gföhl gneisses, the west Moravian granulite belt, and ends up 7 km south of the Náměšť fault where Suess (1912) placed the thrust plane of the Moldanubian overthrust.

It can clearly be seen in time section (Fig. 2.25) that the dynamics and frequency of reflections in all parts of the crust are very good. In the interval 0.6 – 1.8 s TWT, three prominent groups of nearly horizontal reflections occur. The most prominent one is at 1.5 s TWT. These reflections do not appear to be associated with the superficial positions of granulites, amphibolites, and Gföhl gneisses but run without interruption beneath both the complexes. They could represent either the original lower-crust reflections of the planar structures which originated in the course of plastic flow at high metamorphosis under the lower crust conditions, or horizontally deposited magmatic rocks.

The most significant group of reflections consists of reflections dipping S at 1.9 – – 2.5 s TWT. This nearly 2 km thick system could represent a dislocation zone of the Moldanubian overthrust (Suess 1912). The reflections are probably due to mylonitized bands of rocks each of them at least 80 m thick, in order to act as a prominent reflector (Fountain et al. 1984). In the idealized geologic model (Fig. 2.26), where all seismic events are assumed to be in the profile plane, these reflections can be seen to be tilted towards the surface outcrop of the Náměšť fault. These results are in accordance with the Suess' conception of the Moldanubicum as an overthrust complex resting on the less

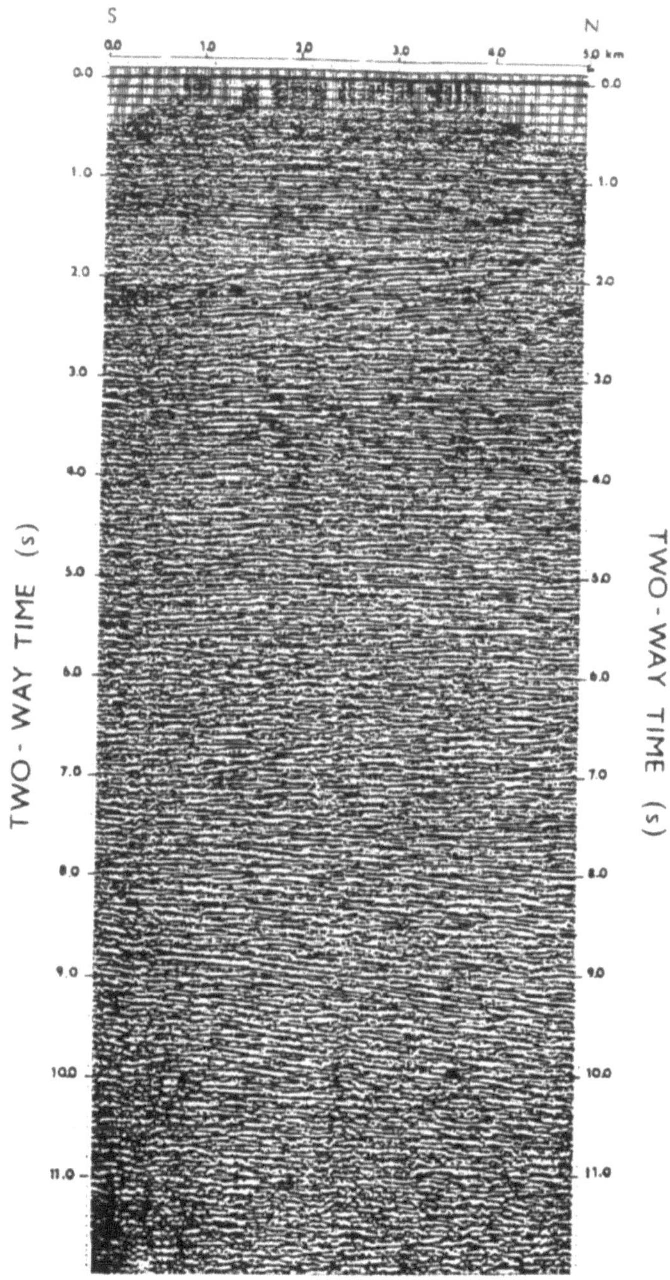

Fig. 2.25 Time section of profile 3/83.

metamorphosed rocks of the Moravicum. The thrust of the Moravicum over the Brunovistulicum interpreted by Jaroš and Mísař (1974), could be represented by prominent subhorizontal reflections between 3.5 – 3.7 s TWT. This interpretation, however, is highly speculative with regard to the length of profile 3. It cannot be excluded that stratification of deeper parts of the Moldanubicum in this region is far more intensive than it is assumed. It is well-known that at depths around 10 km and more even very complex structures can cause subhorizontal reflections similar to sedimentary basin reflections.

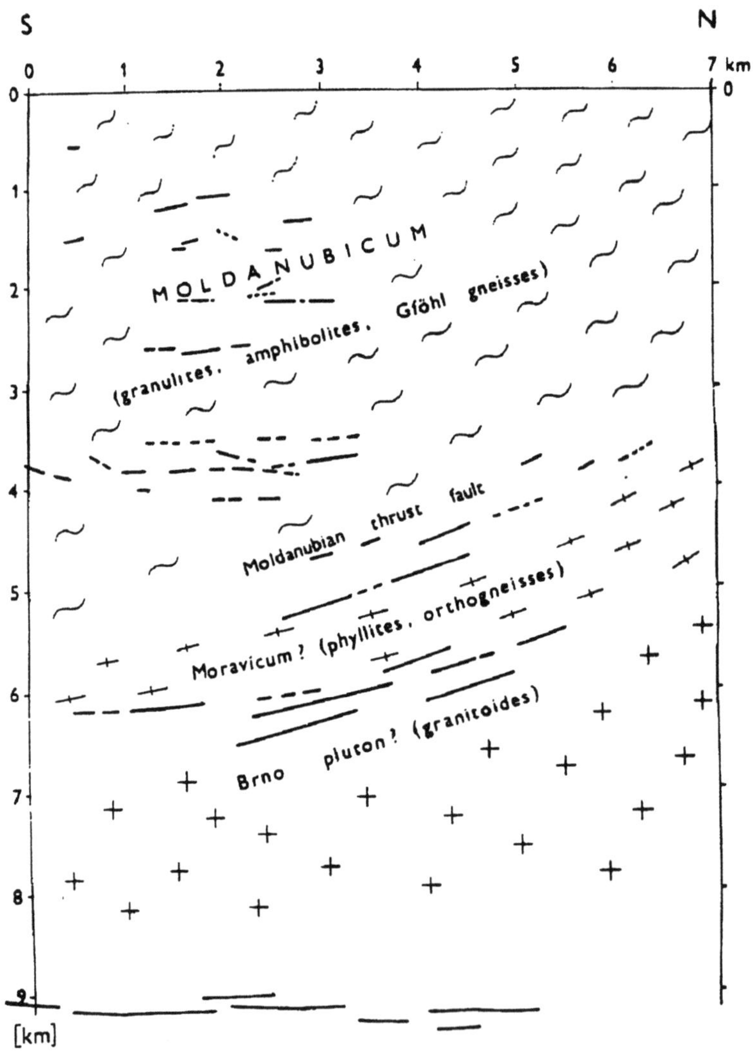

Fig. 2.26 Idealized geologic model section of profile 3/83.

Within the zone at 3.7 – 4.9 s TWT only incomplete reflections occur which cannot be explicitly interpreted. More expressive is the horizontal boundary at 4.9 s TWT. Beneath the corresponding depth of about 15 km numerous diversely tilted groups of reflections can be observed which are quite dynamically striking. Similar reflections are frequent in the lower crust. The majority of reflections in section 3, namely the prominent N–dipping reflections in the interval at 8.8 – 9.4 s TWT, are, after a conversion to depth, shifted several kilometres towards the south. It shows a complex structure of the lower crust at depths of 25 – 30 km beneath the eastern margin of the Moldanubicum. The less prominent and discontinuous horizontal reflections at 11 – 12 s TWT, namely those in the lowest part of the section, probably represent the transient crust–mantle zone.

2.5 Model of the Deep Lithosphere Structure

V. Babuška[1] , J. Plomerová[1], J. Šílený[1]

From the seismological point of view, the lithosphere can be understood as an upper envelope of the Earth's body, which includes the crust and the uppermost part of the mantle down to the low-velocity layer, where a change in physical properties of rocks, caused by their partial melting, occurs. That is why seismic wave velocities are higher in the lower lithosphere than those in the asthenosphere. While numerous contributions in this monograph deal with the structure of the crustal part of the lithosphere, here we try to augment the understanding of its subcrustal part.

With the use of the method for studying the subcrustal lithosphere, elaborated by Babuška et al. (1984a), the values of mean relative residuals in a region and their spatial variations are examined. Arrival times of *P*–waves at a network of seismic stations for a set of earthquakes represent basic data. Foci of the set must be distributed uniformly around the region from the viewpoint of their azimuths and epicentral distances, so that the rays of incident *P*–waves could "scan" the structure beneath the region under various directions.

In order to eliminate all effects that originate out of the subcrustal lithosphere, relative (normalized) residuals are used. At first absolute residuals (i.e. the observed arrival times minus the theoretical arrival times of *P*–waves) are reduced to the Moho discontinuity at a reference depth of 33 km by introducing corrections for the Earth's crust (according to the results of deep seismic sounding) and for unconsolidated sediments. Then, effects of inhomogeneities originated in deep parts of the Earth's mantle and in foci regions as consequences of their structures , inaccuracies in locating the foci and in determining the origin times, are minimized by the normalization. For this purpose we selected 15 reference stations which are evenly distributed in the region and represent individual tectonic units. For each earthquake the mean residual was

1 Geophysical Institute, Boční II/1401, 141 31 Praha 4, Czech Republic

determined, as the value averaged over the residuals of the reference stations. This normalizing value was then subtracted from absolute residuals at individual stations.

Generally, variations of relative residuals are caused by combined effects of lateral heterogeneities and anisotropic structures. To separate both effects, we express the residuals at individual stations of the region by a directionally independent term – the representative mean value and by azimuth – incidence angle dependent terms, i.e. by relative residual related to the representative mean value of a station (Babuška et al. 1988). The lateral changes of the representive mean residuals reflect the variations of average velocities in the upper mantle. The distribution of relatively low velocities (positive residuals) and high velocities (negative residuals) is not random and can be correlated with the tectonics of the region.

In spite of the fact that corrections for sediments were introduced, the regions of subsidence, i.e. the Pannonian Basin, the Po Basin, and the German – Polish Platform, are characterized by positive residuals and thus by relatively low *P*–wave velocities. However, the Rhenish Massif which, on the contrary, represents a region with constantly rising tendency (Illies and Fuchs 1983), exhibit also positive residuals.

Negative residuals, characterizing regions of relatively high *P*–wave velocities, can be observed in the SW part of the East European Platform and in three relatively narrow zones in the Alpine – Carpathian system: in the western Alps, in the eastern part of the Austrian Alps, and in the South Carpathians at the contact with the Moesian Platform. These are probably zones of Alpine palaeosubductions which manifest themselves as inhomogeneities exhibiting relatively higher velocities than those in the surrounding upper mantle down to depths of around 200 km (Baer 1980, Miller et al. 1984).

The model of the lithosphere thickness is based on the lithosphere definition (Sacks and Snoke 1984) according to which, velocities of seismic waves are higher in the rigid subcrustal lithosphere than in the plastic asthenosphere. Similarly as in the work by Poupinet (1979) we assume in our model that the lithosphere thickness, after introducing the correction for the crust, is directly proportional to the representative mean residuals (Babuška et al. 1987) which reflect average velocities beneath individual stations. We transform variations of the mean residuals into a model of lithosphere thickness assuming the effects of lateral changes of the average velocities in the subcrustal lithosphere are much smaller than the effects of the lithosphere thickness variations. Regions with a relatively thin lithosphere are characterized by positive residuals of the *P*–waves (relatively low velocities beneath the stations due to a lack of the high velocity lithospheric material) and vice versa, negative residuals are typical for regions with a thick lithosphere (i.e. high *P*–wave velocities beneath the stations due to abundance of the high velocity lithospheric material).

The inhomogeneity exhibiting higher velocities in the inner part of the western Alps plunges SE beneath the Po Basin (Babuška et al. 1990), the inhomogeneity in the eastern Alps plunges steeply into the mantle (see diagrams of spatial variations of residuals in Fig. 2.27). In accordance with Miller et al. (1984) and with Lidster (1980) we have interpreted these inhomogeneities as remnants of a "subducted" lithosphere which is still cooler than the surrounding mantle. The subduction results from the continent –

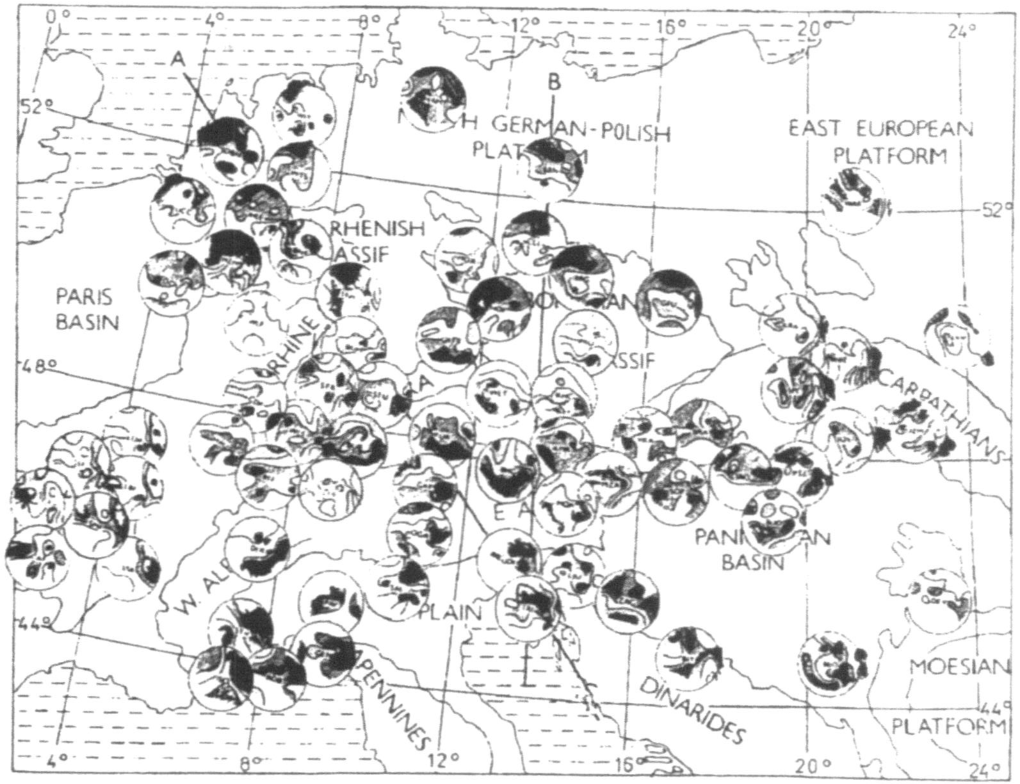

Fig. 2.27 Diagrams of spatial distribution of relative residuals of *P*–waves related to the representative mean residuals of individual stations (Fig. 2.28), expressed in the polar projection for the majority of stations in the region. Steep incident rays with angles around 20° are projected into the central part of diagrams, rays with angles around 50° are projected near the perimeter of the diagrams. Contours are constructed with a step of 0.4 s. The areas of diagrams representing directions of relatively high velocities are hatched, those representing directions of relatively low velocities are stippled.

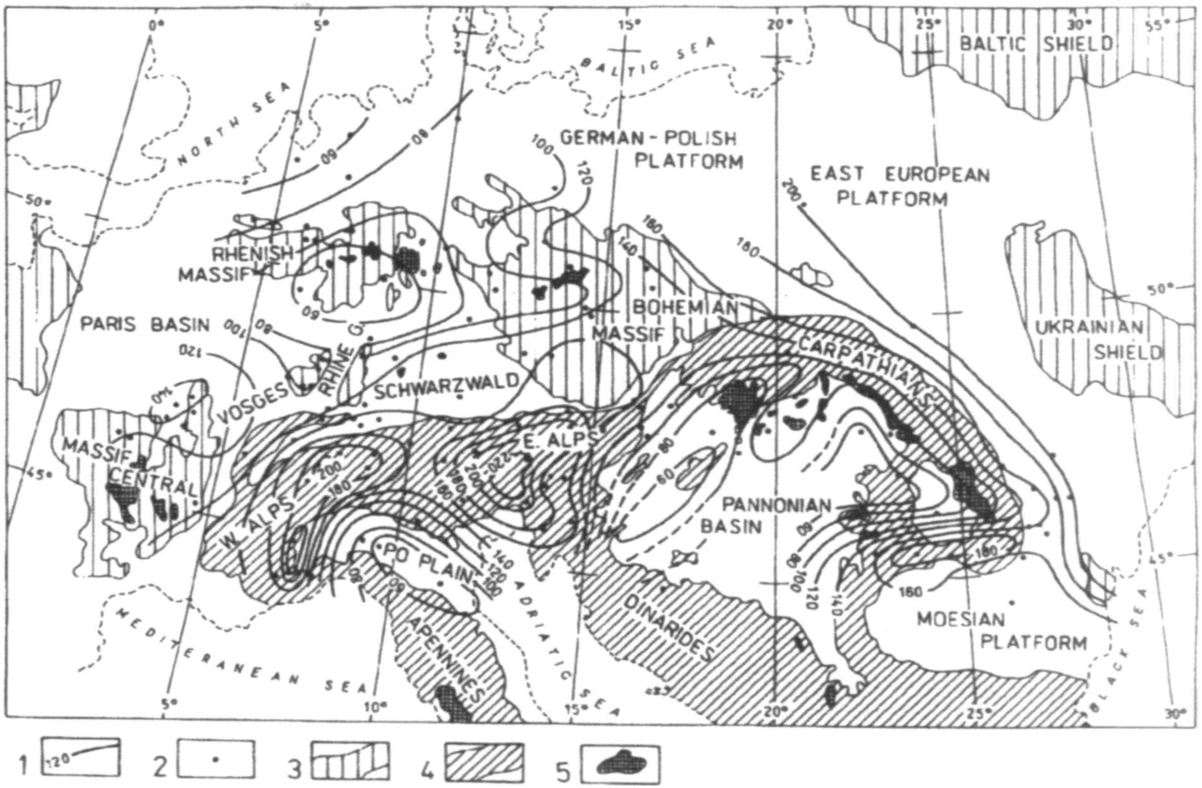

Fig. 2.28 Model of the lithosphere thickness in central Europe. 1 – depth contours of the lithosphere–asthenosphere transition with a step of 20 km derived from the representative mean residuals of *P*–waves, 2 – positions of seismological stations, 3 – Variscan units, 4 – Alpine units, 5 – neovulcanites.

– continent collision between the Eurasian and African Plates. The model thickness of the lithosphere in the central part of the Alps, ranging from 140 to 160 km, is comparable with the lithosphere thickness of the Moldanubian part of the Bohemian Massif.

The lithosphere thickness beneath the Bohemian Massif ranges in our model from 90 to 140 km, which corresponds well to the results of magnetotelluric soundings. These measurements determined the depth of a higher conductivity layer in the upper mantle corresponding to the low-velocity layer in the range from 100 to 150 km (Pěčová et al. 1974, 1976, Praus et al. 1990). From the propagation of surface waves, Panza et al. (1980) derived the lithosphere thickness in the central European Variscides as being around 90 km.

Very good correspondence between the lithosphere thickness determined from the representative residuals of *P*–waves and that obtained from the results of magnetotelluric soundings is observed also in the Pannonian Basin. Ádám (1976) sets values of the depth of higher-conductivity layer in the upper mantle at 60 – 85 km for the central part of the basin and of around 100 km at its margins. The relief of the lithosphere – asthenosphere boundary, derived from *P*–wave residuals, forms here narrow bands of NE–SW direction which correlate with the tectonics. The tectonic line Rába – Rožňava corresponds to the thinnest lithosphere band and the band of a thickened lithosphere (90 – 120 km) detected along the Balaton – Darnó tectonic line correlates with remnants of the ocean lithosphere of the Igal – Bükk zone and with the volcanites on its NW margin, which were interpreted by Csillag et al. (1983) as a part of a fossil island arc. The zone of a reduced lithosphere thickness extends from the Pannonian Basin as far as the inner Carpathians (Fig. 2.28). A gradual increase of the lithosphere thickness (140 – 180 km) is observed in the inner Carpathians and also along the margin of the East European platform.

In order to be able to investigate systematically spatial variations of *P*–wave velocities we constructed polar diagrams of relative residuals, which were always related to the representative mean residual of the station. In the polar diagrams, the directionally dependent components of relative residuals are plotted as a function of azimuths and of incidence angles of the waves approaching the Moho discontinuity. These directional components were obtained from the mean relative residual for a focal region from which the direction-independent representative residual of a station was subtracted. The values of relative residuals determined in this way express their spatial variations. As the zero level is formed by the representative mean residuals of the individual stations, variations of residual patterns can be mutually compared.

Figure 2.27 shows diagrams of the directional dependence of *P*–wave residuals at seismic stations in the region. The distribution of positive and negative residuals in the diagrams is not random. Stations with a similar residual pattern form groups in regions extending several hundreds of kilometres. Besides, it can be observed in numerous diagrams that seismic waves exhibit from particular directions systematically earlier arrivals, and from the opposite directions a systematical delay when compared with theoretical *P*–wave arrival times at individual stations.

Whereas stations in the NW part of the region – i.e. in the Rhenish Massif, Brabant Massif, and in their northern surroundings as well as stations in the Saxothuringicum (Fig. 2.27) are characterized by relatively high *P*–wave velocities that arrive from N, NW and NE, and vice versa, by relatively low *P*–wave velocities of those waves arriving from S, SW and SE. This dependence changes towards the south. On the contrary, stations in central Bohemia and in the Moldanubicum, as well as stations in the Alps south of the Bohemian Massif, are characterized by high velocities of waves arriving from the south. The boundary between the two regions runs in the Bohemian Massif between the Saxothuringicum and the Moldanubicum in the ENE–WSW direction in the vicinity of the GRF and HOF stations, whose diagrams show a transient character between the two types of residual patterns. The border region is characterized by a lithosphere thinning (Fig. 2.28), by increased heat flow (Čermák 1979b), and by an occurrence of the Tertiary to Quaternary volcanism in the surroundings of the Ohře rift. It also coincides with the boundary between the positive and negative region in the map of regional gravity anomalies of Czechoslovakia (Ibrmajer 1981).

The orientation of relatively high velocities of *P*–waves arriving from the south is preserved also at seismic stations in the eastern Alps as far as the Insubric and Gail lines. In the southern Alps, the residual patterns again change. This dependence and thus also the *P*–wave velocities are, in the entire Alps, considerably affected by the lithosphere roots (Fig. 2.28).

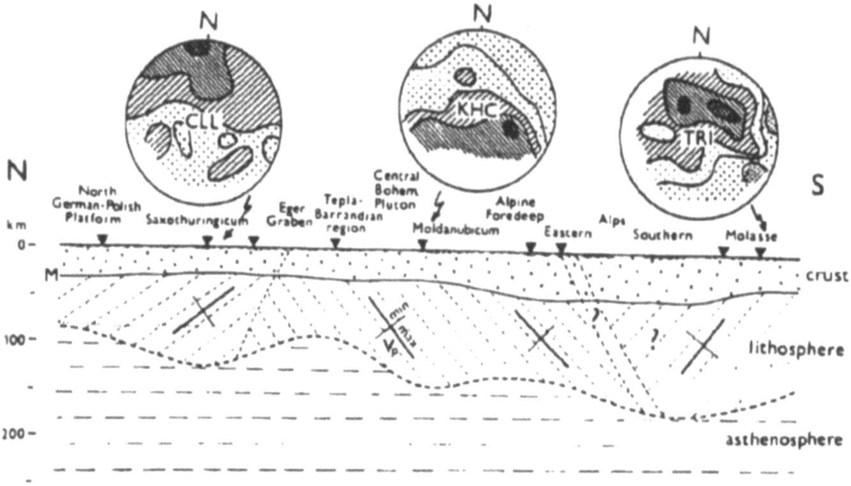

Fig. 2.29 Scheme of an anisotropic structure of the subcrustal lithosphere along profile B (see Fig. 2.27) intersecting the German–Polish Platform, the Bohemian Massif, and the eastern part of the Austrian Alps in the N–S direction. The lithosphere thickness marked with dashed curve was derived from the representative mean residuals (Fig. 2.28), the lower boundary of the lithosphere according to Panza et al. (1980) is dotted. Stations situated in the vicinity of profile B: from the left – BRN, CLL, BRG, PRU, KHC, KMR, BHG, UDI, and TRI are marked with triangles. In the upper part of the figure, there are polar projections of spatial variations of relative residuals of stations CLL, KHC, and TRI reflecting the counter dipping anisotropic structures in the deep lithosphere along profile B (for more details see the caption of Fig. 2.27).

Differences in residuals between the opposite sides of the diagrams commonly reach 1.5 – 2.0 s, and in some cases, they exceed 2.0 s. Such great differences cannot be accounted for by inhomogeneities or by anisotropy structures in the crust, but their causes must be sought in the upper mantle. For example, in order to explain the difference of 1.5 s in the residuals, we need about 260 km of the ray path in an inhomogeneity exhibiting a velocity $v_p = 8.6 \text{ km s}^{-1}$ surrounded by a mantle material with $v_p = 8.2 \text{ km s}^{-1}$. If we admit the existence of such an extensive inhomogeneity,

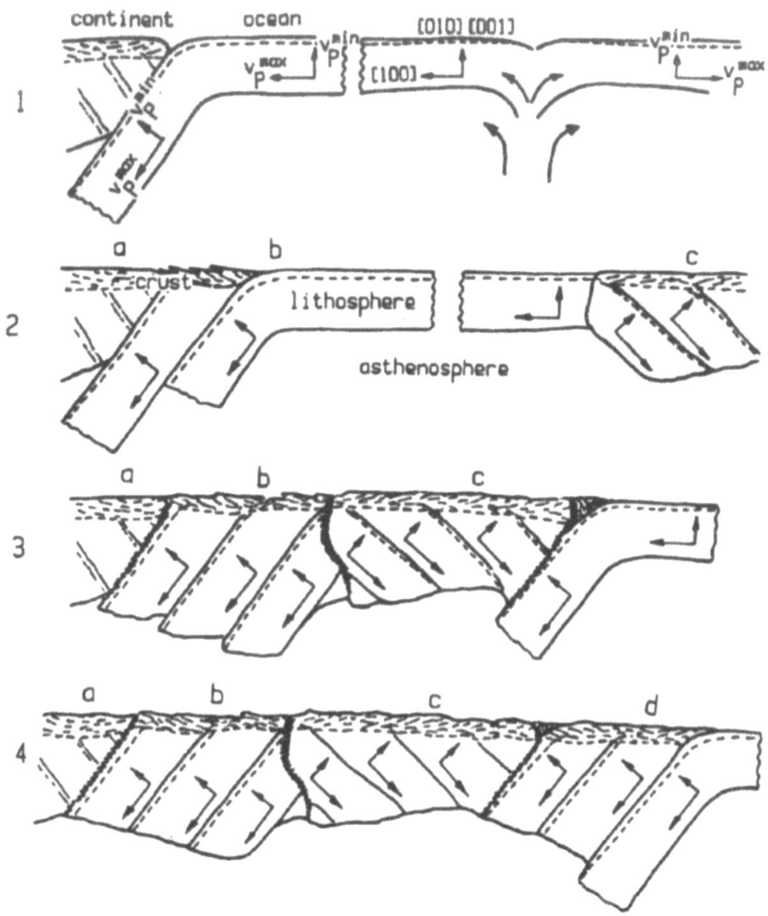

Fig. 2.30 Cartoon illustrating four stages of a possible continent growth by a succession of accretions of the oceanic lithosphere and an origin of a system of inclined anisotropic structures in the continental lithosphere: 1 – anisotropic structure of the oceanic lithosphere due to a preferred orientation of olivine (the axis [100] is in the direction of the oceanic plate motion) is preserved in its subducting part, 2 – extent of the old continental core (a) is augmented by a system of consecutive subductions (b), 3 – after closing-up the ocean, an allochthonous continental fragment (c) which can exhibit a generally diverse structure of the lithosphere adjoins the enlarged continental core, 4 – a series of new subductions of the oceanic lithosphere (d) contributes to further continental growth and to a formation of a new system of inclined anisotropic structures.

which is, as far petrology is concerned, rather improbable, it cannot explain a systematic directional dependence of the residuals. The most probable explanation of this phenomenon is the anisotropy of seismic wave propagation in the uppermost mantle, most likely in the subcrustal lithosphere.

Figure 2.29 shows a schematic section of the lithosphere along approximately N–S profile connecting stations BRN (Berlin) and TRI (Trieste). The section is oriented in the direction of the greatest differences in residuals and approximately normal to the general trend of the central European Variscides (Babuška and Plomerová 1991). In this section, three parts of different orientation of inclined anisotropic structures can be distinguished. These structures are derived from the directional dependence of *P*–wave residuals. Boundaries of diversely oriented structures coincide with important tectonic sutures – the Insubric line which represents the contact of the European and Adriatic plates (Mueller 1982), and the Saxothuringicum–Moldanubicum boundary as a principal deep suture which was determined on the basis of deep seismic reflection investigations (DEKORP Research Group 1985). Orientation of the plunging of structures in the subcrustal lithosphere of the Saxothuringicum and the Moldanubicum, derived from the directional dependence of the residuals (Fig. 2.29), corresponds to the orogene polarity of structures in the upper part of the Earth's crust (Franke 1986). The velocity anisotropy of *P*–waves in the subcrustal lithosphere, calculated from the residuals of waves propagating approximately along the profile in Fig. 2.29 ranges from 6 to 9% (Babuška et al. 1984b). These values correspond well with the anisotropy of olivine ultrabasic rocks determined in laboratory (Babuška 1984).

The system of fossil palaeosubductions of the oceanic lithosphere which maintained its original anisotropic structure in the accretion process during which the continental lithosphere grew, can represent one of the possible explanations of anisotropic structures in the subcrustal lithosphere in central Europe (Babuška et al. 1984b). Figure 2.30 gives an example of such a process which simultaneously goes hand in hand with the formation of anisotropic structures (Babuška and Plomerová 1989).

In conclusion we would like to stress that the estimated thickness of the lithosphere is only a model derived from one of several geophysical parameters. However, it corresponds well e.g. with results of magnetotelluric soundings. The effects of inhomogeneities are represented mainly by lateral changes in the lithosphere thickness and in its temperature. Also, effects of oriented structures of regional extent are strongly reflected in *P*–wave velocities. Distinguishing these structures and determining their orientations can contribute to the formation of realistic anisotropic models of the deep structure and of the geodynamic evolution of the continental lithosphere.

2.6 Deep Structure of the Bohemian Massif from Phase Velocities of Seismic Surface Waves

A. Plešinger[1], H. Neunhöfer[2], E. Wielandt[3]

Introduction

Although several broadband or long-period seismograph stations have been in operation in and near the Bohemian Massif for many years, the number of surface waves dispersion studies for this region is still relatively small. Neunhöfer et al. (1981) analyzed and interpreted the Rayleigh wave dispersion between Moxa (MOX) and Vienna (VIE), a profile that crosses the Bohemian Massif from NW to SE (see Fig. 2.31). A similar study was carried out for the western marginal region of the Bohemian Massif, the profile Moxa – Kašperské Hory (KHC) (Neunhöfer et al. 1983). Theoretical dispersion curves for the profile Kašperské Hory – Ksiaz (KSP), the course of which nearly coincides with that of international DSS profile VII, were derived by Novotný (1983). Wielandt et al. (1987, 1988) measured phase velocities of both Rayleigh and Love waves between KHC and KSP, and inverted the data separately for both wave types to models of shear-wave velocity vs. depth. In the present contribution we briefly summarize the results of the quoted works.

Geologic and Geophysical Characteristics

The path between MOX and VIE essentially crosses three tectonic provinces: in the NW around 25% of the whole path belongs to the Saxothuringicum, in the central part the Bohemian Massif is crossed, and towards SE the Vienna Basin is touched. The profile extends mostly in the Bohemian Massif. According to the results of surface wave investigations in the adjoining areas published by Knopoff et al. (1966) and Neunhöfer and Güth (1975), there probably exists a low-velocity channel in the upper mantle. Neither the crustal structure, nor the subcrustal velocities are homogeneous along the profile. The Moho discontinuity is found at the depth of around 30 km beneath the Saxothuringicum in the NW (Knothe and Schroeder 1972), of 38 km beneath the Bohemian Massif, and of 34 km beneath the Vienna Basin (Beránek and Zounková 1979). The mean *P*–wave velocity in the upper crust obtained from measurements taken along the nearest parallel path – the international DSS profile VI – amounts to 6.3 – – 6.4 km s^{-1}. The depth of the granite – gabbro transition has not yet been defined sufficiently. The subcrustal velocities apparently increase towards SE. The mean upper mantle velocity beneath the profile has been estimated at 8.2 km s^{-1}. The DSS data available suggest the existence of a low-velocity zone in the crust which, however,

1 Geophysical Institute, Boční II/1401, 141 31 Praha 4, Czech Republic
2 Institut für Geowissenschaften, Burgweg 11, Jena, Germany
3 Institut für Geophysik, Richard-Wagner Str. 44, 7000 Stuttgart 1, Germany

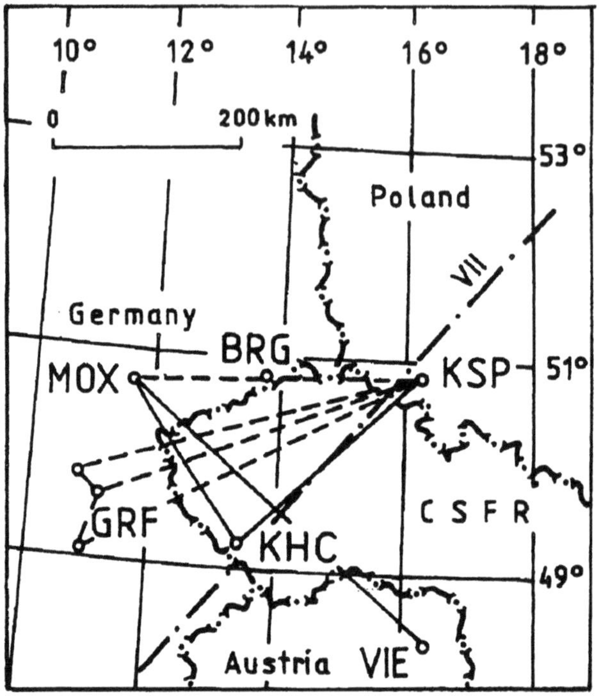

Fig. 2.31 Geographical situation of the investigated (solid lines) and perspective (dashes lines) surface wave profiles. MOX – Moxa, VIE – Vienna, KHC – Kašperské Hory, KSP – Ksiaz, GRF – Gräfenberg, BRG – Berggiesshübel.

cannot be identified as a well-defined consistent layer. After Čermák (1979c) the Bohemian Massif is characterized by relatively low heat-flow values so that the general tendency may be compared to that of a shield-like structure.

The path between MOX and KHC geologically belongs to the Saxothuringicum in the north (around 33%), to the western margin of the Bohemian Massif in the south, and to the Vogtland – Ore Mts. region which is well-known for its swarm-like earthquake activity (Kárník et al. 1987; Neunhöfer and Güth 1988). The DSS results published by Beránek and Zounková (1979) and those obtained by Grummt (1988) by inversion of teleseismic body waves recorded at KHC enable us to estimate the mean depths of the Moho discontinuity at about 32 km and 45 km in the northern and southern parts of the profile, respectively. The heat flow values given by Čermák (1979c) support the assumption that the profile runs in the marginal parts of the Vienna Basin.

The Earth's crust along the path between KHC and KSP has a rather complicated block structure. From SW to NE the profile passes through the Moldanubicum (17% of the total length of 272 km), the Central Bohemian pluton (36%), the Bohemian Cretaceous Basin (21%), and the Sudeten (26%). For the construction of *P*–wave velocity vs. depth models for the individual blocks Novotný (1983) used all respective results of DSS measurements along international profile VII, and adopted Brune's and Dorman's (1963) Canadian shield model for the upper mantle beneath the blocks. Grummt (1988) derived crust velocity models for KHC and KSP by inversion of teleseismic *P*–waves. A recapitulation of Grummt's and Novotný's considerations leads to the following integral image of the crust along the profile KHC–KSP: ascent of the Moho discontinuity from about 45 km beneath KHC to about 29 km beneath the Bohemian Cretaceous Basin, and a slightly descending trend towards 32 km beneath KSP; the mean crustal *P*–wave velocity of about 7.2 km s^{-1} near SW and about 6.5 km s^{-1} around the NE end of the profile; subcrustal velocities around 8.5 and 7.8 km s^{-1} for KHC and KSP, respectively; and an inexpressive low-velocity channel at depths ranging from 18 to 32 km beneath KSP.

Data Analysis and Inversion

The quality of the data and the applied methods of their analysis and inversion varied along the individual profiles as follows:

Profile MOX–VIE

Only two suitable events with complementary azimuths (N. Atlantic Ocean and Turkey) were found for this profile in the time span of operation of the temporary long-period seismograph station VIE. The records of VIE were adapted to the response of the MOX

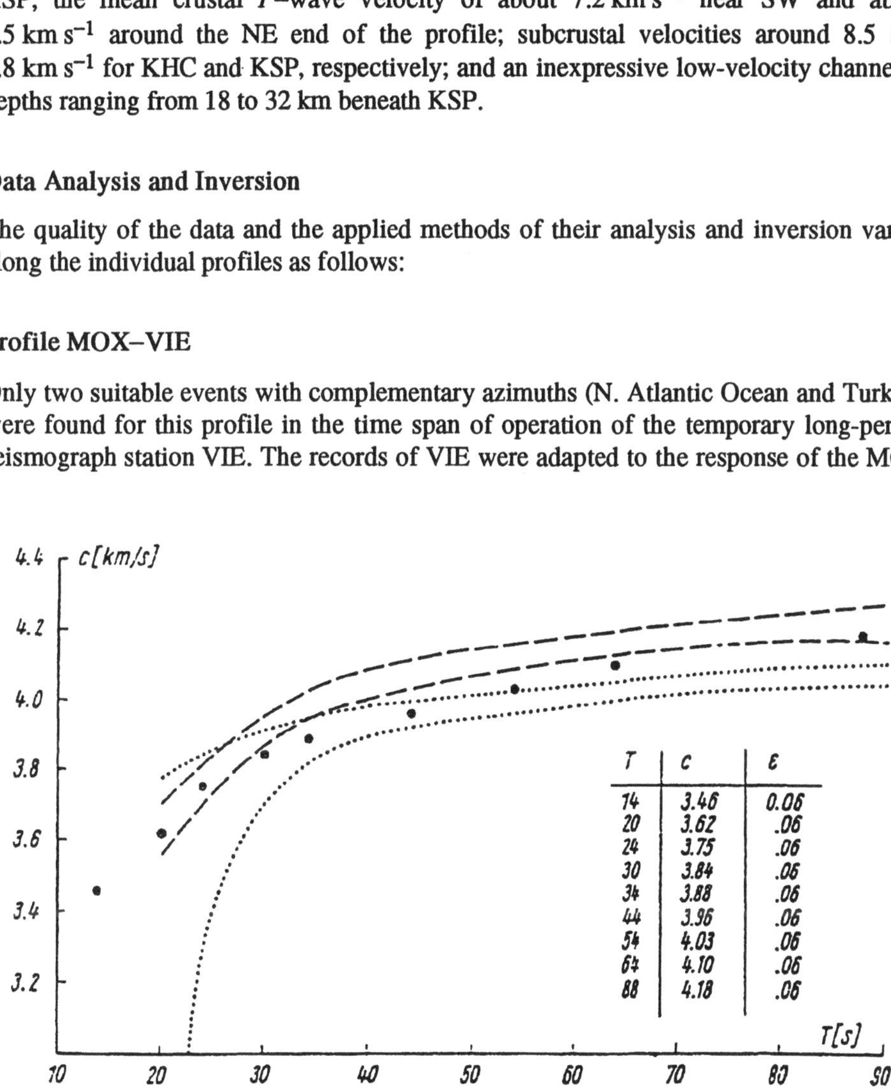

T	c	ε
14	3.46	0.06
20	3.62	.06
24	3.75	.06
30	3.84	.06
34	3.88	.06
44	3.96	.06
54	4.03	.06
64	4.10	.06
88	4.18	.06

Fig. 2.32 Smoothed experimental phase velocity data for the profile MOX–VIE (solid circles), and ranges of phase velocity curves for stable continents (dotted) and shields (dashed) – after Knopoff (1972).

seismograph by recursive correction filtering. The methods described by Panza et al. (1978a) were used to determine the dispersion of the Rayleigh waves recorded at the two stations in the period-range from 14 to 88 s. Since the phase velocity curve obtained for the Atlantic Ocean event exhibited oscillations probably due to multipathing, the data for this event were smoothed. The smoothed data were in good agreement with those obtained for the Turkey event. The resulting mean phase velocities for both events are displayed in Fig. 2.32. The classical hedgehog method was applied for the inversion of the data into shear-wave velocity vs. depth models. The parameters of the used start model are given in Table 2.2. Of the total number of 1 728 calculated models, 22 fitted the observations within the given accuracy range (single point errors $\varepsilon < 0.06$, r.m.s. deviation < 0.03 km s^{-1}). The most probable model was found by introducing three additional conditions based on other geophysical results: the Bohemian Massif has no rift-like properties, the mean Moho discontinuity depth along the profile is about 34 km, and no low-velocity layer exists in the lower crust.

Profile MOX–KHC

The magnetic tape library of the "very broadband" (VBB) seismograph system FBV at KHC (Plešinger and Horálek 1976) and photo-records of the standard-class B seismograph SSJ–1 at MOX served as sources of the observational material for this study. Five earthquakes with complementary azimuths (three from Greece, one from California and one from Idaho) were selected, and the Rayleigh wave dispersion in the period range from 19 to 65 s was determined for each event using the multiple recursive band-pass filtering technique described in (Neunhöfer 1985, par. 2.2.2). The resulting average phase velocities for the whole set of analyzed data are depicted in Fig. 2.33. The single point errors can be assumed not exceeding 0.03 km s^{-1}. As "start-model" for the inversion the final model found for the MOX–VIE profile was used. The partial derivatives of the phase velocity by the shear-wave velocity and depth of the layers were used to fit the model in the experimental data. Three steps in the change of the start-model parameters proved to be sufficient for explaining the observed phase dispersion. The first correction consisted in a reduction of the Moho discontinuity depth, the second

Table 2.2 Crust and upper-mantle model for *S*–waves used in the inversion of the MOX–VIE data. d – layer thickness, β – *S*–wave velocity.

d [km]	step [km]	β [km/s]	step [km/s]
6.7		3.4	
16.3		3.5 – 3.9	0.2
9 – 17	4	3.7 – 4.3	– 0.2
15 – 105	30	4.35 – 4.8	0.15
156 – 270	30	4.4 – 4.7	0.1
134		4.75	
		5.3	

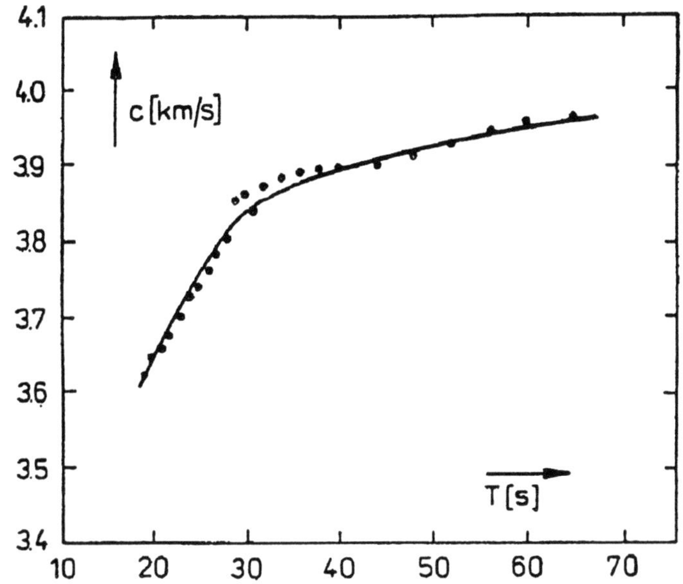

Fig. 2.33 Observed Rayleigh wave phase velocity dispersion (solid circles) and theoretical dispersion curve for the MOX–KHC (MOKH) model.

in fitting the model in the observation data in the long-period range (introduction of a thin low-velocity channel in the upper mantle), and the third in a modification of the model according to the short-period part of the experimental dispersion curve (diminishing of the shear-wave velocity in the lower crust).

Profile KHC–KSP

Since both KHC and KSP had been in operation as very broadband (VBB) stations for a long time (Plešinger 1981), extensive observational material of high quality was available for a detailed investigation of this profile. Events with a surface wave magnitude greater than 5.5, an epicentral distance smaller than 100°, and an azimuth within 10° of that of the profile were retrieved from the tape library. After visual inspection of the seismograms and of their amplitude and phase spectra, 10 events were selected for the analysis. Six of these events had their epicentres in the Honshu – Bonin region (NE azimuth), two in Algeria, and two near the Mid-Atlantic Ridge (SW azimuth). All seismograms were normalized to a standard VBB response, and both Rayleigh and Love waves were analyzed. To extract the Love waves, the two horizontal components were rotated into the direction transverse to the wave propagation. With regard to the high calibration accuracy, equal weight was assigned to each event. The phase delay of each type of waves between the two stations was determined by the cross-correlation method which results in smoother phase velocity curves than those obtained by a separate phase analysis of the two seismograms. Results associated with

Table 2.3 Rayleigh and Love wave phase velocities observed at the KHC–KSP profile.

period [s]	Rayleigh [km/s]	Love [km/s]
10.0	3.44 ± 0.05	3.76 ± 0.09
12.1	3.44 ± 0.06	3.82 ± 0.08
14.7	3.50 ± 0.05	3.88 ± 0.09
17.8	3.61 ± 0.06	3.97 ± 0.08
21.5	3.70 ± 0.07	4.02 ± 0.08
26.1	3.80 ± 0.05	4.11 ± 0.06
31.6	3.89 ± 0.05	4.24 ± 0.17
38.3	3.92 ± 0.06	4.32 ± 0.16
46.6	3.96 ± 0.10	4.46 ± 0.07
56.2	3.96 ± 0.07	4.54 ± 0.08
68.1	4.03 ± 0.07	4.50 ± 0.10
82.5	4.05 ± 0.10	4.67 ± 0.17
100.0	4.02 ± 0.11	4.66 ± 0.09

spectral amplitude minima were rejected in the analysis, and the procedure was repeated with different sets of time and lag windows in order to test the stability of the results. Due to the large bandwidth of the original data, satisfactory results were obtained for periods from 10 to 100 s. The resulting phase velocities are given in Table 2.3 and plotted in Fig. 2.34. Each value represents an average of five to eight observations; no additional smoothing was applied. The stabilized least-squares pattern described by Wielandt and Knopoff (1982) was used to invert separately the phase velocities obtained for the Rayleigh and Love waves. The start-model was a combination of Novotný's model KHKS82 down to 315 km, with a strongly simplified version of the model 1066B for the rest of the mantle (the mantle section of Novotný's model did not appear to be quite appropriate for the profile). All phase velocities were computed for a spherical Earth, and the data were weighted according to their variance.

Final Models

The final shear–wave velocity vs. depth models obtained in the described ways are plotted in Figs. 2.35 and 2.36. The pertaining parameters are listed in Tables 2.4 and 2.5.

In comparison with the models obtained for the western region including the Alpine piedmont and the Alps, and for the northern part of the Saxothuringicum (Knopoff et al. 1966, Neunhöfer and Güth 1975; Panza et al. 1978b), the main differences of the MOVI (Moxa–Vienna) model are: 1. the Moho discontinuity is at a greater depth, 2. no low-velocity channel in the upper mantle, but shear-wave velocity slightly increasing with depth; 3. comparatively low shear-wave velocity in the lid; 4. shear-wave velocity comparatively high in the upper crust and comparatively low in the lower crust. The shield range of the phase velocities is achieved only for long or relatively very short periods. The data with periods of 50 and 70 s correspond to those typical of shields and stable continents. For shorter periods they indicate a stable continental character. These

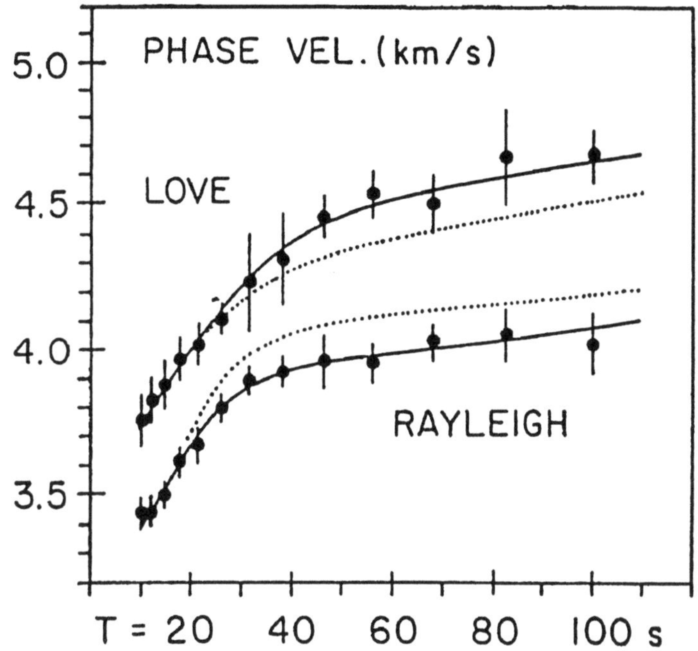

Fig. 2.34 Observed (solid circle-bar symbols) and modelled (solid and dotted lines) phase velocities for the KHC–KSP profile. The error bars indicate the standard deviation of the measurements. Dotted lines indicate velocities predicted from the inappropriate model (β_R for Rayleigh waves and vice versa).

results are in good agreement with Čermák's (1979c) heat flow data which have lower values than those for the surrounding regions but higher values than those of the Baltic shield.

Compared with MOVI, the model MOKH obtained for the Moxa – Kašperské Hory profile is characterized by: 1. a smaller Moho discontinuity depth (i.e. 32 km instead of 36 km); 2. a slightly lower shear-wave velocity in the lower crust; and 3. a probable existence of an inexpressive low-velocity channel in the depth range of the usual upper mantle channel (55 to 140 km in the MOKH model). In comparison with the results obtained for Western Europe the low-velocity channel is very inexpressive and may be interpreted as a transition from the absence of such a channel in the Bohemian Massif (cf. Neunhöfer and Güth 1975) to its clear development towards the west.

The results obtained for the crustal section of the KHC–KSP profile (see Fig. 2.36 and Table 2.5) agree with Novotný's model. Since surface waves data are sensitive to the average shear-wave velocity in the lower crust, of which Novotný had no direct evidence, they provide a desirable confirmation of his crust model. In the mantle, the Rayleigh-wave model employs significantly lower shear-wave velocities down to 215 km; the data require the average velocity in the top 180 km of the mantle to be about 0.24 km s^{-1} lower than in the initial model. This is not surprising since the mantle

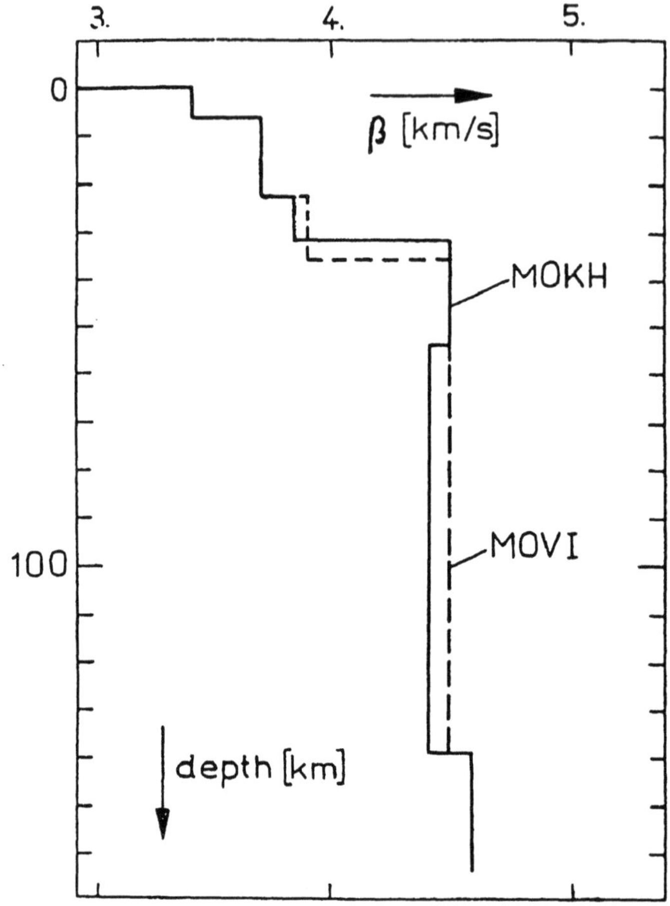

Fig. 2.35 Final shear-wave velocity vs. depth models for the MOX–VIE (MOVI) and MOX–KHC (MOKH) profiles.

section of Novotný's model represents a shield structure. In the Love-wave model the average velocity reduction in the top 180 km of the mantle is 0.06 km s^{-1}. The discrepancy between the two surface-wave models is thus 0.18 km s^{-1} or 4% of *S*–wave velocity. The significance of the discrepancy is evident from the dotted dispersion curves in Fig. 2.34 which indicate Love-wave phase velocities predicted from the Rayleigh-wave model and vice versa. This sort of discrepancy has repeatedly been noticed in the past 20 years (see e.g. Schlue and Knopoff 1976; Lévêque and Cara 1983). Three alternative explanations have been proposed:

1. contamination of the fundamental Love-wave mode with higher modes,
2. existence of an Earth structure containing unresolved low-velocity layers,
3. anisotropy of the elastic wave velocities.

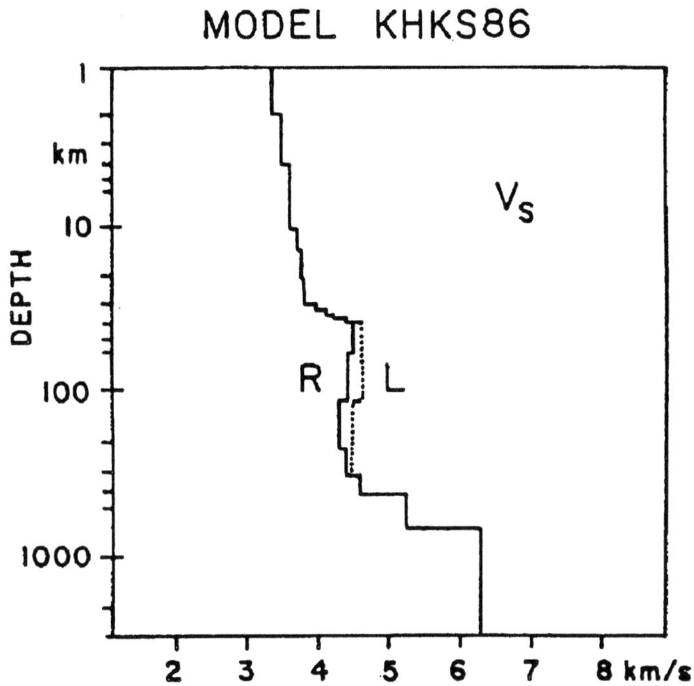

Fig. 2.36 Final shear-wave velocity vs. depth models for the KHC–KSP (KHKS) profile. R – Rayleigh-wave model, L – Love-wave model.

As no mode interference has been obvious in the data (cf. Wielandt et al. 1987), the first alternative can be excluded.

Special inversion experiments in which both the data sets were fitted in a single earth model consisting of relatively thin (10 to 50 km) alternating layers of high and low velocities have shown that an excellent fit can be obtained if velocity contrasts in the

Table 2.4 Parameters of the MOVI and MOKH models; *d* – layer thickness, *z* – depth, β_R – shear waves velocity derived from Rayleigh waves.

MOVI			MOKH		
d [km]	*z* [km]	β_R [km/s]	*d* [km]	*z* [km]	β_R [km/s]
6.7	0	3.4	6.7	0	3.4
16.3	6.7	3.7	16.3	6.7	3.7
13.0	23.0	3.9	9.0	23.0	3.83
105.0	36.0	4.5	23.0	32.0	4.5
156.0	141.0	4.6	85.0	55.0	4.41
134.0	297.0	4.75		140.0	4.6
	431.0	5.3			

Table 2.5 Shear-wave velocity vs. depth models for the KHC–KSP profile; β_s – start model, β_R – final Rayleigh-wave model, β_L – final Love-wave model.

	d [km]	z [km]	density [Mg/m^3]	α [km/s]	β_s [km/s]	β_R [km/s]	β_L [km/s]
1	1.0	0.0	2.81	5.55	3.20	3.10	3.20
2	1.0	1.0	2.86	5.80	3.35	3.35	3.35
3	2.0	2.0	2.90	6.01	3.47	3.47	3.50
4	6.0	4.0	2.93	6.17	3.56	3.62	3.61
5	4.0	10.0	2.95	6.26	3.61	3.75	3.65
6	4.0	14.0	2.98	6.38	3.68	3.79	3.70
7	3.0	18.0	3.02	6.62	3.82	3.80	3.78
8	4.0	21.0	3.05	6.77	3.91	3.81	3.84
9	5.0	25.0	3.08	6.90	3.98	3.83	3.89
10	2.0	30.0	3.12	7.14	4.13	3.98	4.04
11	2.0	32.0	3.17	7.39	4.28	4.13	4.19
12	2.0	34.0	3.21	7.57	4.38	4.23	4.29
13	2.0	36.0	3.26	7.86	4.56	4.41	4.47
14	22.0	38.0	3.30	8.10	4.72	4.51	4.64
15	55.0	60.0	3.30	8.10	4.72	4.42	4.66
16	100.0	115.0	3.44	8.20	4.54	4.32	4.49
17	100.0	215.0	3.53	8.30	4.51	4.44	4.48
18	100.0	315.0	3.60	8.50	5.60	4.60	4.60
19	250.0	415.0	3.90	9.70	6.25	5.25	5.25
20	∞	665.0	5.50	11.30	6.30	6.30	6.30

order of 1 km s^{-1} are accepted between subsequent layers (Wielandt et al. 1988). However, the positions and thicknesses of individual layers cannot be determined from the data. The dispersion curves remain nearly unchanged when high- and low-velocity layers are interchanged. This ambiguity makes it impossible to accept or reject such a model on the basis of surface waves observations.

Earth models with a smooth velocity–depth distribution satisfy the dispersion data for both wave types only if anisotropy of the shear-wave velocity is introduced. Such an anisotropy is expected to occur if anisotropic crystals have a preferred orientation in the mantle. Depending on the assumed axis of symmetry the anisotropy may express itself in a difference between the velocities of horizontally and vertically polarized surface waves, or additionally in dependence of the wave velocity on the direction of propagation.

Conclusion

It is not possible to distinguish and decide unambiguously between the numerous possibilities of interpreting the observed surface wave dispersion data from a single profile. Future investigations will require a two-dimensional network of broadband seismic stations. In this sense a completion of the investigations including the fan of KSP – GRF profiles (see Fig. 2.31) and the KHC – BRG profile is desirable.

3 Seismicity and Seismotectonics

3.1 Earthquakes in Central Europe

D. Procházková[1], V. Schenk[1], Z. Schenková[1]

Introduction

Earthquake activity in Central Europe is determined by its geologic history and tectonics. Investigations of seismicity in Czechoslovakia, however, are not confined to the Bohemian Massif and the West Carpathians only. It is necessary to investigate also earthquakes in the East Alps, the Karnian Alps, Jurassic units of the Alps, and the northern part of the Pannonian Basin which affect Czechoslovak territory.

The investigations of earthquakes provide general knowledge of the structure of the Earth, of the dynamics of contemporary movements within the Earth's crust and upper mantle. In this respect, prime attention is paid to the structural aspects of seismological investigations in Central Europe, particularly to the information obtained in the period 1980–1990. The previous results were summarized in the monograph "Geophysical Syntheses in Czechoslovakia" (Kárník et al. 1981a) and in other works mentioned further on.

Spatial Distribution of Earthquake Foci

Location of earthquake foci in the crust indicate weakened zones where the tension accumulated by recent tectonic processes is released, i.e. the active boundaries of blocks or plates of the lithosphere. The situation in Central Europe has been demonstrated in the map of earthquake epicentres (Kárník et al. 1981b, 1984a; Procházková 1984a, b; Grünthal et al. 1985; Schenk et al. 1989a). The latest version of the maps of earthquake epicentres in Czechoslovakia and its surroundings compiled by Schenk and Schenková (1990, unpubl. work, Geoph. Inst. Czechosl. Acad. Sci.) is given in Fig. 3.1. For all the maps available catalogues complemented with observations from recent years were used.

1 Geophysical Institute, Boční II/1401, 141 31 Praha 4, Czech Republic

The general seismic pattern of Central Europe has not been substantially changed by the latest results. It has only been corrected in some places. As the seismic activity is low, it is not possible to enrich the observations during one or two decades. Also, it can hardly be expected to reveal new information in the archives. So far, the only way to obtain new knowledge of focal zones is the observation of microearthquakes ($M \leq 2.5$).

The earthquakes in the Bohemian Massif took place mainly in its peripheral areas. The intensities of the strongest of them were 7.5° MSK (6 March 1872, Gera in the Saxony), 7° MSK (10 January 1901, in the vicinity of Trutnov in the Sudeten, the Hronov – Poříčí fault). The intensity of the strongest earthquakes in swarms occurring frequently in the western part of the Bohemian Massif near the towns of Kraslice and Aš has not exceeded 7° MSK (Kárník et al. 1987).

The most distinct, seismically active unit of Central Europe is the belt stretching along the rivers Mur and Leitha in the East Alps and along the Peripienian lineament of the West Carpathians. The epicentres between the towns Těšín and Krakow ($I_0 = 7° - 8°$ MSK) also belong to this belt.

Individual parts of the West Carpathians and the adjoining basins have been the scene of rather strong earthquakes ($I_0 = 8°$ MSK) also in this century. The origin of the most disastrous earthquake in Czechoslovakia, i.e. that of 28 June 1763 with the epicentre east of Komárno, is probably connected with the subsidence of the Lesser Danubian Basin.

Its intensity attained 8° – 9° MSK and it caused the death of 63 people. The church and several other buildings in Komárno fell down as well as churches in villages east of Komárno where the earthquake must have had its focus. The activity continued for a hundred years and was shown by six shocks originating regularly at about 20-year intervals until 1869. The Komárno epicentre is now quiet and occasional shocks are very weak, attaining $I_{max} = 4°$ MSK.

A detailed description of seismicity of Czechoslovakia can be found in Schenková et al. (1983), Kárník et al. (1984a), and Procházková (1984a, b, 1990).

As a rule, the earthquake foci in the study area occur within the upper part of the Earth's crust (for details see Procházková and Dudek 1982; Schenková and Kárník 1985), i.e. $h \leq 15$ km. Exceptions are in Bavaria ($h = 16$ km), the Beskydy Mts. ($h = 20 - 40$ km), Central Slovakia ($h = 20$ km) and Northern Hungary ($h = 19 - 27$ km).

Maps of epicentres serve as a basis for delineations of epicentral areas which should take into account also the geological situation and the geophysical fields associated with the origin of earthquakes. Schemes of the source areas were also compiled for Central Europe, from simple separation of large seismotectonic units (the Bohemian Massif, the East Alps, etc.) to contour of isolated zones or clusters of epicentres (Bune et al. 1983; Kárník et al. 1984a, b; Procházková 1984a, b; Schenk et al. 1989a). There are several schemes for Central Europe. They were constructed for different purposes and therefore differ mainly in details, which is further influenced by the latest conceptions of tectonic zoning (cf. Chapter 3.2).

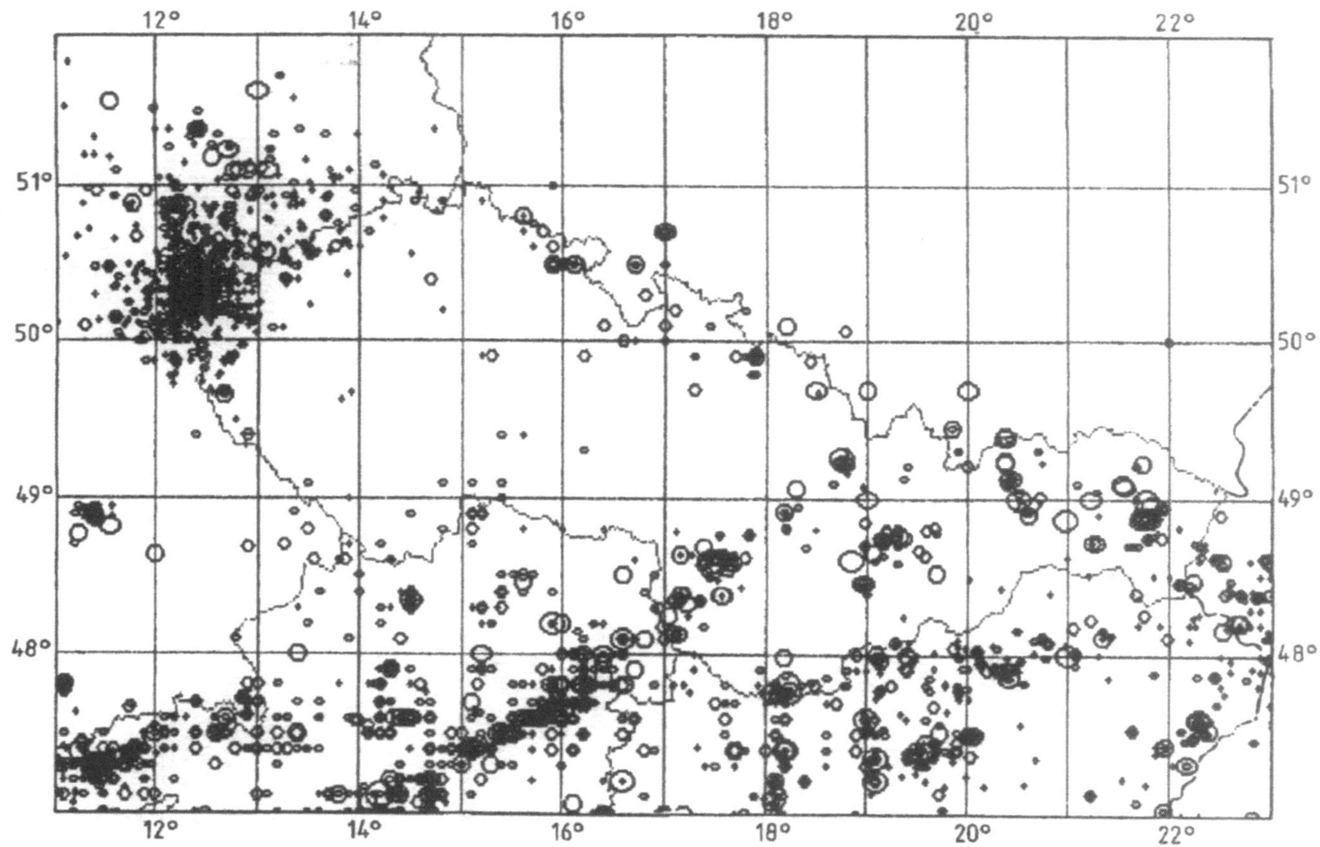

Fig. 3.1 Epicentres of earthquakes in Czechoslovakia and its surroundings - the 1990 version (V. Schenk, Z. Schenková). Circles correspond to the earthquake surface-wave magnitude [epicentral intensity in °MSK] from 2.5-2.9 [III], 3.0-3.5 [IV], 3.6-4.0 [V], 4.1-4.6 [VI], 4.7-5.1 [VII], 5.2-5.6 [VIII] to 5.7-6.1 [IX].

At the present time, the earthquake source regions are defined by their border lines, by the maximum possible earthquakes, and by the magnitude (intensity) – frequency relationships, sometimes by types of earthquakes, by Benioff and space–time diagrams (Bune et al. 1983; Kárník et al. 1984a, b; Procházková 1984a, b; Schenk et al. 1989a). Threshold values I_{max} or M_{max} for individual source regions in Central Europe correspond generally to the maximum values observed in each region during a period of approximately 500 years. We consider them as the possible limit, valid not only for the known epicentre, but for the entire region regarded as a seismotectonic unit and defining its seismogenetic potential.

Statistical estimations of threshold values I_{max} or M_{max} (Kárník et al. 1984a; Schenk et al. 1989b) are frequent. They are effective, however, only in case of numerous and homogeneous sets of observations.

These procedures, as a rule, do not take into account the fluctuation of earthquake activity in future or activation of new foci.

An analysis of focal parameters of European earthquakes (Procházková 1983) confirmed that earthquakes differ from each other by the value of magnitude M or seismic moment M_0 ($\log M_0 = 9.95 + 1.4\,M$), by the size of the source or by the stress drop $\Delta\sigma$.

A large number of relationships between magnitude M and the fault length L have been published, having different validity. They were calculated either for a particular area or for a particular type of fault and slope along the fault plane. The dispersion of observed values is great because the relationship is rather simplified by the given formula.

For an approximate estimation of earthquake magnitude Schenk and Schenková (1983) derived the relation $M = (1.93 \pm 0.08) \log L + (2.62 \pm 0.62)$, where L is the length of the seismoactive part of fault in km. Regarding analogous relations (e.g. Shebalin 1971) the preceding relation can be used for its modification with regard to the earthquake activity of the area. It can then be written

1. for low activity (the Bohemian Massif and the less active part of the West Carpathians) $M = 1.85 \log L + 2.0$,
2. for intermediate activity (active part of the Carpathians, the Alps, etc.) $M = 1.93 \log L + 2.62$.

All the resultant values of M_{max} or I_{max} depend entirely on the reliability of determining the active fault length and on correct correlation between earthquakes and faults. It is generally held that if an earthquake originated at a certain place along a fault or in a fault zone, an earthquake with identical M can originate along its entire length, most probably at points of intersection with other faults.

Schenk et al. (1982) correlated velocity values of recent vertical movements with earthquake occurrence on given faults. With the use of the map of vertical movements velocities V_v in the Carpathian – Balkan region values of gradient of movements along the faults $G_v^{\|}$ and across the fault $G_v^{\perp}$ were calculated. With the help of magnitudes M of the known strongest earthquakes on the investigated faults (Fig. 3.2) it was possible to derive, using the correlation analysis, the following mean dependences:

$$|V_v| = 0.980\,M - 4.657;$$

$$\log G_v^{||} = 0.654\,M - 5.863;$$

$$\log G_v^{\perp} = 0.606\,M - 6.024;$$

$$|V_v| = (1.15 \pm 0.05)\,I_0 - (8.5 \pm 0.5).$$

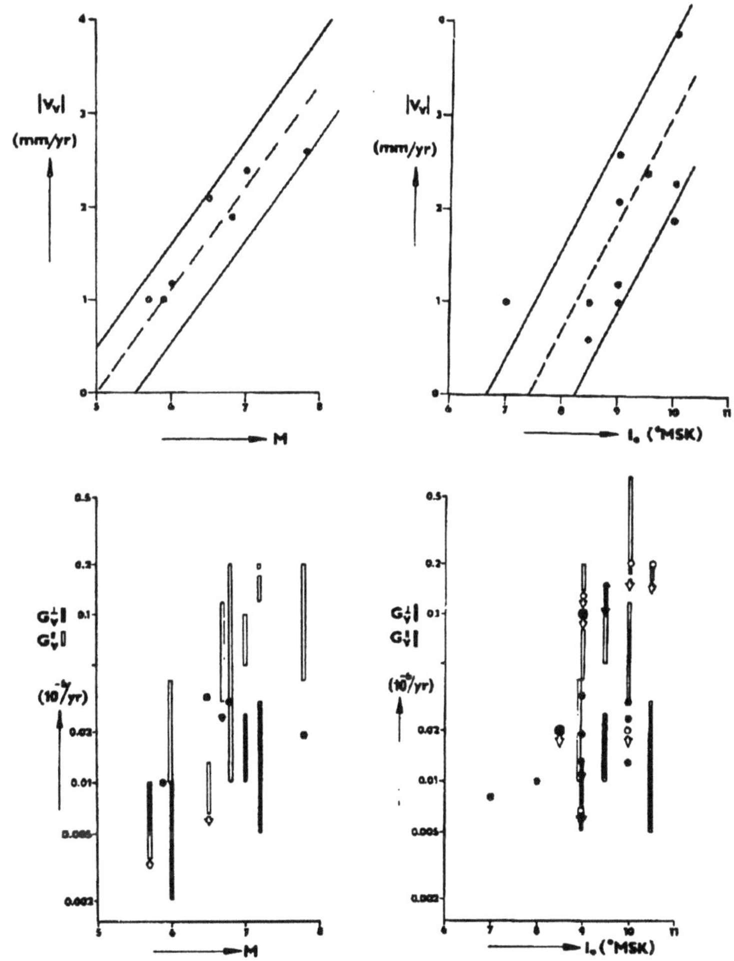

Fig. 3.2 Absolute values of velocity $|V_v|$ of the recent vertical movements and gradient of these movements along the fault $G_v^{||}$ and across the fault $G_v^{\perp}$ versus magnitude M and epicentral intensity I_0 for large earthquakes in Eastern and Central Europe.

The analysis of the Carpathian – Balkan region showed (Schenk et al. 1982) that:

1. strong earthquakes ($M \geq 6.5°$, $I_0 \geq 9°$) originate at faults, vertical movements of which are $|V_v| \geq 2$ mm per year, the velocity gradients $G_v^{\parallel} \geq 2.5 \times 10^{-8}$ per year and $G_v^{\perp} \geq 1 \times 10^{-7}$ per year;
2. for intermediate earthquakes ($5 \leq M < 6.5$, $6.5° \leq I_0 < 9°$ MSK) the following ranges of values can be found for the above given parameters: 0.5 mm per year $\leq |V_v| <$ < 2 mm per year, 5×10^{-9} per year $\leq G_v^{\parallel} < 2.5 \times 10^{-8}$ per year, and 1×10^{-8} per year $\leq G_v^{\perp} < 1 \times 10^{-7}$ per year;
3. for weak earthquakes ($M < 5$ and $I_0 < 6.5°$ MSK) the parameters are as follows: $|V_v| < 0.5$ mm per year, $G_v^{\parallel} < 5 \times 10^{-9}$ per year, and $G_v^{\perp} < 1 \times 10^{-8}$ per year.

Even through these values are the first approximations, they agree relatively well with the values published earlier for areas of decreased seismic activity, for which the following values were assumed: $G_v = 1 \times 10^{-8}$ to 1×10^{-10} per year.

For all epicentral areas it holds that seismic information does not enable us to provide a reliable prediction and determination of the level of earthquake activity and on its tendency. Seismic data analysis must be complemented with analyses of all available information associated with the earthquake origin. It includes, of course, all data about the geodynamics and structure of the Earth's crust, about the position, sizes, and present movements of faults, about the boundaries and magnitudes of movements of individual blocks found by geologic investigations and geodetic measurements, about the depths of density discontinuities, and about mechanical properties of layers as well as about the relations between geophysical fields and seismic activity (e.g. Borissoff et al. 1975, 1976, 1978).

Earthquake Mechanism

The direction of the main acting tectonic stresses can be determined by analysis of the directions of P–wave arrivals at a greater number of seismic stations. The basic mechanism and position of two fault planes can be obtained by stereographic projection of the first motion of P–wave displacement. The method, however, does not indicate which of the two planes is directly associated with the fault plane. Although there is only a small number of earthquake observations in Central Europe, two attempts have been made to determine the focal mechanism. Relatively well documented is the earthquake from 14 April, 1983 near Scheibbs (the East Alps) (Drimmel and Procházková 1985). The attempt of Pospíšil et al. (1985) regarding three earthquakes in the West Carpathians and one in the East Alps was a bit problematic as there were only 4 to 9 first arrivals of P–wave available. The resultant mechanisms are analogous. The optimum solution is a fault plane of N60°E geographic azimuth direction and of 75° dip to the SE. It can be assumed that the northwestern block moved along a dipping plane downwards to the SW or, on the contrary, the southeastern block moved along a dipping plane upwards to the NE. The resultant information, however, indicates only the mutual movements of geologic blocks in the Little Carpatians.

Fault plane solutions of single events of the 1985/1986 swarm, completed from a sufficient number of data, have been described by Špičák (1987), Procházková et al. (1987a), Antonini (1988), Grosser and Köhler (1988), Procházková (1988a), Zahradník et al. (1988). Grünthal et al. (1990) compared all solutions and drew the following conclusions. Except for one event, all the other 14 strikes are in the range from N170°E to N200°E with a mean strike of N187°E, the dip does not exceed 75°. The slip components are generally sinistral with the mean azimuth N136°E. From the fault plane solutions only two consistent features are the essential ones for the derivation of the generalized model of the focal swarm area: N–S to NNE–SSW azimuth of the majority of fault planes, generally sinistral strike–slip components.

Occurrences of Very Weak Earthquakes

Seismic activity in the West Carpathians is comparatively low and even lower is the frequency of earthquake occurrence in the Bohemian Massif. Of course, both regions differ in the intensity of the strongest observed (and possible) earthquakes. The only way to obtain new information is, therefore, localization of very weak shocks of smaller magnitudes than those of catalogued earthquakes. The experiments in Slovakia in 1956 and 1957 when four field seismographs were installed for a month in the vicinity of the towns of Komárno, Žilina, and Dobrá Voda did not yield useful results as the recording time was rather short (Kárník and Vaněk 1957).

After a sufficient amount of information had been gathered at regular stations, localization of seismic activity of the order of $M = 2.5 - 3.5$ started in 1973 (Mrázová and Ruprechtová 1982). In the following years the data were processed through a cooperation of the Geophysical Institute of the Czechoslovak Academy of Sciences with the Geophysical Institute of the Slovak Academy of Sciences. Another experiment was carried out in 1982 (Knaislová et al. 1988) in which recordings of Central European stations (i.e. Průhonice, Kašperské Hory, Ksiaz', Bratislava, Collm, Berggieshübel, etc.) have been used. Approximately 3 000 nearby, weak events are recorded annually of which only a part (30 – 50 %) are localized. Among those events, explosions in quarries prevail, followed by shocks in mines. About 1% of events only, can be regarded as tectonic earthquakes. The only criterion so far to identify natural seismic events is their occurrence during night hours (outside mining areas). In 1982, this classification identified 24 events in Czechoslovakia and a total of 72 events in Central Europe. Some of them occurred in known, seismically active areas, while others were recorded outside them and testify to permanent though very low activity. The accuracy of localization is very low. The epicentres located by the first experiments have so far no systematic geographic distribution which would help to identify active faults.

At the time of the earthquake swarm 1985/1986, strictly speaking after December 20, 1985, several recording systems (Vogtland network, Bavarian network, station Vysoká Pec, Oloví, Vackov, Tisová and Schönberg) were in operation in the swarm epicentral region and its close vicinity (Neunhöfer and Güth 1983, 1984; Neunhöfer et al. 1989; Procházková et al. 1987b). For example, the Vogtland network of stations situated in the territory of the former GDR (MOX, PLN, BDE, KLI and WRG) registered 8423 events, the station Vackov 2542 (22.–27. 12. 1985) and 857 (after 28. 12. 1985) shocks, respectively.

Frequency of Earthquakes

Among the basic characteristics of seismicity is the frequency of occurrence of earthquakes in a particular region and within a given time interval. There is a general empirical relationship between the number N and the magnitude M of earthquakes: $\log N = a - bM$, where a, b are constants valid for a particular region and for a given time interval, b ranges from 0.5 to 1.5. There are also theoretical models of earthquakes distributions in time, models of time intervals between earthquakes, models of the probability of occurrences of extreme events, etc. They, however, are used in earthquake statistics and in seismic hazard assessment and thus they are not closely related to the subject of the present work, i.e. to deep structure. It should be added, however, that according to laboratory experiments parameter b is directly proportional to the degree of medium heterogeneity and indirectly proportional to the increase of stress. Parameter a is directly proportional to the level of activity, length of interval and size of the area. In this regard the frequency diagrams $N(M)$ have a qualitative significance for discovering the properties of the medium in epicentre areas.

Frequency diagrams $N(M)$ or $N(I)$ were constructed for active regions in Central Europe (e.g. Kárník et al. 1984a, b, 1987; Procházková 1984a, b; Schenk et al. 1989a).

Release of Seismic Energy in Time and Migration in Space

Earthquakes are distributed regularly neither in space nor in time. There are periods of increased or decreased seismic activity. In most regions, no regularity in time, i.e. no regular cycles can be detected. The observations to date testify to different intervals between periods of increased activity though there are clear differences in the activity pattern in individual regions, i.e. in the level of activity. The activity variation in time is well demonstrated on the Benioff graphs. They express the sum of quantity $E^{1/2}$ (E is seismic energy) which is proportional to deformation. A detailed description was given by Procházková (1984a,b, 1988b). Some examples show indications of two active periods (the Swabian Jura, Innsbruck), others indicate only one. The East Alps show only one with certain oscillations (the data from the years before 1300 is not complete even in case of strong earthquakes $I_0 \geq 8°$).

Within one source region as well as within a broader area covering several regions, seismic activity can sometimes migrate, which is a proof of re-distribution of stresses in the lithosphere. While one area is seismically active, the other may be at quiescence. This phenomenon can be proved to a certain extent as it can be seen in samples of presumed migration in Central Europe (Procházková 1984a,b).

Earthquakes are not isolated events. They tend to group both in time and in space. Groups of shocks with different structures and properties originate (Procházková 1976, 1984a,b, 1988a). Earthquake swarms in the Krušné hory Mts. have two or more periods of increased activity. The active period may last from three to six days and hundreds of shocks may occur daily. This phenomenon has been ascribed to activization of the system of two main faults, the Krušné hory Mts. fault and Mariánské Lázně fault which intersect in the focal region. In some swarms stronger shocks were associated with the Krušné hory Mts. fault only (e.g. the 1903 and 1936) while in other cases (e.g. the 1897 and 1908, 1985/1986) the shocks were linked with the Mariánské Lázně or Tachov faults (Grünthal et al. 1990).

From studies of microearthquakes with $M = 0.5 - 1.5$ (Neunhöfer and Güth 1983, 1984) in the Vogtland region in the period 1968/1982, it follows (Procházková 1984a,b) that in the study area several partial swarms of microearthquakes can be distinguished. In the period before and after larger earthquake, swarm-clusters of several shocks resembling small swarms occur from time to time.

In the East Alps and in the area Mur, Mürz, and Leitha two types of aftershocks can be observed. The first type prevails and corresponds with mean dependence derived from aftershocks by statistical equalization of an empirical set of data. As regards the second type, the strongest aftershock is much weaker and occurs with a long delay (Procházková 1984a,b, 1990). The reason for this difference must be associated with the processes in earthquake foci, which are different as it is indicated by different values of stress drop in the focus (Procházková 1983).

The results from previous chapters constitute the basis for earthquake hazard calculation. The simplest map of earthquake hazard is the map of maximum observed intensities, or its modifications. There are numerous examples from Czechoslovak territory for the last decade (Kárník et al. 1981b, 1984a; Procházková et al. 1978; Procházková and Brouček 1986; Schenk 1984a; Schenk and Mantlík 1985; Kárník et al. 1988). The seismostatistical model of computing the earthquake hazard in the given locality, which was applied to earthquake hazard assessment of NPP sites in Czechoslovakia was described in (Schenková et al. 1981; Kárník et al. 1982; Schenk et al. 1989b). Further, Schenk et al. (1981) modified the program EQRISK of McGuire (1976) and mapped the earthquake hazard in the Bohemian Massif, in the former German Democratic Republic (Schenk et al. 1984) and in Central Europe (Schenk et al. 1989a). The results were the contour maps of maximum expected macroseismic intensities that will not be exceeded during T years ($T = 10$, 20, 50, 100, 200, 1000, 2000, and 10000 years).

As a rule, earthquake hazard maps generalize the existing information and anticipate the stability of seismic activity. Calculation includes also seismotectonic criteria assessing the earthquake potential of faults as well as data characterizing neotectonic or recent movements. An elaboration of these criteria is tiresome and recent attempts to find the seismotectonic relations in Central Europe will be commented on in the following Chapter.

3.2 Macroseismic Fields and the Main Directions of Structures

D. Procházková[1], V. Schenk[1], Z. Schenková[1]

Macroseismic events on the Earth's surface depend on numerous factors. Among them the response of surface layers, the position of the fault, and the medium which can either attenuate seismic energy or transmit it well play an important role. These characteristics can be deduced from the shape of isoseismals (Procházková 1984a,b; Procházková and Kárník 1978; Procházková and Dudek 1982; Procházková et al. 1986). The generalized course of isoseismals constructed on the basis of earthquakes for which the isoseismal maps exist in selected areas is shown in Fig. 3.3. The known prolongation of isoseismals in the direction of 3° and 4° far from the East Alps foci into the Bohemian Massif together with the very strong attenuation towards the Pannonian Basin are again found to be the most conspicuous phenomena. The shape of typical isoseismals shows that between individual regions of Central Europe differences in shape and size of individual macroseismic fields exist. The characteristic elongations or shortenings of isoseismals depend on a system of faults in the focal zone and on the earthquake mechanism. Figure 3.3 shows the differences in the elongation of isoseismals in the near and far zones. The boundary between these zones in Central Europe corresponds to the isoseismal of the diameter $r \approx 2.5\,h$, where h is the focal depth in kilometres (Procházková 1985a). Practically the same value was obtained for the territory of the Balkan (Procházková 1985b; Procházková and Dudek 1982) and is in agreement with results of Ananin (1983) who derived $r \approx (2-3)\,h$ for seismically active regions in the USSR.

Areas S_n of individual isoseismals are directly proportional to the depth of the focus and to the earthquake magnitude and inversely proportional to the coefficient of attenuation of isoseismals. For Central Europe (Procházková and Dudek 1982; Procházková 1984a,b) $\log S_n = 2.66 + 0.53\,I_0 + 0.01\,h - 0.60\,I_n - 13.00\,\beta_2$, (standard deviation = 0.14), $I_n = 3° - 9°$ MSK, where S_n is the area of isoseismal with intensity I_n, I_0 is the intensity at the point of epicentre, h is the depth of the focus in km, and β_2 is the attenuation coefficient in the distant zone. I_n is a formal quantity, in the given formula it enables expression of the dependence for values S_n corresponding to individual values $I_n = 3° - 9°$ MSK by means of a single formula.

Pospíšil et al. (1985) presented a map of principal directions of seismic waves propagation derived from the shapes of isoseismals of West Carpathian earthquakes. Studies of attenuation differences of these isoseismals due to asymmetry of macroseismic fields reveal regularities in seismic energy propagation caused by directions of structures at depth (elongation of isoseismals in the NW–SE direction) and by surface ENE–WSW trending structures, which can be observed from the Little Carpathians through Central Slovakia towards the east.

[1] Geophysical Institute, Boční II/1401, 141 31 Praha 4, Czech Republic

Studies of Schenk (1984b), Procházková (1984a,b) and of Procházková et al. (1986) concerned with a correlation of shapes of isoseismals, distribution of epicentres, and with the principal structural directions of the Bohemian Massif, are added to this work. It was found that over 50 % of earthquakes are directly associated with the directions NW–SE.

Each focal zone is characterized by attenuation of intensity values which do not differ to a great degree (Procházková and Dudek 1982; Procházková 1981, 1984a,b, 1985 a). It follows from their analyses that the relatively small values of attenuation coefficient $\beta_2 \leq 0.010$ occur in focal zones in Bohemia, Moravia, Austria, and Northern Yugoslavia. The boundaries between the Alps and the Pannonian Basin on the one hand and between the Bohemian Massif and the Carpathians on the other hand are characterized by a notable increase of β_2 values. It can also be explained by a change in Earth's crustal thickness which is notably greater in the Alps and in the Bohemian Massif than in the western part of the Pannonian Basin and in the West Carpathians (Beránek and Zátopek 1981). In the West Carpathians $\beta_2 \leq 0.010$ can be observed only

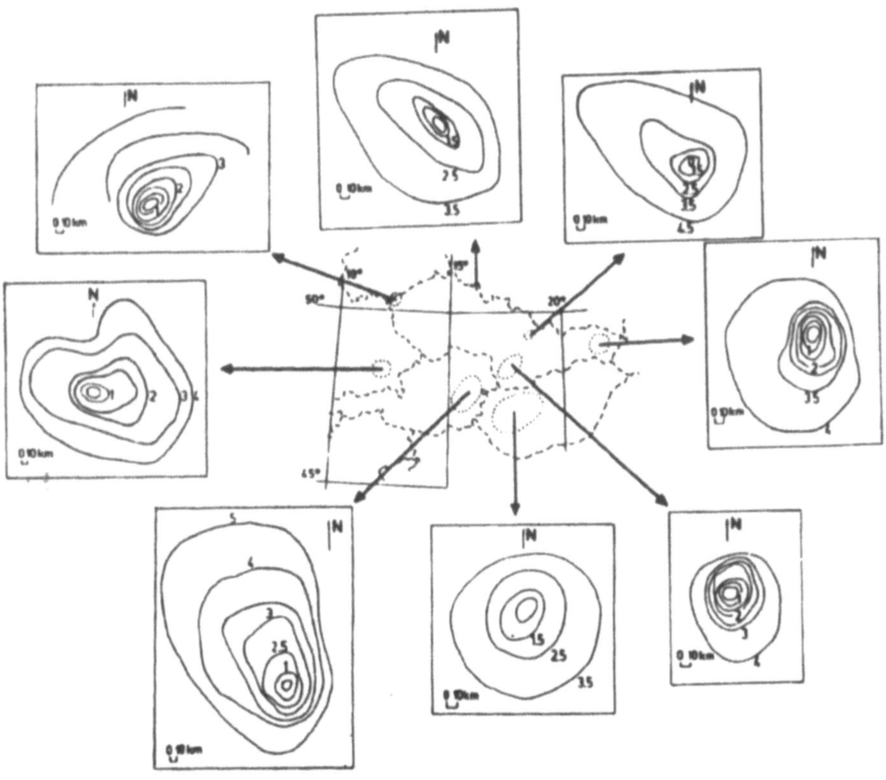

Fig. 3.3 Typical isoseismals of earthquakes in main epicentral regions of Central Europe (figures near isoseismals denote the difference $I_0 - I_n$, I_0 is the epicentral intensity, I_n corresponds to the given isoseismal).

for shocks with their foci at greater depths, i.e. for $h \geq 10$ km. According to distribution of β_2 values in the area of Žilina the estimated depth of Carpathian nappes is only about 7 km (Procházková et al. 1986).

The distribution of attenuation values β_1 (the near zone) shows a difference between the Bohemian Massif on the one hand (relatively small values β_1) and the Alps, the West Carpathians, and the Pannonian Basin on the other hand. Almost the same values β_1 for Mur – Mürz – Leitha area and for the Little Carpatians support the assumption that these areas are mutually linked. In the Carpathians, for deep shocks smaller β_1 are observed than for shallow ones.

It shows that deeper foci occur in a medium characterized by a smaller attenuation coefficient. These smaller attenuation coefficients for deep foci are comparable with values for the Bohemian Massif and, therefore, it can be assumed that the deep structure beneath the nappes of the East Alps and of the West Carpathians is linked with old and deep structures of the Moldanubicum and Brunovistulicum and mainly with Variscan tectonic elements in the Moravicum. This indicates the shallow nappe structure of the Alps and confirms the results of Beránek and Zátopek (1981). An analysis of values β_1 and β_2 in the East Alps showed that earthquake depths of $h \approx 8$ km occur in an area or beneath an area characterized by high attenuation, which also confirms the small thickness of Alpine nappes. This fact was also confirmed by data from boreholes Bärendorf–1 and Urmansau–1 which attain the crystalline rocks of the Bohemian Massif in the depth of 6 km and 3 km, respectively.

In the Pannonian Basin high attenuation of earthquake intensity can be observed neither for medium-depth earthquakes with epicentres in Vrancea (Procházková 1985a,b) nor for earthquakes in the vicinity of Komárno ($h \approx 20$ km), while high intensity attenuation can be observed for shallow shocks in the Pannonian Basin, mainly in its west part. Taking into consideration the depth $h \approx 8$ km it can analogously be assumed that foci in the Pannonian Basin are found in the area or quite nearby beneath it characterized by great intensity attenuation (Procházková 1984a,b, 1985a,b).

In the Bohemian Massif, the fault zones Hronov – Poříčí, Šumperk – Kouty nad Desnou, and near the river Ohře appear anomalous (characterized by high values of β_1).

Maps of maximum observed intensities show the total contribution of earthquakes manifested in the considered area by macroseismic effects (Kárník et al. 1981b, 1984a; Procházková 1984a). The last version of the seismic zoning map for Czechoslovakia is presented in Fig. 3.4 (Kárník et al. 1988). These maps based only on observations do not reflect the actual situation in places where macroseismic data are incomplete or where they are missing for some of the strongest historical shocks. That is why a map of maximum expected earthquake intensities in Czechoslovakia (Schenk and Mantlík 1985; Procházková and Brouček 1986) was compiled on the basis of distribution of maximum observed intensities, isoseismals maps of individual earthquakes, and empiric isoseismals models for individual epicentre areas. It must be noted, however, that the largest area of high intensities in Central Slovakia is based on a model compiled for a strong earthquake in 1443 without the focus position being known reliably.

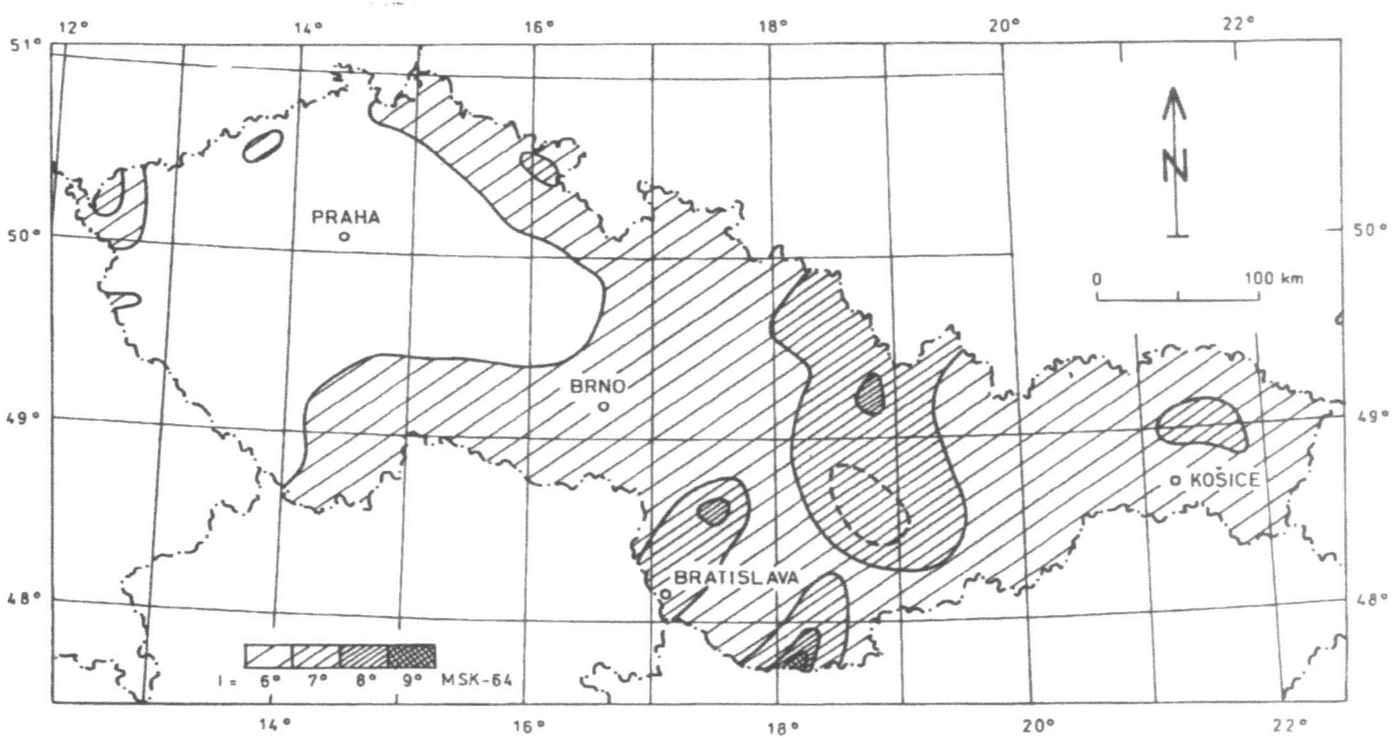

Fig. 3.4 Seismic zoning map of Czechoslovakia - the 1987 version (I. Brouček, V. Kárník, D. Procházková, V. Schenk, Z. Schenková 1988). Dashed area in Central Slovakia marks a possible occurrence of earthquake towards to 8° MSK.

3.3 Seismotectonic Studies of the Bohemian Massif and the West Carpathians

V. Schenk[1], D. Procházková[1], Z. Schenková[1]

A study of spatial distribution of earthquake foci in relation to geologic structure is uncertain by the time-incomparability of geologic and seismological evidence. From the structural-geologic understanding of an evolution of a particular region our data on earthquakes correspond to one or several "seconds of geologic hour". Despite the time-incomparability of the compared quantities being real it is obvious that a causative relation exists between these phenomena. That is why we have tried to find corresponding correlations between earthquake foci distribution and tectonics, recent crustal movements, geophysical fields, etc.

On the basis of existing seismological data, tectonic maps, geophysical fields, and geodetic measurements several seismotectonic schemes have been worked out. They are regional schemes for Central Europe and Southeastern Europe (Mísař and Procházková 1982; Procházková and Zeman 1982a,b; Schenk et al. 1982; Grünthal et al. 1985) as well as seismotectonic analyses and models of smaller areas or seismoactive zones of the Bohemian Massif and the West Carpathians (Procházková et al. 1986; Schenk et al. 1986, 1989c; Grünthal et al. 1990). The above-mentioned studies were argued by analyses of relations of macroseismic fields lengthening to structural-geologic units of Central Europe with special attention paid to structures of the Bohemian Massif and the West Carpathians (Mísař and Procházková 1982; Procházková and Dudek 1982; Procházková and Zeman 1982a,b; Procházková 1984a,b; Procházková et al. 1986). Without seismotectonic studies of extensive regions, the relationships in smaller foci or zones are hard to understand. In the next part we thus recapitulate the results obtained, starting with regional aspects and ending with selected smaller seismoactive zones.

Regional scheme for Central and Southeastern Europe (Schenk and Schenková 1982; Schenk et al. 1982) was elaborated on the basis of present knowledge of tectonic structure of the region, which were correlated with other available geologic, geophysical, and geodetic data. It considers deep faults interpreted from gravimetric, seismic, and magnetometric measurements. It has been shown that a simple correlation of earthquake epicentres with faults identified by e.g. Shatsky (1962) and Maheľ (1974) is not, in this region, unique from the point of view of difference in position (or preference) of fault structures at depth, and inaccuracy in locating earthquake foci (± 5 to 20 km) (Kárník et al. 1984b). Only the most striking lineaments of the first rank and regional faults separating adjacent large structural units by geologic and geophysical criteria were selected for the analysis. In this way a compiled map of main fault lines was used to plot foci of strong earthquakes ($M \geq 5.5$ or $I_0 \geq 8.5°$ MSK) and only those deep faults which were verified by gravimetric or seismic measurements (Schenk et al. 1982).

---

[1] Geophysical Institute, Boční II/1401, 141 31 Praha 4, Czech Republic

On the basis of gravimetric and other geophysical data the sub-Carpathian, sub-Balkan, Marica, and Jambol deep faults were given priorities over the remaining tectonic lines, because gravity anomalies detected over these deep faults can be interpreted as due to step-like structure, the deeper part of which rests beneath mountain ranges. The contact of the West Rhodopean Massif with the Balkan Mts. is an exception. Analysis of neotectonic movements known from the West Carpathians show some evidence for old structures of the West European Platform being "copied" through younger Alpine-type units; see e.g. the Upper Moravian Basin (Schenk et al. 1982,1986; Roth and Procházková 1988).

Beside the regional seismotectonic study of the Carpathian – Balkan region (Schenk et al. 1982) geologic data and satellite images of adjacent eastern part of the West European Platform (Grünthal et al. 1985) were also interpreted from the earthquake occurrence point of view. In the first study we have paid attention to Alpine-like structures of the Carpathian–Balkan system, the second one aimed at the Bohemian Massif. It appears that the majority of seismic activity of the Thuringian and the Krušné hory Mts. regions of the Bohemian Massif correlate well with photolineations oriented in the N–S direction. In the western part of the region earthquake foci are less frequent. The Tachov fault (the Cheb – Domažlice trough) appears as a southern continuation of these photolineations. On the contrary the earthquake epicentres occurring along the lines oriented in the E–W direction could be correlated only with analogous photolineations of the same direction. These E–W oriented tectonic elements are known and attributed to the main Proterozoic megastructures of Central Europe. Later, they were masked by younger tectonic lines of NW–SE and NE–SW directions intersecting geologic structures of the upper crust. In some cases, the E–W structural elements could, however, still define neotectonic movements of blocks in the intermediate part of the Earth's crust in the places where earthquake foci occur (the vast majority of the foci occur at depths of 7 – 10 km). In the eastern part of the West European platform the following seismotectonic regions were distinguished (Grünthal et al. 1985):

1. Thuringian – Saxonian region and the Krušné hory Mts. block of the Bohemian Massif,
2. Lužice region with the Krkonoše Mts. – Jesenice block,
3. South Moldanubian region characterized by weak earthquakes in the Vltava – Dyje block,
4. Tachov deep-fault zone (the Cheb – Domažlice trough),
5. Jizerské hory Mts. seismic zone which could include the earthquakes occurring between the Thuringian – Saxonian and Lužice regions.

The Elbe lineament zone and the zone of the Alpine and Hessen foredeep are tectonically active, but no strong earthquake occurrence was evidenced. The seismic zones of the Franc line and the zones of Delitszch – Brandenburg are still unresolved because there is a lack of earthquake activity. The Delitszch – Brandenburg zone could be considered as a northern continuation of the central part of the Thuringian – Saxonian region. The increased seismicity in the area 40 km SE of the Bohemian Massif – the

West Carpathian contact corresponds to the tectonically, very active boundary of the outer Carpathians which is likely due to the West Carpathian overthrusting the Bohemian Massif. Deeper parts of the overthrust boundary, where earthquake foci probably occur, correspond to the seismic zone "Little Carpathians – Vienna thermal line". It is generally known that every seismotectonic model describing relatively small regions must include also the known geologic facts. Their interpretation obviously offers a compilation of several different models. A problem thus arises which structural-geologic interpretation is actually the one closest to reality. Two models were compiled for the Bohemian Massif and the West Carpathians. The first one resembles the classical structural-geologic interpretation of the geologic structures of the Bohemian Massif and the West Carpathians (Procházková et al. 1986), the second one, on the contrary, is based on neotectonic and geomorphologic investigations argued by the results obtained from geophysical interpretations of tectonic structures (Schenk et al. 1986).

In the following, we have tried to summarize all the results available. We took the second model as a base and included conclusions of the first one in it.

The Neoid model of block structure of the Bohemian Massif (Balatka et al. 1983) and the structural scheme of the West Carpathians by Fusán et al. (1981, 1987) were used for a comparison. This structural scheme was adapted using the data obtained from the interpretation of satellite images (Pospíšil et al. 1986) and adjusted to the newly processed gravimetric data (Šefara et al. 1985). This approach enables us to analyze both the main geologic units parallel (Fig. 3.5).

The Bohemian Massif is an extensive elevation which forms a complicated horst-like anteclise of the Epivariscan platform. In the Neoid epoch, marginal parts of the Bohemian Massif exhibit a rising tendency while the core of the Bohemian Massif maintained a subsiding tendency (Procházková et al. 1986). The Neoid structural model of the Bohemian Massif (Balatka et al. 1983) which is result of an interdisciplinary approach delimited several striking tectonic blocks (Fig. 3.5), i.e. Vltava – Dyje (A), Elbe – Svratka (B), Krušné hory Mts. (C), Český les Mts. and Smrčiny Mts. (D), and Krkonoše Mts. – Jeseníky Mts. (E), and several fault zones (Fig. 3.6), i.e.: Cheb – – Domažlice trough (a), Žatec – Litoměřice (b), Klatovy – Závist (c), Lužice – Jílovice (d), Kyšperk – Zábřeh (e), Poděbrady – Železné hory Mts. (f).

Based on analysis of the spatial distribution of earthquake foci and tectonic activity in the Vltava – Dyje block (A) we can make an assumption that movements still take place along the Late Variscan dislocations (i.e. the Blanice furrow (2), the Přibyslav zone (3), the Rudolec trough (4), etc.). There, however, probability of earthquake occurrence is very low. The foci of weak earthquakes in separate parts of the Bohemian Massif can be associated with a disintegration of the Massif into a great number of minute structures, which according to Kopecký and Vyskočil (1972), are the main characteristic feature of the neotectonic evolution of the Bohemian Massif. The slight increase of earthquake occurrence in Southern Bohemia seems to be joined with an origin of the Tertiary basins, although an effect of tectonic movements along the Jáchymov fault (1) cannot be neglected. The Neoid boundary between the Vltava – Dyje

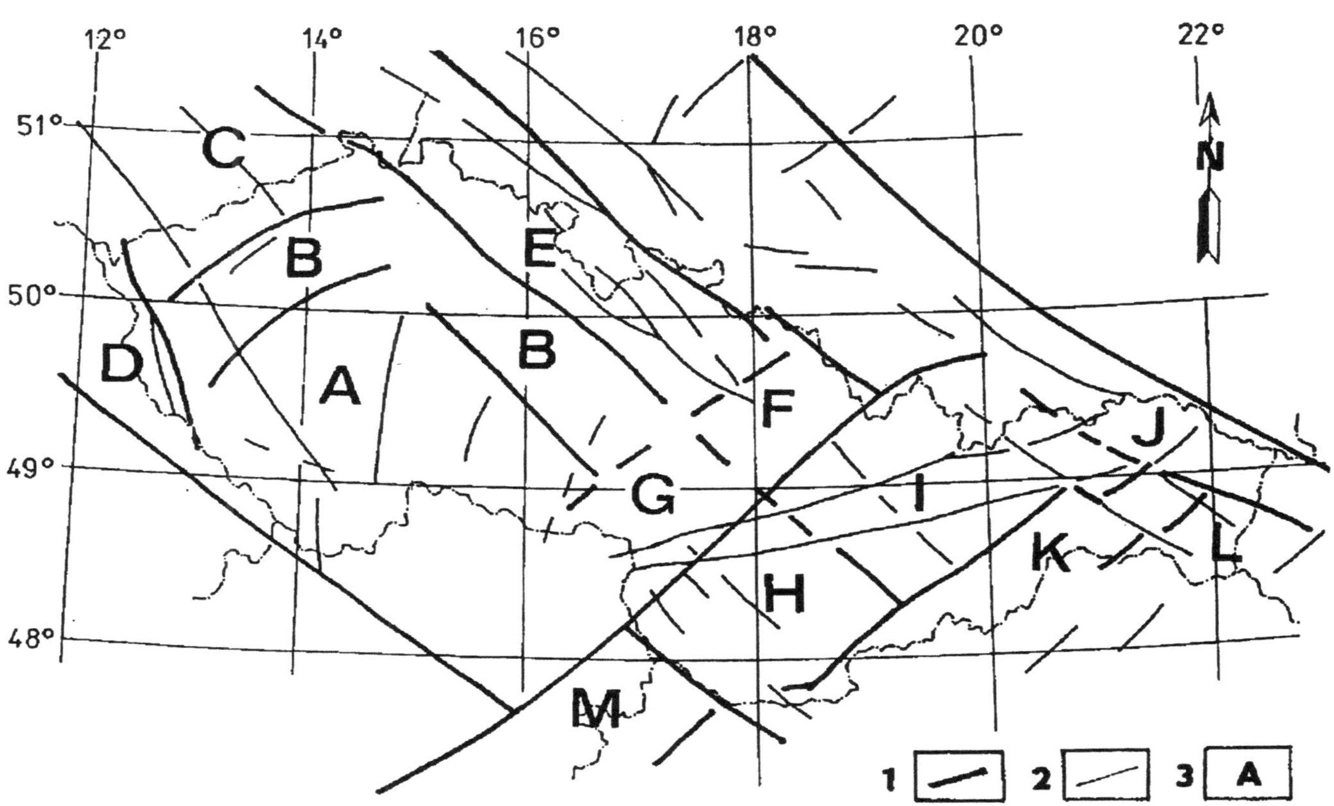

Fig. 3.5 Seismotectonic block pattern of Czechoslovakia (V. Schenk, Z. Schenková, L. Pospíšil, A. Zeman). Materials by Balatka et al. (1983) and Fusán et al. (1979, 1981) were used: 1 – block boundary, 2 – other fault lines, 3 – block designation. A – Vltava – Dyje, B – Elbe – Svratka, C – Krušné hory Mts., D – Český les Mts. and Smrčiny Mts., E – Krkonoše Mts. – Jeseníky Mts., F – Slovakia – Silesia, G – Moravia – Slovakia, H – Danube, I – Fatra – Tatra Mts., J – Beskydy Mts. – Bukovina, K – Southern Slovakia, L – Eastern Slovakia, M – Sopron.

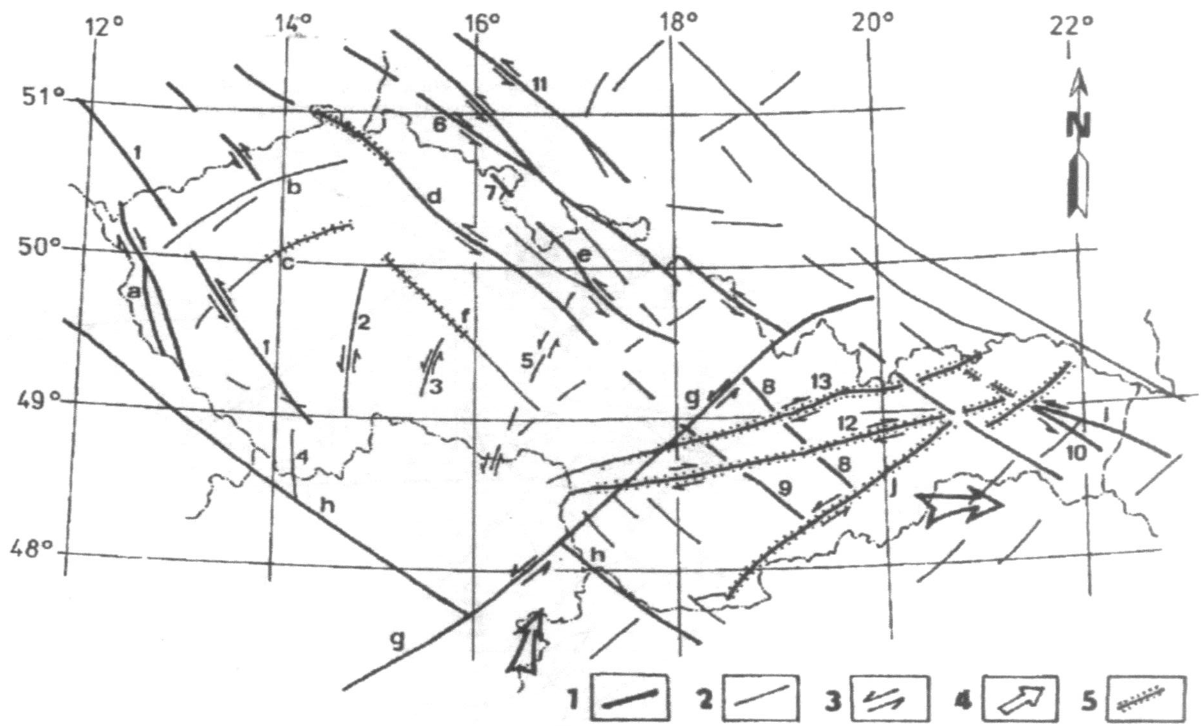

Fig. 3.6 Seismotectonic fault zones and boundaries in Czechoslovakia (V. Schenk, Z. Schenková, L. Pospíšil, A. Zeman). 1 – main striking boundaries, 2 – other striking boundaries, 3 – assumed horizontal displacement, 4 – block movements, 5 – overthrust character of the movement; **Bohemian Massif**: 1 – Jáchymov dislocation zones, 2 – Blanice dislocation zone, 3 – Přibyslav dislocation zone, 4 – Rudolec trough, 5 – Boskovice dislocation zone, 6 – Intrasudetic dislocation zone, 7 – Hronov – Poříčí zone, a) Cheb – Domažlice trench, b) Žatec – Litoměřice fault zone, c) Klatovy – Závist fault zone, d) Lužice – Jílovice fault system, e) Kyšperk – Zábřeh fault system, f) Poděbrady – Železné hory Mts. fault zone; **West Carpathians**: 8 – Revúca dislocation zone, 9 – Přerov – Štiavnica dislocation zone, 10 – Topľany – Močarany dislocation zone, 11 – Odra dislocation zone, 12 – Hron dislocation zone, 13 – Myjava – sub-Tatra dislocation zone, g) Verona – – Semmering – Váh fault system, h) Danube fault system, i) Parečiny fault system, j) Muráň fault system.

block (A) and the Elbe – Svratka block (B), i.e. the Klatovy – Závist fault zone (c), has, mainly in the north and partly in the NW, an overthrust character with the dip towards NW (the Závist overthrust). It is clearly observable between Zbraslav and Mníšek pod Brdy. The Poděbrady – Železné hory Mts. fault zone (f) delimiting block A in the NE has an overthrust character with the dip towards NE, in the Železné hory Mts..

In the Elbe – Svratka block (B) only two remarkable earthquakes are known, one near the town of Kutná Hora (in 1588) and the other one near Mělník (in 1898, the "Mělník blow"). The earthquake which occurred near Kutná Hora has often been explained as due to local mining activity.

The majority of earthquake occurrences in the Bohemian Massif are known from the Krušné hory Mts. block (C). Two systems of faults extend through the block: oriented in WSW–ENE and NW–SE directions. In some places, faults of N–S or E–W directions can also be found. Earthquake activity of the Krušné hory Mts. block (C) is limited by the Žatec – Litoměřice fault zone (b) in the south. On its other side, in the Elbe – Svratka block (B), no earthquakes have been known. The regional character of earthquake occurrences can be explained e.g. by an arching of the entire region. Closer explanation of the higher concentration of earthquake foci in Kraslice and Vogtland areas, including earthquake swarms, remains open. New data on earthquake occurrences in this region document high seismic activity of faults of the NNW–SSE, or N–S directions as compared with the main tectonic lines of the Krušné hory Mts. direction. The seismotectonic analysis of the last swarm of 1985/1986 (Procházková 1988c; Grünthal et al. 1990) confirmed this conclusion unambiguously.

In the block of the Český les Mts. and Smrčiny Mts. (D) earthquakes are confined to the Cheb – Domažlice trough (a). Saxonian orogenic processes have caused repeated shifts along old dislocations. The region between the main faults relatively subsided (Tertiary sediments are preserved in some parts of the region). Recent earthquake occurrences indicate that movements along these faults have not ceased yet.

Seismotectonic characteristics of the Krkonoše Mts. – Jeseníky Mts. block (E) are not quite clear. Earthquakes occur more or less in the whole region but significant ones can be observed only between the eastern part of the Krkonoše Mts. and the Carpathians (Kárník et al. 1984a; Procházková 1984a,b; Schenk et al. 1989a). Lower frequency of earthquake foci occurrences in the western part of the block can be explained by the presence of the resistant Krkonoše – Jizera Massif. Numerous striking faults intersect the block, which originated in the course of the Variscan and Caledonian orogenes and which have remained active ever since the Recent. The faults extend mostly in the NW–SE direction: the Lužice fault (d), the Intrasudetic fault (6), etc. The Neoid block boundary between the Elbe – Svratka (B), Krušné hory Mts. (C), and Krkonoše Mts. – – Jeseníky Mts. (E) blocks, the Lužice – Jílovice dislocation (d), in the north of the Bohemian Massif, exhibits an overthrust character with the dip towards NE. The pre-Neoid complexes are elevated along this line forming a horst above the Cretaceous sediments. It has been known (Vyskočil P. 1983) that the above-mentioned block (E) has a rising tendency (ca. 1 mm per year) in the NE part, and up to 1.5 mm per year in part of the Jizerské hory Mts. and the Lužické hory Mts. These movements are explained by a compensation of horizontal stress affecting the Bohemian Massif from the south and SE (Balatka et al. 1983). These facts agree with earthquake foci occurrences in the Krkonoše Mts. – Jeseníky Mts. block. It appears that this seismicity zone plunges underneath the Carpathians continuing towards the SE over the known earthquake epicentre near Žilina to foci near Rimavská Sobota.

In the West Carpathians, in accordance with the opinions of Fusán et al. (1981, 1987), the following structural-geologic blocks (Fig. 3.5) and fault zones (Fig. 3.6), respectively, can be distinguished: Slovakia–Silesia (F), Moravia–Slovakia (G), Danube (H), Fatra – Tatra Mts. (I), Beskydy Mts. – Bukovina (J), Southern Slovakia (K), and Eastern Slovakia (L), Revúca (8), Přerov – Štiavnica (9), Topľany – Močarany (10), Odra (11), Hron (12), Myjava – sub-Tatra (13), Verona – Semmering – Váh (g), Danube (h), Parečina (i), and Muráň (j).

Earthquake foci in the West Carpathians are confined to weakened tectonic zones more tightly than those in the Bohemian Massif. From the regional point of view, earthquake occurrence is confined to the three main structural-geologic directions:

1. Carpathian faults (SW–NE), related to the origin of sliding surfaces between the main blocks,
2. transverse faults (NW–SE) reflecting the basement structure of the West Carpathians (Schenk 1984b), and
3. shear and "gravity" thrust planes along which the blocks move in an easterly direction (Schenk et al. 1986).

The best known activity of the West Carpathians is associated with the Peripienian lineament zone (the Záhorie – Humenné fault of Procházková et al. (1986), which appears as a tectonic continuation of the Verona – Semmering lineament (g) over the East Alps towards NE. Contact of the units of the Bohemian Massif with blocks of the East Alps is not accompanied with a striking fault zone which would show itself by earthquake foci. An assumption of transcurrent fault existing between the East Alps and the West Carpathians (Unrung 1984), which could explain the difference of Tertiary kinematics and tectogenesis of both the mountain ranges, together with the conclusions of Biely (1975) concerning the wrong and problem-creating use of the term "Peripienian lineament" in the sense of Máška (1961), has forced us to designate this boundary as the Váh fault (g). On the basis of results of remote sensing and geophysical data this fault can be considered as transcurrent.

It follows from the geodynamic analysis that, at present, movements take place probably along this fault zone. Here, SE blocks move towards NE and, vice versa, NW blocks move towards SW. This fact has also been confirmed by known movement tendencies in the Bohemian Massif and in the West Carpathians which appear to be associated with the primary movement of the Sopron block (M) towards the north or NE.

More frequent occurrences of earthquake foci in this region can also be explained by the above-mentioned tendency of slow movements of active Carpathian blocks towards NE. An elongated shape of macroseismic isoline closures of fields of strong earthquakes indicate a shallow basement of Carpathian structures. The relation of earthquake foci to lines of the Danube dislocation zone (h) extending over the Komárno region, the Revúca fault zone (8) from Žilina to Rimavská Sobota, southern border (the Parečiany fault zone) (i), Beskydy Mts. – Bukovina block (J), etc. also follows from a spatial distribution of earthquake epicentres. Seismic manifestations of smaller extent are observed along the Přerov – Štiavnica fault (9), Topľany – Močarany fault zone (10), transverse faults of the Little Carpatians, etc.

If we consider that on one side of the West Carpathian system the Sopron block (M) shifted considerably towards NE what is clearly documented by the course of the Danube fault zone (h) as well as by gravimetric data obtained in this region, and that on the other side the Tornquist – Teisseyre line resistance (the western border of the Russian Platform) has not enabled a continuation of these movements, we can then assume that due to a compression a mutual shift of transverse dislocation zones took place towards NE (Fig. 3.6) which brought them closer. It follows from this seismotectonic model that the Přerov – Štiavnica fault zone (9), separating the Danube block (H) from the Fatra – Tatra Mts. block (I), could be a continuation of the Poděbrady – Železné hory Mts. zone (f). A continuation of the Lužice – Jílovice (d) and the Kyšperk – Zábřeh (e) fault zones could analogously correspond to intensive earthquake activity of the Revúca fault zone (8). The Odra fault zone (11) could continue into the Topľany – Močarany fault zone (10). The presence of tectonic zones extending over a long distance in the WSW–ENE direction (Pospíšil et al. 1986) appears as a significant element in this evolution. Along the tectonic zones blocks could be thrusted over other blocks as well as the northern blocks shifted to the east. The presented idea is confirmed to some extent by seismoactive intervals of transverse tectonic lines in the West Carpathians: their northern termination does not coincide with the Klippen Belt position (in the past it was considered to be a continuation of the Peripienian lineament towards E) but with the above-mentioned "shear" transcurrent fault zones of WSW–ENE direction. The bending of gravity and conductivity anomalies known from the Fatra – Tatra Mts. block (Ibrmajer 1981; Praus et al. 1981) supports this idea.

The overthrust character of the movement can be observed not only on actual boundaries, e.g. the Myjava – sub-Tatra fault system (13) in the Strážovské vrchy Mts. (Janků et al. 1984), in the High Tatras, or in the Ružbachy island (Pospíšil et al. 1986), but also on boundaries delimiting head fronts of the moving blocks to the east, e.g. the Klippen Belt with the Magura unit in the Plaveč – Lipany part (Němčok and Rudinec 1979). A compressional origin of some structures, e.g. the synclinorium of the North Gemericum in the proximity of the Muráň fault (j), or the pattern of the Past-Paleogene anticlines and synclines in the West Carpathians (Maheľ 1974) indicates a pressure propagation from SW (Unrung 1984).

If earthquake occurrences in the West Carpathians are evaluated from this point of view, the majority of them lie then in weakened zones. Deeper foci are found in the Little Carpatians, Revúca line (8), Topľany – Močarany zone (10), Hron zone (12), and the Myjava – sub-Tatra (13) transcurrent fault system, and of course in the Danube lineament (h) passing over the foci of the Komárno earthquakes. From the seismological point of view the Muráň line (j) appears extremely stimulating, which delimits the continuation of transverse seismoactive tectonic elements further to SE. Beside the Southern Slovakia block (K) the Moravia–Slovakia block (G) appears also aseismic.

In conclusion it is possible to say that both the models (Procházková et al. 1986, Schenk et al. 1986) yield similar global conclusions. The models differ only in the detail to which they were elaborated. The first model starts from the fact that the position of earthquake epicentres is inaccurate and that courses of faults at depth are known from boreholes only in few places. That is why the model considers fault zones of up to 10 km in width and does not distinguish individual faults presented in regional geologic maps belonging to the particular zone. When joining epicentres and fault zones, the first model utilizes only data on earthquake mechanism and geologic and geodetic data on the fault zones movements. On the contrary the second model discusses in detail the seismogenic character of individual tectonic structures of known geologic units in their Neoid evolution. In this model, the recent activity of fault zones was determined by an interdisciplinary evaluation of presently existing information on earthquake occurrences (or on earthquake mechanisms), on geophysical interpretations of fault structures, neotectonics, geomorphology, recent movements of the Earth's crust, and for the West Carpathians, on remote sensing data also. The resulting model shows higher degree of dynamics in the sub-Recent to Recent epoch than it was assumed until now. This fact is also confirmed by the coincidence of earthquake epicentre locations with the boundaries obtained from interpretations of satellite images and geomorphologic data. It cannot be excluded, however, that in some zones intersecting the entire part of the Earth's crust (Jáchymov, Lužice – Jílovice, Odra, Danube, Hron, etc.) character of movements could vary with time. The overthrusts on young fault boundaries generally dipping to north in the Bohemian Massif, and to south in the West Carpathians give evidence that also in the Recent epoch both geologic units are affected by pressures from the south, from the East Alps. The overthrust character of faults documents mutual compression of blocks, when relative shortening of lengths takes place. This fact confirms a recent mobility of the whole investigated regions. Based on present knowledge it can also be considered that a large amount of stress is not accumulated.

The models show an areal division of Czechoslovakia into regional Neoid blocks, and represent a simplified two-dimensional expression of tectonic and seismological phenomena between one another. In the near future, the models will have to be further elaborated from point of view of the spatial distribution of individual structural blocks and specifically the genetic superiority of fault zone activity among themselves.

4 Results of Heat Flow Studies in Czechoslovakia

V. Čermák[1]

Introduction

A heat flow project has been under way in Czechoslovakia since 1964, its main objective to obtain a sufficient data base for interpreting the deep crustal and lithospheric structures.

In the mid-1970s the growing interest in acquiring alternative sources of energy, including geothermal energy, substantially accelerated the programme of fundamental research into heat flow. The early heat flow investigations started with particular efforts to construct suitable field and laboratory equipment so as to improve temperature logging and to measure the thermo-physical parameters of rocks. In this period all available older temperature logging records were re-assessed and the first 34 heat flow values were published (Čermák 1967a). This work resulted in the clear recognition that the mean geothermal activity decreased from younger to older tectonic elements and disclosed the different geothermal structures of both main tectonic units in Czechoslovakian territory, i.e. the Bohemian Massif (Hercynian) and the West Carpathians (Alpine) (Čermák 1977c).

The principal institution taking care of the heat flow research programme since the beginning has been the Geophysical Institute of the Czechoslovak Academy of Sciences in Prague. Since 1970 a similar project has been under way in Geofyzika Brno (the branch in Bratislava). While the former has been orientated more academically, the latter has been a part of a regional applied survey serving to assess usable sources of geothermal energy.

The manner in which the programme effort has been directed towards heat flow data acquisition has been reflected on and off in several reports, see e.g. Čermák (1979a, 1981), Čermák et al. (1991a), and Lizoň (1982). This chapter is focused on a comprehensive review of the heat flow research in Czechoslovakia and on discussing the main results in terms of the lithospheric structure. In view of the rather complicated tectonic setting and of a relatively small territory, Czechoslovak data were complemented with the analogous results in the neighbouring countries. The territory investigated and discussed thus occupies not only Czechoslovakia but also sizeable parts of all adjacent countries in Central Europe. Formally this territory is limited by the coordinates 11° to 24° E in longitude and 46° to 52° N in latitude.

1 Geophysical Institute, Boční II/1401, 141 31 Praha 4, Czech Republic

4.1 Previous Major Results

After the introductory heat flow determinations were made in several locations, which tested both the methodology and apparatuses used, standard measurements continued to improve the national data coverage. The data obtained were completed with more detailed investigations in various specific geologic regions: in several sites of the oldest, tectonically consolidated units of the Bohemian Massif (Čermák and Krčmář 1967, 1968a; Čermák et al. 1968b; Čermák 1975a), in the Bohemian Cretaceous Basin (Čermák et al. 1968a; Čermák 1977a), in Permocarboniferous coal basin of Central Bohemia (Čermák 1967b), and of Lower Silesia (Čermák 1968c), and of Upper Silesia (Čermák 1968b, 1976a,b,c), in the Carpathian Foredeep (Čermák 1968d, 1975b; Novák 1971), in the Neogene depressions of the West Carpathians in SW Slovakia (Čermák and Krčmář 1968b; Marušiak and Lizoň 1976), in east Slovakia (Čermák 1968e), as well as in the faulted units of the West Carpathians (Čermák 1977b; Čermák et al. 1977; Lizoň 1982). The first review of more than a hundred original data together with the attempted first heat flow map of Czechoslovakia were prepared in 1976 (Čermák 1978). The local geologic environment and the effects of the crustal structure rendered the general heat flow pattern quite complicated. Anomalous heat flow (over 120 mWm^{-2}) was found in the area of thermal waters of Teplice (Čermák 1967c) demonstrating the hydrothermally disturbed temperature field in the uppermost layer of about a few hundred metres. An increased heat flow in the Bohemian Cretaceous Basin was interpreted in terms of a possible thinning of the Earth's crust (Čermák 1972; Zahradník and Čermák 1979), but the heat flow there may have been also regionally affected by extensive groundwater circulation in permeable Mesozoic sediments (Čermák and Jetel 1985; Čermák 1989b). The effect of mud circulation during drilling boreholes on natural temperature field was discussed (Čermák 1968b; Čermák et al. 1968c), and it was proved that some of the old temperature records could have been used as additional information. The Vienna Basin data (Čermák and Krčmář 1968b) were later reworked (Marušiak and Lizoň 1976) and lowered by about 20 – 30 % because of a high clay content recorded by the SP logging, new thermal conductivity data were proposed according to relative clay-versus-sand composition. Taking into account the detailed lithostratigraphic profile, we re-assessed the conductivity data from the Upper Silesian coal basin (Kubík and Čermák 1986) and corrected them for relative abundances of the individual rock types. The 1978 heat flow map was later checked and the courses of contours were adapted to new heat flow data (Čermák et al. 1991a,b). The most recent version of the heat flow distribution pattern in Czechoslovakia is shown in Fig. 4.1.

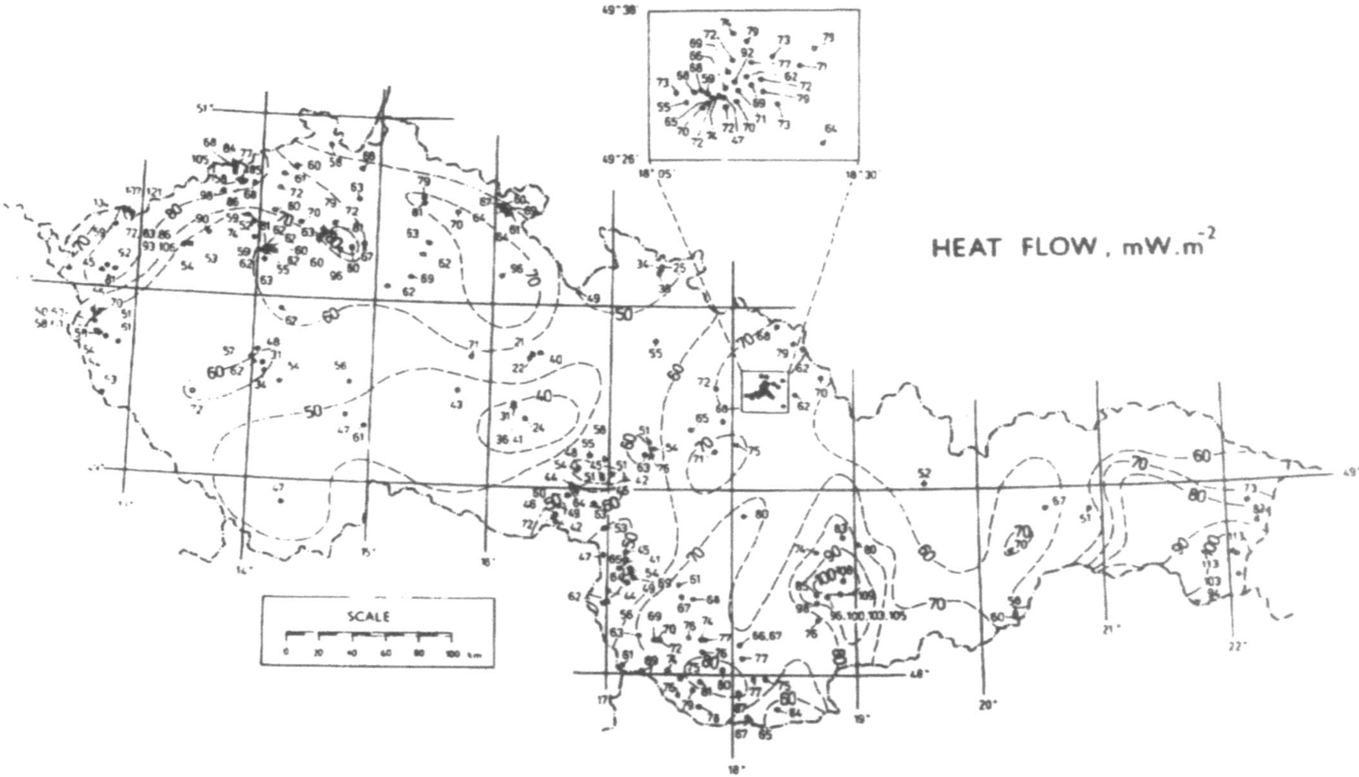

Fig. 4.1 Heat flow stations and the heat flow map of Czechoslovakia, after Čermák et al. (1991b).

Since the beginning of the 1970s there has been an emphasis on the heat flow data correlation with other geophysical parameters. Preliminary information concerning the relationship between the geothermal activity and the gravity field represented by the geoid shape (Čermák 1972) facilitated the complex interpretation of the heat flow field to the deep crustal structure (Čermák et al. 1976). Results of the deep seismic sounding were then utilized, together with the extrapolation of the surface heat flow distribution pattern to a depth. The deep crust temperatures were then calculated for all major tectonic units (Čermák 1975c) and also along the cross-section across the West Carpathians (Čermák 1974). The heat flow-versus-age relationship was described (Čermák et al. 1975) and the linear relationship between the heat flow and the near surface heat production was studied for the Bohemian Massif (Čermák 1975a, 1981).

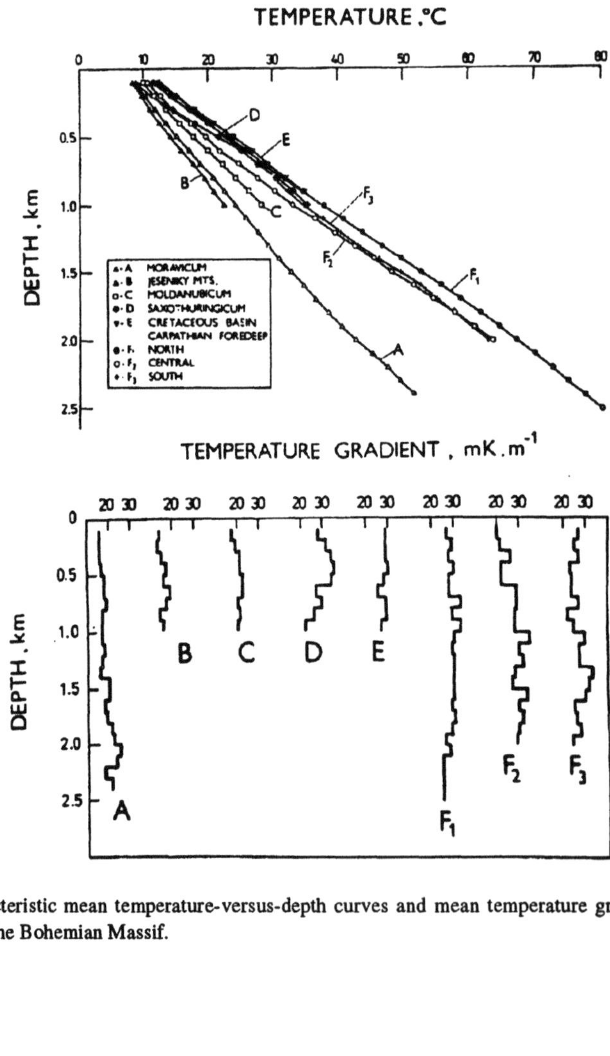

Fig. 4.2a Characteristic mean temperature-versus-depth curves and mean temperature gradients for the main tectonic units of the Bohemian Massif.

Information about the thermal history is often required for purposes of coal, gas and oil prospecting, as well as for tectonic studies. The coalification data of 12 boreholes in the Central Bohemian basin were used to evaluate the palaeotemperature gradients for the Upper Carboniferous period (Šafanda et al. 1990). By combining the present thermal conductivity and palaeogradients, the estimated Upper Carboniferous heat flow was about 40% higher than its present value.

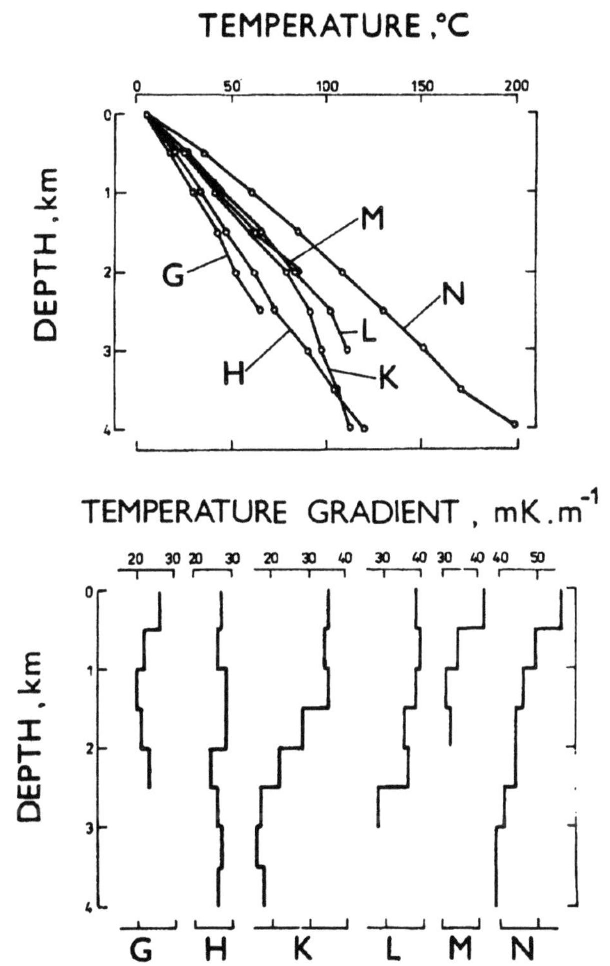

Fig. 4.2b Characteristic mean temperature-versus-depth curves and mean temperature gradients for the main tectonic units of the West Carpathians: G – Slovenské Rudohorie Mts. (inner units), H – Outer West Carpathians (Flysch Belt), K – Vienna Basin, L – Danube Basin, M – neovolcanites in Central Slovakia, N – East Slovakian (Neogene) Basin.

4.2 Technique of Measurements

The thermometer used in the majority of the early measurements was a common logging thermometer, whose basic configuration was a temperature-sensitive element. Variations of voltage proportional to a temperature increase were recorded continuously as a function of depth. At selected check points the recorded temperatures were checked with the help of a set of mercury thermometers. The accuracy of these early data was not better than several tenths of a degree of centigrade. Since 1972, a light portable thermistor thermometer has been used for the measurements which has helped to increase the measurement accuracy considerably (Krešl 1976, 1981). The probe thermistors were calibrated against laboratory precise, resistance thermometers. The estimated accuracy of the probe is better than ± 0.05 K , and the relative accuracy, corresponding to a temperature uncertainty, amounts to a few millidegrees. Temperature readings are usually taken point by point at 5 or 10 m intervals, the depth between the readings interpolated by a depth counter with the accuracy of a few centimetres. For gradient calculations the data are processed to determine the geothermal gradient by the least-squares method, either for regular depth sections or for individual lithostratigraphic units. The gradient can be estimated with 2 – 5 % accuracy considering the logging record and with accuracy better than 1 % considering the thermistor probe.

Nearly all conductivities reported were obtained with the use of the divided-bar apparatus (Čermák and Krešl 1968) on specimens saturated with water under vacuum. Lately the original device was modified into a semi-automated version (Krešl and Veselý 1973) which has enabled us to measure routinely large quantities of samples. Crystalline and fused quartz discs were used as primary standards to calibrate the instrument. With the use of the automated version certain problems appeared with saturating the samples. In addition to this, some loose sediments (mudstones and claystones) disintegrated when re-saturated and could not have therefore be measured. That is why the conductivity of a large number of selected samples was measured twice, i.e. in a "dry" state and after the re-saturation (in a "wet" state), and the effect of water content on the ratio "wet" to "dry" conductivity was examined (Čermák 1967d, 1977a). The experimental relationship obtained were then used in all cases if the sample could not be resaturated. Since 1980 additional measurements have also been obtained with the help of the commercially produced Showa–Denko QTM–2 conductivity meter (transient method) and since 1986, the needle probe has been used for measuring the thermal conductivity of sedimentary rocks. The reproducibility of the conductivity results of all the above-mentioned laboratory techniques is better than 3 – 5 % except for a few cases of either poor-quality samples or of extremely inhomogeneous rocks. The absolute accuracy of the thermal conductivity determination depends above all on the grain size, porosity, inhomogeneity, and sample size, and can be estimated as being between 5 and 15 %.

4.3 Data Set

Subsurface Temperatures

All available results of the temperature loggings performed in boreholes were recently collected and a catalogue of data prepared for the Geothermal Atlas of Europe (Čermák et al. 1991b). While in the Bohemian Massif temperature conditions are fairly well studied only in the upper 1 500 m (there is only one borehole in which measurements were taken from the depth of 2 400 m), in the Carpathian Foredeep and in the West Carpathians the regional temperature field is known better at a greater depth (to 3000 m), and there are a few boreholes which penetrated rocks at depths of 4000 m (11 boreholes) and 5000 m (6 boreholes). For construction of the geotemperature maps, all measured values were used together with the extrapolated temperatures, the extrapolation, however, did not exceed 50 % of the depth interval actually measured. In addition to the observed data, knowledge of the tectonic setting, gross-scale geologic structures, surface morphology, and heat flow distribution pattern was taken into consideration in projection of the geo-isotherms.

Due to principally different tectonic structures, the geothermal conditions are studied separately for the Bohemian Massif (Variscan province) and for the West Carpathians (Alpine province). Characteristic temperature-versus-depth profiles and the geothermal gradients in special tectonic units and sub-units of both the tectonic system are shown in Figs. 4.2a and 4.2b. The temperature and gradient profiles represent mean values calculated for all available measurements within corresponding 200 m depth intervals (Bohemian Massif) and/or 500 m depth intervals (Carpathians), respectively. The regional distributions of the subsurface temperatures are shown in the maps of geoisotherms at depths of 500, 1000, 2000, and 3000 m below the Earth's surface (Fig. 4.3).

On basis of temperature measurements in almost 200 holes the map of temperature distribution at a depth of 100 m below the terrain level was constructed for the Bohemian Massif and then interpreted in term of hydrogeological, geomorphological and microclimatic factors, the corresponding effect on the heat flow data obtained in shallow holes was assessed (Kubík 1990).

Heat Flow

The locations of 246 sites used for the 208 heat flow determinations in Czechoslovakia together with the values obtained and with the proposed contours of heat flow density was shown in Fig. 4.1. Practically all measurements were performed in boreholes, and only 3 data are from mines (Čermák et al. 1968b). The borehole depths ranged from minimum of about 100 m to the extreme 6000 m, and were originally drilled for various purposes, including basic geologic studies and mapping, hydrogeological studies, as well as prospecting for oil, gas, coal, and minerals. The borehole inclination, terrain

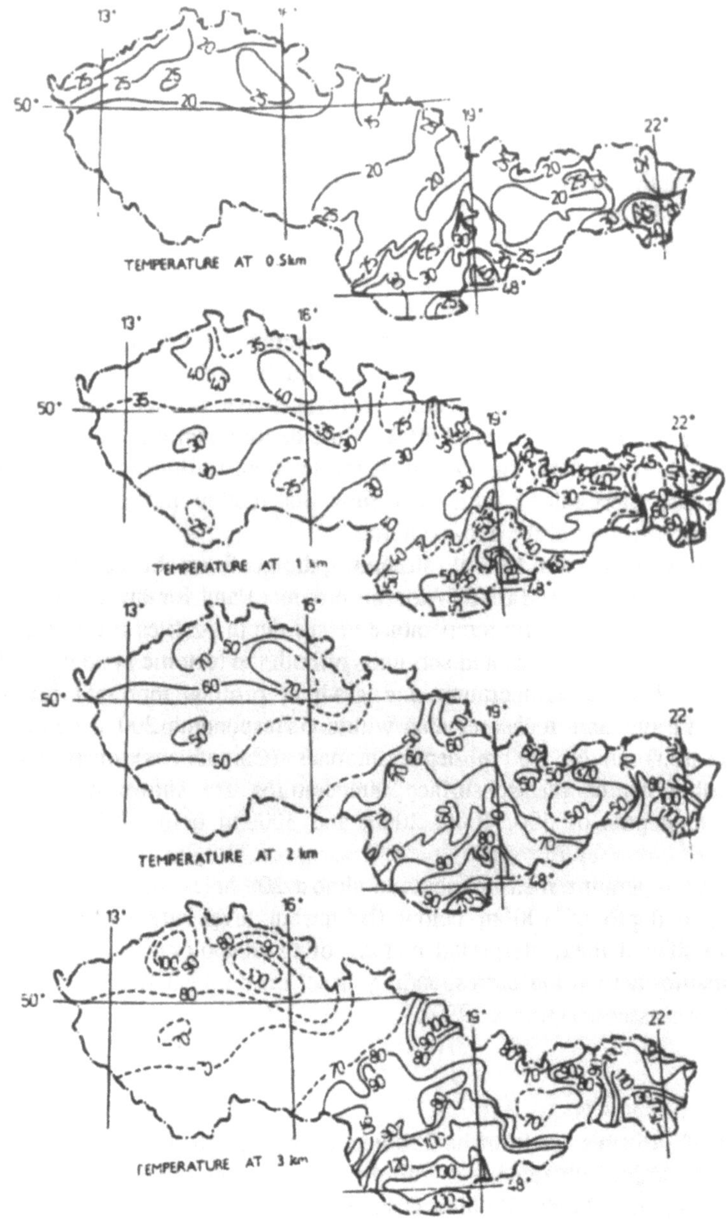

Fig. 4.3 Subsurface temperature distributions (in degrees Celsiurs) at the depths of 500 m, 1 km, 2 km, and 3 km.

Table 4.1 Mean heat-flow values in Czechoslovakia

Tectonic unit	Number of data	Heat flow [mWm^{-2}]	s.d. [mWm^{-2}]	s.e. [mWm^{-2}]
Bohemian Massif	92	67.8	24.3	2.4
Carpathian Foredeep	60	72.2	19.4	2.5
West Carpathians	56	78.0	22.9	3.9
Total	208	71.0	22.8	1.6

topography and in some cases also the insufficient temperature equalization were taken into account and corresponding corrections were proposed. As the depths of the boreholes exceeded 300 – 500 m (except for a few cases) and quite often the boreholes were deeper than 1000 m, the magnitude of topo-correction was generally small. No attempts have been made yet to assess the climatic disturbances due to the last ice age for any specific location, but the land was not covered with ice, being just to the south of the maximum extent of the continental glacier in the Quaternary. With the use of the theoretical climatic step model (Čermák 1977d) the maximum correction for the thermal diffusivity of 0.01 $cm^2\ s^{-1}$, could be estimated to be about 1 – 2 $mK\ m^{-1}$ for the depth interval of 300 – 800 m, and would increase the measured heat flow by 3 – 5 $mW\ m^{-2}$ on average. Below 1 000 –15 000 m the disturbance changes its sign, the climatic correction being negative. It amounts to 0.5 – 1.0 $mK\ m^{-1}$.

To distinguish the data quality, quite strict criteria were applied to express the maximum uncertainty of each value following the three categories proposed by Jessop et al. (1976) that differentiated the reliability of data into "good", "fair", and "unreliable" groups. All data were recently listed and arranged in the IHFC format (Čermák et al. 1991b). The mean heat flow and its standard deviation is $71 \pm 22.8\ mW\ m^{-2}$. Due to its complicated tectonic structure and the fact that the territory belong to two essentially different tectonic provinces, it is reasonable to evaluate the geothermal activity separately, see Table 4.1. There is not very much difference in the mean heat flow between both tectonic system, but when individual smaller tectonic units and subunits are compared, the characteristic heat flow values can vary from 40 – 50 to 80 $mW\ m^{-2}$, and more. To complete the heat flow distribution pattern for a larger area (see further), data from Austria (Clark 1961; Boldizsár 1968) Federal Republic of Germany (Bram 1979; Hurtig and Oelsner 1979), Hungary (Dövényi et al. 1983), Poland (Majorowicz and Plewa 1979), and West Ukraine (Lubimova et al. 1973; Kutas and Gordienko, pers. commun.) were used, see Figs. 4.4 and 4.13.

Heat Flow Map

The lowest relatively heat flow values in Czechoslovakia were observed in the southeastern and central parts of the Bohemian Massif, i.e. in areas where crustal thickness attains its maximum close to 40 km (Beránek and Dudek 1972). The mean heat flow in the stable intramontane block which forms the nucleus of the whole massif is 50 – 60 mW m^{-2}. The heat flow generally increases in all directions as the crustal thickness decreases. To the NW and to the NE, the central part of the Bohemian Massif is bound by two zones of relatively thinner crust (about 4 – 6 km), which separate the stable part from the areas of intensive Variscan tectonogenesis. An increased geothermal activity is typical of both these zones which are assumed to be old rift structures (Kopecký 1972); the Labe zone corresponds to the Cretaceous Basin, the Ohře zone is parallel to the southern rim of the Krušné hory Mts. Relatively high heat flow values (70 – 80 mW m^{-2}, locally as high as 90 mW m^{-2}) observed here are generally being explained by an increased heat supply from below the crust associated with deep seated processes of rejuvenation during the Alpine–Carpathian orogenesis (Pačes and Čermák 1976). This explanation seems to be supported further by the occurrence of Tertiary volcanism, numerous thermal springs and CO_2 discharges along both these zones. The eastern anomaly of high heat flow zone, the Labe zone, approximately follows the characteristic Sudetic direction of Saxonian faults and coincides with the axis of the sedimentary basin. The igneous rocks forming the roots of the Krušné hory Mts. and some Late Variscan granitic plutons belong to the most radioactive rocks of the Bohemian Massif (Matolín 1970). The radiogenic heat contributes here to the observed high heat flow by about 30 – 40 mW m^{-2}.

There is a certain increase of heat flow along the Carpathian Foredeep, from 50 – – 60 mW m^{-2} in the SW to about 70 mW m^{-2} in the NE. The geothermal field in the Upper Silesian Basin may have been affected by the formation of the West Carpathians

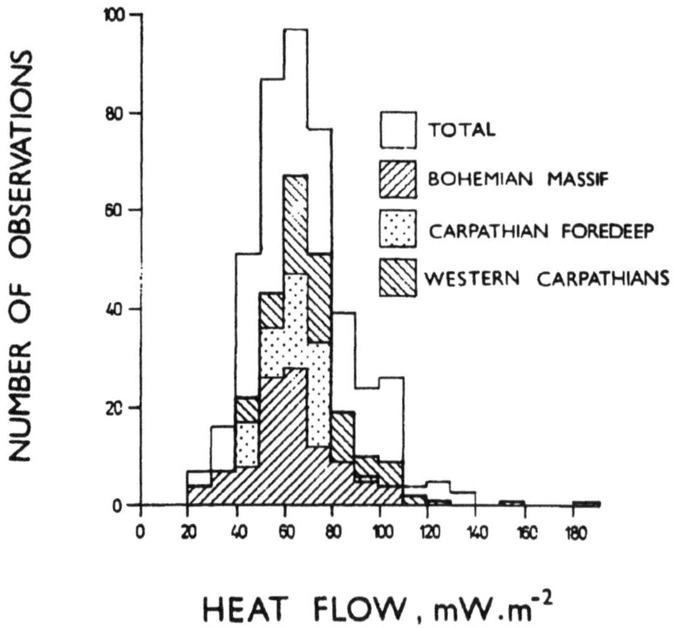

Fig. 4.4 Histogram of heat flow data in Czechoslovakia and its close vicinity limited by the coordinates and 11° to 24°E in longitude and 46° to 52°N in latitude.

(Kubík and Čermák 1986). The general increase of the heat flow in the direction from outer zones of the Carpathian belt towards inner structural units and intramontane depressions is dominant in the whole Carpathian system. Contrary to the Alps, where the zone of high heat flow roughly corresponds to the range of folded units, the Carpathian crest area is a zone of the strongest horizontal gradient in the heat flow field. The heat flow in the Palaeozoic platform and the outer Flysch Belt in south Poland is only 50 – – 60 mW m^{-2}, but it attains over 100 mW m^{-2} in the Neogene basins (Tisa Lowland) in east Slovakia, and even more in central parts of the Pannonian Basin in Hungary (Dövényi et al. 1983). The near-surface temperature gradient increases correspondingly from less than 20 mK m^{-1} to almost 50 mK m^{-1}.

4.4 Crustal Geothermics

Heat Flow and Crustal Thickness

The knowledge of the crustal thickness and of the contribution by crustal heat sources can be of some importance in regional heat flow studies. As the crustal radioactivity generates, on average, more than 40 % of the surface heat flow, one might expect that a thick crust would correspond considerably to regions of high heat flow. While in some places it may be true, it has not been generally observed in many places in Europe (Čermák 1979b). The results of deep seismic sounding in Czechoslovakia (Beránek and Dudek 1972) when correlated with the surface heat flow distribution pattern (Zahradník and Čermák 1979) revealed a negative character of such a correlation, i.e. a high heat flow was detected over local Moho discontinuity elevations, while low heat flow was detected over local Moho discontinuity depressions. Similar relationships were also reported for the Ukrainian shield (Kutas 1979), for the East Carpathians (Veliciu and Demetrescu 1979), and were treated in detail for a number of selected locations in Europe as well as in other continents (Bodri and Bodri 1985). In the latter study the authors investigated the heat flow – crustal thickness relationship in several different categories, defined according to the thickness of the consolidated crust (i.e. total crustal thickness minus the thickness of the sedimentary cover) and while the negative correlation was generally confirmed in areas with normal (30 – 40 km) and thin consolidated crust (less than 30 km), they claimed that the positive correlation existed in tectonic formations of the thick consolidated crust (over 40 km). Such structures usually include uplifted massifs and folded mountain ranges, such as e.g. the Alps, where heat flow is typically high (70 – 80 mW m^{-2}) and the crustal thickness is 40 – 50 km. The correction for uplift and rapid erosion can amount here to 30 % of the measured heat flow (Haenel and Zoth 1973) and the reported positive correlation thus becomes less convincing.

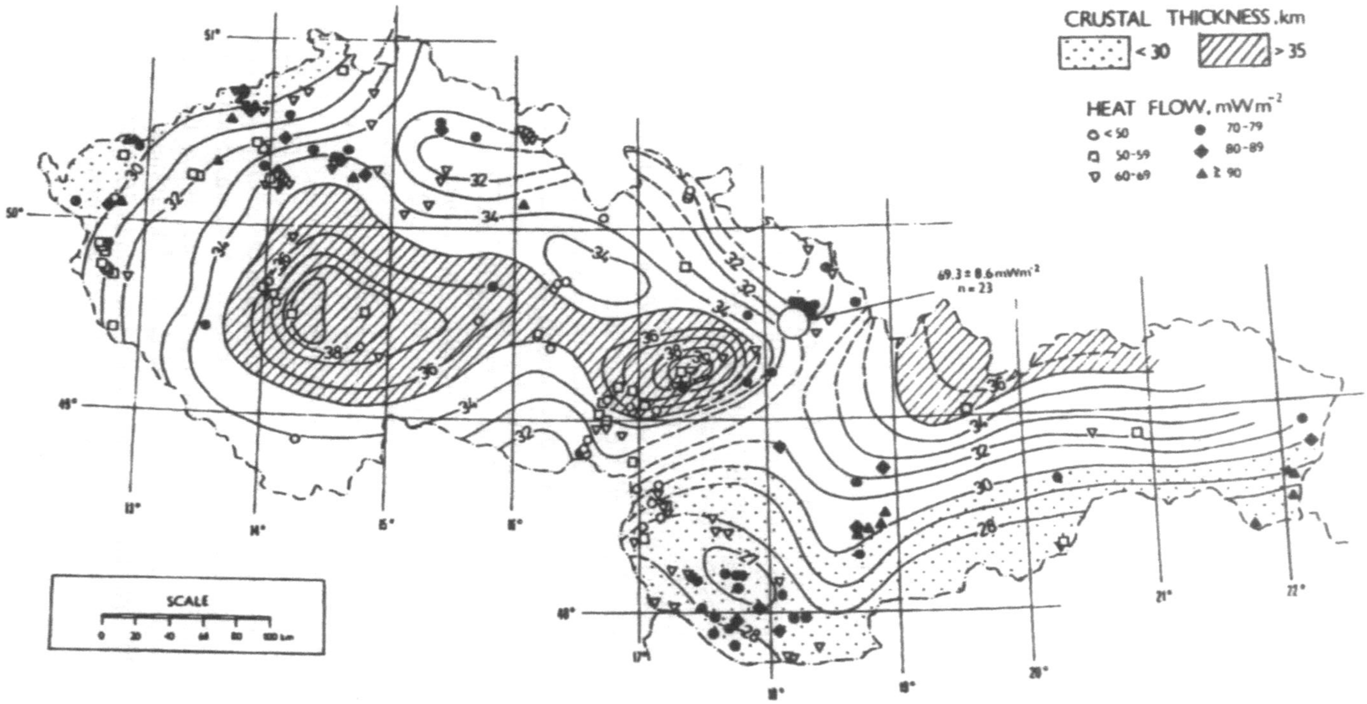

Fig. 4.5 Map of crustal thicknesses in Czechoslovakia together with the heat flow data.

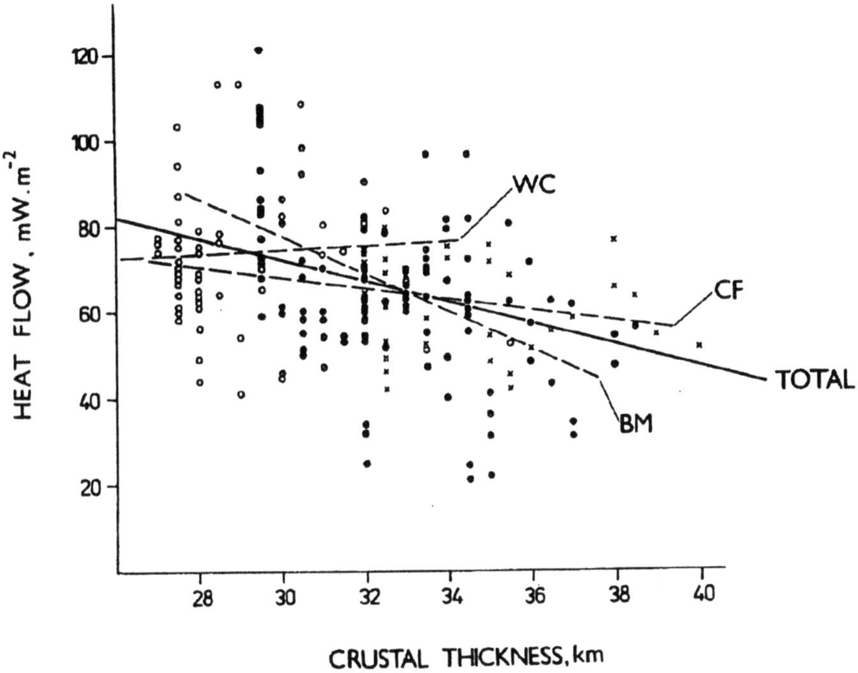

Fig. 4.6a Heat flow versus crustal thickness: solid dots – Bohemian Massif, crosses – Carpathian Foredeep, open circles – West Carpathians.

The problem of interpreting the negative correlation between the heat flow and crustal thickness has not yet been solved satisfactorily, but there is little doubt that it is directly connected with the heat flow from deep parts of the upper mantle. The crustal heat contribution due to radioactivity probably does not exhibit such large regional variations and thus the Moho heat flow, ranging from less than 20 to more than 50 mW m^{-2} (Čermák 1989a) is more definite for the variation of the surface heat flow density. The areas of thin crust are usually younger in their origin (except for young mountain ranges), while relatively "colder" areas with normal to thicker crust belong to the oldest tectonic units such as shields and continental platforms. In the regions like the Pannonian basin, high crustal temperatures, which account for the increased heat flow from the depth, can cause "subcrustal erosion", i.e. phase changes of crustal rocks into denser ones of the upper mantle properties. The process is accompanied by a general thinning of the crust, subsidence, and sedimentation on the surface.

In tectonically old regions with a low heat flow that have been cooling down for the past few millions of years, the conditions of the crust–mantle boundary have also been consequently changing and the crust may have gradually grown thicker. At its early stages, a typical present-day shield rarely had its crustal base at a depth of 40 – 50 km (or even more if some thickness reduction by erosion is taken into account). In the late Cenozoic orogenetic belts, the present thick crust is the product of downwarping

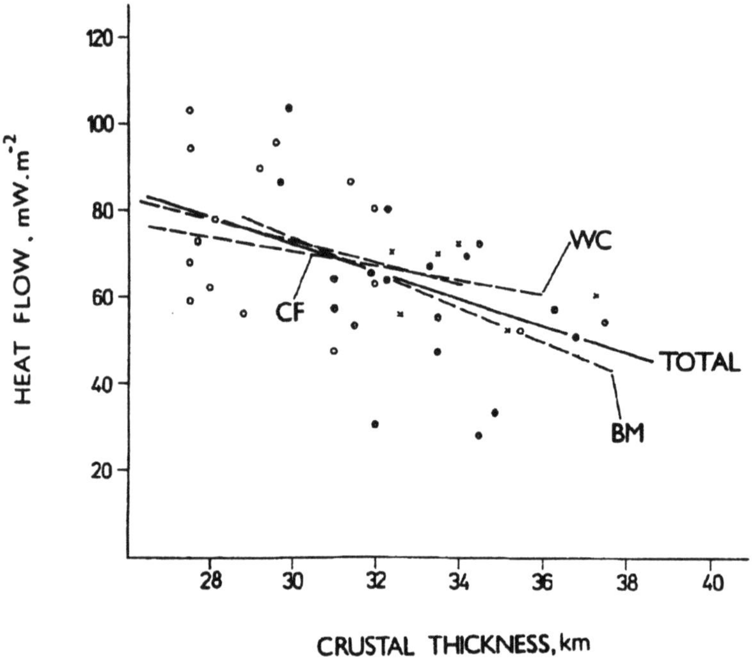

Fig. 4.6b Heat flow versus crustal thickness, individual symbols corresponds to the mean data in 1° × 0.5° longitude/latitude grid elements.

processes when the mountain roots were created and depressed into the upper mantle. Not only were the heat transfer conditions and steady state temperatures disturbed during an orogeny, but also the heat flow – crustal thickness relation could have been anomalous. If the position of the Moho discontinuity is time dependent it may be assumed that the light crustal material being suddenly exposed to extreme *p–T* conditions may also gradually alter and that the local crustal bulges below the mountains will slowly diminish by the process of basification. Contrary to the zones of lithosphere extension, as most continental areas of an abnormal thin crust, where enhanced heat flow and crustal thinning are a consequence of the extension regime, the abnormal thermal regime of orogenes is generated by collision tectonics, which can also involve some upward mass transfer from the mantle, but causes a thickening of the crust.

Results of the deep seismic sounding performed in Czechoslovakia were used to construct a crustal thickness pattern of the country (Beránek and Zátopek 1981, Mayerová et al. 1985), which was used for correlating the depth to the Moho discontinuity with the heat flow. Figure 4.5 shows the individual heat flow data plotted on the crustal thickness map and the corresponding heat flow - versus - crustal thickness data are given in Fig. 4.6a. These data show a considerable scatter, which is mostly due to local crustal inhomogeneities.

To get a smoother relationship, both patterns were subdivided into a regular 0.5° × 1° latitude–longitude grid and the corresponding characteristic heat flow and crustal thickness values in each grid element were plotted (Fig. 4.6b). The general negative tendency of such a relationship is more convincing in all three major tectonic units Bohemian Massif, Carpathian Foredeep, and West Carpathians. To be able to study this relation in more detail and to investigate a larger territory, the procedure was repeated for a greater part of Central Europe, and for a 10' × 10' grid system (Čermák et al. 1991a); the heat flow - crustal thickness relationship was investigated independently in all tectonic units. It is interesting to mention that practically in all cases, except for the Fore-Sudetic monocline and the neovolcanites of Central Slovakia (with only 5 data points and data coverage with a low reliability), the studied correlation is negative. This result should not be given too much credit, as the scatter is sometimes too great and correlation coefficient varies from – 0.3 to – 0.7. The general negative tendency is well confirmed, however, regardless of the large span of geological ages of the individual structures and their diverse tectonic history.

The present correlation study thus agrees with the earlier work (Zahradník and Čermák 1979) which investigated the regional behaviour of the correlation coefficient between both parameters. The positive correlation found now for the Fore-Sudetic monocline may have some physical meaning and may correspond to the previously

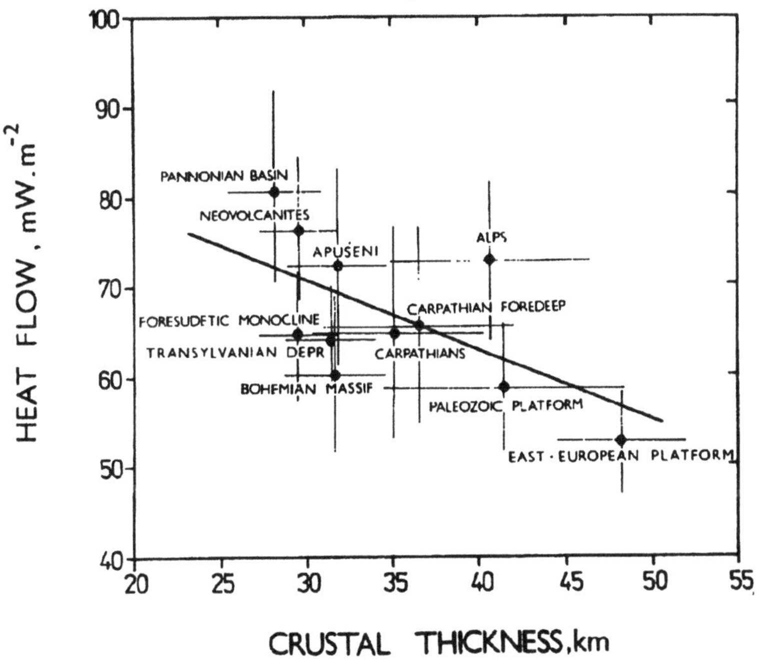

Fig. 4.7 Mean heat flow data versus mean crustal thickness data calculated for 10' × 10' longitude/latitude grid elements in specific tectonic units of Central Europe. Error bars correspond to standard deviations of the mean.

established belt of positive LR (linear regression) values stretching from the north Bohemia to south Poland (coinciding with the area of intensive Variscan tectonogenesis). Higher values of heat flow accompanying increased crustal thickness were explained by higher contributions of crustal heat sources, some of the plutons of late Variscan granitoids belong to rocks of the relatively highest radioactive content (Matolín 1970). Bodri and Bodri's (1985) explanation proposed for the anomalous positive relationship between heat flow and crustal thickness (i.e. their 1B sub-group) can also be applied here, namely that the increase of the total crustal thickness is mainly the result of the presence of a thicker sedimentary layer, the thickness of the consolidated crust being practically constant. High heat flow should then correspond to increased radiogenic heating of the sediments, the thickness of which gradually increases from west to east.

Mean heat flow versus mean crustal thickness in each individual tectonic unit is plotted in Fig. 4.7 together with the corresponding standard deviations of the respective means and with the resulting regression line. It is interesting to mention that the older geologic provinces are generally below the regression line, while the younger ones are above it. This fact must be related to higher heat supply from deeper (subcrustal lithosphere) heat sources in tectonically younger units. The difference between the mantle heat flow in the Precambrian/Palaeozoic terrains compared with that of the Tertiary terrains of about 10 – 15 mW m^{-2} corresponds well to the data by Vitorello and Pollack (1980) who reported it as the heat from transient thermal perturbation associated with tectogenesis.

Heat Flow and Heat Generation

Crustal radioactivity contributes to the observed heat flow, and the linear relationship between observed surface heat flow (Q), and the near-surface heat generation (A_0) (Roy et al. 1968), (Lachenbruch 1968): $Q = q_0 + DA_0$, defining heat flow provinces, was reported from various tectonic settings all over the world. Heat flow provinces correspond to large tectonic areas characterized by similar geologic evolution within which the value of the reduced heat flow (q_0) and the depth parameter (D) are constant. A knowledge of them can help in understanding the vertical distribution of heat producing elements within the crust. The above linear relationship requires combined heat flow and heat production observations in the basement rocks forming a sizeable crustal section. Taking the depth range of current drilling into account within the area of Central Europe studied, only two regions were suitable for such studies, namely the Bohemian Massif and the closest part of the East European platform. The remaining territories are mostly overlaid with either sedimentary cover or with allochthonous formations. The underlying basement has not been entered by any borehole.

Majorowicz (1984) published 9 combined heat flow–heat production data from east Poland (Mazury – Belorussian anticlise), which revealed $q_0 = 35$ mW m^{-2} and $D = 4.2$ km (Fig. 4.8). The value of the reduced heat flow of 35 mW m^{-2} may appear too high, but taking the low value of D into account, and the heat contribution of 6 – – 8 mW m^{-2} from the crust below 10 km depth calculated for two nearby crustal sections (Čermák et al. 1989), its agreement with the estimated heat flow from below the crust (Moho heat flow) of 26 mW m^{-2} (Čermák 1989a) is not bad. The value of D is nearly identical with that given by Kutas (1984) for the granitoid terrains of the Ukrainian shield, the low value of D reflects a strong vertical differentiation of radiogenic elements in the uppermost crust.

Čermák (1975a) reported two combined heat flow–heat generation values from the Bohemian Massif, which were later completed with seven additional data (Čermák 1981), see Fig. 4.8, which revealed $Q = 35 + 6.6\,A_0$. These data also showed a considerable scatter, mostly due to not very suitable locations of the boreholes which

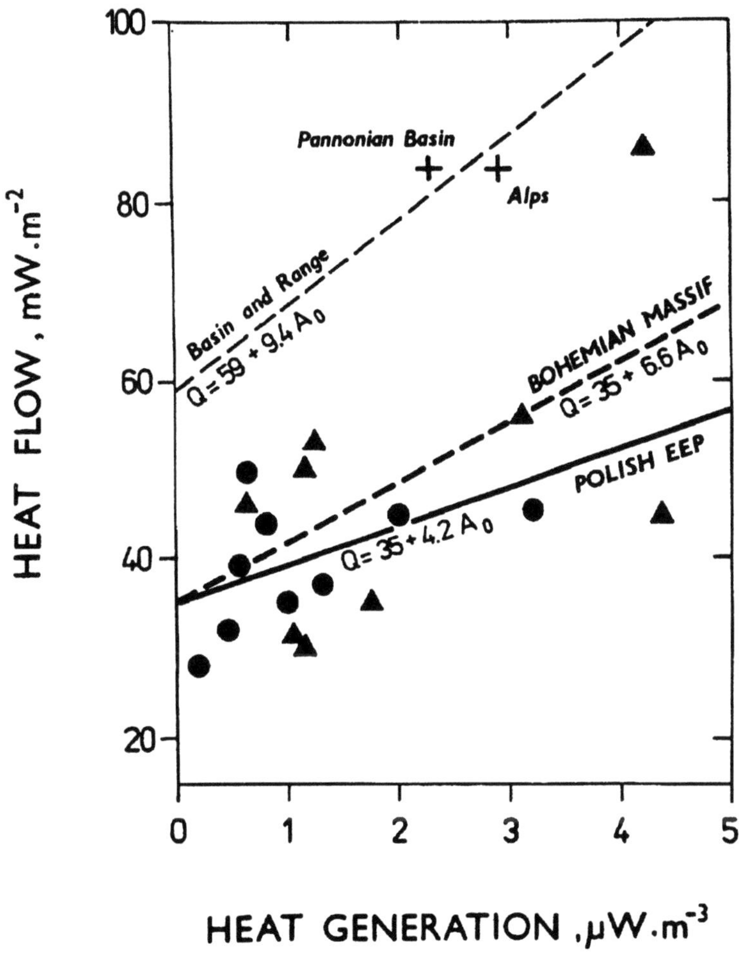

Fig. 4.8 Heat flow versus heat generation: triangles correspond to the data obtained in the Bohemian Massif, solid dots – Precambrian platform on the Polish territory (Majorowicz 1984), crosses – Pannonian Basin (Stegena et al. 1975) and Alps (Rybach 1973). For a comparison the Basin and Range province relationship is also shown (Roy et al. 1968).

were drilled in the course of an applied survey and were situated mostly in zones of industrial interest. As the tectonic evolution of the Bohemian Massif was generally terminated in the late Palaeozoic, the above relationship corresponds reasonably well to similar data reported for areas of the Palaeozoic consolidation, such as e.g. eastern part of the USA ($Q = 33 + 7.5\,A_0$, Roy et al. 1968).

A more detailed interpretation of heat flow–heat generation relationship was enabled by the multi-component analysis (Jaupart et al. 1981). The authors supposed vertical distribution of each of the radiogenic elements contributing to the heat production within the crust (U, Th, and K^{40}). The surface heat flow should then be examined as the linear combination: $Q = q_0 + D_U A_U + D_{Th} A_{Th} + D_K A_K$, where A_i (i = U, Th, K) is the surface heat production of the i-th element and D_i the parameter of its vertical distribution:

$$D_i = \int_0^h \left[\frac{C_i(z)}{C_i(0)}\right] \mathrm{d}z \quad ,$$

where h is thickness of the near surface layer containing the bulk of crustal radioactivity and $C_i(z)$ is the concentration of the i-th element.

This technique was applied to the ample material obtained for numerous rock samples from different localities in the north and west parts of the Bohemian Massif (the region of intensive Variscan tectogenesis), the area of generally high heat flow and also of relatively high radioactive content in rocks. Original data based on gamma-spectrometric determinations of the U, Th, and K contents (Bartošek, Chlupáčová, Kašparec and Manová, unpubl. reports) were used to estimate characteristic heat production values in several geologic units (Table 4.2) and showed systematically lower D_U and D_{Th} values compared with D_K. The q_0 and D values calculated for the above data amount 47.5 ± 5.1 mW m^{-2} and 8.4 ± 1.0 km, respectively. The value of the

Table 4.2 Concentrations of radioactive elements (C_U, C_{Th}, and C_K), total heat production and heat flow in several selected regions of the Bohemian Massif. Least square fit: $Q = 47.5 + 8.4\,A_0$, $Q = 42.7 + 7.0\,A_U + 11.9\,A_{Th} + 25.7\,A_K$

region	approx. coordinates	C_U [ppm]	C_{Th} [ppm]	C_K [ppm]	A total [Wm^{-3}]	heat flow Q [mWm^{-2}]
Slavkovský les region	50°09' N 12°47' E	7.9	14.2	3.3	3.4	73
Teplice region	50°39' N 13°50' E	14.9	35.1	3.9	6.7	109
Zlatý kopec locality	50°25' N 12°54' E	23.4	5.6	4.0	6.9	100
Krušné hory Mts.	50°30' N 13°20' E	11.0	15.5	3.7	4.3	85
Rakovník–Kladno basin	50°15' N 14°06' E	3.8	11.7	1.2	1.9	63

reduced heat flow differs from the previously assessed value for the whole Bohemian Massif. This difference can be explained by substantially higher Moho heat flow in northern and eastern rims of the Bohemian Massif (Čermák 1975c, 1989a) and generally higher radioactivity of crustal rocks in this area (Matolín 1970). The values of $q_0 = 42.7$ mW m^{-2} and $D_U = 7$ km, $D_{Th} = 11.9$ km, and $D_K = 25.7$ km are in good relation with similar values based on geochemical estimates (Haack 1982). They confirmed the more striking differentiation of U and Th in the uppermost crust, while K does not show much vertical variation.

4.5 Crustal Temperatures

One-Dimensional Models

Characteristic deep temperatures of individual tectonic sub-units of the Bohemian Massif and of the West Carpathians were calculated (Čermák 1975c) by means of multilayered step models of the crust composed of layers of surface sediments and/or metamorphosed rocks and of layers of granitic and basic rocks. For this purpose, the results of deep seismic sounding were used to define the crustal structure. Some assumptions of the distribution of thermal conductivity and heat production in crustal rocks were based on probable petrologic model. The solution of the equation of heat conduction was facilitated by the presumption of steady-state conditions. The problem was treated as one-dimensional.

For thermal conductivity $k = k_0$, and heat production $A = A_0$, both uniform within a surface layer of thickness z_1, temperature $T(z)$ as a function of depth z is given by :

$$k_0 T = Qz - \left(\frac{A_0}{2}\right) z^2 \ ,$$

where Q is the surface heat flow. If the conductivity and heat production change abruptly to $k = k_1$ and $A = A_1$ at depth $z = z_1$, temperature $T(z)$ for $z > z_1$ is expressed as:

$$T = T(z_1) + \left(\frac{1}{k_1}\right)(Q - A_0 z_1)(z - z_1) - \left(\frac{A_1}{2k_1}\right)(z - z_1)^2 \ ,$$

where $T(z_1)$ is the temperature at depth z_1 calculated from the previous equation. Similarly, the temperature can be calculated for any arbitrary layered model representing the Earth's crust composed of a number of layers.

The coefficient of thermal conductivity k usually decreases with increasing temperature and this dependence roughly follows the law $k_0/k_T = (1 + CT)$, where k_T is the conductivity at temperature T, and k_0 is the conductivity at surface conditions, C is the experimentally determined constant (0.0005 – 0.0015 K^{-1}) (Čermák and Rybach 1982). If the relation for k_0/k_T is introduced into the solution for temperature, the latter formula remains valid when the $(k_0/C)\ \lg(1 + CT)$ is substituted for the left side. The temperature dependence of the thermal conductivity can be of importance in the upper part of the crust composed of crystalline schists and of granites, the conductivity of which can drop from about $k_0 = 3$ W m^{-1} K^{-1} to 2.3 – 2.5 W m^{-1} K^{-1} at 200° C and less than 2 W m^{-1} K^{-1} at 400° C, the conductivity of mafic rocks forming the lower part of the crust is usually only little dependent on temperature.

The sites for which the temperature–depth profiles were calculated were chosen at those places where the crustal structure was relatively well known and where reliable heat flow data existed, and above all in places where the crustal $T(z)$ profiles may have represented the specific tectonic section: (a) Moldanubicum – the oldest, most rigid, and stable part of the Bohemian Massif, where the thickness of the Earth's crust attains almost 40 km, and where heat flow is generally low (40 – 50 mW m^{-2}); (b) Krušné hory Mts. – characterized by thinner crust (30 – 32 km), increased heat flow of over 70 mW m^{-2}, and of typically high radioactivity of rocks forming the basement structure; (c) Cretaceous Basin – suture zone framing the stable nucleus of the Bohemian Massif, typically with increased heat flow, reduced crustal thickness, and probably with increased heat supply from the upper mantle; (d) Vienna Basin – extending just off the SE rim of the Bohemian Massif, where crystalline basement rocks are deeply submerged under the Carpathian Flysch Belt, an area with generally low heat flow (50 – – 60 mWm^{-2}); (e) Danube Basin – filled with the Neogene sediments, forming a protruding edge from the Pannonian Basin, with relatively thinner crust (28 km) and increased heat flow (70 – 80 mW m^{-2}); (f) Outer Flysch Zone of the West Carpathians – contact zone with the Palaeozoic platform to the north, typically with thick crust (over 40 km) and low heat flow; and (g) Pannonian Basin – hyperthermal zone characterized by thick sedimentary cover, high gravity and hydrothermal activity, thin crust (25 – – 27 km), and very high heat flow (over 80 mW m^{-2}).

The most important conclusion from the crustal temperature assessments is the existing temperature difference at the base of the crust between the Bohemian Massif and the Neogene depressions of the Carpathian system, which may attain up to 300° C for the range of surface heat flow of 40 to 90 mW m^{-2}. Equally important is the fact that between both tectonic systems regional variations in the Moho heat flow may exceed 30 – 40 mW m^{-2}.

Figure 4.9 shows a comparison of the calculated temperatures. The characteristic Moho temperatures under the Bohemian Massif are within the range of 500 to 550° C at depths 35 – 40 km with local excesses manifesting temperatures of 600 – 700° C in regions of enhanced surface heat flow, such as the Cretaceous basin or the Krušné hory Mts. The Carpathian Foredeep is characterized by Moho temperatures between 450 and 600° C, however, the low estimated upper mantle heat flow in its southern part of only 15 mW m^{-2} or less is surprising. The hyperthermal region of the Pannonian Basin has extreme subsurface temperatures, the mantle heat flow can be of 40 – 60 mW m^{-2} compared with 20 – 30 mW m^{-2} existing beneath the Bohemian Massif. Corresponding Moho temperatures in Neogene depressions can exceed 800° C. The downward

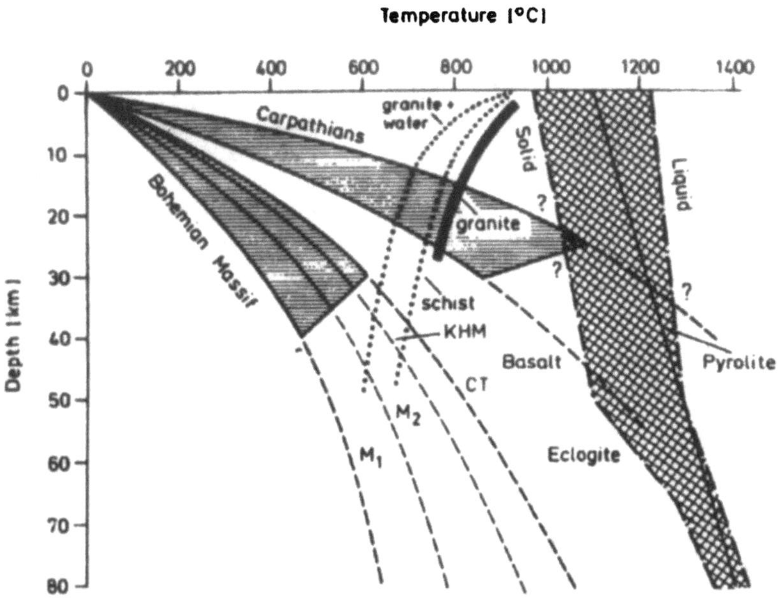

Fig. 4.9 Range of crustal temperatures beneath the Bohemian Massif and the Neogene depressions of the West Carpathians. M_1 and M_2 are the minimum and maximum temperature curves for the Moldanubian (oldest) part of the Bohemian Massif, KHM – mean temperature for the Krušné hory Mts. Cross-hatched area shows the melting relation for basalt and eclogite (Yoder and Tilley 1962).

extrapolation of the temperature curves in the upper mantle is uncertain here, but the calculated geoisotherms can intersect the solidus for the upper mantle rocks at depths in the range of 40 – 50 km. A local zone of the partial melting below the Pannonian Basin cannot be excluded.

Two-Dimensional Sections

The idea of a horizontally homogeneous medium may be a suitable simplification of reality only in some cases, e.g. when characteristic temperature conditions of different large tectonic units are to be compared. The major disadvantage of 1-D solutions is the problem of modelling the temperature effect of lateral inhomogeneities of thermal conductivity or of heat sources. In these cases the solution of the heat conduction equation requires a two-dimensional approach. As 2-D solution can be only very rarely expressed in an analytical form, no practical advancement was achieved before the use of numerical solutions became facilitated by computer techniques in recent years.

An application of the 2-D steady-state solution of the heat conduction equation by the finite difference method was performed for Czechoslovakia first by Šafanda (1985) who generally followed the technique proposed by Stromeyer (1984). As the boundary conditions the following were used: the known surface conditions (for simplicity taken as $T_0 = 0$ °C), the symmetry of the temperature field at vertical boundaries, i.e.

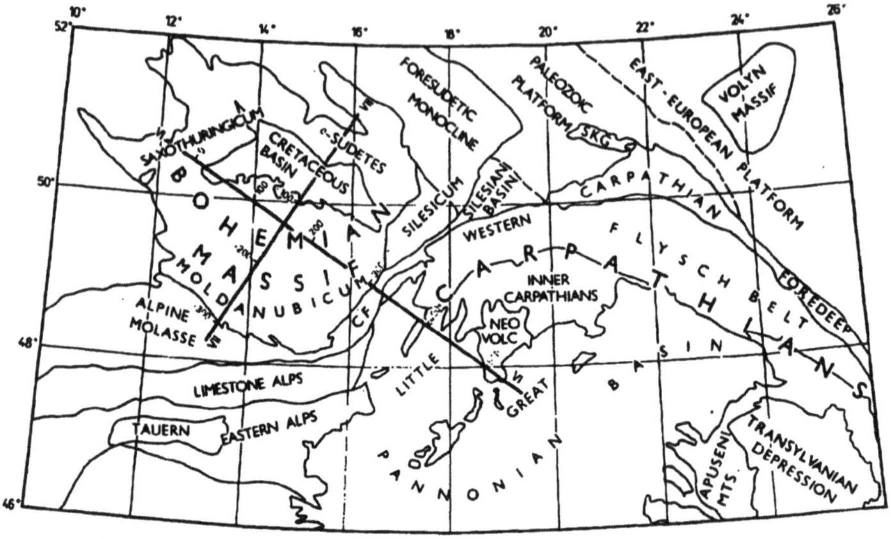

Fig. 4.10 Schematic tectonic setting of Czechoslovakia and its surrounding areas together with the position of two deep seismic sounding profiles along which 2-D deep temperature distribution was calculated.

$dT/dx = 0$ at $x = 0$ and $x = L$ (L being the length of the profile), and the known data on the heat flow at the base of the model, Q_M. As Q_M is usually not known, the solution of the problem is ill-posed and requires a first solving of the inverse problem, in which the observed heat flow on the surface (Q_0) is used to evaluate the heat flow from depth. The surface heat flow is assumed to be the sum of two components, the crustal contribution due to crustal radiogenic heat sources (Q), and the mantle heat flow (Q_M). By repeated calculation of the surface heat flow corresponding to the model (Q_B), starting with an arbitrary value of Q_M and successively correcting the calculated Q_M – value by the difference ($Q_0 - Q_B$), it is possible to assess the conditions at the lower boundary of the model within a certain limit of accuracy.

The 2-D temperature-depth distribution was computed for two deep seismic sounding profiles, DSS VI, and DSS VII, crossing Czechoslovak territory. Crustal temperatures were calculated to a depth of 40 km. Both profiles are parts of geotraverses, the East European Geotraverse EEGT 5 and/or EEGT 1, respectively, along which the temperature field was studied by slightly different techniques (Čermák and Bodri 1986). Positions of both profiles are shown in Fig. 4.10 together with major tectonic units.

The region of solution of the heat conduction equation was approximated by oblongs with dimensions of 620 × 40 km for profile DSS VI, and 360 × 40 km for DSS VII. To include the temperature dependence of the thermal conductivity coefficient into the model, a simple stratified model was applied with values of $k = 2.5$, 2.2, 2.0, and 2.5 W m^{-1} K^{-1} in the depth intervals of 0 – 10, 10 – 20, 20 – 30, and 30 – 40 km. The position of the Moho discontinuity was adopted from Mayerová et al. (1985), the surface heat flow from Čermák (1981).

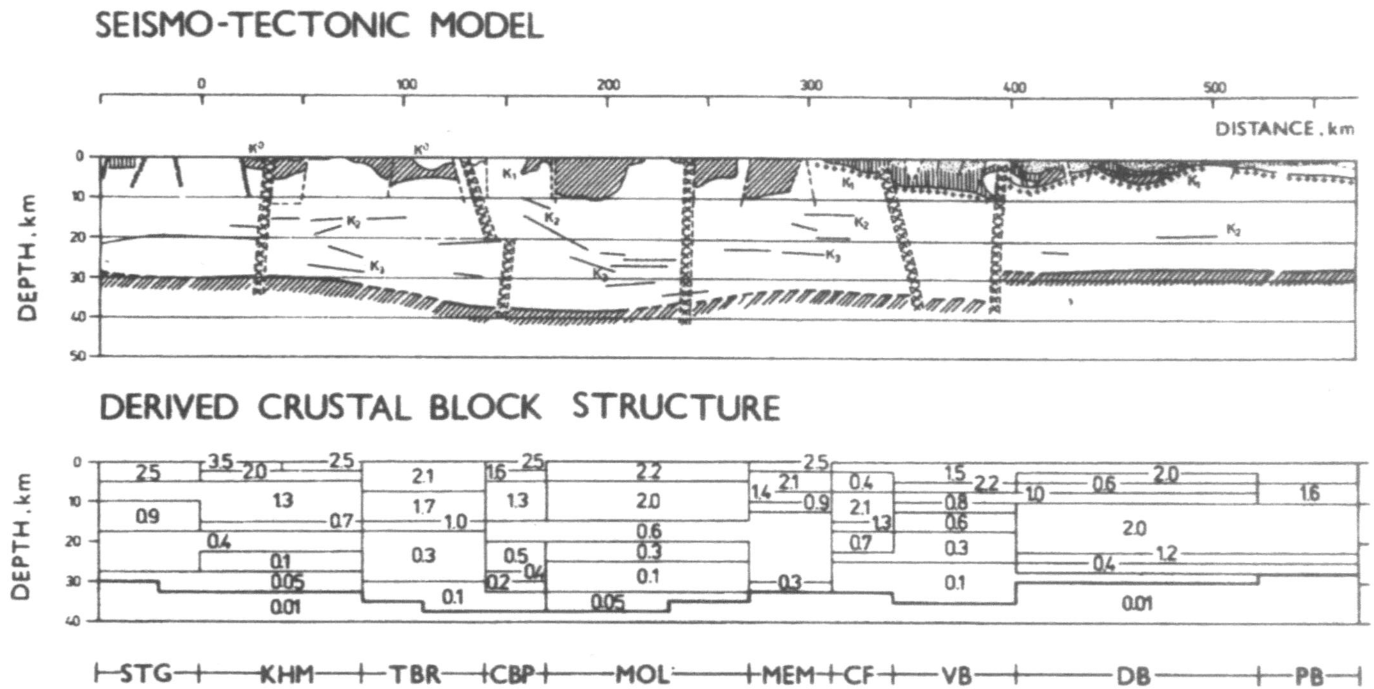

Fig. 4.11a Crustal cross-section along DSS profile VI: upper diagram – seismo-tectonic model based mainly on seismological data, lower diagram – derived block structure giving the characteristic heat production values within each block, in $\mu W\ m^{-3}$ (see the text).

The heat released by the radioactive decay was considered to be the only source of heat within the model. Below the Moho discontinuity the uniform heat production of the upper mantle rocks was adopted to value given by Haack (1982), i.e. $A = 0.01\ \mu Wm^{-3}$. To assess the distribution of the radiogenic heat sources within the crust, the experimental relationship between seismic velocity and heat production proposed by Rybach and Buntebarth (1984) complemented by the corrections for the effect of temperature and pressure on the seismic velocity (Čermák 1989a) was employed. The heat production values were assigned to the large blocks of crustal rocks according to their characteristic seismic velocities.

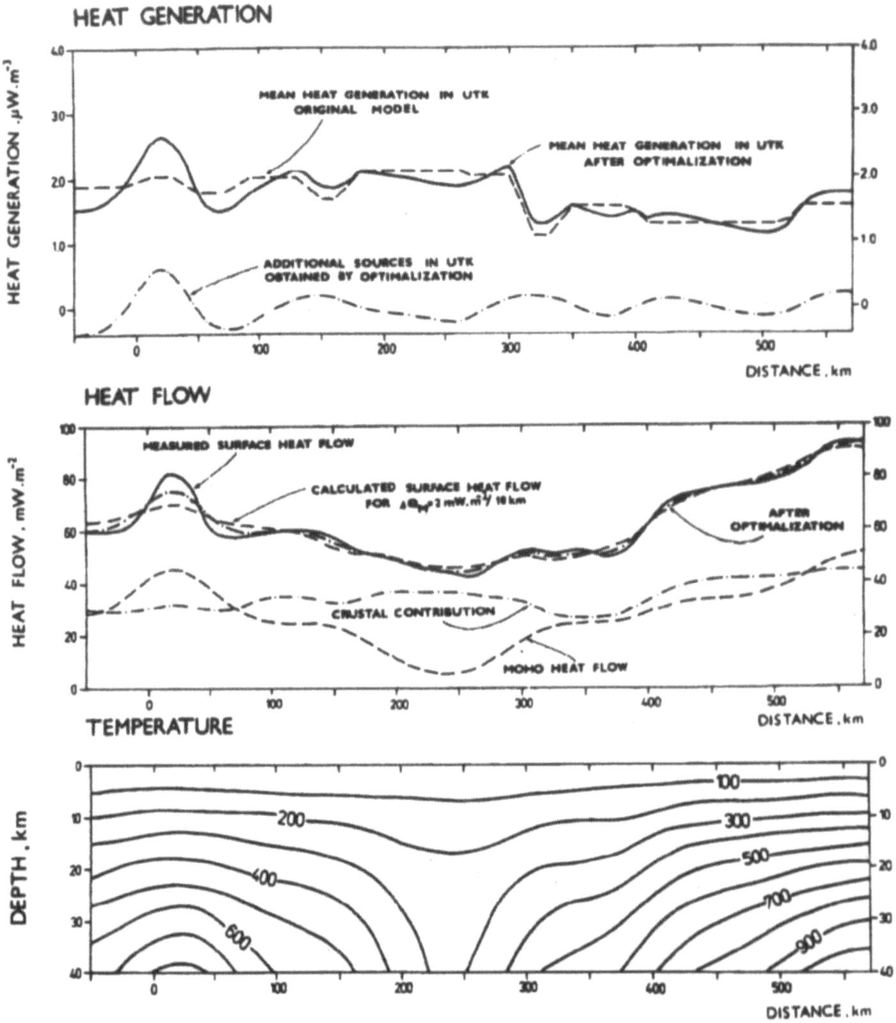

DISTANCE, km

Fig. 4.11b Depth distribution of heat generation, heat flow and calculated crustal temperatures along DSS profile VI.

The initial seismo-tectonic structures along both profiles investigated together with the derived block model structure and characteristic values of heat production within each of the blocks are shown in Figs. 4.11a and 4.12a.

To solve the inverse problem, in principle, it is possible to attain a Q_M-distribution corresponding to any surface heat flow distribution pattern for any required accuracy, however, at the expense of either unrealistic horizontal heat flow gradients or an unrealistic distribution of additional heat sources. It is thus necessary to evaluate carefully the variation of mantle heat flow per unit distance with respect to the local tectonic structure and also to assess the quality of data which serve as a criteria of the solution, and to seek an acceptable compromise. This problem was discussed in detail by Stromeyer (1984) and Šafanda (1985, 1986). It was also shown that the value of the Q_M-variation of 2 mW m^{-2}/10 km solved the problem.

Laboratory measurements (Vaňková et al. 1981) revealed a variability of heat production in surface rocks in the range of 0.4 – 10 μW m^{-3}. Such a wide interval of values justifies the assumption that the variations of the observed surface flow,

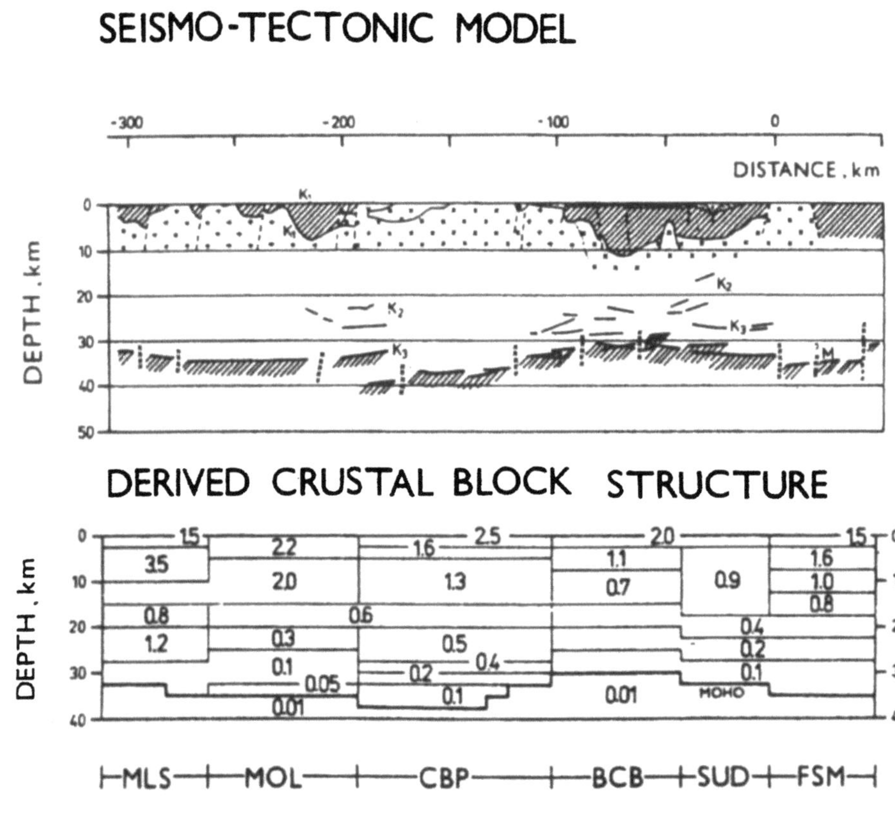

Fig. 4.12aCrustal cross-section and derived block structure along DSS profile VII.

especially those of small-scale, are caused by lateral inhomogeneities of heat sources in the uppermost part of the crust. The method of optimization of heat sources described by Stromeyer (1984) was therefore employed to further improve the fit between the measured (observed) and calculated surface heat flow values (Šafanda 1985). The optimization procedure was applied to the original distribution of heat sources within the crust, when additional sources, uniform in the vertical direction, were sought in order to reduce the remaining difference in the surface heat flow after the $(Q_0 - Q_B)$ value had fallen below the chosen limit. Analogously to the maximum Q_M-variation allowed, also the maximum horizontal variation of the additional heat sources was limited, in our case this value amounts 0.1 μW m^{-3}/10 km and the additional heat sources were distributed in the upper 10 km (UTK-layer). The distribution patterns of the surface heat flow and of the crustal heat sources along both profiles documenting the algorithm used are shown in Figs. 4.11b and 4.12b together with the calculated temperature fields. The calculated temperatures correspond to the final solution after optimization of heat sources.

The calculated temperatures generally well correspond to the 1-D results. They confirmed the existing differences in the deep temperature distribution between both major tectonic systems. Along profile DSS VI (Fig. 4.11b) the Moho temperatures beneath the Saxothuringicum amount 600 – 650 °C and sharply decrease under the Bohemian Massif. Further to the SE the crustal temperatures increase again and attain 550 – 600 °C beneath the Vienna Basin, 700 °C beneath the Danube Basin, and over 800 °C beneath the Pannonian Basin in Hungary. Along profile DSS VII (Fig. 4.12b) high temperatures exist beneath the Alpine molasse (700 – 800 °C), low temperatures beneath the Bohemian Massif (450 – 500 °C), and T_M increases to 650 °C beneath the Cretaceous Basin.

The only problematic (and not yet fully explained) point is the extremely low temperature $T_M \approx 300$ °C at 40 km depth beneath the southeastern rim of the Bohemian Massif (km 220 – 260, profile DSS VI, Fig. 4.11b) evidently connected with a fictitious, low calculated Q_M of 6 mW m^{-2} only. This section of the profile corresponds to block "D" with a zone of low seismic velocity ($v_p < 6.0$ km s^{-1}) in its upper part, reaching to depth of 15 km. Such low velocity v_p, when converted to heat production, gives 2.0 – – 2.2 μW m^{-3}. This would mean that the crustal radioactivity itself can explain most of the surface heat flow with little left to deeper heat sources. Crustal radioactivity in this case cannot be real and either the seismic velocities reported for the block "D" have to be checked or the conversion technique, in this case, failed. Another explanation can be connected with an untrue very low value of the observed heat flow on the surface ($Q_0 \approx 40$ mW m^{-2}). If a higher value of about 50 mW m^{-2} is supposed (more probable in view of the gross-scale heat flow pattern), Q_M and also the Moho temperature will be adequately higher.

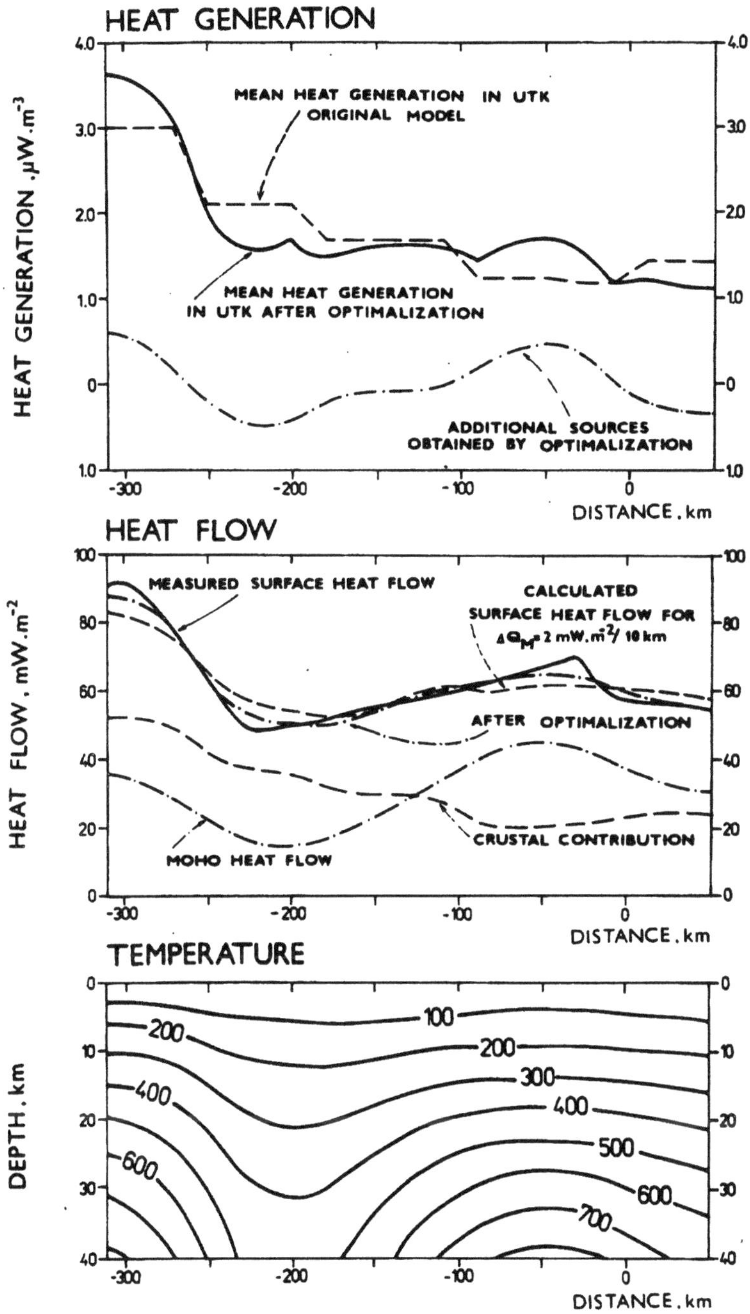

Fig. 4.12b Depth distribution of heat generation, heat flow and calculated crustal temperatures along DSS profile VII.

4.6 Regional Geothermics

To be able to study the regional distribution of the geothermal parameters on a larger scale, the Czechoslovak data were complemented with heat flow observations from the surrounding territories. The corresponding heat flow and the crustal thickness patterns of the area investigated are shown in Figs. 4.13 and 4.14 and cover the part of Central Europe bounded by the coordinates 11° – 24° E, and 46° – 52° N.

Moho Temperatures

The previous 1-D and 2-D temperature profiling was completed with the regional study of the distribution of the temperature on the Mohorovičic discontinuity. For this purpose, the simple two-layer crustal model was used to calculate 1-D T_M-temperature in each 30' × 30' square element of the grid system.

The model represents the crust composed of the upper and lower parts, the boundary between them corresponding to the seismic velocity of 6.6 – 6.8 km s^{-1}.

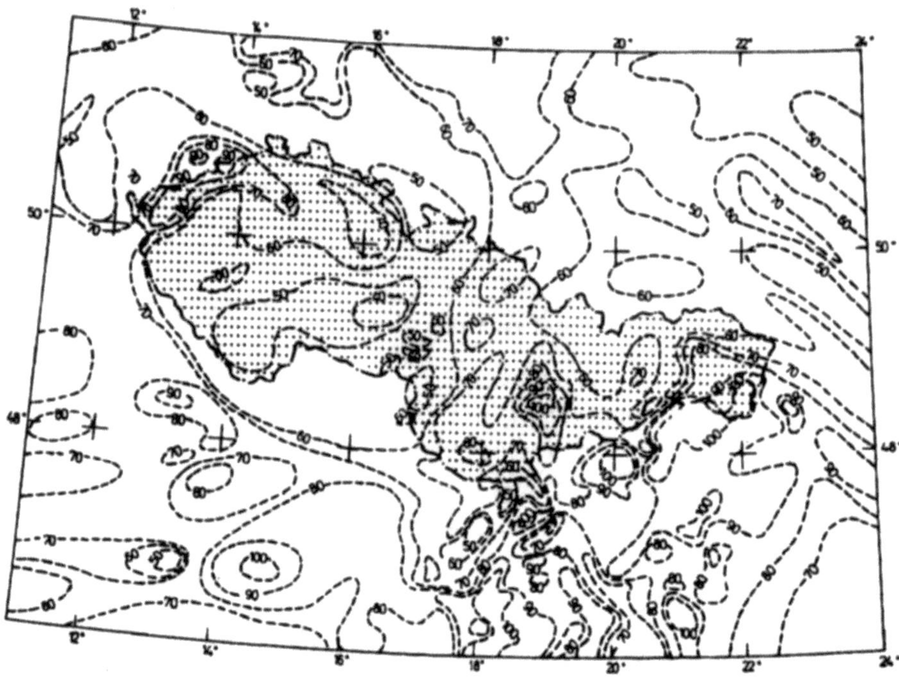

HEAT FLOW, mW.m^{-2}

Fig. 4.13 Heat flow distribution pattern in Czechoslovakia and in its close vicinity.

The steady-state one-dimensional step-like model with exponentially distributed heat sources $A(z) = A_0 \exp(-z/D)$ within each layer and with temperature dependent thermal conductivity $k_T = k_0(1 + CT)^{-1}$, gives the temperature $T(z)$ in each layer (Čermák 1982a):

$$\left(\frac{1}{C_i}\right) \lg\left[1 + C_i T(z)\right] = \left(\frac{1}{C_i}\right) \lg\left[1 + C_i T_i\right] + Q_i \frac{z - z_i}{k_{0i}} - \frac{A_{0i} D_i}{k_{0i}} \left[D_i \exp\left(\frac{z_i - z}{D_i}\right) + z - z_i - D_i \right]$$

with $Q_{i+1} = Q_i - DQ_i$,

where $DQ_i = A_{0i} D_i \left[1 - \exp\left(\frac{z_i - z_{i+1}}{D_i} \right) \right]$.

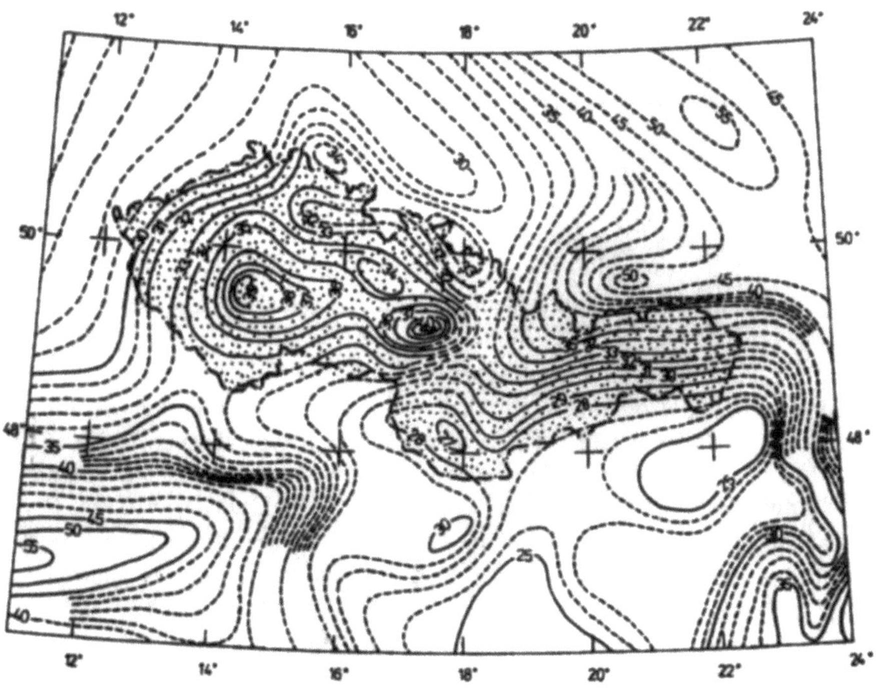

MOHO DEPTH, km

Fig. 4.14 The Moho depth pattern in Czechoslovakia and in its close vicinity.

In the above solution, T_i and/or Q_i ($i = 1, 2$) are the temperature and heat flow, respectively at depth z_i, and k_{0i}, C_i, A_{0i}, D_i are the corresponding parameters in the i-th layer, bounded by depths z_i and z_{i+1}. For k independent of temperature or A constant in a certain layer, the solution remains valid, if $Q_i \to 0$ and/or $D \to \infty$ in the respective layers.

In order to assess the values of the heat flow production A_{0i}, two relationships were used:

1. the concept of heat flow provinces, which relates the surface heat flow Q_0 and the near surface heat production A_0: $Q_0 = q + D\, A_0$, the intercept value q is called the reduced heat flow, and the slope value D characterizes the vertical distribution of heat sources (Roy et al. 1968; Lachenbruch 1968),
2. empirical relation between mean surface heat flow (Q) and the reduced heat flow q: $q = 0.6\, Q$ (Pollack and Chapman 1977).

If both the above relationships are generalized and within a certain surface area the mean heat production A_0 in the near surface layer of thickness D can be related to the mean surface heat flow: $A_0 = (Q - q)/D$, and there is no substantional difference between Q and Q_0 (which can be assumed as in our case the heat flow values were taken from a heat flow distribution pattern smoothed in 10' × 10' grid elements), the value of A_0 can be formally expressed as $A_0 = 0.4\, Q_0/D$.

For exponentially decreasing radioactivity with depth, the mean heat production A_0 in a layer bounded by depths z_1 and z_2 is given by

$$A_0 = \frac{1}{z_2 - z_1} \int_{z_1}^{z_2} A_{01} \exp\left(-\frac{z}{D}\right) dz \quad ,$$

where A_{01} is the heat production at depth z_1. For $z_1 = 0$ and $z_2 = z_K$ (the depth of the upper/lower crust boundary) it follows that $A_{01} = A_0/(1 - \exp(-z_K))$ and $A_{02} =$ $= A_{01} \exp(-z_K)$ is the heat production at depth z_K.

D-values reported from the above heat flow-heat production relationship from various tectonic settings varied between 4 and 16 km, a typical value being of about 10 km (Morgan and Sass 1984). The value of 10 km was therefore used to characterize the logarithmic decrement in the upper crust of our model, i.e. $D_1 = 10$ km. For this value and the observed interval of surface heat flow values, 40 to 80 mW m^{-2}, the interval of A_0 amounts to 1.6 – 3.2 μW m^{-3}, the interval of A_{01} is approximately 2 – 4 μW m^{-3} and the interval of A_{02} is 0.45 – 0.9 μW m^{-2}, which corresponds relatively well to the data based on petrologic assumptions of the crustal structure.

The radioactivity of the lower crust is even less known, but in order to complete the model, the exponential function $A(z) = A_{02} \exp(-z/D_2)$ was applied with $D_2 = 5$ km (Čermák 1982a). The mean heat production of the lower crust amounts thus to less than 0.1 – 0.4 μW m^{-3} according to the crustal thickness and surface heat flow. The heat production at the crustal base varies from 0.001 to 0.1 μW m^{-3}. The lower crust contributes to the surface heat flow by about 5 %. The Moho heat flow varies from less than 20 to almost 50 mW m^{-2} (see below).

Thermal conductivity of upper crustal rocks was supposed to decrease with temperature from $k_0 = 3$ W m^{-1} K^{-1} to 2.3 – 2.6 W m^{-1} K^{-1} at the depth of about 15 km, which roughly corresponds to z_K, i.e. $C = 0.0008$ °C^{-1}. The conductivity of the lower crust was taken 2 W m^{-1} K^{-1}, independent of temperature, i.e. $C = 0$.

Figure 4.15 shows the regional distribution of the calculated Moho temperatures within the area of study. The relatively coldest are the central and eastern parts of the Bohemian Massif and also its northern rim and southern part of the Carpathian Foredeep, with T_M below 500 °C. Both thinner-crust zones of the Bohemian Massif, i.e. the Krušné hory Mts. (the Ohře graben) and the Cretaceous Basin, are characterized by increased values of T_M (slightly over 600 °C). Most of the Inner Carpathians and the Carpathian Flysch Belt units are characterized by T_M between 500 and 600 °C as well.

Crustal temperatures are increasing beneath the Alps, up to 800 °C at a depth of 40 – 45 km, and beneath the Pannonian Basin, 700 – 800 °C at a depth of 25 – 28 km. Isolated anomaly of locally increased Moho temperature (over 700 °C) in the southern

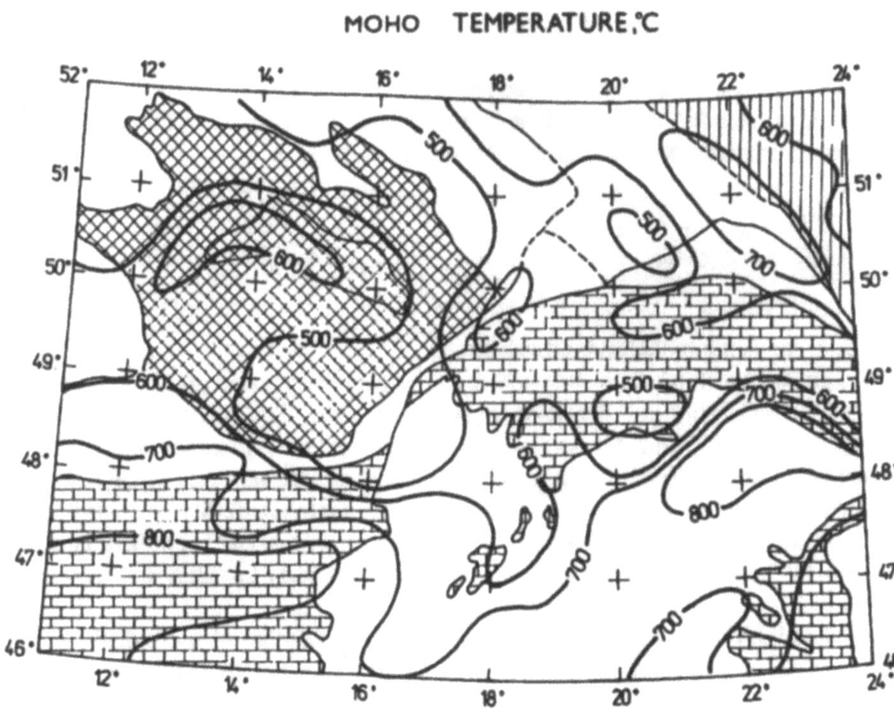

Fig. 4.15 Regional distribution of the calculated Moho temperature.

part of the Teisseyere–Tornquist zone in Poland is partly due to enormous crustal thickness attaining 50 km or more, but the increased surface heat flow in this area can be the product of salt tectonics, and the anomaly is difficult to be interpreted in terms of deep structure. The southwestern slopes of the East European platform have T_M of about 600 °C and further to the NE crustal temperatures decrease to less than 500 °C (Čermák and Bodri 1986).

Moho Heat Flow

The heat flow measured in a stable continental crust can be described by two components: i) heat from the upper mantle or from below a relatively greater depth, beneath which heat sources are generally uniform, and ii) the heat generated by radioactive heat sources within the crust. The local variations in the observed geothermal activity on the surface can mostly be explained by variations in the heat production of crustal rocks, in addition to the variations due to local hydrogeology and local geologic effects. The mantle or Moho heat flow represents the energy outflow from greater depth, which is further the sum of the background heat flow and the heat from a transient thermal perturbation associated with tectogenesis (Vitorello and Pollack 1980). Moho heat flow can thus be more dependent on the tectonic and magmatic activity than on the age of the overlying crust. It is more important to know the Moho heat flow for understanding the distribution of heat flow on a global or regional scale. Moreover, the recognition of the outflow of heat from depth is an essential boundary constraint for crustal two-dimensional temperature modelling.

In order to evaluate the Moho heat flow Q_M it is necessary to subtract the crustal contribution from the observed surface geothermal activity, i.e.

$$Q_M = Q_0 - \int_{z_0}^{z_M} A(z)\,\mathrm{d}z \quad ,$$

where z_M is the depth to the Moho discontinuity.

To estimate the total crust radioactivity it is necessary to create a corresponding petrologic model what can be quite problematic task. The radioactivity certainly decreases with depth, and the exponential model can well illustrate this fact. The exponential model, which is widely used for approximate assessment of deep-crust temperatures, however, does not give an equally acceptable possibility of evaluating the Moho heat flow. The radioactivity of the lower crust, if this model is applied, is usually underestimated, and the calculated Q_M is then correspondingly too high to represent real conditions.

The step model defining the crust as a multilayered medium of individual layers of specific radioactivity corresponding to typical crustal rocks, is schematic, but provides a good basis for the Moho heat flow estimates. The problem of determining a reasonable characteristic value of the radiogenic heat generation for each layer is then crucial. It is then useful to apply the deep seismic sounding data and the experimental relationship between the heat production and seismic velocity v_p (decreasing radioactivity with increasing velocity), originally reported by Rybach (1973). Explosion seismology data thus provide a good basis for both the specification of the crustal structure and its subdivision into individual layers, and for the rough estimate of the characteristic values of heat generation within each layer.

The preliminary map of Moho heat flow distribution in Europe (Čermák 1982b) was based on the compilation of the estimates of Q_M published by various authors completed with the evaluation of crustal heat production by converting the mean seismic velocity to heat production data on gross scale, i.e. 2° × 2° grid network. This map confirmed the variation of the Moho heat flow which, within the study area ranges from 15 to 50 mW m^{-2}. Such procedure, however, is too coarse to be applied to a smaller scale area, unless more detailed information is available on seismic velocity distribution. No regional distribution of Q_M value has been therefore proposed yet and only a few irregularly distributed crustal sections $v_p(z)$ could have been used (Čermák 1989a). The data which fall into the area discussed here are summarized in Table 4.3.

These data are based on the conversion of a series of 1-D $v_p(z)$ profiles reported along five East European geotraverses (Sollogub et al. 1980). The conversion technique $v_p - A$ used, was based on the data published by Rybach and Buntebarth (1984) together with the corrections for the effect of in situ conditions (temperature and pressure) on seismic velocity (Čermák 1989a).

The values of the Moho heat flow obtained vary in a broad range from less than 20 mW m^{-2} to more than 50 mW m^{-2}, the mean value (after the unreliable value for

Table 4.3 Heat flow, its crustal components, crustal thickness and Moho heat flow. Note: code indicates the geotraverse and serial number of station (Čermák 1989a). [a] problematic or less reliable value, [b] unreliable value.

code	approximate location longitude	latitude	surface heat flow [mWm^{-2}]	Moho depth [km]	sedimentary cover [km]	heat flow UTK layer [mWm^{-2}]	contribution under UTK [mWm^{-2}]	Moho heat flow [mWm^{-2}]
1–2	14–00	49–00	60	37.5	0	12.7	24.8	13[a]
1–3	14–30	49–30	56	37.5	0	14.1	22.3	20
5–1	15–00	49–30	50	38	0	12.7	18.2	19
1–4	15–10	50–20	66	30	1	16.6	9.5	40[a]
1–5	15–30	50–50	60	32	0	15.2	17.4	27
1–6	16–10	51–20	62	30	2	15.6	12.2	34
4–2	15–50	51–20	66	34	2	16.6	21.0	28
1–7	16–50	51–40	68	34	2.5	16.8	23.3	29
4–3	21–00	50–40	(48)	36	0	12.1	41.1[a]	0[b]
2–2	23–30	49–10	50	55	11	10.0	23.1	17[a]
5–2	17–00	48–40	70	30	4.5	16.4	16.5	37
1–1	13–10	48–20	75	29.5	2.5	18.3	17.9	39
2–1	21–00	47–50	90	26	2.5	21.5	11.3	57
5–3	20–00	47–20	90	28	6.5	17.1	7.1	66

4–3 profile was excluded) is 32.8 ± 15.5 (s.d.) mW m^{-2}. Usually only one or a few $A(z)$-profiles are available for each specific tectonic unit, thus any detailed statistical conclusions are out of the question, the mean value is, however, substantially higher than the mean value obtained for the Precambrian realm of the East European platform, i.e. 22.6 ± 4.4 (s.d.) mW m^{-2} (Čermák 1989a). The outer East Carpathians profile (No. 2–2) yielded low Moho heat flow of only 17 mW m^{-2}, which has not been explained correspondingly yet, but all other profiles clearly confirmed the increasing outflow of heat from the upper mantle in younger terrains. The Variscan structures in Central Europe, which includes the Bohemian Massif (profiles nos. 1–2, 1–3, 5–1, and 1–4, $Q_M = 23 \pm 11.7$ mW m^{-2}), and northern slopes of the massif plus the Fore-Sudetic monocline (profiles Nos. 1–5, 1–6, 4–2, and 1–7, $Q_M = 29.5 \pm 3.1$ mW m^{-2}) are characterized by the mantle heat outflow of about 26 mW m^{-2}. Alpine molasse (profile 1–1) is characterized by the Moho heat flow of 39 mW m^{-2}, the inner West Carpathians – Danube Basin (profile 5–2) by $Q_M = 37$ mW m^{-2}. The Pannonian Basin (profiles nos. 2–1 and 5–3), wedged between two branches of the Alpine ranges, has an anomalous structure and also a high Moho heat flow of over 50 mW m^{-2}. Profile 4–3 (Holly Cross Mts.) yielded zero value for mantle heat flow, obviously due to high crustal contribution thus representing completely unreliable datum, which has to be explained in the future.

Heat Flow and Lithospheric Thickness

Seismic velocities in the upper mantle increase with depth, but at a certain depth, there is a sharp velocity decrease and from here, a zone of low velocity is observed. This zone is a general terrestrial phenomenon. It characterizes the transition from a solid lithosphere to a viscous asthenosphere. Since the seismic velocity is temperature dependent, the velocity decrease obviously corresponds to the zone where the existing temperature approaches the melting temperature of upper mantle rocks. It was the idea of Pollack and Chapman (1977) to use the calculated geotherms in combination with the mantle melting relations to estimate the top of the low velocity zone and to map regional variations of the lithospheric thickness.

By following the idea of the above cited authors, a more detailed model of the continental lithosphere was proposed and a map of the lithosphere thickness in Europe was compiled (Čermák 1982a). In contrast to the original work, a more detailed vertical distribution of the coefficient of thermal conductivity, especially its temperature dependence, was used. A two-layer crustal model of heat production was used, heat production being proportional to the observed surface heat flow. Since the actual crustal thickness affects the total radiogenic heat contribution, i.e. also the computed temperatures, the surface heat flow pattern was completed by using the map of regional variations of crustal thickness. Finally, unlike Chapman et al. (1979) who based their calculations on the spherical harmonic representation, advantage was taken of the existing heat flow map of Europe completed with the tectonic background (Čermák and Hurtig 1979). Both maps were divided in 4° × 4° latitude–longitude grid elements. The mean values of heat flow and of crustal thickness were taken for each such element and the corresponding lithosphere thickness was then calculated. Using this technique, dependence on the uneven data distribution was partly removed and the resolution of the map was improved by a factor of 2.5, i.e. from less than 1100 km in the former map (Chapman et al. 1979) to about 400 km (Čermák 1982a).

Similar to the latter work, using the same technique and the same model of the crustal thermal conductivity and heat generation, the detailed lithospheric thickness distribution pattern was proposed for the study area (Fig. 4.16). This pattern is based on contouring the data calculated for the 0.5° × 0.5° grid elements. The resolution of the map is thus approximately 50 km. Characteristic heat flow and crustal thickness data were taken from Figs. 4.13 and 4.14. The pattern obtained seems to correspond to the expected deep lithospheric structure. Except for the deep-reaching lithospheric roots beneath the East European platform, which extend down more than 130 km beneath the platform margins (upper right hand corner of Fig. 4.16) and amount more than 200 km beneath the major part of the craton (Čermák 1982a), the deepest lithosphere/asthenosphere boundary is beneath the Bohemian Massif. The central and eastern parts of the massif are characterized by the lithosphere thickness of 120 – 140 km, which corresponds to its submerged portion beneath the overthrust Carpathian system. The lithosphere is thinner beneath the northwestern part of the

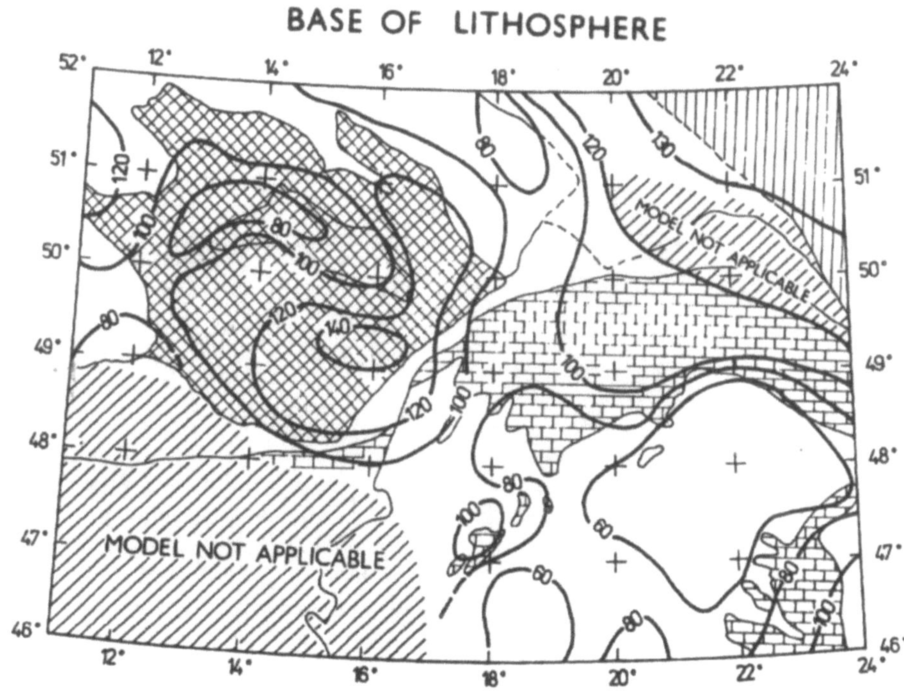

Fig. 4.16 Regional pattern of the depth to the lithosphere base.

massif, the zone of minimum lithosphere thickness (80 km) follows the crooked-shaped belt of an intensive Variscan tectogenesis, framing the central stable block. Under a great part of the Pannonian Basin the lithosphere/asthenosphere boundary is only 60 km deep (Ádám 1978; Horváth et al. 1979).

The model used for the calculation is based on the steady-state solution of the heat conduction equation. The results cannot be applied in the Alps, where the geothermally determined lithosphere/asthenosphere boundary would be at a depth of only 80 km. Here, the crust as well as the lithosphere attain substantional depth (Calcagnile and Panza 1980). The thermal regime at the depth of 100 km and more has not reached its stationary state yet. The area of the Teisseyere–Tornquist zone in south Poland is also a problematic one. Here, the surface heat flow may be affected by the salt tectonics and can hardly be used as a boundary constrain for assessment of the lithosphere thickness.

5 Zones of Geoelectrical Inhomogeneities

5.1 Earth's Magnetic Field Variations and Geomagnetic Induction Characteristics

V. Červ[1], J. Pek[1], V. Petr[1], K. Pěč[2], J. Pěčová[1], O. Praus[1], M. Tobyášová[1]

The results of systematic research into the peculiarities of the geomagnetic field variations and their interpretation in terms of internal electrical conductivity distribution over Czechoslovak territory, with regard to the results in the neighbouring countries, are addressed in this chapter. Main conclusions concerning the investigations carried out at the boundaries of the West Carpathians and the surrounding units are presented. Investigations of the anomalous zones at the eastern margin of the Bohemian Massif are discussed. Together with the spatial delimitation of anomalous zones, the depth estimates of anomalous field sources, obtained by various methods, are presented and models of internal geoelectrical structures of the provinces under investigation are suggested.

Surface distribution of induction characteristics deduced from geomagnetic field variations has been used to study the internal geoelectrical structure on both the global and regional scales for several years. For regional studies, the "Wiese induction vector" represents one of the most descriptive characteristics. It follows, from the theory of electromagnetic fields in a horizontally homogeneous medium, that the vertical component of the geomagnetic field equals zero. In actual structures, however, a non-zero vertical component of the geomagnetic field variations has always been observed and in specific regions, it reaches significant values and shows a large site to site variability.

A lateral inhomogeneity in the electrical conductivity distribution is the source of the vertical component variations. It generates the anomalous geomagnetic fields of internal origin, observed and measured at the Earth's surface together with the primary external inducing field. Wiese vectors and/or other induction characteristics clearly express the spatial distribution of fields of the internal origin. By analysing these characteristics, information can be obtained on spatial/depth parameters of the inhomogeneous distribution of the electrical conductivity.

---

1 Geophysical Institute, Boční II/1401, 141 31 Praha 4, Czech Republic
2 Department of Geophysics and Meteorology, Faculty of Mathematics and Physics, Charles University, Ke Karlovu 3, 121 16 Praha 2, Czech Republic

The Wiese induction vector can be determined from the relation:

$$\Delta Z_i = A\Delta H_i + B\Delta D_i\,, \quad i = 1,2,\ldots,N \quad , \tag{5.1}$$

where ΔZ_i(+ down), ΔH_i(+ north), ΔD_i(+ east) are the actual variation magnitudes of the geomagnetic field components commonly used in observatory practice. In estimating the parameters A, B, we formed an overdetermined set of equations by substituting into (5.1) the variation magnitudes of the field components for each specific event, scaled at the moment of the maximum in the Z–component. The equation system is solved in the least squares sense for A and B as components of a vector (the Wiese induction vector). The quantities $|C| = (A^2 + B^2)^{1/2}$ and $\Theta = \text{arctg}\,(B/A)$ define its magnitude (module) and azimuth (argument), respectively. For a 2-D inhomogeneous medium we assume, on the basis of model calculations, that in the vicinity of the inhomogeneity the induction vector is directed perpendicularly to its strike and points away from a zone of increased electrical conductivity.

More refined information on the surface distribution, the period dependence and on other parameters of internal fields can be obtained by analysing generalized induction vectors estimated from the complex transfer functions of a system "Earth conductor/external inducing EM source". From the estimates of transfer functions (*TF*) we define the vector of their real parts (the in-phase vector) and the vector of their imaginary parts (the out-of-phase vector). These complex vectors have frequently been referred to as the "Schmucker vectors" and their maps characterize the variability of the Z–component across a region. The maps can be generated for a set of periods. The Wiese induction vector represents special case of a complex vector for the zero or negligibly small imaginary component. This takes place when the induction process is assumed to be a pure self-induction (induced and inducing fields are in phase).

Additional information furnishing further details on internal geoelectrical inhomogeneities can be inferred from inter-station transfer functions and corresponding vector characteristics related to a common reference station situated at a sufficient distance from the inhomogeneities under investigation, where approximately "normal" field can be expected. Induction characteristics estimated in this way also define anomalous zones and, in addition, they can be used to synchronize the measurements performed at stations along individual profiles during different time-windows. Then, the properties of the internal field can be studied, as if the measurements were conducted over a 2-D synchronous array of field stations. Two independent methods were developed and refined for estimating complex transfer functions together with corresponding induction vectors at each individual station (single-station induction characteristics) and the inter-station characteristics from time-series of D, H, Z of the geomagnetic field components recorded at both the individual stations and the reference station (the Budkov observatory) simultaneously. The first of them is referred to as the multivariate coherence analysis (MCA). According to the physical character of the mechanism, we divide the field components into those representing a driving mechanism and those corresponding to a dependent one. Their mutual relations can be expressed by the transfer matrix the components of which form auto- and cross-power spectra of the field components. Corresponding induction characteristics are estimated for each individual sample. Suitable statistical data treatment can be applied to get averaged induction characteristics. The method provides a good measure of the residual part which cannot be associated with the driving mechanism (quantity R^2).

The second method (TDA – Time Domain Analysis) considers relation (5.1) as a convolution equation with general time-series of the field components to be solved in the time domain for the whole set of samples simultaneously. Suitable algorithms are available (e.g. Wieladek, Ernst 1977). Sets of discrete operators representing an impulse response of the system are the solution to the convolution formula. By applying the Fourier transform, we obtain estimates of transfer functions for a series of periods.

The theory of the two methods is explained in detail in Pěč et al. (1988a,b). Analysing the actual data from profiles P–83 and P–82, we estimated the sets of induction characteristics by both methods and assessed the differences between the results. It can be concluded that the agreement is good for main induction characteristics, and that the results reasonably match within a few per cent.

Treating the variations of the geomagnetic field components together with the induction characteristics, we can estimate the depths of the anomalous field source, i.e. geo-electrical inhomogeneities, and its position beneath the profile. Assuming that the anomalous field is generated by a current carrying line conductor of infinitesimally small cross-section in a homogeneous half-space, the source position can be estimated in following steps:

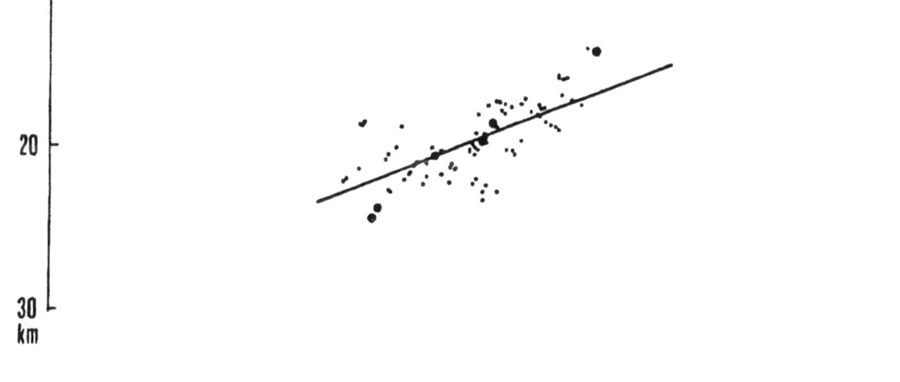

Fig. 5.1 Set of intersection points according to the D–P method for the data obtained along profile P–81 (correction with H_U) and their interpolation. Circles mark repeated occurrence of the value.

1. Let the field of the horizontal component H_i in the direction of a profile and the field of the vertical component Z for selected variation events be divided into internal and external parts (H_i, Z_i) by applying a mathematical transform (the D–P method, Pěčová 1985). If normals are constructed from various points of the profile to the vector of the internal field, they should theoretically intersect, according to Biot–Savart's law, in one point corresponding to the schematized point-source. In practice, the numerous normal intersects for a set of selected events scatter around the source. Treating individual estimates statistically, we obtain approximate co-ordinates of the anomalous field source (Fig. 5.1).
2. From numerous variation events we estimate the depths and positions of anomalous field sources beneath the profiles by applying a physical–statistical approach and the methods of observation calculus (the F–S method, Petr 1980).
3. Using the Wiese and in-phase vector estimates at observing sites of a specific profile, we construct sets of characteristic functions for various periods. From parameters c_{max} (amplitude of the characteristic function maximum), and x_{max} (the distance of the maximum from point $c = 0$ in km) of characteristic function we can estimate the depths and the source position beneath the profile (Petr et al. 1985).

It is necessary to emphasize that the depths obtained by applying these methods represent the maximum estimates due to schematic assumptions concerning the source and medium. Consequently, the actual inhomogeneities of finite dimensions corresponding to geological structures can be expected closer to the Earth' surface.

5.2 Main Zones of Geoelectrical Anomalies in Europe

V. Petr[1], K. Pěč[2], J. Pěčová[1], O. Praus[1]

The Wiese induction vectors map zones of anomalous internal fields, the sources of which are geoelectrical inhomogeneities of the medium. Exploiting our own results and the data published prior to 1975 (Calota et al. 1971; Ádám et al. 1972), we compiled a map of induction vectors for the periods ranging from 20 to 60 min in Central, Southeastern, and in a part of Eastern Europe. From anomalous induction vectors pattern, especially from reversals of their orientation or sudden changes of their orientation and lengths, it is possible to mark the zones with internal electrical inhomogeneities (Fig. 5.2).

1 Geophysical Institute, Boční II/1401, 141 31 Praha 4, Czech Republic
2 Department of Geophysics and Meteorology, Faculty of Mathematics and Physics, Charles University, Ke Karlovu 3, 121 16 Praha 2, Czech Republic

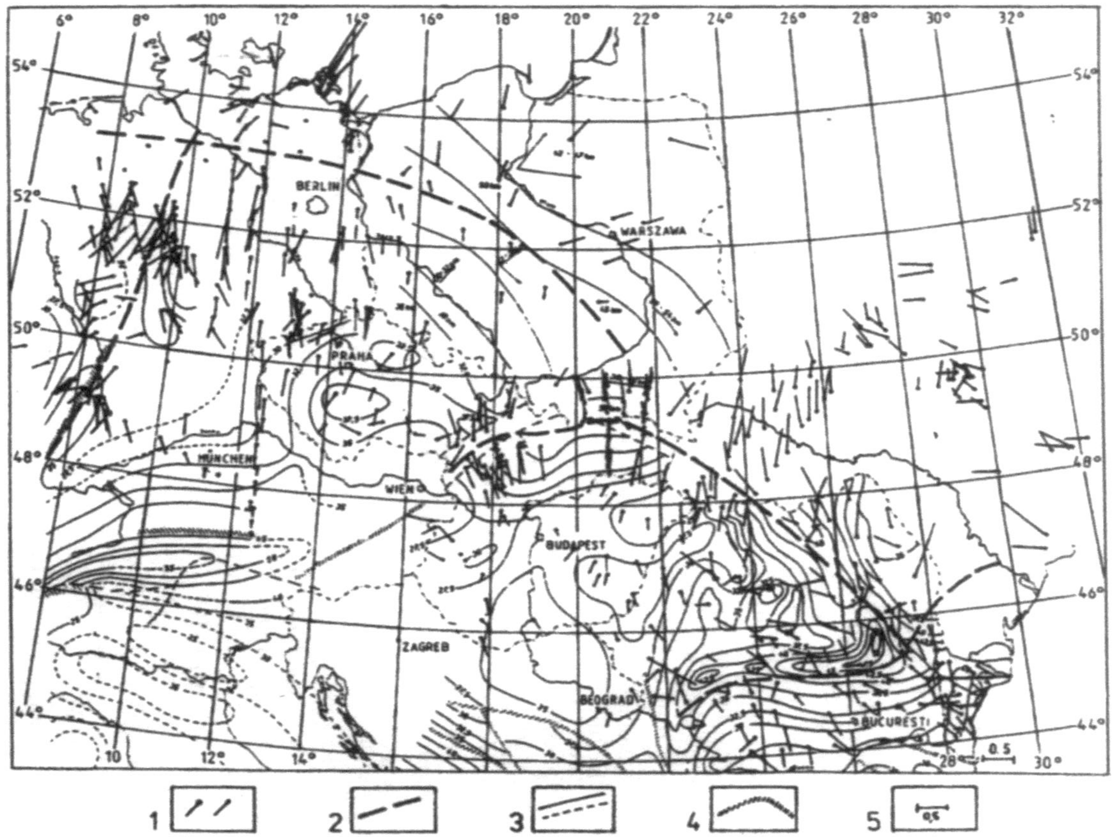

Fig. 5.2 Distribution of the Wiese induction vectors in Central, Southeast, and East Europe (results to 1975 included). 1 – Wiese induction vectors, 2 – zones of geoelectric anomalies, 3 – assumed depths to the Moho discontinuity in km (according to Beránek and Dudek 1972), 4 – tectonic elements assumed from the DSS results, 5 – scale of the induction vectors lengths.

The Carpathian geoelectrical anomaly is probably the most significant feature of the European region. It can be traced starting from the Vienna Basin and extending along the Carpathian arc in the west, east to the south Carpathians, where its further continuation is rather uncertain for the MV data sparsity. In the Czechoslovak – Polish sector of the West Carpathians the position of the anomaly has been spotted in detail analysing the data of stations along several synchronous profiles traversing the Carpathian arc in key regions (Praus et al. 1981; Jankowski et al. 1985).

In SE Romania the anomaly is expressed vaguely by the vectors the azimuths of which scatter considerably, but are oriented north or southward (Ritter 1975). The tendency of the vectors to reverse their azimuths and to point from the Carpathian arc, however, remains. Considering these results together with the data from the Black Sea and its vicinity, from the Crimea peninsula (which are not presented in Fig. 5.2), and from the Ukraine, we can assume an anomalous geoelectrical structure which branches off the Carpathian anomaly roughly in the bend of the Southern Carpathians (the Vrancea region), and which extends in a NEE direction. It has been assumed that it delimits the northern margin of the Dobrudža massif.

In NW Hungary (the Bakony Mts.) a trans-Danubian conductivity anomaly has been revealed (Ádám and Pospíšil 1984). It is explained by increased electrical conductivity in a system of deep-rooted faults at an average depth of 7 – 10 km.

An important anomalous zone which traverses the northern part of Central Europe is referred to as the northern German – Polish conductivity anomaly. Its characteristic feature is the phase change of the Z variation by 180° and, thus, the reversal of induction vector azimuths from northern to southern ones. The southern margin of the anomaly, roughly defined by the maximum value of the Z variation, marks a transition from highly conductive sediments of the North German basin to low-conductivity rocks of the Hercynian zone in Central Europe. The anomaly seems to be generated by a 2-D inhomogeneous source only on Polish territory (Porstendorfer et al. 1976). Towards the west, the zone widens and its core probably branches off west of Hamburg. One part extends towards the west, the other strikes towards the NW into the North Sea.

The previous interpretation assumed a schematized-line current-source the maximum depth of which was approximately 85 km. Recent interpretations suggested the great thickness of surface sediments to be the source of the anomaly. At present, an additional highly conductive zone separated from the surface by a high-resistivity layer is assumed in the lower crust. It is based on model calculations (Ritter 1975).

In the western part of the FRG a geoelectrical anomaly of the Rhine graben was delineated. Studying the anomaly, Ritter (1975) shows that the induction vectors on the profile, oriented in WNW – ESE direction, clearly indicate the effect of currents induced in the direction parallel to the graben. The anomalous zone extends through the graben valley of the Leina river NNW to the region of the main anomaly. A part of this system of anomalies is represented by the Göttingen D–anomaly, characterized by a maximum deviation of induction vectors from the N–S direction for the variations with periods of 40 – 50 minutes. All these anomalies seem to be a part of a specific system which is confirmed by a close relationship between electrical conductivity and positive heat flow anomalies (Hurtig et al. 1975). The anomaly is probably caused by a combined effect of highly conductive surface sediments and a conductive sequence in the lower crust and upper mantle.

Carpathian Geoelectrical Anomaly

The MV measurements in the West Carpathians resulted in numerous data on variations of the geomagnetic field components at 90 field observatories. For each individual field station both the Wiese induction vector and the transfer functions for the periods ranging from 700 to 7 000 s were computed together with the in- and out-of-phase vectors (computed from real and imaginary parts of the transfer functions, respectively). They will be alternatively referred to as the complex induction vectors. Synchronous records of the field components at all stations of a specific profile made it possible to divide the selected pronounced variation events into internal and external components and analyze their internal component (the D–P method, Pěčová 1985), or process the variation events in a physical–statistical way (the F–S method). From a set of the characteristics thus obtained, we could estimate the position and depth of the conductive zone beneath the given profile.

With regard to these results the position of anomalous zone centre (the Z–component variation = 0) was estimated with the accuracy of ± 2 km on all profiles. The anomalous zone seems to extend over the whole sector of the Czechoslovak – Polish West Carpathians. The results are displayed on the map in Fig. 5.3, where the Wiese induction vectors are schematically plotted over the Czechoslovak territory and its immediate vicinity. Here other zones are revealed as well. A detailed mapping of the anomaly resulted in a considerably better knowledge of its regional trend in the West Carpathians and its position with regard to the fundamental geological and tectonic features (i.e. the Klippen Belt, the Peripieninian lineament, the Odra line, etc.).

In the Vienna Basin and in the Záhorská nížina lowlands additional MV measurements were conducted altogether at 11 stations, distributed along two profiles. The estimates of induction vectors are shown on the map in Fig. 5.3. They contribute to solving the problem of further extension of the anomaly in the close vicinity of and across the Czechoslovak – Austrian border. According to these results we can state that the position of the anomaly axis, extrapolated to the state border from the Wiese vectors in the vicinity of the profile P–VI, requires only minor corrections as shown in Fig. 5.3 (where a and b indicate the original and the refined positions, respectively). The anomaly seems to continue to the region east of Vienna because, according to the data obtained at the geomagnetic observatory Aufhof (Ádám et al. 1972), a very small induction vector modulus (practically zero) indicates a position directly on or close to the anomaly axis.

Further characteristics of the Carpathian geoelectrical anomaly were deduced from the results of transfer functions computations and related in- and out-of-phase vectors (the complex or the Schmucker vectors). For details, the reader is referred to our publication (Jankowski et al. 1984). An example of the complex vectors distribution in the West Carpathians is displayed in Fig. 5.4 for the period of 3800 s.

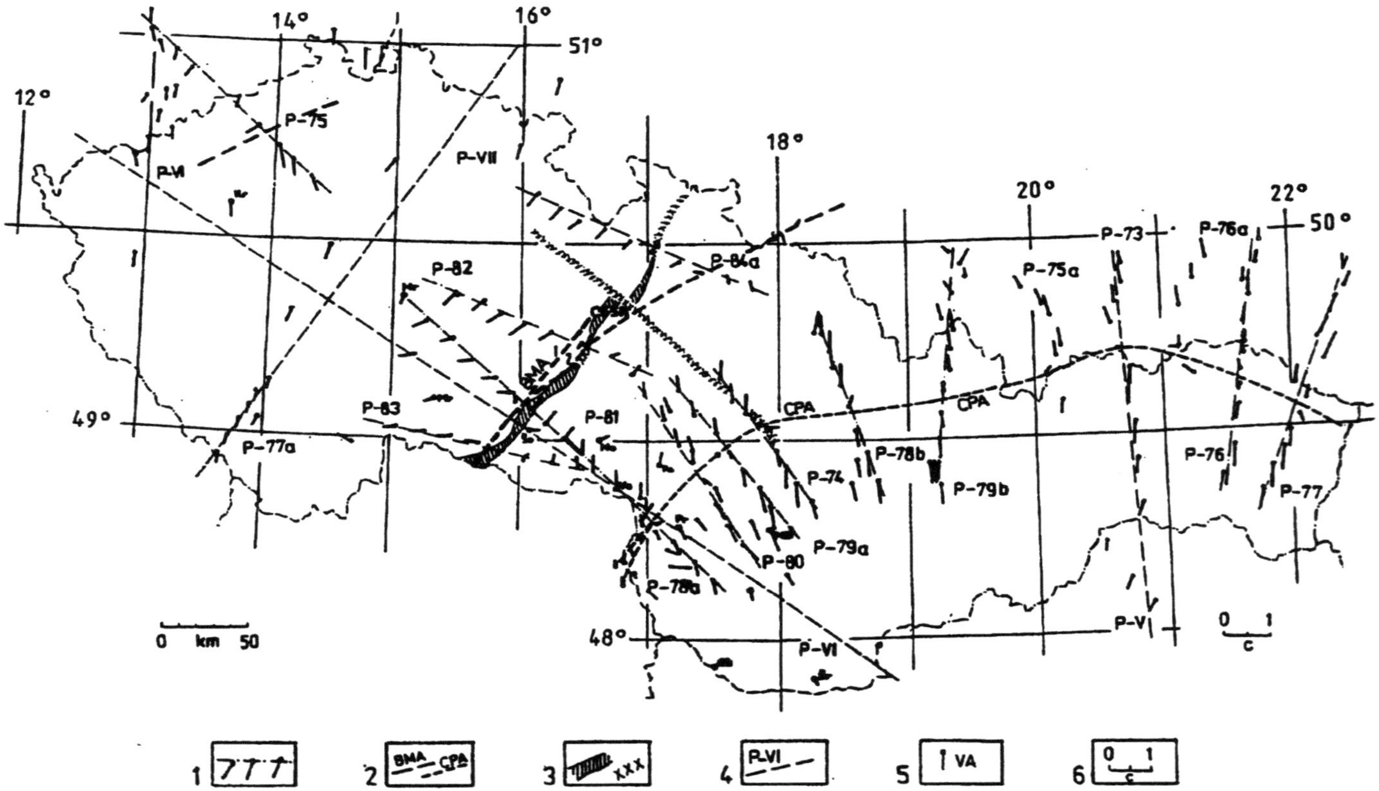

Fig. 5.3 Wiese induction vectors in Czechoslovakia and surrounding regions (results to 1986 included). 1 – Wiese induction vectors and the MV profiles, 2 – geoelectrical anomalies (BMA – the Bohemian–Moravian, CPA – the Carpathian), 3 – elements of the deep structure of the Earth's crust (after Suk in Suk and Vacek 1984), 4 – deep seismic sounding profiles (DSS), 5 – magnetotelluric stations, 6 – scale of the induction vector lengths.

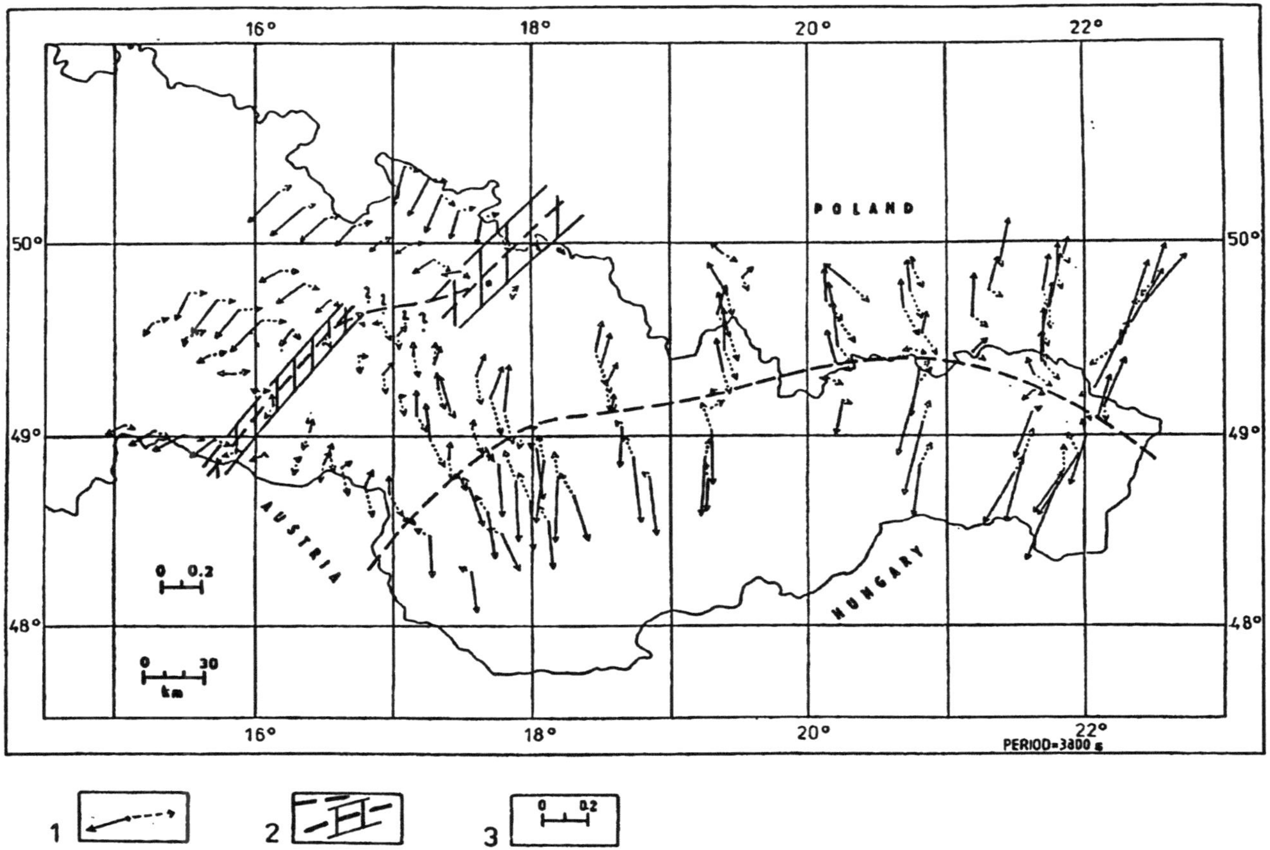

Fig. 5.4 Complex induction vectors (in- and out-of phase vectors, full and dashed arrows, respectively) in the West Carpathians and across the eastern margin of the Bohemian Massif. 1 – the vectors, 2 – zones of geoelectrical anomalies, 3 – scale of vector length

Analysing the induction vector results in the foreland of the West Carpathians, on the eastern slopes of the Bohemian Massif, and in the Polish Palaeozoic platform, we detect an antiparallel orientation of in- and out-of-phase vectors for periods from 3600 s to about 1500 s. It is a qualitative indication that the medium in the surroundings of the anomaly axis has approximately the character of a 2-D inhomogeneous structure. For periods shorter than 1200 s the out-of phase vectors show a tendency to rotate into the perpendicular orientation. This tendency then strongly prevails at many stations in the inner Carpathian region. We assume that it is an expression of induction in superficial sediments of the marginal parts of the Carpathian basins and an increasing general 3-D inhomogeneous character of the geoelectrical structure of the sedimentary basin in the close vicinity of the Inner Carpathian mountain ranges.

Recently, the MV data of 143 field stations distributed over the West Carpathian sector, over the transitional block of the Brno unit (the Brunovistulicum) and over the eastern margin of the Bohemian Massif were analyzed by improved MCA and TDA methods to obtain different kinds of transfer functions of the geomagnetic field and related complex induction vectors. Contour maps of transfer functions were computer-generated to map objectively the zones of anomalous induction. The most expressive representation of the Carpathian geoelectrical anomaly is provided by the contour map of the component TF1 in Fig. 5.5 (for details and notation see Praus and Pěčová 1991). Thick and "0" labelled heavy dash-dot contours separating the regions of positive and negative TF1 values clearly mark the central zone of the Carpathian anomaly and the anomalous zone in the eastern margin of the Bohemian Massif, discused further.

In estimating the depth of the anomalous field source, we exploited the results of separating the field components of several selected variation events into internal and external components (the D–P method), the sets obtained by the F–S method and from the parameters of the characteristic functions defined by the Wiese and in-phase vector module on each individual profile. In the Czechoslovak – Polish part of the West Carpathians the depths range from 15 to 25 km, maximum depths being in the Vienna Basin on profile P–78a and on profile P–76 traversing SE Poland and East Slovakia (see Fig. 5.3). The depths of the anomalous field source are shown schematically in Fig. 5.6 as function of profile position on the Carpathian arc.

Further information on the depth of the anomalous field source is obtained from the magnetotelluric sounding (MTS) results. The MTS measurements at or in a close vicinity of the anomaly axis were conducted in various parts of the West Carpathians. Three MTS curves in Czechoslovak territory, 2 MTS, curves in Poland and several sounding curves obtained in the west Ukraine were used. For details reference is made to our paper by Jankowski et al. (1984), only main conclusions are summed up here. Interpreting the MTS curves, we detected a layer exhibiting low resistivity of 2 – 5 ohm-meters at depths from 7 km to 12 km. The thickness of the layer ranges from 1 km to 7 km, the variability of depths and thicknesses is due to the electrical equivalence of layers, and, therefore, to ambiguity in the interpretation. According to the MTS curves at stations approximately at the anomaly axis, the geoelectrical section is characterized by a layer of high integrated conductance in the range of 1 000 – 3 000 Siemens. As the

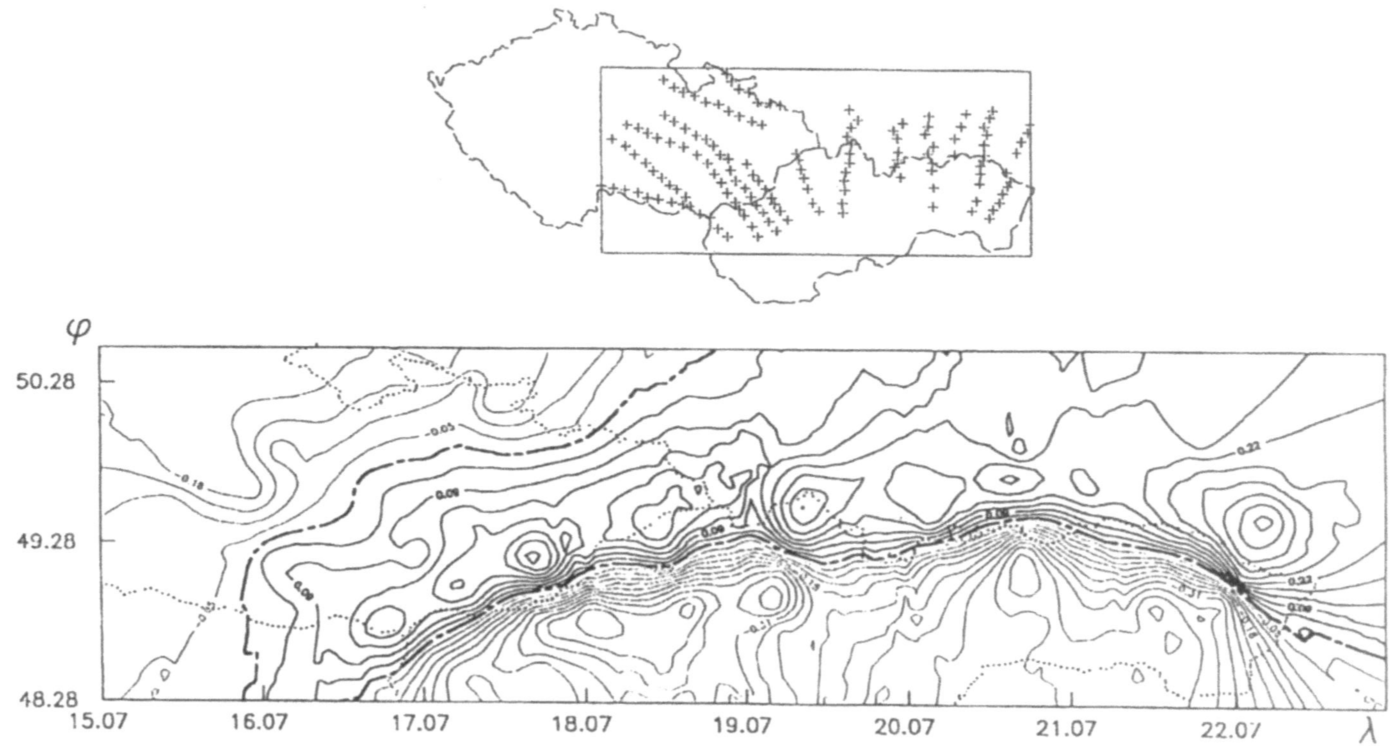

Fig. 5.5 Contour map of the TF1 component for the set of stations over the marginal part of the Bohemian Massif and the West Carpathians (k and z are the geographic coordinates). The sector of investigations and the observing sites are apparent from the inset in the upper part of the figure. Heavy dash-dot curves mark the zero contours (central part of anomalous zone), heavy solid contours enhance the region of the positive TF1 values, light solid contours denote the negative TF1 contours.

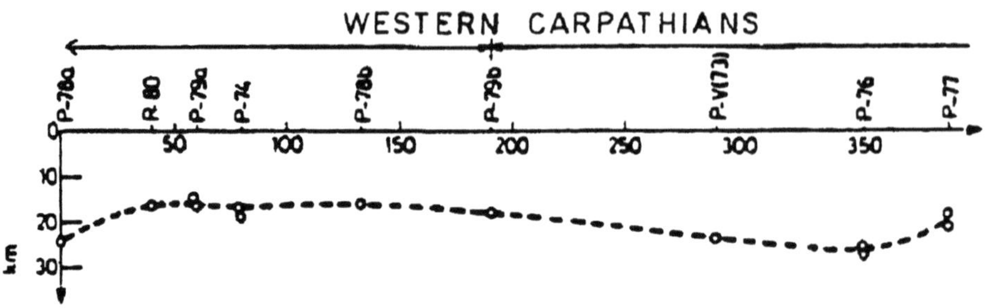

Fig. 5.6 Summary of the source depths for the Carpathian geoelectrical anomaly extending along the Czechoslovak and Polish – Czechoslovak section of the West Carpathians (from the Vienna Basin to the USSR – Czechoslovakia frontier). Averages are shown with specific profiles traversing the anomaly (profile P–78a, etc.).

distance of the station from the anomaly axis increases, the integrated conductance decreases to about 200 Siemens at a distance of 15 – 18 km from the anomaly axis. The layer with high conductance is identified with an internal anomalous field source. In spite of the fact that the MTS results yielded the anomalous field source depths smaller than the MV ones, the two estimates are not in contradiction as the interpretation of MV data based on the line current approximation are theoretically shown to provide maximum estimates. Consequently, the actual geological structure can be expected at a smaller depth depending on its shape .

From the results of research into the anomalous geo-electrical zone in the Czechoslovak – Polish part of the West Carpathians and from those obtained in other regions of Carpathian countries it is clear that the geo-electrical anomaly is one of the geophysical characteristics distinguishing the Carpathian unit as a whole. The anomaly is located in the proximity of important geological boundaries, is related to significant deep-rooted tectonic zones (the Klippen Belt, the Peripieninian lineament, the Odra line), and is found to be closely correlated with anomalies of other geophysical fields (regional trend of the gravity minimum, anomalies in the permanent magnetic field). Specifically, the anomaly is found to coincide with the DSS sections where a vertical jump amounting up to 15 km of the Moho discontinuity depths is indicated by DSS. These zones are assumed to distinguish the blocks with different thicknesses of the Earth's crust. All these facts support the assumption that the geoelectrical anomaly corresponds to a regional structure at the immediate contact of the Carpathian plate with the surrounding blocks of the Bohemian Massif, the East European platform, and of the Ukrainian shield. An increased electrical conductivity of the zone can be explained by the following factors or by their combinations:

1. the presence of thick sediments deposited in the Carpathian foreland and their saturation with mineral waters;
2. temperature increase and consequent warming-up of mineralized waters, or partial melting of rocks drawn down into crustal depths in the process of subduction;
3. the heating-up of rocks and reduction effects in rocks containing in various amounts hydrocarbons according to the mechanism suggested by Bucha (1980).

Ádám and Pospíšil (1984) present also some other possible mechanisms such as e.g. a process of regional metamorphism of the flysch complex at depths below 10 km, or a dehydration process which can be responsible for the increase of rocks conductivity at temperatures of 600 – 700 °C when overheated water cannot escape from the rocks, or water-saturated silicate solutions can ascent through deep-rooted faults towards the Earth's surface in zones of partial or total melting (see corresponding references in the paper by Ádám and Pospíšil (1984).

Considering all previous and recent results on our territory with those from neighbouring countries, the Carpathian geo-electrical anomaly is suggested as marking a deep-rooted collision zone of the Carpathian lithospheric plate with stable blocks of the Meso- and Palaeo-Europe (the Bohemian Massif, the Polish Palaeozoic platform as a part of the East European platform). These units came into contact during the collision of the European and African continental lithospheric plates.

The Anomalous Geoelectrical Zone in the Eastern Margin of the Bohemian Massif

The geoelectrical measurements conducted in 1962 – 1972, namely the MTS measurements carried out along the DSS profile VI, implied an existence of another geoelectrical inhomogeneity in the eastern margin of the Bohemian Massif (Pěčová et al. 1976). Therefore, the MV measurements were organized in 1981 along the profile P–81 which extended the trans-Carpathian profile P–78a into the Bohemian Massif in the direction of the DSS profile No. VI. (see Fig. 5.3). The measurements along six profiles P–81 to P–86, which traversed the eastern margin of the Bohemian Massif in key regions between the Austrian – Czechoslovak and Polish – Czechoslovak borders and the results of subsequent data processing confirmed the existence of geoelectrical inhomogeneity. The organization of the measurements and instrumentation used were the same as in the West Carpathians.

In processing the variation data, we employed the approaches explained in Section 5.1. In the first step, the set of equations (5.1) was solved, and for each individual station, the Wiese induction vectors were estimated. The vector parameters were estimated with a relative error of about 6 % for their module and about 5° to 6° for the angles at stations outside the anomalous zone. In the zone itself, the relative errors increased to 20% and to 15° – 18° for amplitudes and angles, respectively. Large relative errors in determining the induction vector parameters are, thus, an additional indication of anomalous conditions at an observing site of a given profile. A general distribution pattern of the Wiese vectors on our territory is displayed in Fig. 5.3 and specifically for the eastern margin of the Bohemian Massif in Fig. 5.7.

In the next step, a series of 10 – 15 samples with increased geomagnetic activity were digitized and used for calculating the complex transfer functions and related single- and inter-station induction vectors. The computations were performed for all stations along profiles P–81 through P–86 using the two TDA and MCA methods. In estimating the inter-station induction characteristics, we used as reference the records of the permanent observatory of the Geophysical Institute of the Czechoslovak Academy of Sciences (GFÚ ČSAV) in Budkov.

The period dependence of the induction process is shown in Fig. 5.8 for the results at the profile P–82. With decreasing period, we usually observe an increase of in-phase vector module, one or two maxima are reached at certain periods and a decline of module magnitudes takes place again for short periods. On the contrary, the modulus of the out-of-phase vector decreases with decreasing period. Minimum values are reached approximately at periods corresponding to the in-phase vector maximum, i.e. in this particular range of periods the complex vector approximately equals to the real Wiese vector. For the profile P–81 the module maxima occur at periods around 1300 - 1500 s and 800 s, in the SE part of the profile, these maxima become

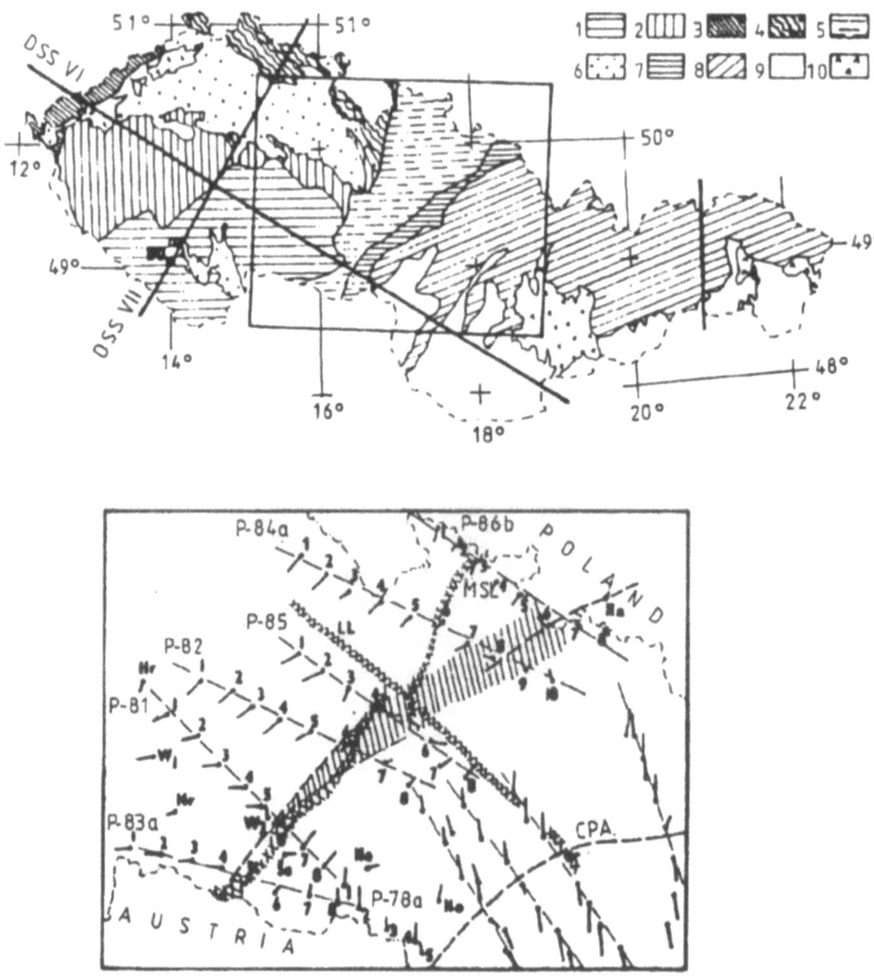

Fig. 5.7

inexpressive. In the close vicinity of the anomaly centre they disappear. An analogous tendency is apparent in the results of the profile P–82, where distinct maxima at periods around 1800 – 1900 s and 800–900 s are observed. They also diminish and vanish at localities closer to the inhomogeneity centre. Further to SE only one maximum occurs. Finally, at all stations along the profile P–83, one maximum at a period of about 1000 s is observed. Comparing the periods of maximum induction effect on profile P–82 with that from P–81 and P–83, we find a slight shift towards longer periods. Therefore, it can be qualitatively deduced that the source of an internal anomalous field (the geo-electrical inhomogeneity) dips to greater depths towards NE which does not contradict the results of depth estimates.

In order to understand better the character of the medium as to its geo-electrical inhomogeneity, we also plotted the in- and out-of-phase vectors for a series of identical periods at all stations of profile P–81 together with those obtained on profile P–78a (continuation of the traverse across the Carpathian arc into the inner Carpathian region – Fig. 5.9). The results obtained on profile P–83 are plotted analogously together with those from the profile P–78a. Comparing the lengths of in- and out-of-phase vectors and their mutual orientations, we can quantitatively distinguish areas and specify ranges of periods within which the medium can be approximated by a 2-D inhomogeneous structure (roughly parallel or antiparallel orientations of the vectors). This presentation

---

Fig. 5.7 a) Generalized geological map of Czechoslovakia showing Deep Seismic Sounding profiles Nos. V, VI, VII. The main geologic units are marked by the following symbols. The Bohemian Massif: 1 – Moldanubian block, 2 – Central Bohemian block, 4 – West Sudeten block, 5 – Moravian–Silesian zone, 6 – Cretaceous and Tertiary platform deposits, 7 – Alpine–Carpathians foredeep. The Carpathian zone: 8 – pre-Neogene folding of the Carpathian unit, 9 – Neogene basin of the Carpathian region, 10 – Tertiary eruptive rocks, Bu – reference observatory. b) Enlarged inset of Czechoslovakia where MV measurements were analyzed. Wiese vectors obtained by treating visually scaled amplitudes of variations for periods from 6 to 40 min are shown with individual locations grouped into profiles (arranged from SW to NE). Localities of MV measurements: P–83: 1 – Číměř (Čí), 2 – Stálkov (ST), 3 – Báňovice (BÁŇ), 4 – Dešov (DE), 5 – Šumná (ŠU), 6 – Únanov (UN), 7 – Čejkovice (ČEJ), 8 – Drnholec (DRN), P–78a: 1 – Nová Ves (NOV), 2 – Bavory (BAV), 3 – Valtice (VA), 4 – Polesí Lanžhot (PLA), 5 – Kúty (KU), other locations of the profile are beyond the limit of the sector. P–81: 1 – Koberovice (KOB), 2 – Úsobí (ÚS), 3 – Velký Beranov (VB), 4 – Svatoslav (SV), 5 – Hostákov (HO), 6 – Popůvky (PO), 7 – Jamolice (JA), 8 – Vedrovice (VD), P–82: 1 – Mířátky (MI), 2 – Bílek (Bíl), 3 – Radostín (RAD), 4 – Vlachovice (VLA), 5 – Ždánice (ŽDA), 6 – Sychotín (SY), 7 – Sloup (SL), 8 – Rychtářov (RY), P–85: 1 – Chotovice (CHO), 2 – Trstěnice (TRS), 3 – Sklenné (SKL), 4 – Jevíčko (JV), 5 – Konice (KON), 6 – Lutotín (LUT) 7 – Výšovice (VYS), 8 – Uhřičice (UHR), P–84a: 1 – Semechnice (SM), 2 – Javornice (JVR), 3 – Kunvald (KNV), 4 – Dolní Boříkovice (DB), 5 – Bohdíkov (BH), 6 – Rudoltice (RDL), 7 – Ondřejov (ON), 8 – Nová Valteřice (NVA), 9 – Svatoňovice (SVA), 10 – Vrchy (VRC), P–86a: 1 – Bernartice (BE), 2 – Velké Kunětice (VK),3 – Dolní Údolí (DU), 4 – Komora (KOM), 5 – Krásné Loučky (KLO), 6 – Skrochovice (SKR), 7 – Hněvošice (HN), 8 – Píšť (PŠ). Other stations: Ra – Racibórz (Poland); magnetotelluric stations: Mr – Mrákotín, No – Nosislav, Ho – Hodonín. W1, W2 – measurements performed by H. Wiese. Unmarked locations with Wiese vectors belong to the Carpathian region. Shaded zone in the central and northeastern part of the region marks a sector where the centre of the zone is not determined reliably. Symbols (xxxxx) are tectonic-structural elements from the cross-section at the level of –10 km (Suk in Suk and Vacek 1984). Further notations: LL – Labe fault zone, MSL – Moravo–Silesian fault zone, BMA – Bohemian Massif anomaly, CPA – Carpathian anomaly.

of vectors clearly defines the Carpathian geoelectrical anomaly, discussed in Section 5.2. The zone of the geo-electrical inhomogeneity near the SE margin of the Bohemian Massif is characterized by a well-developed reversal of in-phase vectors between the Svatoslav – Popůvky stations on profile P–81, and in the range of periods $T > 1000$ s, also by reversal of out-of-phase vectors. Consequently, the electrical conductivity distribution beneath the localities of Svatoslav – Popůvky and for periods between 1000 – 3000 s seems to show a reasonably well-developed 2-D inhomogeneous structure. The Popůvky station seems to characterize a structure in a transition zone to the Carpathian anomaly.

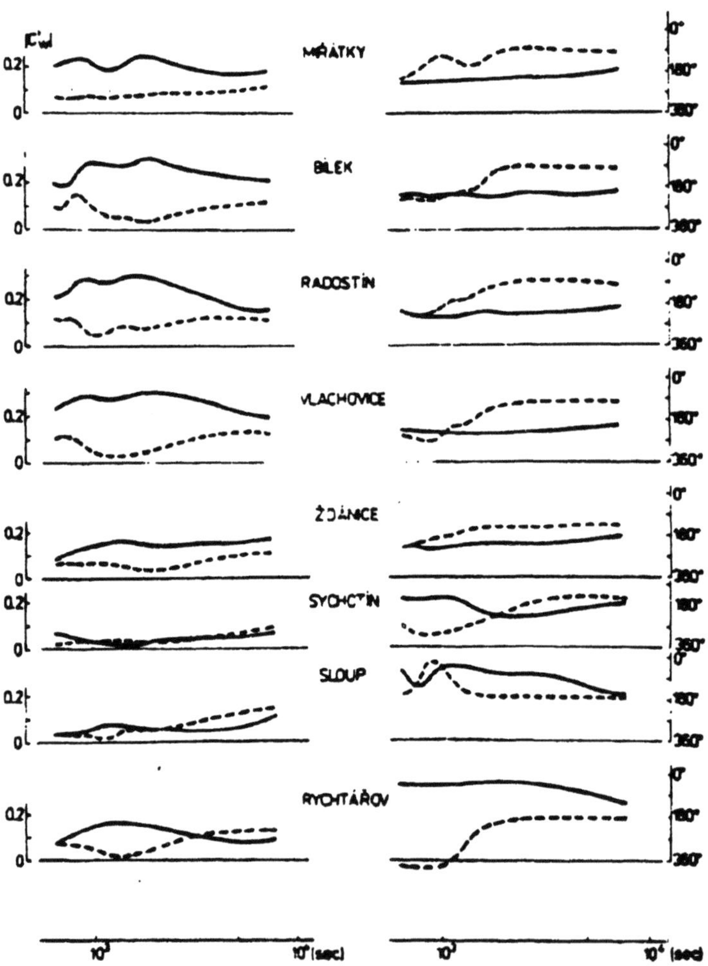

Fig. 5.8 The period dependence of the induction process. The left-hand part: magnitudes of the in- (full) and out-of-phase vectors (dashed). The right-hand part: azimuths of the in- (full line) and out-of-phase vectors (dashed).

Further to the NE, analogous period dependences were analyzed considering the results of data analyses at stations on profile P–85. The maximum induction response according to in-phase vectors (W_R) is found at periods between 1550 – 1220 s in analogy to conditions on profile P–81. According to the minimum module of the in-phase vectors and to changes of their azimuths, the anomalous zone represents an about 40 km long section of the profile P–85 with Jevíčko, Konice, and Lutotín field sites.

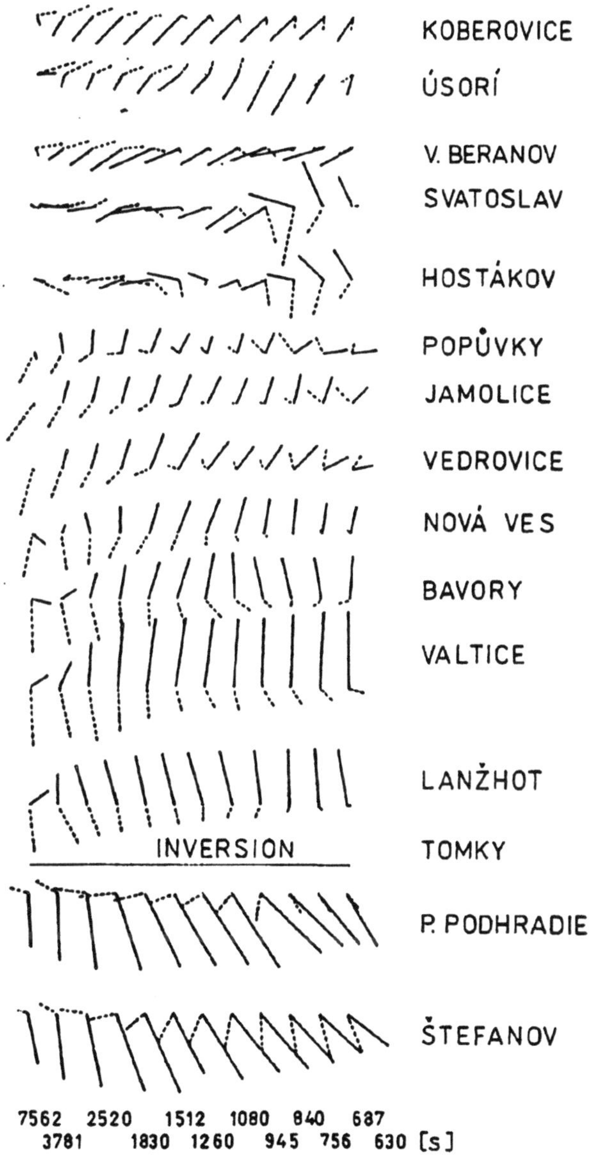

Fig. 5.9 In-phase (full) and out-of-phase (dashed) induction vectors along joint profiles P–81 and P–78 a.

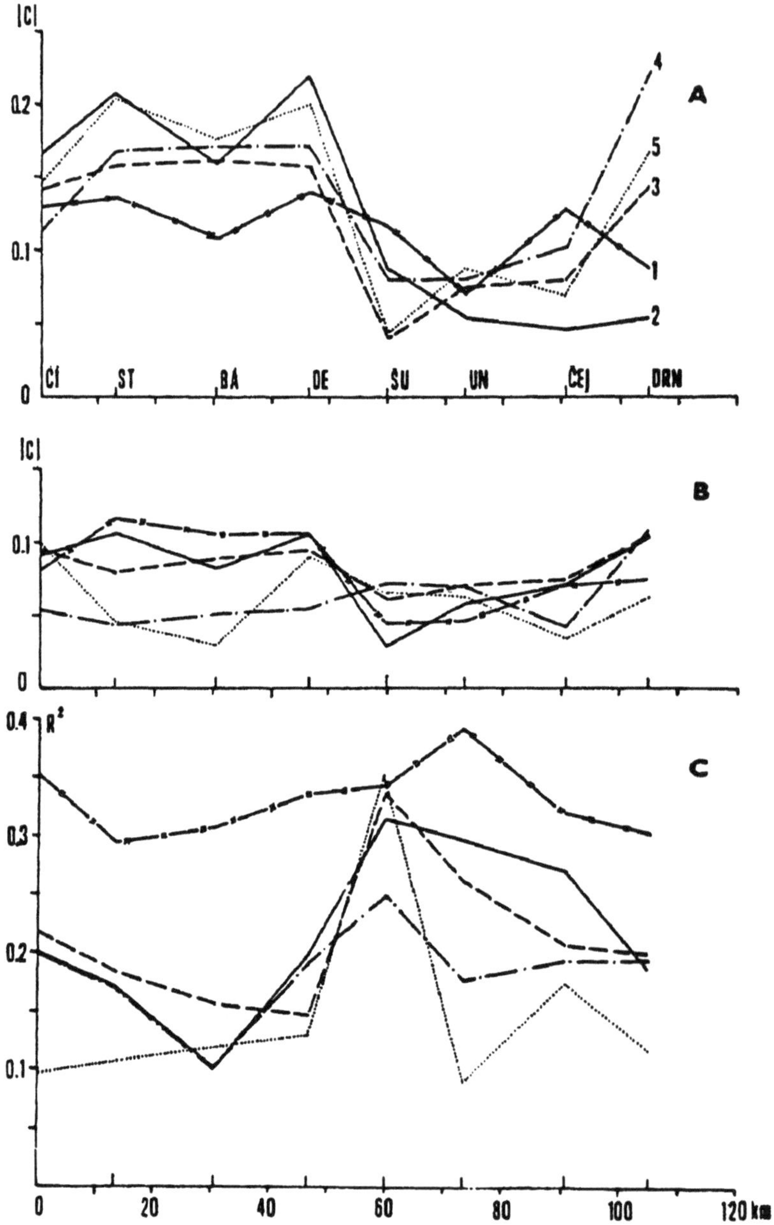

Fig. 5.10 Moduli |C| of in-phase (part A) and of out-of-phase (part B) vectors W_R, W_I at individual stations and magnitudes of R^2 defined in the text (part C) for stations along profile P–83. The figures with the curves in the part A mark the ranges of periods (in minutes): 1 – <96,98>, 2 – <64,65>, 3 – <48,49>, 4 – <36,38>, 5 – <32,33>.

The induction characteristics obtained by applying the MCA and TDA methods to process the variation data substantially widen our information on anomalous features of the region. In Fig. 5.10 (parts A and B), the module $|C|$ of in- and out-of-phase vectors are plotted for individual stations on P–83 and for periods of 98, 65, 45, 38, and 23 minutes. A sudden decrease of in-phase vector magnitudes between Dešov (DE) and Šumná (ŠU) stations is obvious. Minimum modules remain also at the Únanov (UN), and partly also at the Čejkovice (ČEJ) stations. The out-of-phase vectors (Fig. 5.10, part B) yield an analogous pattern although with smaller magnitudes.

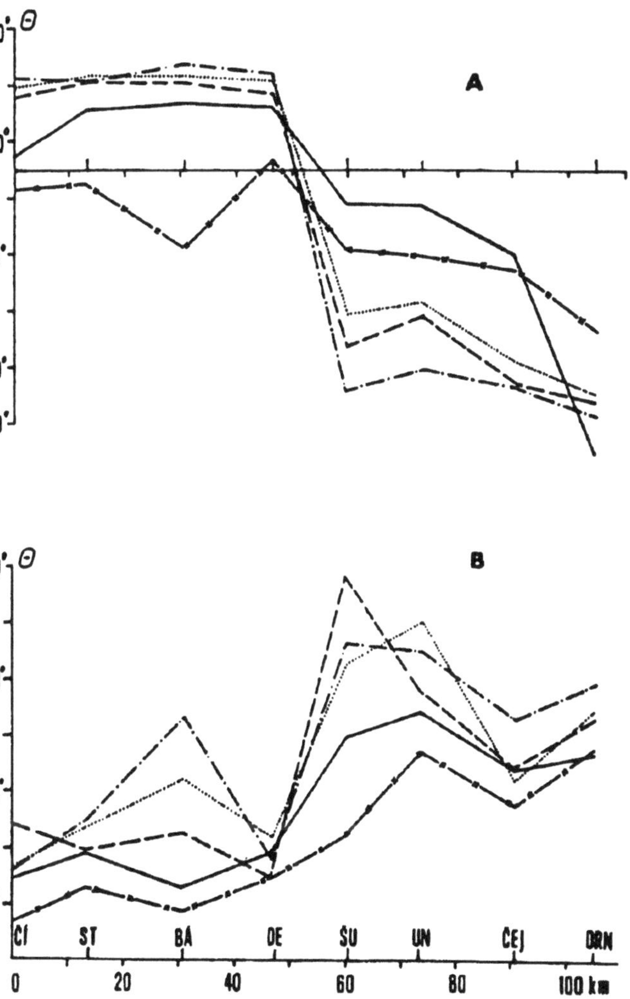

Fig. 5.11 Arguments of in-phase (part A) and out-of phase (part B) vectors W_R, W_I at stations along the profile P–83.

In the lowest part C of Fig. 5.10, the quantity R^2 (see Chapter 5.1) is plotted for stations of profile P–83. Excluding the graph for the longest period from our consideration, we find pronounced maxima of R^2 roughly at the ŠU station and diminishing magnitudes at stations further to SE. Consequently, the residual part of the Z component, which cannot be linearly associated with the horizontal components H, D, seems to increase substantially at observing sites within the anomalous zone. This supports, in spectral terms, the fact that the anomalous zone is also distinguished by maximum errors in determining the parameters of induction vectors. Pronounced changes of in- and out-of-phase vector azimuths in Fig. 5.11 (parts A and B) also clearly define the anomalous zone.

Analogous results were also obtained for other profiles. For the sake of brevity we do not present them here. We complete our survey of induction characteristics by illustrating the possibilities of the inter-station vectors in delimiting the anomalous zone. We focus our attention only on the vectors associating the vertical component Z of the field at each individual station with all field components at the reference station (the Budkov observatory). The calculations were performed using the both methods and Fig. 5.12 shows an example of results obtained by applying the MCA method. Analogous information is also provided by difference induction vectors (the differential

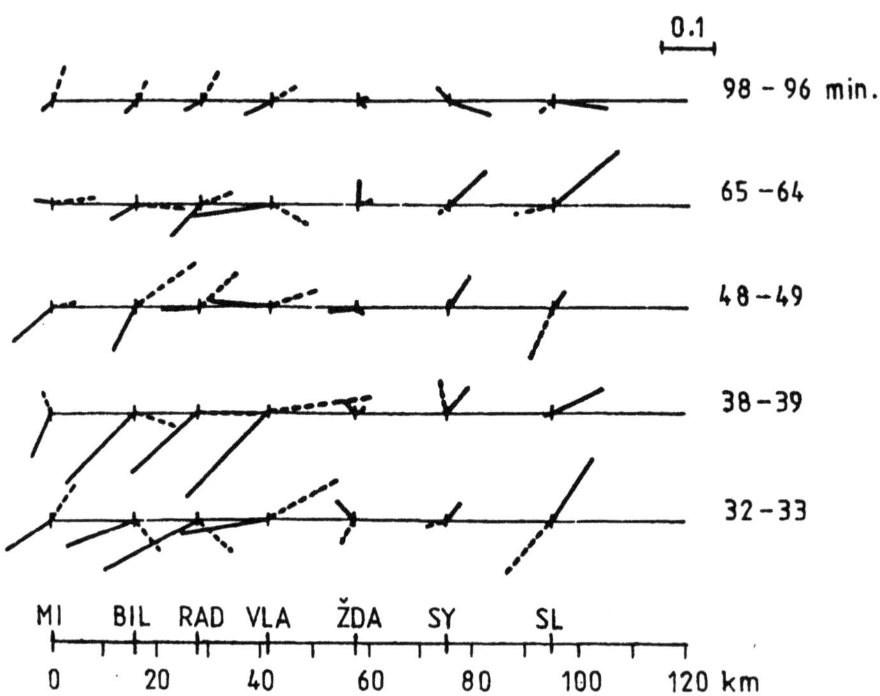

Fig. 5.12 In-phase WZ_R (full) and out-of-phase reference-station vectors WZ_I (dashed) for stations along profile P–82 (MCA results; Budkov observatory as the reference).

field of the Z–component at a field and the reference station is associated with the horizontal field components at the reference station). Such vectors estimated by the TDA method for a wide series of periods are shown in Fig. 5.13. Both types of these characteristics refine the localization of the anomalous zone.

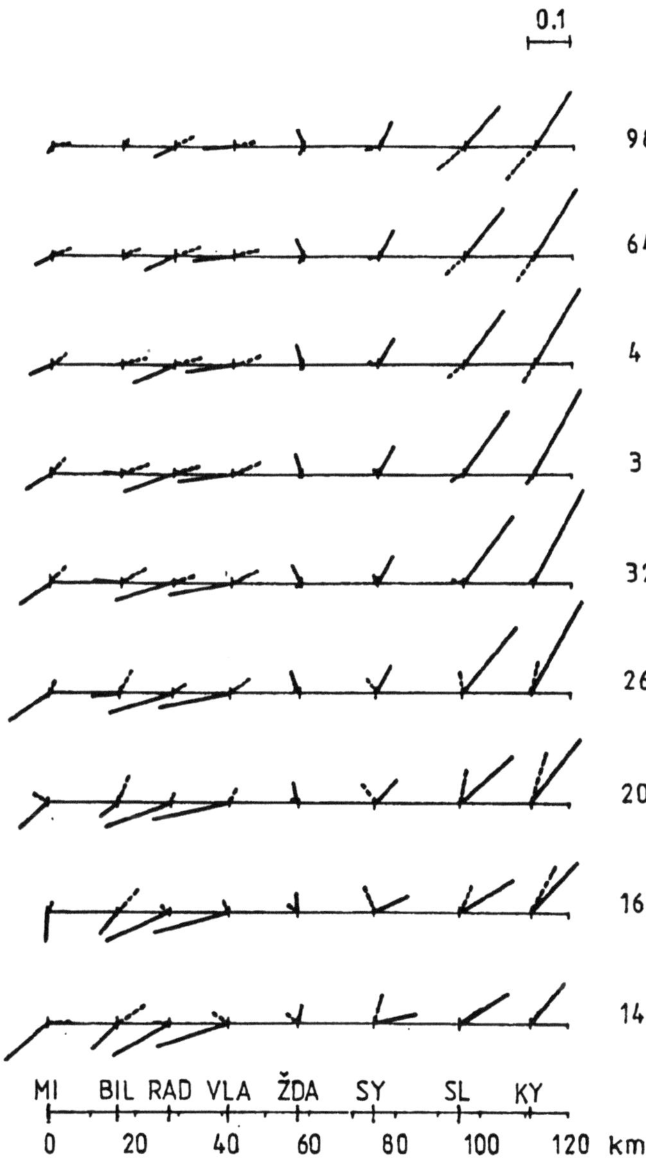

Fig. 5.13 In-phase DWZ_R (full) and out-of-phase DWZ_I (dashed) reference-station vectors of differential fields at the stations along profile P–82 (TDA results; Budkov observatory as the reference).

In estimating the depths and locations of internal sources generating the anomalous fields, we exploited the methods explained in Section 5.1. Assessing the reliability of the results, we have to take into the account the fact that all methods assume a 2-D character of the geo-electrical inhomogeneity (a line current approximation). In our case this assumption is obviously violated. The spatial distribution pattern of the Wiese vectors (Fig. 5.7) over the region under investigation indicates clearly a 3-D character of the electrical inhomogeneity. This is why the results of subsequent depth estimates should be considered only as crude approximation.

By applying systematically the D–P method (field separation into internal/external parts) to the data of each specific profile traversing this region, we obtained the sets of intersection points of the normals to internal field vectors from different points of profile (an example in Fig. 5.1). Averages and their standard deviations were estimated for all sets of the horizontal and vertical coordinates d and h corresponding to the distance from the starting point of a specific profile and to the depth of the electrical inhomogeneity beneath the Earth's surface, respectively. The least squares lines were fitted to the sets of averaged coordinates. In this way we aimed at assessing a possible dipping tendency of an anomalous field source. Some sets, specifically those of profiles P–82 and P–83, show rather great scatter.

The standard deviations for the horizontal coordinate d range up to 11 %, and for depths h up to 22 %, with the exception of profile P–85, where the points are distributed in a special way, as we shall show later.

The results of estimating the depths h and the positions d on the profiles are summarized in Table 5.1 for two variants of introducing a correction for the regional horizontal field H. The first variant uses H_Φ as the average of values at both ends of the

Table 5.1 Anomalous field source positions estimated by the D–P method

profile	correction with H_Φ		correction with H_0	
	d [km]	h [km]	d [km]	h [km]
P–83	63.3 ± 7.3	16.8 ± 3.8	62.6 ± 7.0	19.0 ± 3.9
P–81	66.9 ± 4.1	19.7 ± 2.3	65.9 ± 3.8	23.8 ± 1.6
P–82	66.4 ± 6.5	22.5 ± 4.4	65.7 ± 6.3	25.1 ± 4.4
P–85 left branch	31.9 ± 8.2	21.2 ± 7.2	32.6 ± 6.1	20.9 ± 7.3
P–85 right branch	62.6 ± 5.8	14.8 ± 8.1	58.6 ± 9.7	13.8 ± 6.5
joint of intersection of the branches	51.0	30.0	41.0	25.5

profile, the second variant takes the value H_0, i.e. the starting point of a particular profile for the regional field.

The depth estimates of the internal anomalous field source were supplemented on all profiles with the results obtained by the F–S method and with estimates from the parameters of the characteristic function for a particular profile. The estimates are within the limits shown in Table 5.1.

Summing up the interpretations of MV measurements on profiles across the eastern margin of the Bohemian Massif, we emphasize the following facts:

1. Analysing the variations of the geomagnetic field components at 50 field stations along profiles P–81, ..., P–86, traversing the eastern margin of the Bohemian Massif, we obtained sets of single- and inter-station in- and out-of-phase induction vectors. Total or incomplete reversals of their azimuths and substantial changes of their magnitudes mark the presence of anomalous fields generated by internal geoelectrical inhomogeneity. The anomalous zone seems to be a characteristic feature distinguishing the contact of the Bohemian Massif with the Moravian–Silesian geological units and the West Carpathians.
   Contrary to the Carpathian geoelectrical anomaly, this zone is not marked by a complete reversal of the Wiese induction vectors (change of azimuth by 180°), but only an incomplete change of azimuths by 90–120° is developed. The change and a new vector orientation is observed at 2 – 3 stations on the majority of profiles.
2. The zones of reversals, defined by induction characteristics, characteristic functions and the results of D–P and F–S methods, and recently computer generated maps of the TF1 and TF3 components suggest an anomalous belt striking regionally SW/NE–wards from the Austrian – Czechoslovak border across the Moravian–Silesian sector towards the Czechoslovak – Polish border (Fig. 5.7b), where only the Wiese vectors are plotted with the stations on all profiles, the zone width being marked by hatching.
3. The depth estimates of an anomalous internal field source, obtained by the D–P method are plotted schematically in Fig. 5.14. The source seems to represent a regional contact of two tectogenes, where gradual changes of electrical conductivity of rocks take place. The small stars mark the horizontal coordinates corresponding to D–P results at each specific profile. The double dashed line specifies the centre of the anomalous zone. In spite of some deviations, its position coincides with that in Fig. 5.7 within the accuracy limits. The actual depths are entered in Table 5.1, in Fig. 5.14 only the averages are plotted schematically. It is obvious that the depth values are around 20 km. Segments with profiles define an approximate width of the anomalous zone. On P–83 the zone is about 30 km wide, to NE the width seems to be reduced to about 15 km, and on profile P–82, it is again 26 km. The results from profiles P–85 and P–86 suggest an about 50 km-wide anomalous belt.
   The line segments with arrows at profiles in Fig. 5.14 and the depth estimates suggest a dip of internal source in the vertical plane of the profile. The existence of a boundary dipping to W or NW can be assumed only on profiles P–81 and P–85, while the results from the remaining profiles seem to indicate rather constant depth

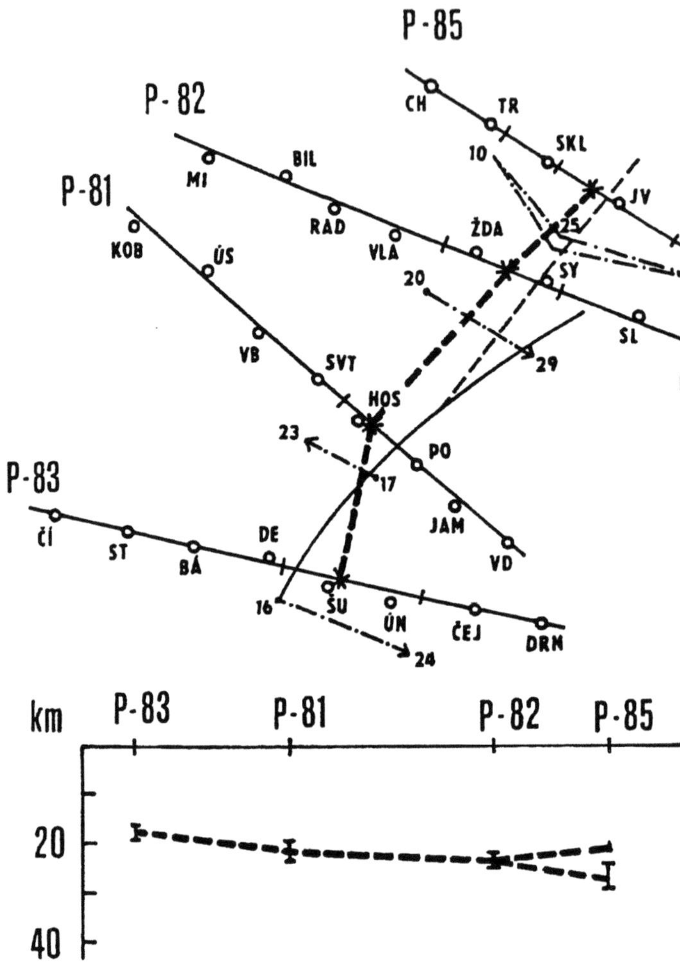

Fig. 5.14 Schematized summary of the anomalous field source depths according to the results of D–P method. Notations: * – projection of the source on the Earth's surface. The thin full and the dashed lines mark the anomaly centre based on earlier estimates, the double dashed line (═══) marks its improved position. Line segments show the anomaly extent on the profile, arrows with numbers point in the directions in which the anomalous field source depths tend to decrease. In the lower part of the Figure the mean depths are schematically shown.

of the source, or even a dip to SE. Thus, only in some regions is it possible to assume the existence of a boundary dipping under the Bohemian Massif. This seems to support geological interpretations assuming the submerging of the Moravian–Silesian region and the Brunovistulicum under the Moldanubicum and the Kutná Hora – Svratka regions (Dudek 1980).

4. Comparing the position and the general trend of the geoelectrical anomaly in Figs. 5.7 and 5.14 with the results of geological investigations, the anomalous zone

appears to mark a region of utmost significance for understanding the geological development and relations between fundamental structural blocks on Czechoslovak territory. The zone seems to delineate a deep-seated boundary of two tectogenes with different histories and consequently with different electrical conductivities. The complex contact seems to be marked by increased electrical conductivity indicating probably the past tectonic activity which might have caused transport and precipitation of conductive matter from carbonate-bearing solutions of deep origin.
The southern and central branches of the anomalous zone are closely related with the Moravian–Silesian lineament of Stille which is a structure separating two different tectogenes. We also compared the distribution pattern of induction vectors and anomalous zones, thus, suggested with the elements of deep geological structure of Czechoslovakia at depth level of – 10 000 m (Suk, Vacek 1984), displayed for this particular region on the map in Fig. 5.7. The southern and central part of the anomalous zone coincides again with the Moravian–Silesian deep -rooted fault zone.
Profile P–82 appears to be a decisive line of block separation. Starting from here, the geoelectrical anomalous zone substantially deviates from the strike of the Moravian–Silesian lineament towards E as proved by the results of MV measurements on profiles P–84 a and P–86 across the Orlice – Jeseník – Opava region. According to these results, the anomalous belt is shown to traverse the Drahany and Silesian culm. In the northern part the belt can be attributed rather to a deep-rooted boundary marked by the belts of the neovolcanites and their conduits in the Šternberk–Horní Benešov zone which is assumed to be a significant linear element in the structure of the Nízký Jeseník Mts. (Mísař et al. 1983). These authors suggest a deep-seated fault zone separating the structurally diverse West-Jeseník synclinorium from the one of the East-Jeseník with the culm deposits.
A close spatial correlation with elements of the deep geological structure may suggest the anomalous zone to mark a specific termination of the Hercynian development in the eastern part of the Bohemian Massif. This problem has been discussed and some variants have been proposed in Mísař et al. (1983). The most acceptable concept seems to be the one that claims all the fundamental zones of the Hercynides in the NE margin of the Bohemian Massif to bend into SSW directions. As yet, the Wiese vector estimates on profiles P–84a and P–86 have provided no indication of that.

5. The complex induction vectors (Fig. 5.4) and a strong eastward trend of the TF contours in the central part of the Moravian–Silesian zone suggest the discontinuous character of the anomalous zone and, therefore, a possible displacement of the internal field source along the Labe lineament (Haná fault zone) towards the SE (Fig 5.3). It is assumed that the lineament continues into the WCP region. In view of this assumption, the bend of the Carpathian anomalous zone into the inner Carpathian region in northern Slovakia (Fig. 5.4) can be alternatively replaced by a discontinuous model assuming a SE displacement of the anomaly source in the region of Považská Bystrica.

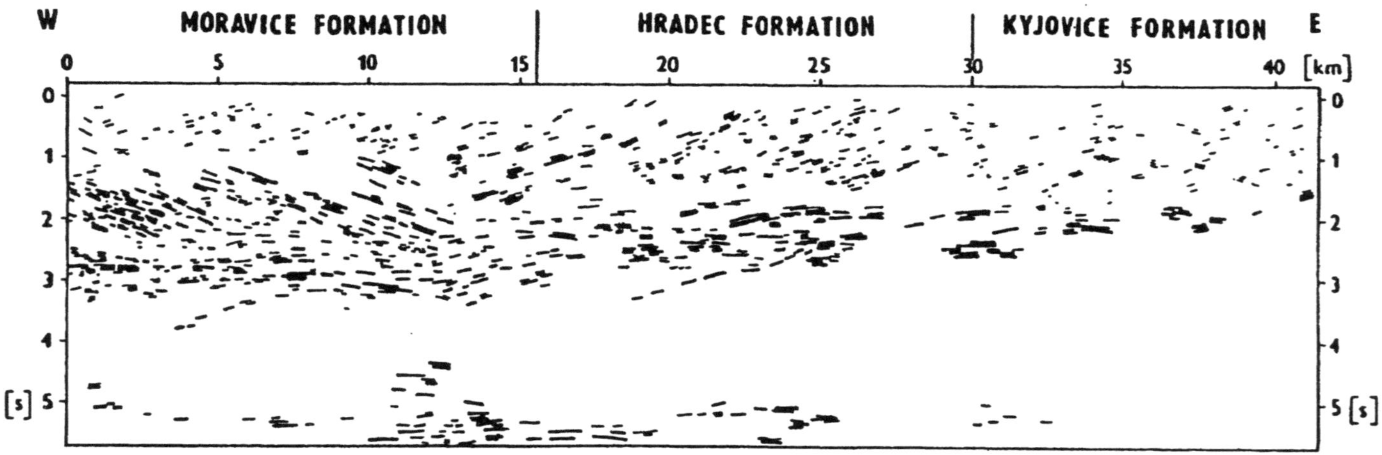

Fig. 5.15 The results of research into the Earth's crust structure obtained by deep seismic reflection profiling (the CDP modification with prolonged period of recording time) along the profile Poruba – Horní Benešov (see Fig. 2.3)

6. The heavy solid contours highlighting the positive values of the TF1 component (Fig. 5.5) seem to define an independent block between the two principal structural units (the BM and the WCP). The block coincides remarkably well with the Brunovistulicum, a unit which has been proved by recent geological/geophysical investigations and their interpretation in terms of the deep structure and by geochronologic studies to be a separate geological unit, very likely belonging to the Fenno-Sarmathian (East European) platform. The distribution of internal geomagnetic fields, therefore, seems to support this interpretation.
7. A remarkable fact is provided by comparing the results of MV measurements with the results of investigations into the deep structure of the Earth's crust by the method of deep reflection seismic profiling (the CDP technique with recording time extended to more than 6 s) (Blížkovský et al. 1986b). The results of measurements, conducted along a 42 km-long profile 6/84 extending in the W–E direction between the towns of Poruba and Horní Benešov, are shown in Fig. 5.15. The profile runs practically parallel with the MV profiles P–84a and P–86 (compare Fig. 5.1). The profile traversed the Kyjovice layers, the Hradec (near Opava) layers and eastern part of the Moravice formation (the Lower Carboniferous of the Nízký Jeseník Mts.). The results reveal very well a fine scaly structure of the culm, with eastern vergency of the reflections. The intensity of the scaliness increases towards the west. In the Moravice formation at the depth of 5 – 8 km within the layered sections the dip of reflections changes towards the east. It cannot be excluded, however, that they could correspond to the overthrust submerged Devonian folds of the Jeseníky Mts. It is worth noticing that the region where the change of the reflection vergence takes place and where strong reflections from the underlying crystalline unit appear at times of 5 – 6 s (i.e. at depths of 15 – 18 km) agrees very well with the central part of the anomalous geo-electrical zone. It seems probable that structural phenomena are accompanied with a significant change of the electrical conductivity.

5.3 Magnetotelluric Models of Inhomogeneity Zones

V. Červ[1], J. Pek[1], B. Pícha[2], O. Praus[1], M. Tobyášová[1]

Models of the internal geoelectrical structure will be addressed in this Chapter, specifically those for SE Moravia, where numerous MTS measurements were performed. In the eastern margin of the Bohemian Massif, distinguished also by an anomalous geoelectrical belt, the MTS measurements are sparse. Therefore, no modelling has been started here, as yet.

In connection with the interpretation of deep electromagnetic soundings numerous authors (e.g.. Berdichevski and Dmitriev 1976) have shown the distortion effects of surface layer inhomogeneities on the results of studying deep levels of the Earth. For

1 Geophysical Institute, Boční II/1401, 141 31 Praha 4, Czech Republic
2 Geofyzika Brno, Ječná 29a, 612 46 Brno, Czech Republic

further interpretation of the MTS results obtained in Czechoslovakia, namely of the data obtained along the international DSS profile No. VI, the effect of the surface layer had to be evaluated. Therefore, a map of electrical conductance over Czechoslovak territory was constructed. In the first step, we concentrated our attention on the transition zone between the Bohemian Massif and the West Carpathians, in the next step we extended the data collection across the whole region of Bohemia and Moravia. Slovak territory has not been investigated, as yet. The integrated longitudinal electrical conductance S (hereafter given in Siemens units) was estimated from the electrical resistivities recorded by geophysical logs. With hydrologic logs for which no electric logging data were available, the electrical resistivities were taken from laboratory measurements of rocks. The conductance S was estimated according to the relation:

$$S = \sum_{i=1}^{n} \frac{h_i}{R_i} \quad ,$$

where n is the number of overlying layers with thicknesses h_i and resistivities R_i. Quantity S corresponds to the actual conductance of the overlying sedimentary complex for those boreholes which have penetrated into the crystalline basement.

In the Bohemian Massif including its contact zone with the West Carpathians, 594 boreholes were selected for estimating the conductance S. In those regions, where boreholes were close together, the mean value of S was determined and attributed to one of the boreholes. When a particular borehole of a group penetrated into the crystalline, its S value was obligatorily accepted for the map. In this way, the number of input data for the map was reduced to 295, 107 boreholes being situated in the transition zone from the Bohemian Massif to the West Carpathians, and the remaining 188 being distributed throughout the whole region (Praus and Tobyášová 1992).

The sets of S estimates were interpolated and contour maps of conductance S were computer-generated using the programs for the geomagnetic data processing. One of the maps with the contour density of 25 Siemens is presented in Fig. 5.16. Owing to the fact that the boreholes are distributed unevenly across the area under investigation, with highest density in regions where deposits of economic raw materials can be expected, the contours tend to form closed patterns around the clusters of boreholes. Consequently, it is difficult to assess quantitatively the accuracy of contour maps.

Despite this, some features in the map of S contours seem to reflect important changes of the thickness in the sedimentary cover. Zero values prevail in the southern province of the Bohemian Massif, where a thin layer of sediments or none at all cover the crystalline basement, while the Bohemian Cretaceous Basin is characterized by conductances ranging up to 100 Siemens. A characteristic contour pattern is developed across the transition zone between the Bohemian Massif and the West Carpathians. The increase of conductance S ranges from 600 to 1 000 Siemens, the values reach about 320 Siemens in the central and about 300 Siemens in the northern parts. Further, the conductance S tends to decrease towards NW due to the effect of emerging crystalline

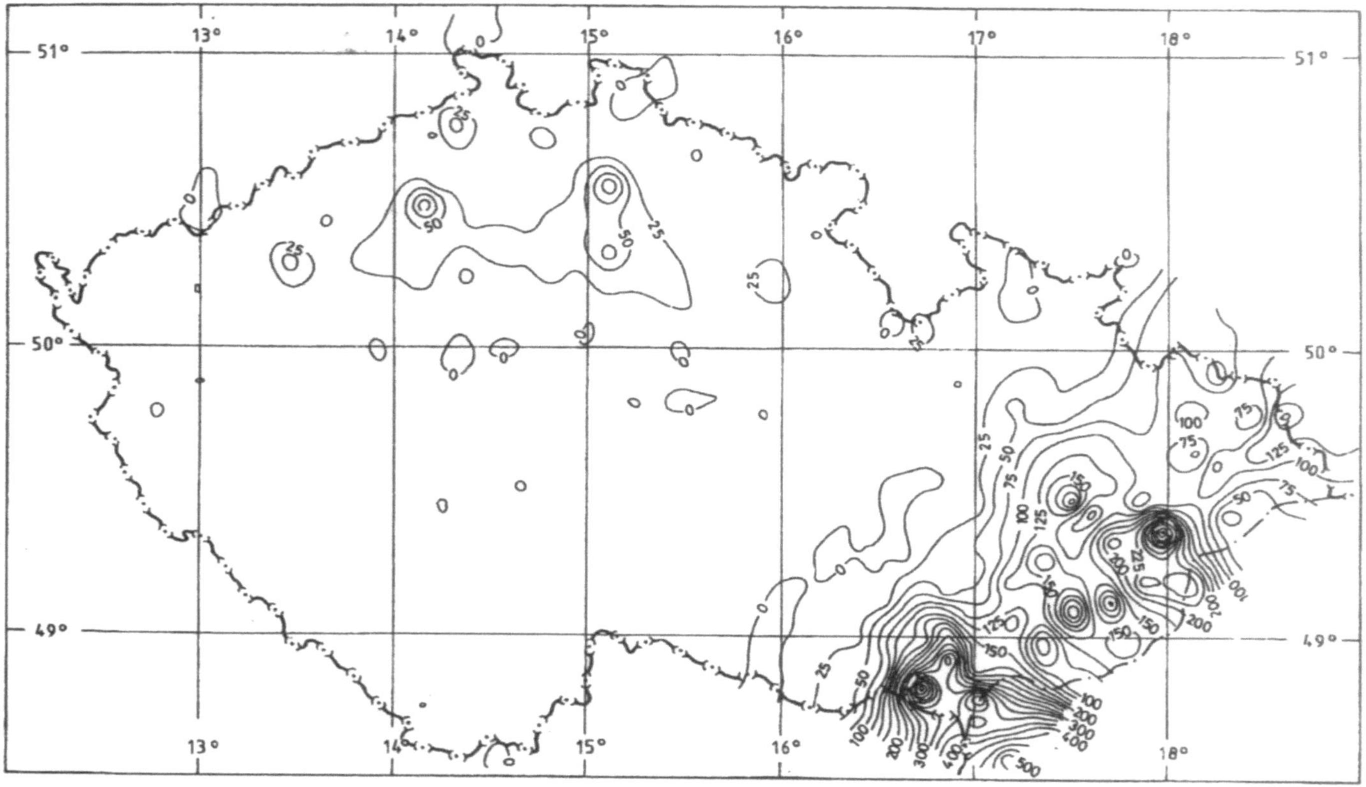

Fig. 5.16 Map of integrated electrical conductance across the Czech republic – contour interval of 25 Siemens.

basement (the Bohemian Massif). The spatial distribution of S contours seems to map also the basement topography of deep basins (the Nasvačilský trough marked by a maximum value of S exceeding 750 Siemens), and further to NE two depressions with maximum values around 300 Siemens. It is obvious from the contour pattern that these depressions are clearly divided by ridges striking across the region in the azimuths parallel to the Labe lineament. The most prominent one runs through the central part of the transition zone (the Přerov – Rusava elevation).

The conductance map served further as a basis for modelling geoelectrical structures on Czechoslovak territory on a regional scale (Červ et al. 1987). In particular, the map has been used for detailed modelling which can yield a closer insight into the geoelectrical structure of the transition zone between the Bohemian Massif and the West Carpathians. Within the zone, the MTS results were available at several sites distributed

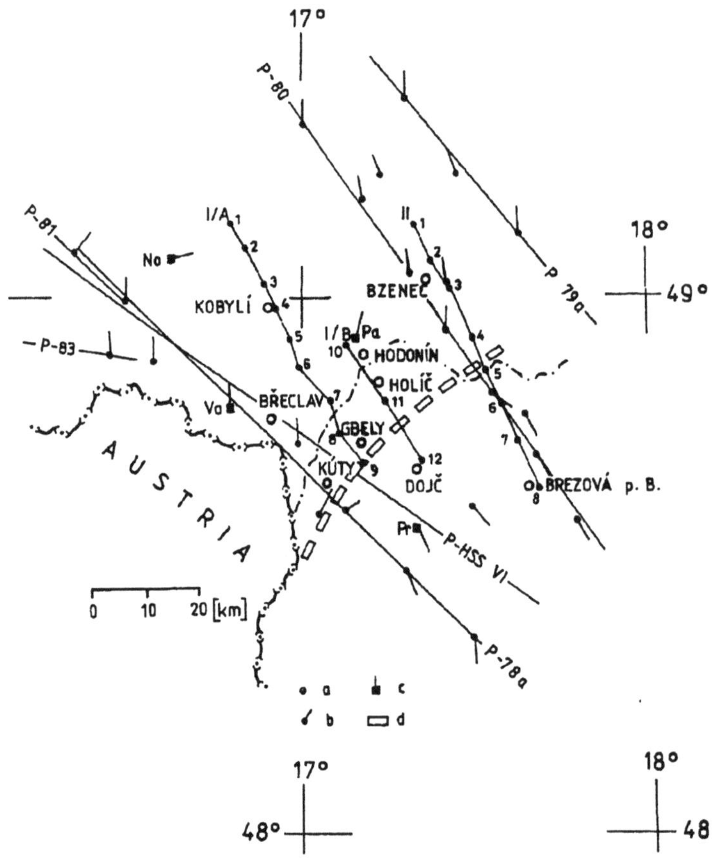

Fig. 5.17 Positions of the MTS sites: a) MTS sites providing intermediate depth estimates) b) MTS sites with induction vectors, c) MTS providing deep structure estimates, d) axis of the Carpathian geoelectrical anomaly. Further notations: P–DSS VI – international DSS profile No. VI. MTS profiles – I: Těšany – Smolinské (point I/A – 1 – deep borehole Těšany I),– II: Osvětimany – Brezová pod Bradlom (point II – 1 – deep borehole Osvětimany I).

along two main profiles with deep boreholes Osvětimany–1 and Těšany–1 and along a parallel auxiliary profile. The variation data recorded were within the range of periods 0.1 – 200 s. All MTS localities were situated in the vicinity of the DSS profile No. VI along which the MTS measurements in the wide range of periods from about 1 s to several hours were performed earlier and the data interpreted (Pěčová et al. 1976; Červ et al. 1984). The location of the profiles and of the sounding sites is shown in Fig. 5.17. The southern branch of the Carpathian geoelectrical anomaly also traverses this region (Figs. 5.2 and 5.3).

In the first step, the directional geoelectrical characteristics of sedimentary layers were analyzed by applying the method of thin sheet approximation (Červ et al. 1987). The integrated conductance S in the southern part of the transition zone between the Bohemian Massif and the West Carpathians (Fig. 5.16) were the input data. Employing a special numerical method by Yegorov et al. (1983), solutions were obtained for two perpendicular polarizations of the electric field on a rectangular mesh with dimensions $10 \times 10\ \text{km}^2$ and generally with different values of S in the mesh cells. The solutions

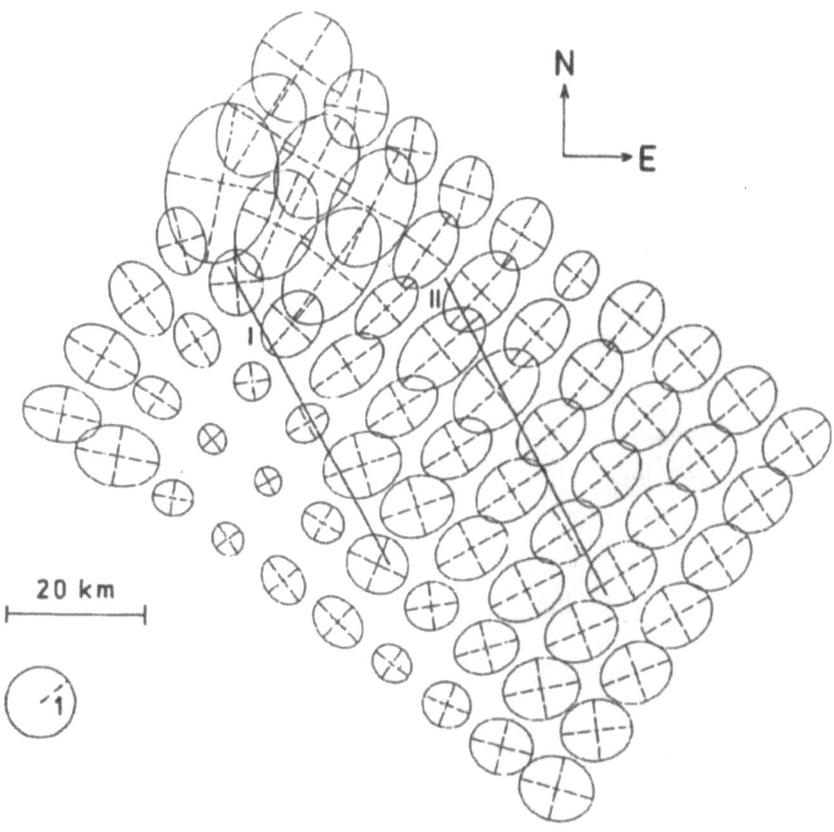

Fig. 5.18 Telluric ellipses obtained by modelling the distortion effects of the sedimentary cover by Price's thin sheet approximation, in the neighbourhood of magnetotelluric profiles I and II.

represented by a system of polarization ellipses in Fig. 5.18 show that the minor semi-axes of all ellipses are oriented in NW–SE directions on profile II. A more complicated situation seems to occur on profile I, where the NW–SE orientation can be observed only in SE part of the area. In the NW part, the orientation seems to be affected by a local, (from 300 to 700 Siemens) maximum of conductance S. Despite the approximate character, the thin sheet modelling confirms that the NE–SW orientation of geoelectrical characteristics prevails over the area in accordance with the geological conclusions concerning the deep structure of the medium.

The geoelectrical directional characteristics of the crystalline basement were estimated by analysing the MTS polar diagrams and, on profile I, also by investigating the magnetic polar diagrams (range of periods 1 – 900 s). The minimum of magnetic polar diagrams, corresponding to the homogeneity axis of the medium, is located approximately in the E–W direction (Fig. 5.19). In the same azimuth also the maxima of

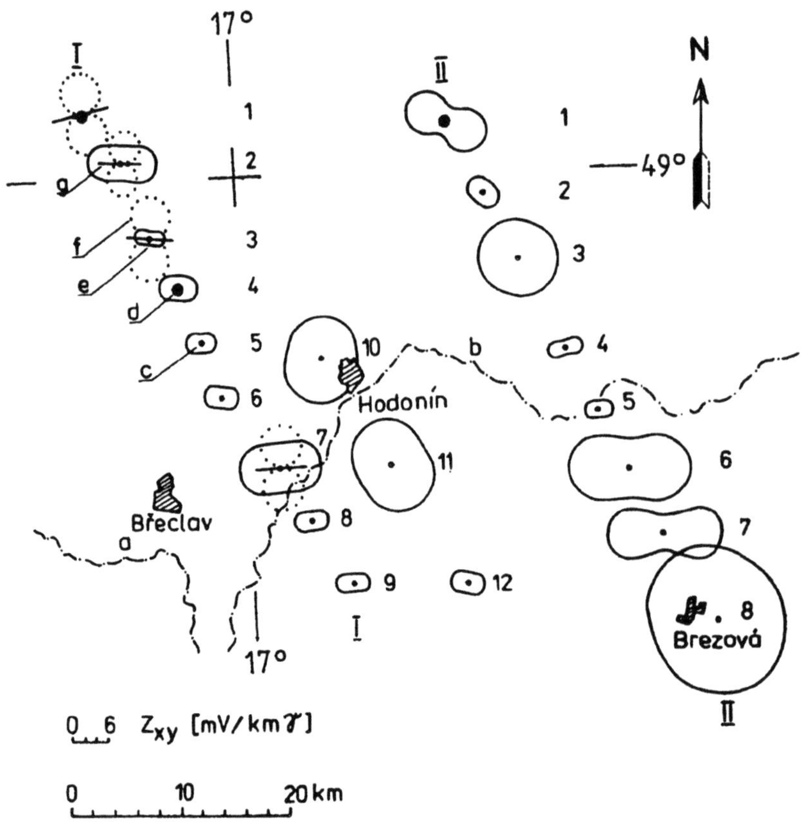

Fig. 5.19 Positions of magnetotelluric stations and notations: a) Czechoslovak – Austrian border, b) Czech – – Slovak Republic border (The Moravian – Slovakian border), c) magnetotelluric sounding sites, d) deep borehole, e) impedance polar diagram $|Z_{yx}(\alpha)|$, f) magnetic polar diagram $|W_x(\alpha)|$, g) homogeneity axis of geologic structures, direction of E polarization.

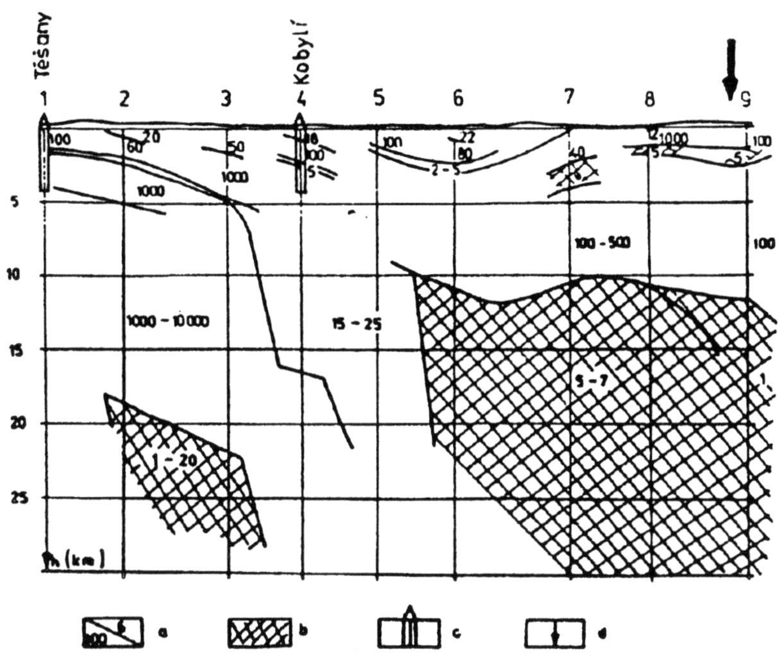

Fig. 5.20 Geoelectrical section suggested from the results of formal 1-D interpretations at the MTS sites of the profile I/A. a) contact of blocks with different conductivities, b) sedimentary and crystalline rocks with resistivities up to 20 ohmmeters, c) boreholes and their penetration depths, d) axis of the Carpathian geoelectrical anomaly.

impedance polar diagrams are observed. These characteristics suggest that the MTS apparent resistivity curves ρ_{yx} correspond to the *E*– and ρ_{xy} to the *H*–polarization, respectively. The directional characteristics comply well with the geological conclusions claiming that the crystalline complexes were strongly affected by the E–W tectonics.

The induction vector orientations estimated from the MV measurements in the transition zone between the Bohemian Massif and the West Carpathians for periods from 900 to 3600 s, are confined to the sector of azimuths between NW–NNW and SE–SSE (see Fig. 5.3). Consequently, the homogeneity axis of the structure seems to strike NEE–SWW- ward. Contrary to the superficial sedimentary complexes, a slight rotation of the homogeneity axis takes place at deeper levels of the structure.

Analysing the directional characteristics, we find that, despite indications for local 3-D inhomogeneities of the medium, the geoelectrical structure of this region can be approximated reasonably well by models of laterally inhomogeneous structures (2-D inhomogeneous medium). Exploiting the results of formal 1-D interpretation of the MTS curves at individual sounding sites on the profiles and taking into account the recent geological/geophysical facts available, we suggested several 2-D models of the internal geoelectrical structure, solved for them forward problems and computed their electromagnetic responses by numerical methods. The responses were then compared with experimental results and in the next step modified by solving the MTS inverse problem. Here, we shall present the results of analysing the MTS data on profile I (Těšany – Smolinské). The direct MTS problem was solved for the model shown in Fig. 5.20 which was further schematized to adjust it for numerical calculations. From the results of calculating the model response, we constructed MTS curves for periods of $T = 10$ s and 100 s. Their comparison with the experimental results showed a principal disagreement. The experimental apparent resistivities for the *H*–polarization are substantially lower than those corresponding to the *E*–polarization, but the results of modelling behave conversely. Calculating and analyzing the responses for numerous structures, we concluded that the disagreement can be removed by embedding the structure into a relatively conductive block, the broader surroundings of which exhibit higher resistivity. Such a model corresponds well to actual situation on the DSS profile No. VI (Fig. 5.21). The Bohemian Massif in the NW and the West Carpathians in the SE simulate those model blocks with high resistivity and increased conductivity, respectively and the whole transient zone between the Bohemian Massif and the West Carpathians represents a highly conductive block with embedded local structures. The results in Fig. 5.22 show an abrupt decrease of apparent resistivities for *E*–polarization in this region. Additional results for *H*–polarization [not presented here but apparent from the MTS model curves in the paper by Červ et al. (1984)] prove the apparent resistivities to be even smaller than those for *E*–polarization for periods longer than several seconds.

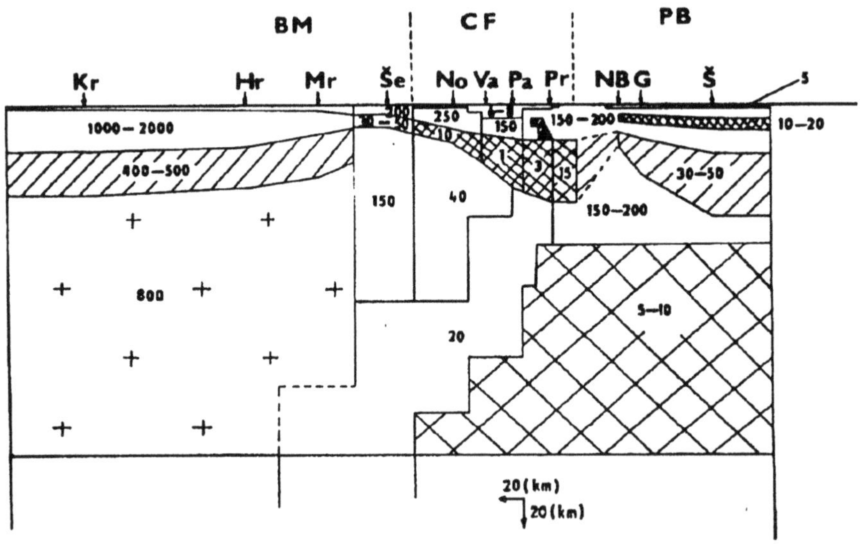

Fig. 5.21 The geoelectrical model in the cross-section along DSS profile No. VI based on the MTS and MVS results. Notations Kr, etc. mark the MTS localities. The principal structural units of Czechoslovakia are shown: BM – Bohemian Massif, CF – Carpathian Foredeep (the transition zone), PB – Pannonian block. Numbers denote resistivities of blocks in ohmmeters.

Though this structure improved agreement between the experiment and the model results substantially, severe discrepancies still persisted, namely in the NW of the profile. Here, the measurements suggest maxima and minima of apparent resistivities at relatively short distances (about 10 km), specifically at sounding sites No. 2 and 3 of the profile, while the model results do not. To explain these features, we analyzed numerous 2-D inhomogeneous models with variety of conductive bodies embedded. In particular, a structure was synthesized with a highly conductive vertical dyke emerging practically at the Earth's surface between the sites No. 2 and 3 (Fig. 5.23). The electromagnetic response of this specific model seems to express relatively well the behaviour of the apparent resistivity curves for the *E*–polarization at periods between 10 and 100 s (Fig. 5.23). The structure is in a general agreement with geological features assumed in the internal structure of the province. The position of the vertical dyke could correspond to the Nasvačilka trough.

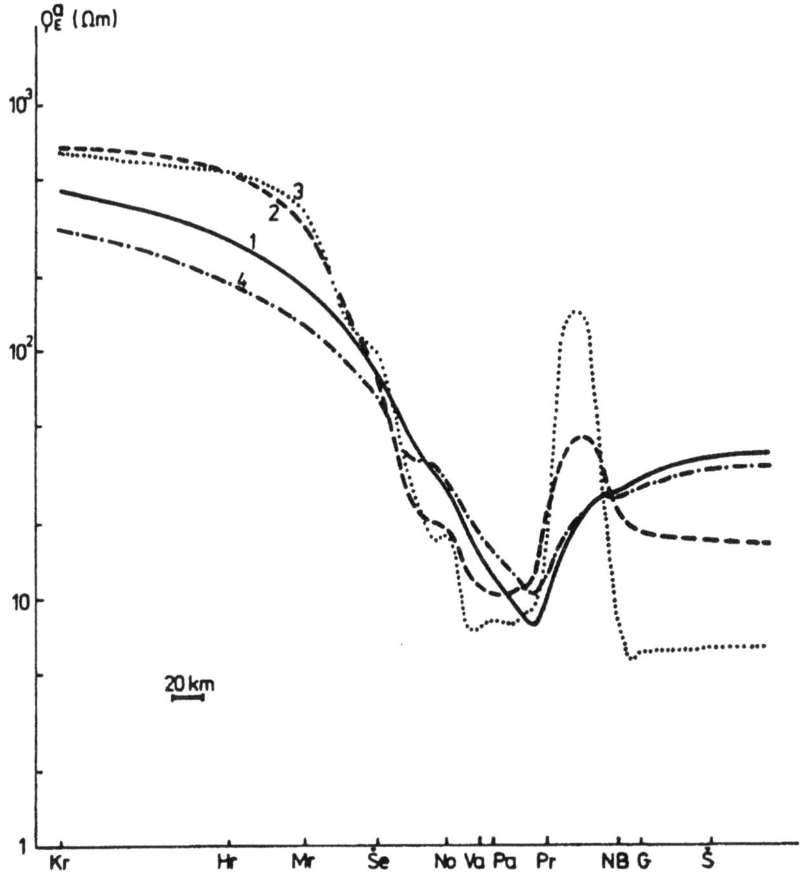

Fig. 5.22 Computed apparent resistivities (*E*–polarization) for the model from Fig. 5.20 for the following periods: 1 – 10 s, 2 – 100 s, 3 – 900 s, 4 – 1 800 s.

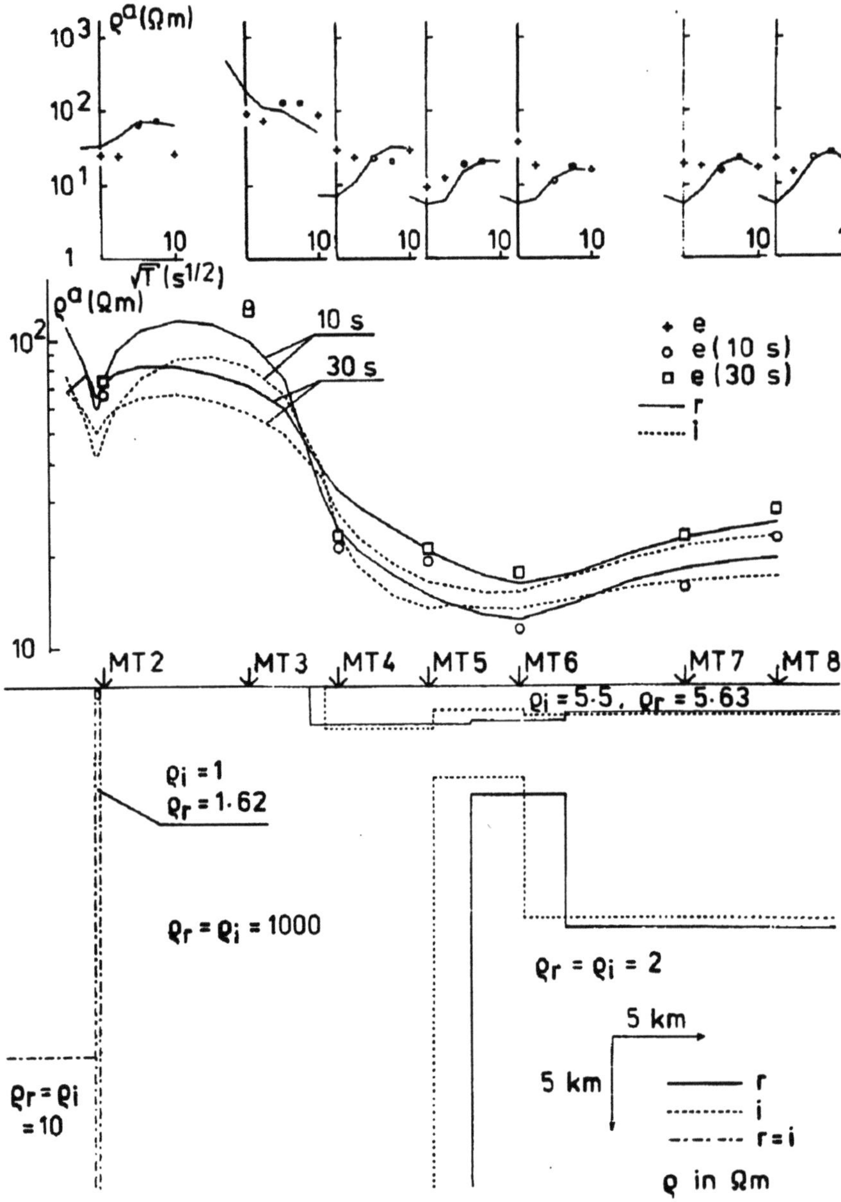

Fig. 5.23 Simplified 2-D geoelectrical model beneath the magnetotelluric stations 2–8 situated along profile I (Těšany–Smolinské) used for 2-D numerical modelling. The apparent resistivity curves obtained along the profile (for two periods T = 10 s and 30 s) at individual sounding sites (for two periods T = 10 s and 30 s), e - experimental results, i – initial approximation, r – the result of the inversion obtained by applying the Marquart's algorithm.

The MTS data along the profile Osvětimany – Brezová pod Bradlom were interpreted using an analogous approach and the results were published earlier (Pícha et al. 1984). It can be claimed that the geoelectrical cross-sections at deeper levels do not differ basically along the two profiles. In both sections the conductive blocks correspond to the Carpathian geoelectrical anomaly in the SE section of the profile I and in the central part of profile II. There are some differences between the sections in the superficial structures, they seem to be of local character.

5.4 Contribution of the MTS to the Study of Lithosphere in Central Europe

V. Babuška[1], J. Plomerová[1], V. Petr[1], J. Pěčová[1], O. Praus[1]

In most geophysical models of the Earth's internal structure the lower boundary of the continental lithosphere is given by the isotherm of partial melting. This process remarkably increases the electrical conductivity of asthenosphere and makes it, thus, substantially different from non-conductive rocks of the upper mantle. It is possible, therefore, to estimate the depth of the lithosphere/asthenosphere transition (L–A transition) by employing the methods of electromagnetic soundings (EM). The MTS and MVS results of our own were supplemented with a set of geoelectrical models based on the MTS curves and MVS measurements given for characteristic provinces of Europe in the professional literature. The results of the EM data interpretation, i.e. the depth estimates to the top of the layer with increased electrical conductivity, attributed to the L–A transition, together with its electrical parameters (i.e. electrical resistivity, integrated conductance) are summarized in a paper by Praus et al. (1990, Table 1).

On the other hand, a partial melting of rocks decreases the seismic wave velocities. The effect can be detected applying a newly developed method of analyzing the seismic data by calculating the representative mean *P*–wave residuals over a province under investigation (Babuška et al. 1984a, 1987). As shown in detail (Chapter 2.5), the positive residuals characterize regions with a relatively thin lithosphere and vice versa, the negative residuals are typical of regions with a relatively thick lithosphere.

The seismological determinations of the lithosphere thickness derived from *P*–wave residuals are graphed by contours (isohypses) in a map covering Central and parts of SE and Western Europe (Fig. 5.24). In the map, the EM depth estimates of the top of a layer with increased electrical conductivity corresponding probably to the L–A transition with the asthenosphere in the basement, are marked by bold figures with the MTS localities.

Comparing the depths obtained by the two independent geophysical methods, we claim their generally good, mutual agreement across the majority of provinces. In a model obtained by analysing the *P*–wave residuals in the Bohemian Massif, the L–A transition depths are suggested in the range from 90 to 140 km. These estimates are in

[1] Geophysical Institute, Boční II/1401, 141 31 Praha 4, Czech Republic

good agreement with the MTS results at Czechoslovak stations distributed roughly along the DSS profile No. VI, as the layer of increased electrical conductivity, attributable to the L–A transition, is interpreted at depths of about 100 – 150 km (Pěčová et al. 1974, 1976). Reinterpretation of MTS results respecting the distortion effects of near-surface inhomogeneities and the EM response of a 2-D inhomogeneous structure, suggests an alternative model with depths of the upper asthenosphere boundary exceeding 150 km (Červ et al. 1984). Buryanov et al. (1983, 1987) assume, on the basis of their own reinterpretation of our original MTS curves, that the asthenosphere does not exist in the central part of the Bohemian Massif. These controversial interpretations imply that the asthenospheric layer is probably not well developed beneath the Bohemian Massif. Its electrical conductivity is not much higher than that of overlying layers and is, therefore, at the limit of its detectability by EM sounding methods.

An exceptionally good agreement between both the seismological and EM data interpretations were obtained in the Pannonian Basin. The most likely depths of the layer with increased electrical conductivity are between 60 and 85 km in the central part of the depression. Towards its margins the depths increase to 100 km. According to *P*-wave residuals the L–A transition zone is assumed to form narrow belts extending from SW to NE and correlating with the tectonics. The Rába – Rožňava tectonic line corresponds to a belt, where the lithosphere thickness is reduced to 60 km or even less according to the recent results by Plomerová and Babuška (1988).

Analysing the seismic *P*–wave residuals, we find that the regions of thinning lithosphere penetrate from the Pannonian Basin to the West Carpathians in two promontories. The eastern one, extending as far as Užhorod, correlates well with the MTS results yielding conductive layer depths of 80 – 100 km. In the western promontory, the L–A transition zone is, according to the *P*–wave residuals, at depths of 90 – 100 km, while the MTS results yield somewhat higher estimates of about 100 – 120 km for the depths of the layer with an increased electrical conductivity. In the outer West Carpathians and further towards the margin of the East European platform a gradual increase of the lithosphere thickness up to 140 – 180 km is observed. It is also very well documented by the MTS results from the Belorussian platform, the Moscow syneclise, and other regions. There, no conductive layer seems to be developed in the upper mantle, or it is questionable at depths over 200 – 400 km.

From the *P*–wave residuals a tendency of the lithosphere is apparent to thicken from 60 – 100 km in the central part of the Pannonian Basin to 180 – 200 km in the transition zone towards the outer East Carpathians, and in the East European and Moesian platforms. The MTS results along the profiles traversing the East Carpathians confirm this model (Stanica et al. 1986). In the Pannonian Basin they yield depths of 70 km increasing to 250 – 300 km in the platform.

An excellent agreement of the EM and seismological results was found across the Rhenish Massif. The *P*–wave residuals indicate a regional thinning of the lithosphere from 100 – 80 km to only 60 km (the central part of the lithosphere elevation is at about 50°N and 8°E). The results of EM sounding conducted along two profiles in this region also indicate an upheaval of a highly conductive layer to depths between 60 and 45 km

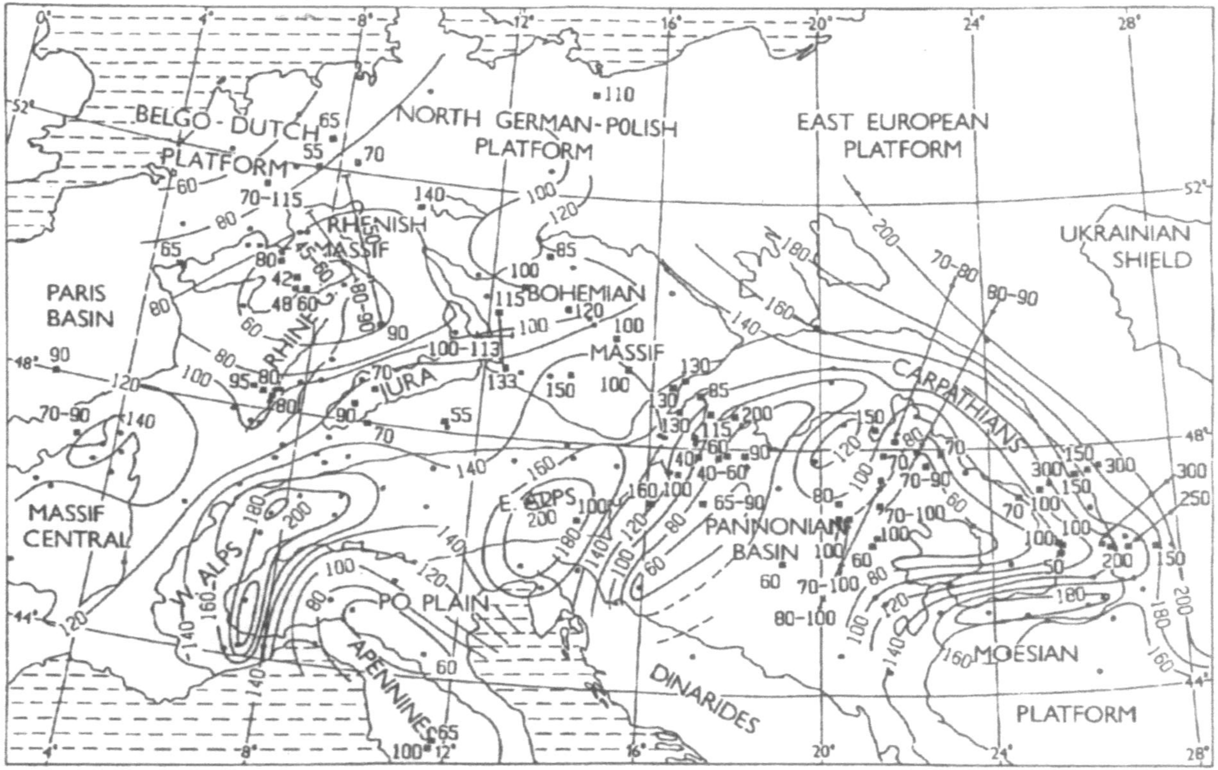

Fig. 5.24 Depths of the lithosphere–asthenosphere transition in Central and Southeastern Europe. The isopath contours constructed at intervals of 20 km are derived from the representative average *P*–wave residuals computed for individual seismological stations (dots). The bold numbers denote depths of the top of the layer with increased electric conductivity determined at individual MTS localities (squares), or along profiles (abscissae in the Rhenish Massif and in southwestern part of the Bohemian Massif). The arrows in the Carpathian region point to the localities, where the depths of conducting layer in the upper mantle were determined and their values had to be placed elsewhere.

(Jödicke et al. 1983). A very good agreement between results has further been detected beyond the western margin of the Bohemian Massif, where the seismological and EM estimates imply a submerging of the L–A transition towards the south (Plomerová and Babuška 1988).

No satisfactory agreement with good seismological results can be claimed from sparse results of EM soundings in the Alps. The EM data are insufficient and their interpretation unreliable due to a strong electrical inhomogeneity of this province, specifically in the area of periclinal junction of the East Alps with the Pannonian Basin.

6 Gravimetric Investigation of Deep Structure of the Earth

6.1 Gravitational Field of Central Europe

M. Blížkovský[1], J. Ibrmajer[1], A. Novotný[1], M. Suk[2], J. Šefara[3]

The synoptic gravity map of Central Europe is presented in Fig. 6.1. The striking negative megastructure in the Alpine – Carpathian region is divided into several independent structures: West and East Alps, West and East Carpathians, West Ukraine, and South Carpathians (in Romania) (Ibrmajer, 1981). While in the Alps and in the West Ukraine the increased Earth's crust thickness contributes to forming the gravity low, a substantial part of the gravity low in the West Carpathians has been explained as due to surficial sources. In the Pannonian Basin, there is a positive gravity anomaly in spite of the significant accumulation (up to 7 km) of low-density Tertiary sediments. The extensive elevation of the Moho discontinuity in the central part of Hungary, whose depths reach around 25 – 27 km, represents its main source.

The negative gravity anomaly of the northern Alpine foreland passes into the Bohemian Massif whose gravitational field is characterized by a zonal arrangement of positive and negative anomalies of NE–SW direction. The gravitational field does not correlate with the morphology of the Moho discontinuity, that can be explained by the presence of density inhomogeneities in the upper part of the crust.

In contrast to the Bohemian Massif, the gravity anomalies of the Polish part of the East European platform are prevailingly oriented in NW–SE direction. Their amplitudes, as well as absolute values, are significantly higher and their structures are more extensive.

Increased values and more undulating gravitational field has been observed in the Saxothuringian and Rhenohercynian regions. While in the Saxothuringian region the Moho discontinuity depths range from 25 km to 30 km, in the Polish part of the East European platform they are in the range of 30 – 50 km. This is why the lithosphere structures are diversified in the East European platform, in the Variscan orogene as well as in the Panonian basin.

1 Geofyzika Brno, Ječná 29a, 612 46 Brno, Czech Republic
2 Faculty of Natural Sciences, Masaryk University, Kotlářská 2, 600 00 Brno, Czech Republic
3 Geological Institute, Comenius University, Mlýnska dolina, 842 15 Bratislava, Slovakia

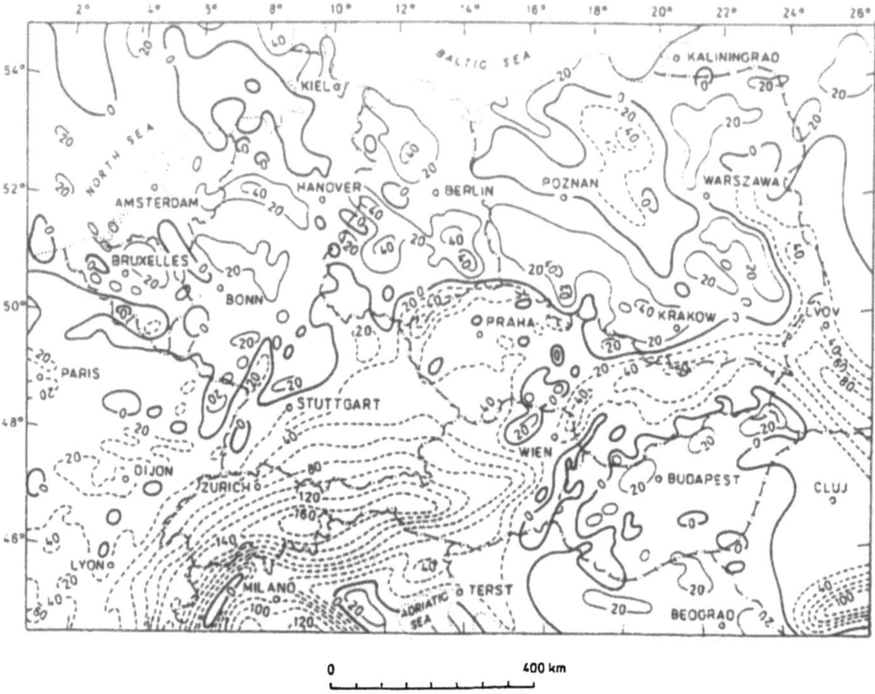

Fig. 6.1 Synoptic gravity map of Central Europe (contour interval $20 \times 10^{-5} m\ s^{-2}$) – Klíma 1965.

6.2 Stripped Gravity Maps in Czechoslovakia

M. Blížkovský[1], J. Šefara[2], M. Burda[3], V. Vyskočil[1]

Constructions of stripped gravity maps initiated the epoch of the geologic filtering of the gravitational field using the principle of "processing from the known to the unknown". For their construction the following assumptions and rules were obeyed:

1. Existence of sufficiently accurate information on the shape and distribution of densities within the body whose gravity geologic correction is calculated,
2. Magnitudes of the gravity geologic corrections of individual bodies are higher than the mean error of the calculation,

1 Geofyzika Brno, Ječná 29a, 612 46 Brno, Czech Republic
2 Geological Institute, Comenius University, Mlýnska dolina, 842 15 Bratislava, Slovakia
3 Geophysical Institute, Boční II/1401, 141 31 Praha 4, Czech Republic

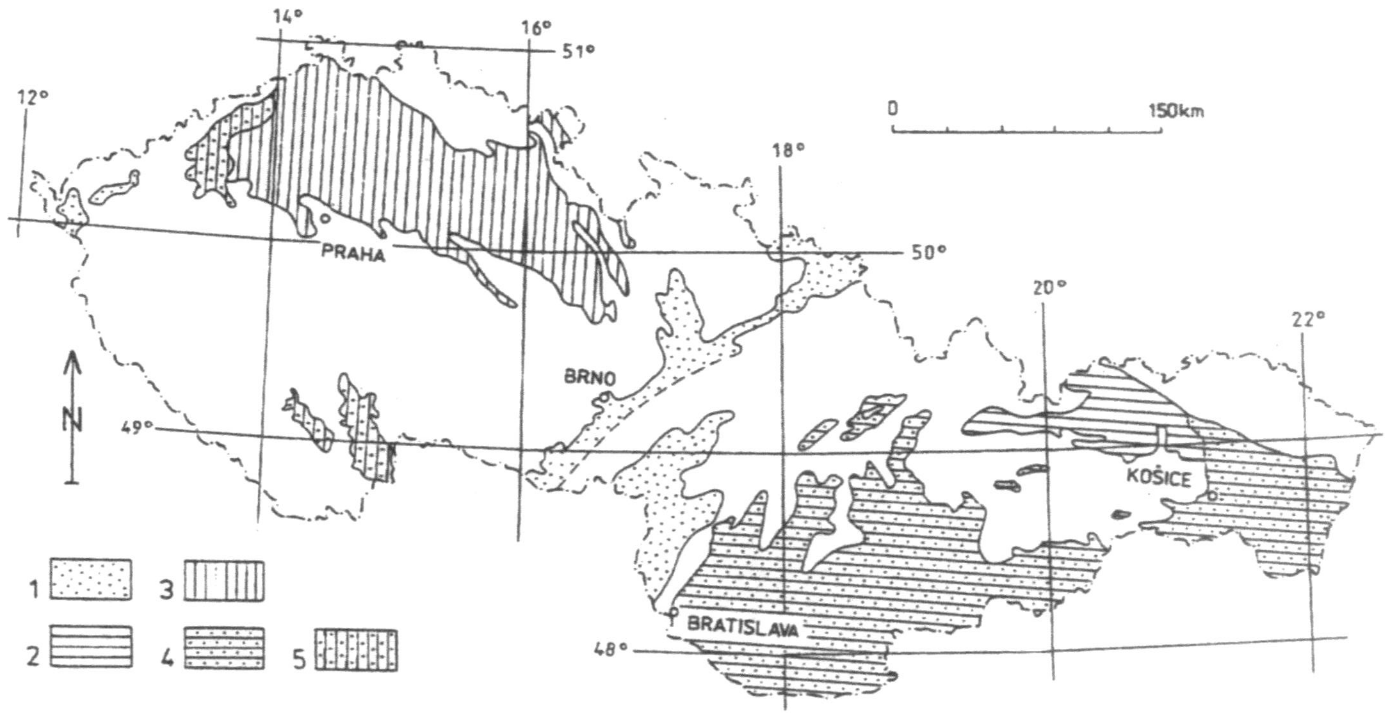

Fig. 6.2 Regions corrected for gravity effects of sediments 1 – Neogene, 2 – Palaeogene, 3 – Cretaceous, 4 – Neogene and Paleogene, 5 – Neogene and Cretaceous.

3. Differential density of the body is derived with respect to the mean basement density 2.67 Mg m^{-3} which also coincides with the reduction density of the Bouguer anomaly map,
4. The correction is being determined at points of calculation at the level of the terrain topography.

Respecting the above rules and using suitable methodology of calculation, a stripped gravity map of defined accuracy can be derived. Such map is able to be used for applications of all known qualitative and quantitative interpretative procedures.

With regard to these requirements the stripped gravity map could be constructed only in the regions of Czechoslovakia shown in Fig. 6.2.

The stripped gravity map of the Bohemian Massif was compiled during 1975–1980 (Blížkovský 1978; Blížkovský and Novotný 1981). In fact, it is the gravity map on a scale of 1:200 000 (Ibrmajer 1963), corrected for the gravity effects of masses representing the sediments of the Bohemian Cretaceous Basin, the system of Neogene basins at the foot of the Krušné hory Mts., the basins of South Bohemia, and partially the Neogene rocks of the Carpathian Foredeep (Fig. 6.2). The corrections were computed for the sedimentary rocks exhibiting densities lower than 2.67 Mg m^{-3}, whose area exceeded 0.5 km^2, and whose expected gravity effect was higher than 5 µm s^{-2}.

The construction of the stripped gravity map required preparatory petrophysical, geologic, and geophysical research. The petrophysical activity consisted of examinations of old reports, preparation of 5000 rock samples, and determination of the rock's physical properties, and compilation of revised maps of mineralogical and bulk densities, and porosities (Ondra and Hanák 1981). During the stage of preparatory geologic work, geometric shapes of density models of individual geologic bodies were constructed. With the use of the petrophysical data, individual bodies were divided into n-sided horizontal "layers" of constant density. The geologic bodies composed of individual layers were limited at their tops by the natural topography, at their bases (with the exceptions of Cretaceous layers of the Most Basin and the volcanic region of the České středohoří Mts.) by the plane of the basement and on their sides by vertical planes intersecting the frames of the map sheets of 1:50 000 of the Gauss-Krüger projection.

The flow-chart of operations for constructing the stripped gravity map is in Fig. 6.3 and the processing programme package is described in Blížkovský (1978). The main input data, i.e. contours of Bouguer gravity, elevation contours, and isohypses of the physical-geologic models were digitized at chosen intervals for individual map areas of 1:50 000, which also represented an archive set. In each particular map, at the grid points of square coordinate network 500 × 500 m, the data was recorded, and partial as well as total gravity effects of the physical-geologic models were calculated. In order to calculate gravity effects of horizontal n-side prisms the formula of Smíšek et al. (1972), which is more effective compared with Talwani – Ewing's formula, was used. The value of the mean quadratic error in the stripped gravity map was determined as ± 11 µm s^{-2}. The map of corrections for the total gravity effects of sediments of the Tertiary and Cretaceous in the Bohemian Massif is shown in Fig. 6.4. The maximum value of the

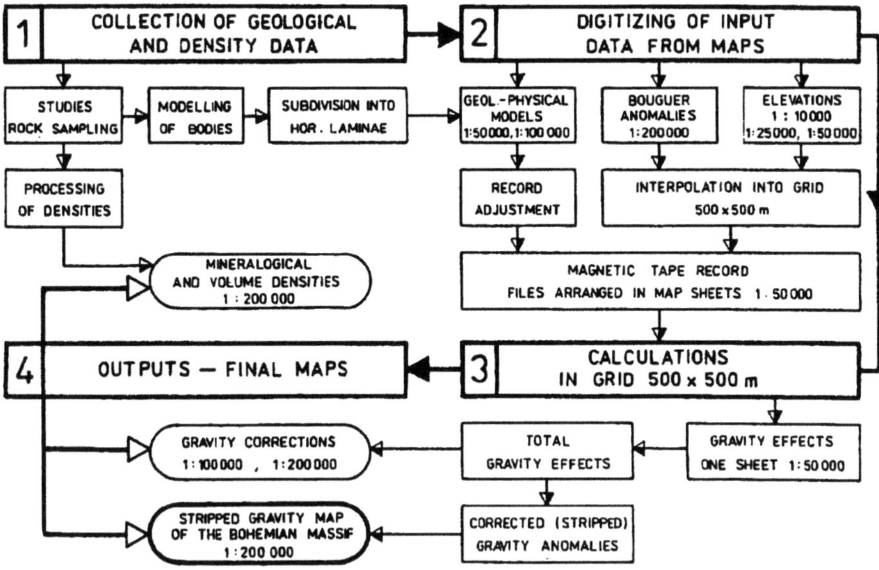

Fig. 6.3 Flow-chart for a construction of the stripped gravity map.

correction exceeds 100 μm s^{-2} in the N part of the Bohemian Cretaceous Basin, and at the SE margin of the Bohemian Massif. The value of the correction elsewhere ranges up to 60 μm s^{-2}.

In contrast to the Bohemian Massif, 63% of the area of the inner West Carpathians is covered by the Tertiary rocks of up to 7 km in thickness but their petrophysical properties as well as their deposition are known with lower accuracy. In deep sedimentary basins, significant changes of densities depending on the depth occur and in neovolcanic regions are also combined with the horizontal changes (Fig. 6.5). This is why the spatial density distribution was approximated by:

1. the mean density–depth function (Fig. 6.5b) in deep basins,
2. the mean density of the entire thickness of the Tertiary with a lateral change (Šefara 1977) in the neovolcanic regions (Fig. 6.5a),
3. the estimate of the mean density, separately in the Neogene and in Palaeogene, in regions with a small number of data.

The methodology of the stripped gravity map computation was as follows:

1. Construction of the initial model of the Tertiary on the basis of borehole data, geophysical (seismic, VES) data, of the type of the above specified dominating density distribution, its approximation by a set of n-sided prisms; and computation of the gravity effects in the grid 1×1 km.
2. Construction of the preliminary stripped gravity map;

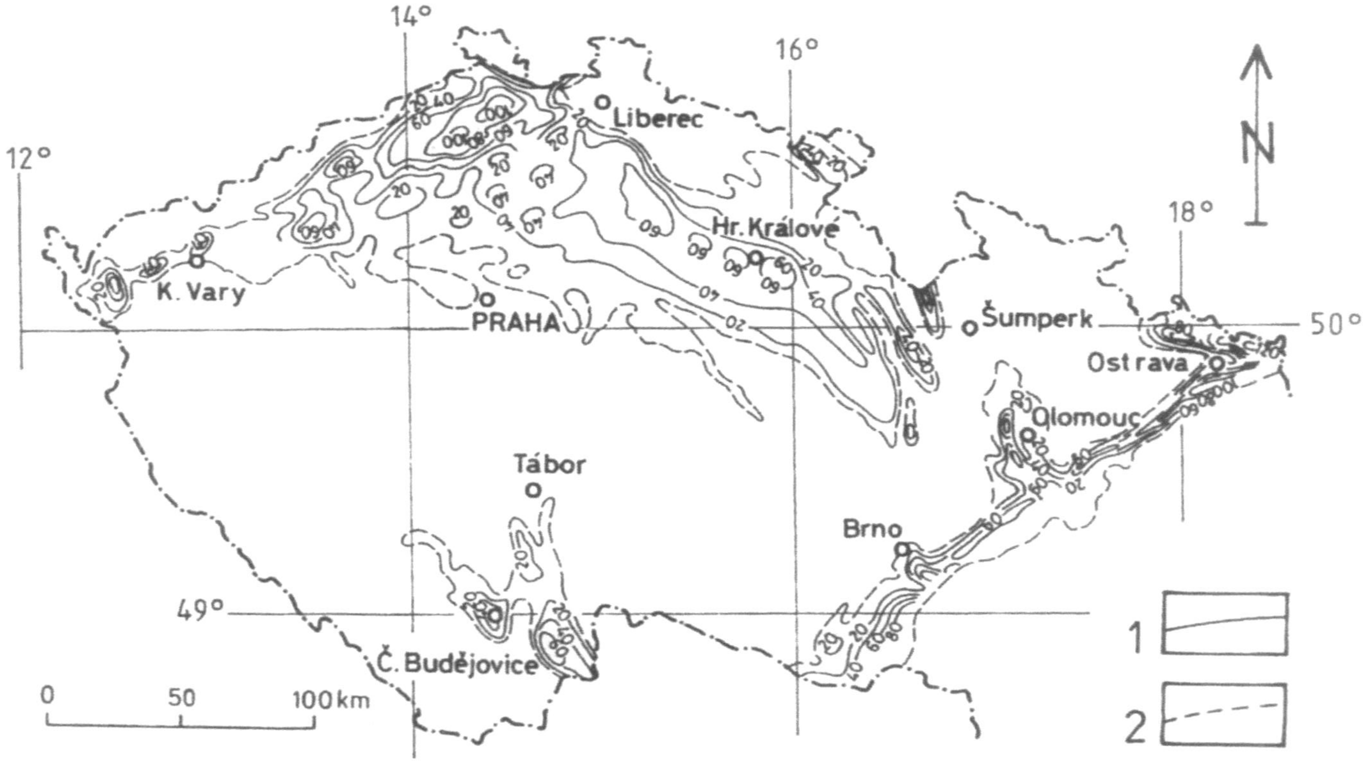

Fig. 6.4 Map of gravity effects of the Tertiary and Cretaceous sediments of the Bohemian Massif. 1 – contour interval 20 $\mu m\, s^{-2}$, 2 – contour interval 5 $\mu m\, s^{-2}$.

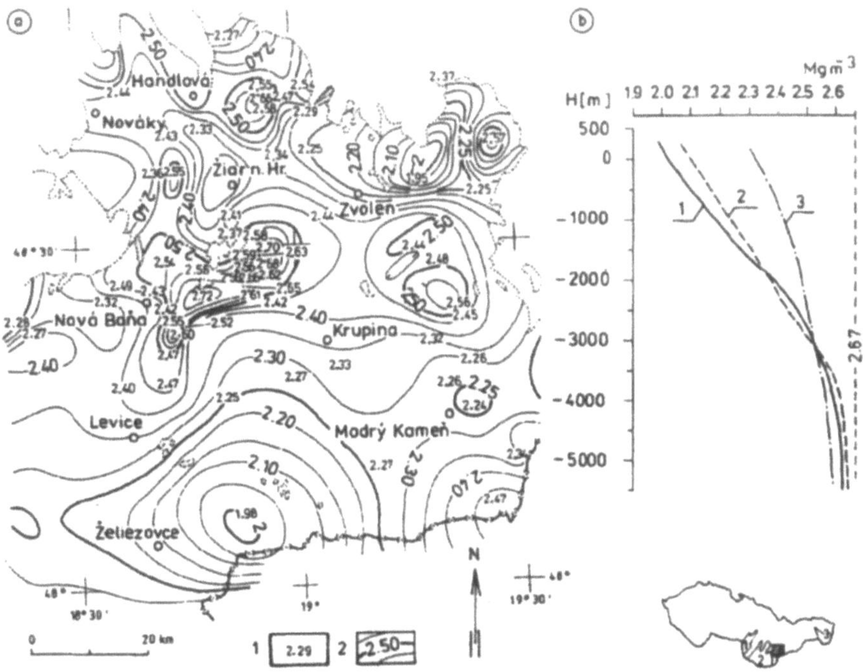

Fig. 6.5 Tertiary rock density distribution in the Inner West Carpathians: a) map of mean densities of the Tertiary volcano–sedimentary complex of Central Slovakia: 1 – mean rock density determined in boreholes in Mg m^{-3} 2 – mean rock density contours in Mg m^{-3}, b) density–depth functions for sedimentary basins 1 – Vienna Basin, 2 – Danube Basin, 3 – East-Slovakian Basin.

3. Estimation of depths to anomalous sources, correction of the model so that its estimated depth was in the range of the thickness of sediments, and computation of the gravity effects of the model;
4. Computation of the stripped gravity map or repetition of point 3.

In those regions where the terrain is flat, the computation was performed in the plane of the mean elevation above sea level. In more undulating terrain the computation was performed at the surface of the terrain.

The stripped gravity map of the inner West Carpathians was constructed during the period 1982 – 1986 (Fig. 6.6). Its mean quadratic error was determined as ±18 μm s^{-2}. The resulting map of corrections for the gravity effects of the Tertiary sediments (Fig. 6.7) reaches the maximum values exceeding 450 μm s^{-2} in the Danube and Vienna basins; smaller corrections of up to 350 μm s^{-2} occur in the East Slovakian Basin. Very small gravity effects of up to only 80 μm s^{-2} are due to the Paleogene rocks of the interior basins of the inner West Carpathians (Šefara 1987).

In order to suppress gravity effects of density boundaries at greater depths (the Moho discontinuity and the lower margin of the lithosphere) geologic filtering can be used.

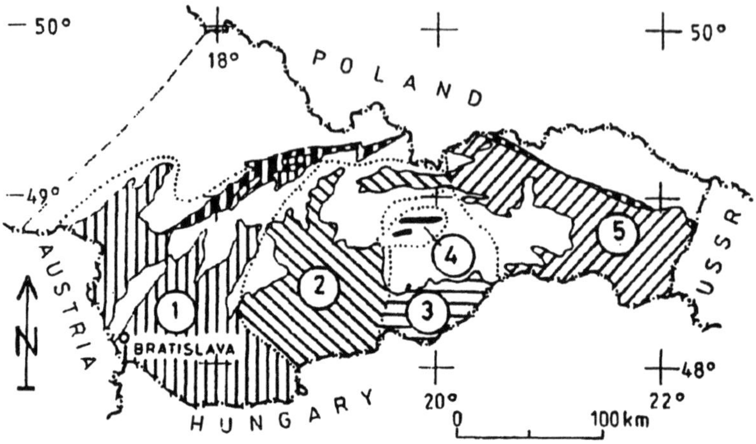

Fig. 6.6 Regions elaborated by individual authors for a compilation of the stripped gravity map of the Inner West Carpathians: 1 – Šutora – Bielik, 2 – Šefara – Velich, 3 – Bodnár, 4 – Kurkin – Mikuška, 5 – Pospíšil – – Velich.

Results obtained by this method cannot be very accurate; one must consider the fact that the data on the depths of these boundaries, as well as on the density contrast, are still a subject of intensive investigations (Burda et al. 1988; Burda and Vyskočil 1991).

For computation of gravity effects of the density boundaries at great depths the methodology presented by Vyskočil and Burda (1976) has been used, which uses "spherical prisms" for modelling the anomalous masses.

The model of the Moho discontinuity (Mayerová at al. 1985 – see Fig. 6.8) was digitized to grid points in the interval of geographic coordinates 10' × 15' ($\Delta\varphi$, $\Delta\lambda$). At each grid point, a gravity effect of the mass from the area of 2° × 3° was computed (Burda and Vyskočil 1991). The depth of 30 km was chosen as the reference base of the "spherical prisms". For the selected differential density of + 0.1 Mg m^{-3} the map of the Moho discontinuity gravity effect was constructed (Fig. 6.9). The model of the lithosphere – asthenosphere boundary was taken from Babuška et al. (1987). It was digitized in the 1° × 1° grid. Its gravity effect at the grid points was computed from the area of 8° × 12° (Burda et al. 1988). Fig. 6.10 shows the resulting gravity effect for the density contrast of – 0.03 Mg m^{-3}, and for the reference level of – 120 km of the lithosphere – asthenosphere boundary.

With regard to the diversity of the opinions concerning the magnitude of the density contrasts at the Moho discontinuity and at the lithosphere – asthenosphere boundary, two versions of the map were constructed:

1. The gravity map corrected for the gravity effects of the Moho discontinuity relief ($\Delta\rho$ = + 0.1 Mg m^{-3}) and of the lithosphere – asthenosphere boundary ($\Delta\rho$ = = – 0.03 Mg m^{-3}) (Fig. 6.11).

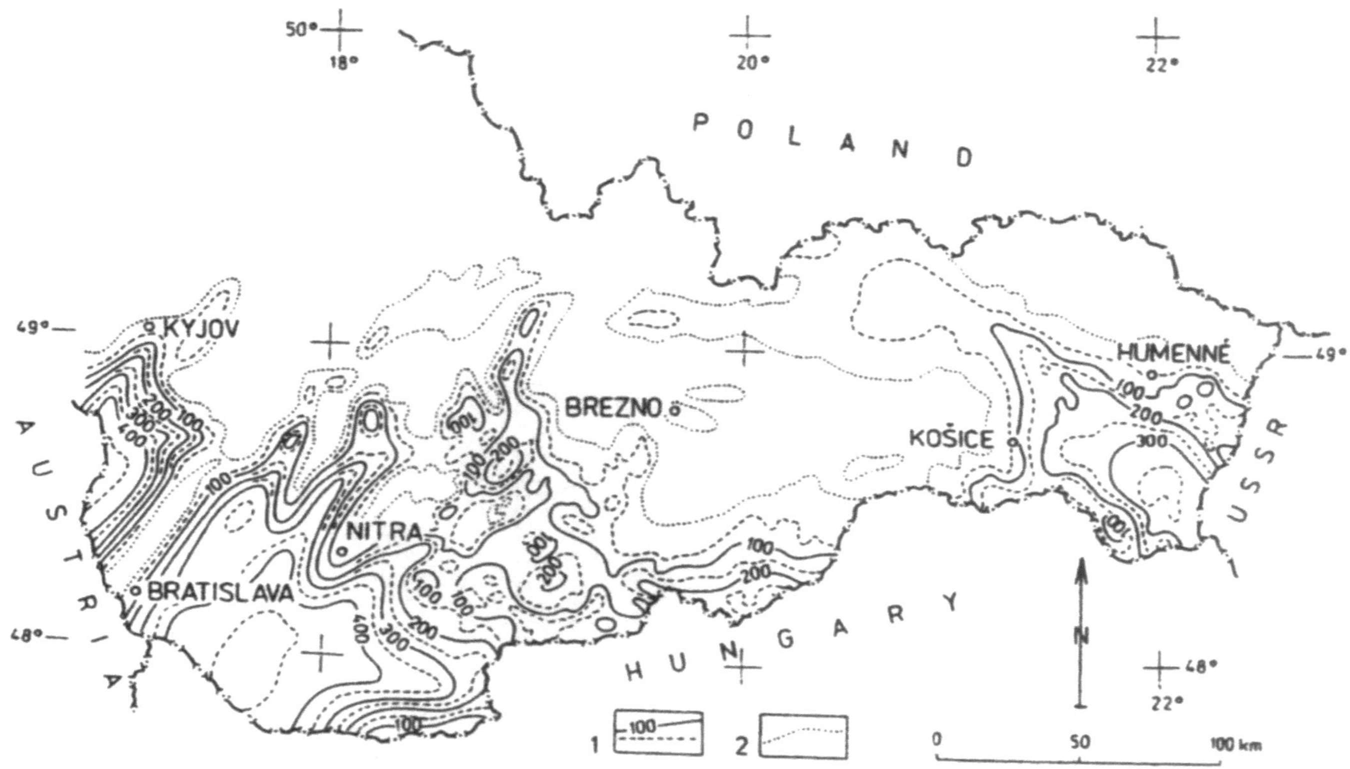

Fig. 6.7 Map of gravity effects of the Paleogene and Neogene sediments of the Inner West Carpathians: 1 – contours in $\mu m\ s^{-2}$, 2 – negligible values of gravity effect (cca 1 $\mu m\ s^{-2}$).

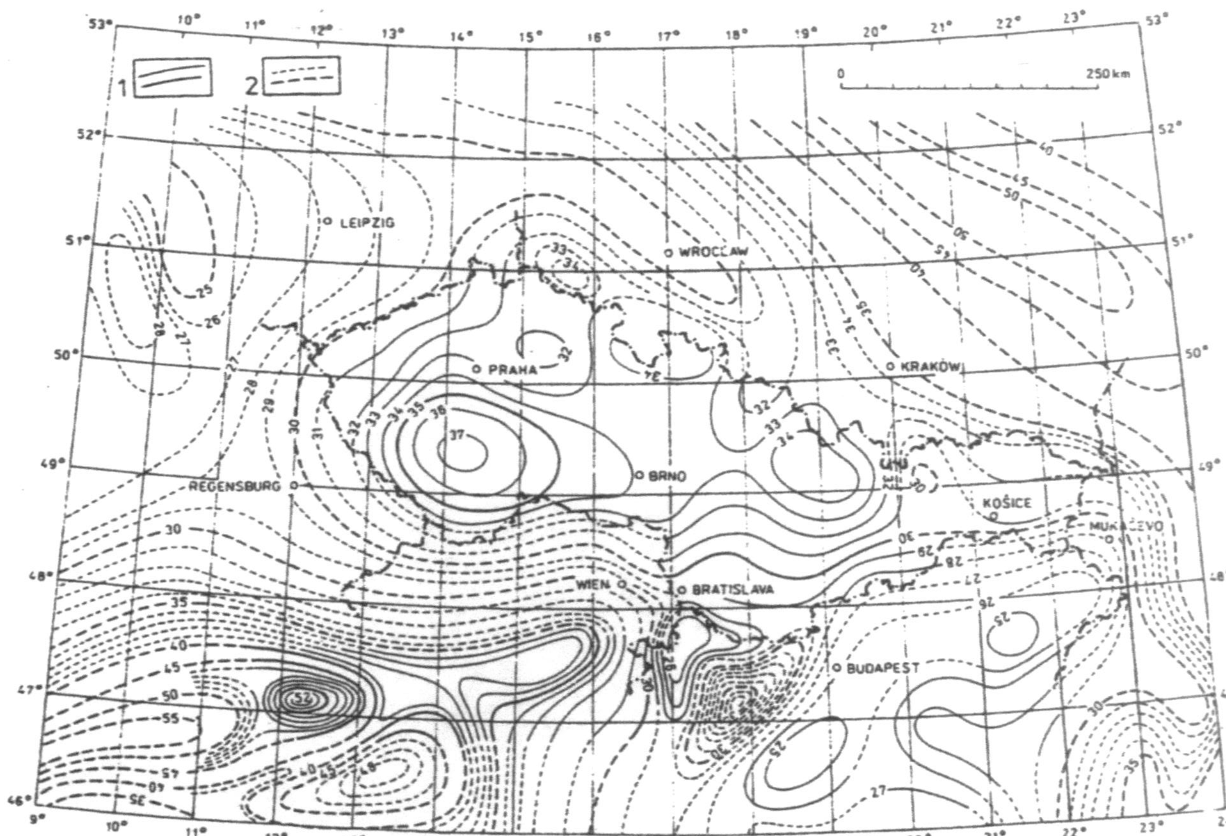

Fig. 6.8 A tectonic scheme of the Moho discontinuity relief in the east part of Central Europe: 1 – Moho discontinuity relief contours in km derived from seismic data, 2 – extrapolated contours.

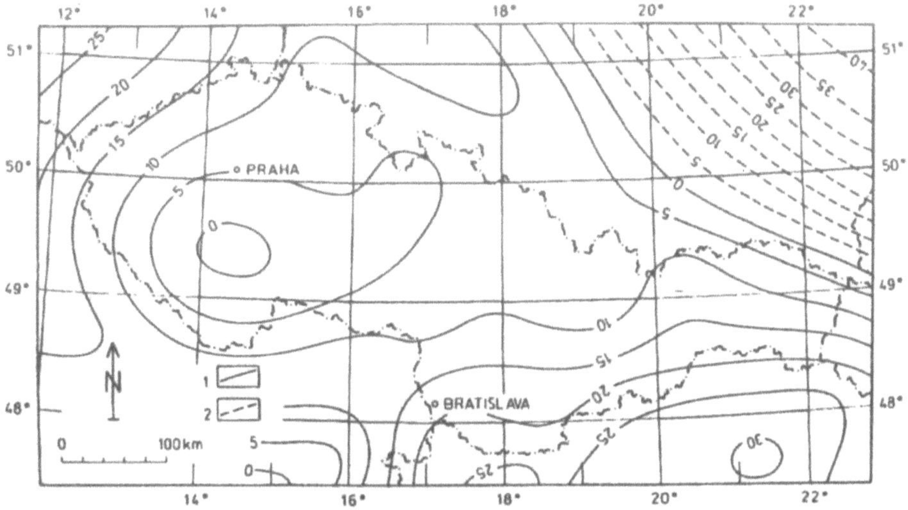

Fig. 6.9 Gravity effect of the Moho discontinuity relief in the east part of Central Europe; density contrast + 0.1 Mg m^{-3}, reference depth 30 km. 1 – contours of positive gravitational field in 10^{-5}m s^{-2}, 2 – contours of negative gravitational field in 10^{-5}m s^{-2}.

2. The gravity map corrected for the gravity effects of the Moho discontinuity relief ($\Delta\rho = + 0.25$ Mg m^{-3}) and of the lithosphere – asthenosphere boundary ($\Delta\rho =$ $= - 0.03$ Mg m^{-3}) (Fig. 6.12).

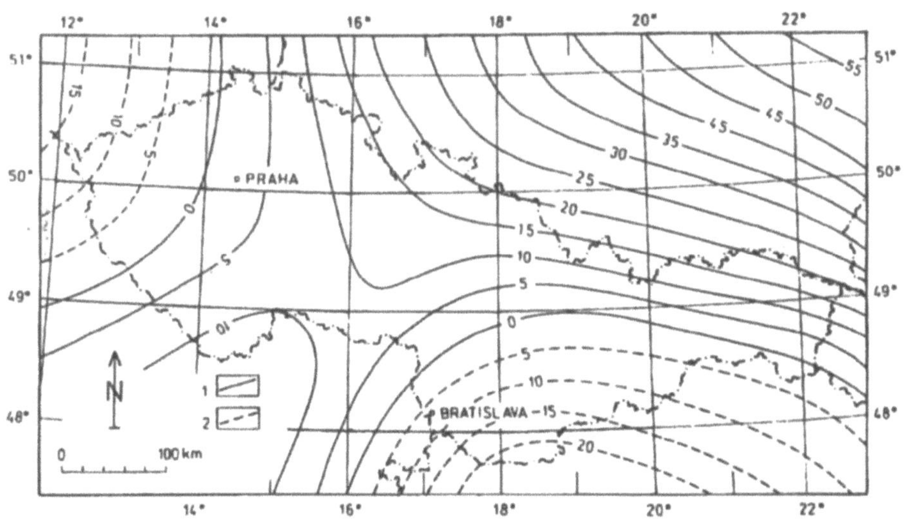

Fig. 6.10 Gravity effect of the lithosphere–asthenosphere boundary in the east part of Central Europe; density contrast – 0.03 Mg m^{-3}, reference depth 120 km. 1 – contours of positive gravitational field in 10^{-5}m s^{-2}, 2 – contours of negative gravitational field in 10^{-5}m s^{-2}.

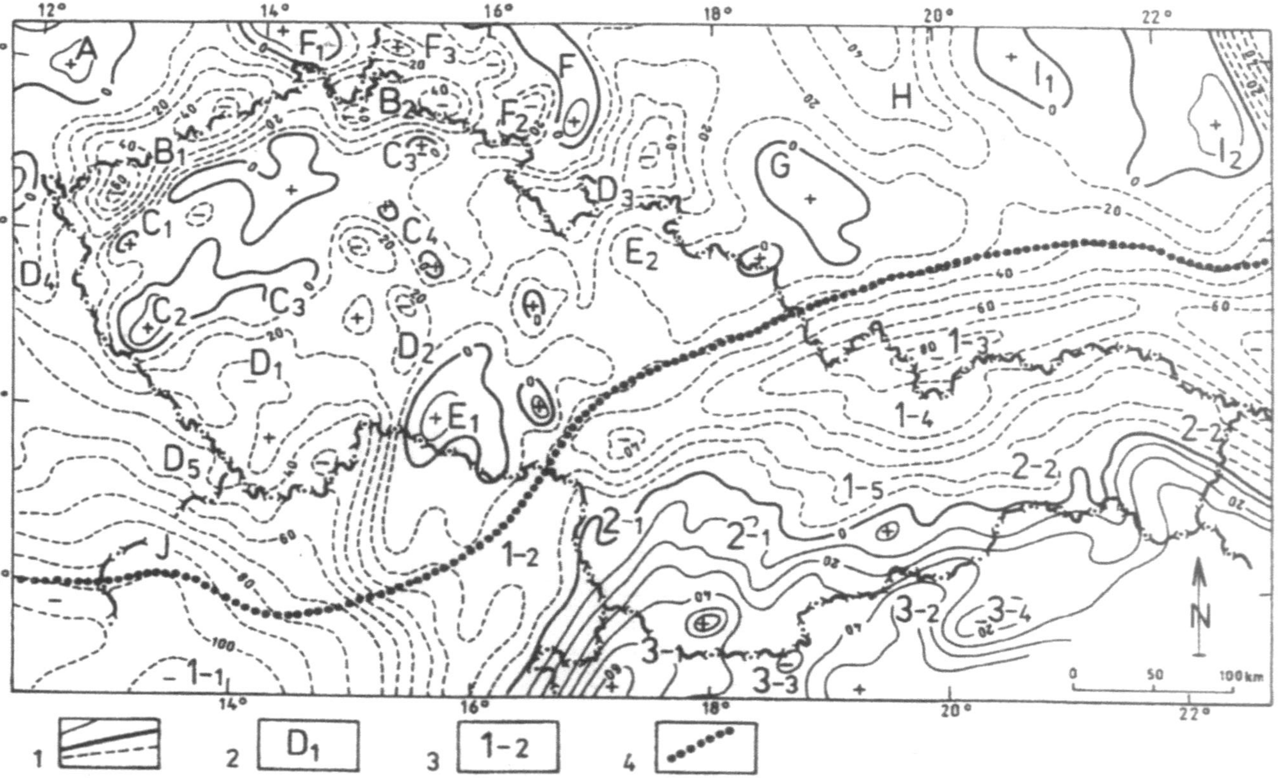

Fig. 6.11 Bouguer gravity map (in Czechoslovakia and Hungary, the stripped gravity map) corrected for the effects of the Moho discontinuity relief ($\Delta\rho = +0.1$ Mg m^{-3}) and of the lithosphere–asthenosphere boundary ($\Delta\rho = -0.03$ Mg m^{-3}). 1 – contours of positive (full line) and negative (dashed line) anomalies in 10^{-5}m s^{-2}, 2 – interpreted structures of the European Hercynides and East European platform (A – J), 3 – interpreted structure of the Alpine, Carpathian and Panonian regions (1 – 3), 4 – border between gravitational field of Hercynides and Alpides.

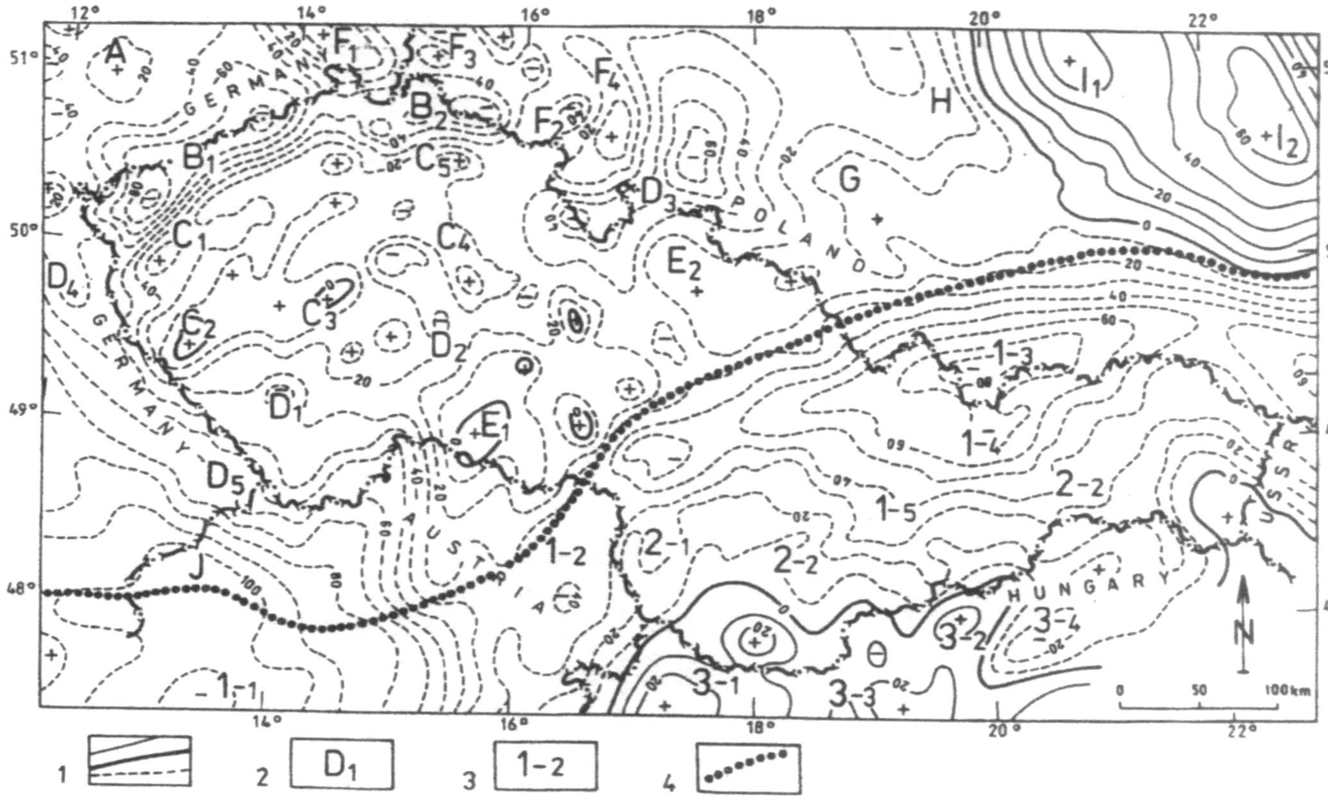

Fig. 6.12 Bouguer gravity map (in Czechoslovakia and Hungary the stripped gravity map) corrected for the effects of the Moho relief ($\Delta\rho = +0.25\ \mathrm{Mg\ m^{-3}}$) and of the lithosphere–asthenosphere boundary ($\Delta\rho = -0.03\ \mathrm{Mg\ m^{-3}}$). 1 – contours of positive (full line) and negative (dashed line) anomalies in $10^{-5}\mathrm{m\ s^{-2}}$, 2 – interpreted structures of the European Hercynides and East European platform, 3 – interpreted structure of the Alpine, Carpathian and Panonian regions, 4 – border between gravitational field of Hercynides and Alpides.

A comparison of these two maps with the original gravity map shows that majority of the amplitudes of anomalies remained, and, therefore, they can be explained as due to sources distributed at depths from the Earth's surface up to 15 km.

6.3 Explanation of the Main Gravity Structures in the Czechoslovak Territory and in Its Close Vicinity

M. Blížkovský[1], J. Ibrmajer[1], A. Novotný[1], M. Suk[2], J. Šefara[3]

The starting basis for the interpretation of the gravitational field distribution in Czechoslovakia were the Bouguer gravity maps of Czechoslovakia on a scale 1:200 000 and on a scale 1:500 000 (Ibrmajer 1963, 1965), the stripped gravity map of the Bohemian Massif 1:200 000 (Blížkovský and Novotný 1981) and the stripped gravity map of the inner West Carpathians compiled from the detailed gravity measurements 1:25 000 (Šefara et al. 1985). The results of the study of density parameters of rocks collected from the surface outcrops as well as from boreholes (Eliáš and Uhmann 1968; Biela 1978; Ondra and Hanák 1981) were used.

A basic characteristic of the gravitational field of Czechoslovakia was given by Buday et al. (1969) who distinguished four main regions in the Bohemian Massif, and two main regions in the West Carpathians. A new gravimetric mapping on a scale 1:25 000 as well as introducing corrections for gravity effects of sedimentary basins made it possible to define with more precision courses of individual structures in the gravitational field for the purposes of geologic interpretation. After introducing corrections for individual basins, the gravitational field pattern changes only locally in the Bohemian Massif but substantially in the West Carpathians.

On the basis of the gravitational field pattern, of the distribution of gravity gradients, of correlations of gravity data with other geophysical methods, and with the use of stripped gravity maps, numerous structural-tectonic schemes were constructed which reflect the geologic structure of a particular region. In the Bohemian Massif, such schemes were compiled by Blížkovský et al. (1975), Pokorný et al. (1985), Blížkovský et al. (1985b), and others. In the West Carpathians, Fusán et al. (1981) distinguished deep fault zones and derived neotectonic blocks which were later defined in more detail by Šefara et al. (1985).

In the map of gravity anomalies in Czechoslovakia and its surroundings, presented in Figs. 6.11, 6.12, the region of European Hercynides together with a part of the Fennosarmatian platform as well as the Alpine – Carpathian region including the Pannonian Basin are clearly distinguished. In Czechoslovak territory and in the northern part of Hungary, after removing the gravity effects of the Cretaceous and Tertiary sediments (with the exception of the Flysch Belt sediments), mainly those elements

1 Geofyzika Brno, Ječná 29a, 612 46 Brno, Czech Republic
2 Faculty of Natural Sciences, Masaryk University, Kotlářská 2, 600 00 Brno, Czech Republic
3 Geological Institute, Comenius University, Mlýnska dolina, 842 15 Bratislava, Slovakia

prevail which reflect magmatic and metamorphic structure. Some gravity negative anomalies outside Czechoslovak territory are mostly caused by sedimentary rock formations.

In the Bohemian Massif, the gravity structures of SW–NE direction are dominant, with the exception in the Český les Mts., Bayrische Wald Mts., and the Lugicum, where directions perpendicular to these structures prevail. The positive gravitational anomalies in the Bohemian part of the Bohemian Massif correspond to abundant basic massifs and volcanic bodies, and these regions tend to subside secularly. On the contrary, the negative gravitational anomalies correspond to regions with a rising tendency associated with an extensive granitic magmatism. Blížkovský et al. (1985b) distinguished fundamental structures of the gravitational field in Czechoslovakia. From NW to SE, the following structures can be differentiated in the gravitational field (Figs. 6.11, 6.12):

The positive anomaly in the Saxothuringian region (A) is caused by volcano–sedimentary Palaeozoic formations and influenced by low-density Triassic and Neogene sediments.

The negative anomaly of the Krušné hory Mts. – Krkonoše Mts. region (B), in which significant "depressions" exist over granitoids and orthogneisses of the pluton of the Krušné hory Mts. (B1) and over low-density granitoids of the Krkonoše – Jizerské hory Mts. pluton (B2). A quantitative interpretation of DSS profile No. VI confirmed the thickness of low-density granitoids of the Karlovy Vary massif to be of at least 10 – 15 km (Polanský 1973; Blížkovský et al. 1985b).

The positive anomaly of the Barrandian – Železné hory – Broumov region (C) is divided into three subregions. In the Teplá – Barrandian subregion, the gravity "elevations" are distributed in a parallel-band pattern of SW–NE direction. They correspond to known basic volcanites as e.g. the metabasite complex of Mariánské Lázně (C1), the Stříbro – Plasy belt, or the Kdyně basic complex (C2). The southernmost "elevation" belt indicates the volcanic zone of the Jílové zone (C3).

In the transverse subregion of the Železné hory Mts., the gravity "elevations" are caused first of all by rocks of the metamorphosed Proterozoic and the Lower Palaeozoic but also by a part of the Nasavrky massif (C4) containing abundant basic bodies.

The gravity structures of the Labe – Broumov subregion have a typical W–E pattern. They are probably caused by basic volcanites of the Proterozoic (C5) in the basement of the Cretaceous and Permo–Carboniferous.

The negative Moldanubian – Klodsko region (D) can be divided into three subregions with different pattern of gravity contours.

In the Moldanubian subregion, the negative structures are confined to the Central-Bohemian (D1) and the Moldanubian (D2) plutons. In the Orlické hory – Klodsko subregion, there is a gravity low (D3) in the Orlické hory – Klodsko arch, continuing into Poland, which indicate bodies of orthogneisses, and possibly also of "low-density" granites in the basement of migmatites. In the subregion of the Český les Mts. and Bayrische Wald Mts., the gravity contours are in a NW–SE direction, which is the direction of the Bohemian quartz lode and of the Danube fault zone. In the NW part of this subregion, the gravitational field pattern is deformed probably by a hidden large Tertiary cover with possible granites in its basement (D5).

The positively disturbed Moravian–Silesian region (E) includes SE part of the Moldanubicum, Moravicum, Brunnia unit, and W part of the Silesicum. The Moravian subregion (E1) is separated by the zone of negative anomalies over the Hornomoravský úval (E3) from the Silesian subregion (E2). The zone of negative anomalies indicates not only a striking fault zone of NW–SE direction but also hidden bodies of granitoids, confirmed by drilling. In the west, the Moravian subregion is bound by a very strong gradient on the gravitational field, which has been associated with the Přibyslav deep zone. In the south, this gradient indicates a thrusting of the Alps over their foreland. This extensive "elevation" of the gravitational field is caused by a great thickness of the Varied Group of the Moldanubicum, basic rocks of the Brno Massif and of the Letovice crystalline unit, and probably also by basic bodies at greater depths.

The Silesian subregion consists of relatively dense rocks of the Devonian and culm. To the positive gravitational field anomaly caused by the Silesian subregion, basic rocks of the Jeseník and Sobotín amphibolite massifs also contribute. It could also be caused, however, by the lower thickness of the Earth's crust (i.e. 32 km according to Mayerová et al. 1985).

In the Lužice (F) region, positive and negative zones of NW–SE direction are strongly manifested, which are associated with the Lužice granite pluton (F1), with the low-density granites of the Strzegom massif (F2), or with a greater thickness of the Tertiary sediments (F3), and with an elevation of the high-density basement rocks beneath the platform cover (F4).

The west part of the Fennosarmatian platform is in the gravity map shown by a positive-anomaly zone connecting Kraków and Ostrzeszow (G), and by the negative-anomaly zone Lodz – Tarnów (H), which both reflect the Odra lineament and the variable thickness of the platform cover, also by the Swietokrzyskie Góry Mts. – Lublin (I) region whose SE projection extends into the Carpathian Foredeep. The region can further be divided into Swietokrzyskie Góry Mts. (I1) subregion and the Lublin subregion (I2) which is a part of the Teisseyre - Tornquist line where the depth of source is about 10 km. In the stripped gravity map, at the SW margin of the Bohemian Massif, the transient region of the Alpine Foredeep (J) can be distinguished, where the gradient corresponds to a gradual increase of the Tertiary sediments thickness towards SW.

The gravitational field in the Alpine – Carpathian region can be divided into the region of negative gravitational anomaly – the Alpine – Carpathian (1) one and the region of a positive gravitational anomaly – the Pannonian (3) one. The area between the two above specified regions, on the gradient of gravitational field is designated as the transient zone (2).

The most striking minimum of the gravitational field is in central East-Alpine subregion (1–1), that is probably caused by a combination of effects of magnitudes and low-density sediments in the basement of Alpine nappes. Effects of deep sources, however, cannot be excluded.

The negative anomaly (1–2) in the area of Alpine–Carpathian gravity ridge indicates the rocks of the Carpathian Flysch Belt and autochthonous cover of the platform. The east part of the Alpine–Carpathian gravitational field is designated as the Carpathian gravity low (1–3) which is caused by porous sediments of the Flysch Belt, or by the underlying molasse (Buday et al. 1969; Tomek et al. 1979; Tomek 1982), or by a great thickness of sediments of the Flysch Belt (Plíva 1980 a,b).

The inner West Carpathian negative gravitational field subregion (1–4) indicates probably greater thickness of the crust. The Vepor negative gravitational field zone (1–5) has been associated with mainly Neo-Alpine granitoids including the neovolcanic plutonites at a deep zone.

In the transient zone can be, on the basis of the gravitational field pattern, distinguished: the Malé Karpaty – Skýcov subregion (2–1) where the Sennonian to Palaeogene sediments in the Vienna Basin basement, or rocks of the Carpathian nappes are the sources of the anomaly, and the Gemer – Vihorlat subregion (2–2) with horizontally distributed sources of negative anomalies, as e.g. granites in the Gemericum or the Palaeo-Alpine and Hercynian granites of the Veporicum.

The positive gravitational field of the Pannonian region can be divided into the following subregions: The Rába – Danube (3–1), the Alföld – trans-Carpathian (3–2), where the anomalies are caused mainly by basic rocks of the lower crust, occurring already at depths of 12 – 16 km. The Bakony – Börzsenyi (3–3) and Mátra – Zemplín (3-4) subregions, where Alpine plutonites are assumed, are characterized by relatively less striking positive gravitational field.

6.4 Density Models of the Earth's Crust Along Seismic Profiles

M. Bielik[1], M. Blížkovský[2], M. Burda[3], O. Fusán[1], M. Hübner[3], H. Herrmann[2], A. Novotný[2], M. Suk[4], Č. Tomek[2], V. Vyskočil[3]

In order to augment the interpretation along the international DSS profiles (i.e. Nos. V, VI, VII in Fig. 6.13), density modelling of the Earth's crust was performed along the Czechoslovak part of the profiles. Density models along other profiles crossing the Carpathian arc CP1, CP2, CP3, and CP4 are published in Bielik et al. 1991; Vyskočil et al. 1991.

The two-dimensional quantitative interpretation along DSS profiles was based on the following input data:

1. information on the geologic structure of the near-surface layer of the Earth's crust to a depth of approximately 5 km,

1 Geophysical Institute, Dúbravská cesta, 842 28 Bratislava, Slovakia
2 Geofyzika Brno, Ječná 29a, 612 46 Brno, Czech Republic
3 Geophysical Institute, Boční II/1401, 141 31 Praha 4, Czech Republic
4 Faculty of Natural Sciences, Masaryk University, Kotlářská 2, 600 00 Brno, Czech Republic

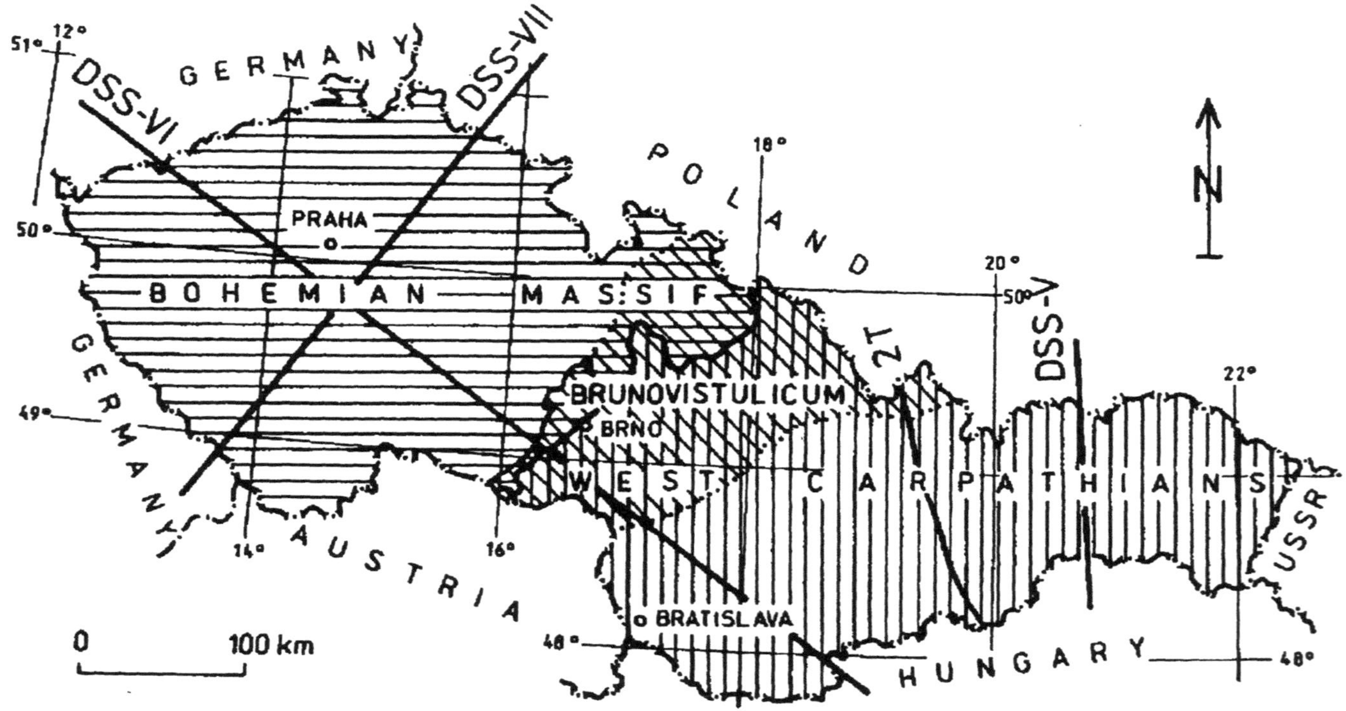

Fig. 6.13 Situation of the density model profiles.

2. densities of rocks determined in laboratory from sets of rock samples (according to Eliáš and Uhmann (1968), Ondra and Hanák (1981)),
3. the Moho discontinuity (according to Beránek 1980),
4. the seismic velocity section (according to Beránek and Zátopek 1981),
5. the assumed contact density of 3.2 Mg m^{-3} below the crust-mantle boundary.

To solve the direct gravimetric problem, the method of Talwani (1973) was used with side corrections for bodies that do not satisfy conditions for being two-dimensional. Calculation points on the profile were distributed in 5 km intervals. Gravity effects of a sedimentary cover were taken from the map of corrections constructed for compiling the stripped gravity map. In order to limit the errors occurring as a result of neglecting the effects of remote bodies, we used the following definition of normal density distribution of the reference model:

Layer No.	Depth extent [km]	Reference density [Mg m^{-3}]
1	surface topography – 0	2.67
2	0 – 10	2.70
3	10 – 25	2.80
4	25 – 40	3.10
5	exceeding 40	3.20

In the graphical presentation the gravity effects of the model of layers 1 and 2 were united, so that the geologic idea of density distribution could be described at three depth levels: top to 10 km, intermediate in the range 10 – 25 km, and bottom in the range 25 – 40 km.

Density Model Along International DSS VII Profile

In Fig. 6.14 a density model along DSS VII profile is presented together with other data. Plots of separate gravity effects at three depth levels enable us to formulate the following conclusions:

1. the gravity effect of the deepest level (25 – 40 km) corresponds to the Moho discontinuity topography,
2. the gravity effect of the intermediate level (10 – 25 km) is mainly due to the course of the layer with density of 2.8 Mg m^{-3}, which beneath the Czechoslovak territory is ascending and thinning from SW to NE,
3. the most striking variations in the course of the Bouguer anomalies are caused by gravity effects of the masses located within the depth of 10 km. Gravity lows with centres at – 100 km and 0 km are caused by low-density differentiates of the Central-Bohemian and the Krkonoše – Jizerské hory plutons. The gravity high at – 25 km is associated with metamorphics of the Proterozoic and Paleozoic volcano-sedimentary formation. In the sector between – 130 km and – 180 km, the granitic layer of density of 2.70 Mg m^{-3} almost outcrops.

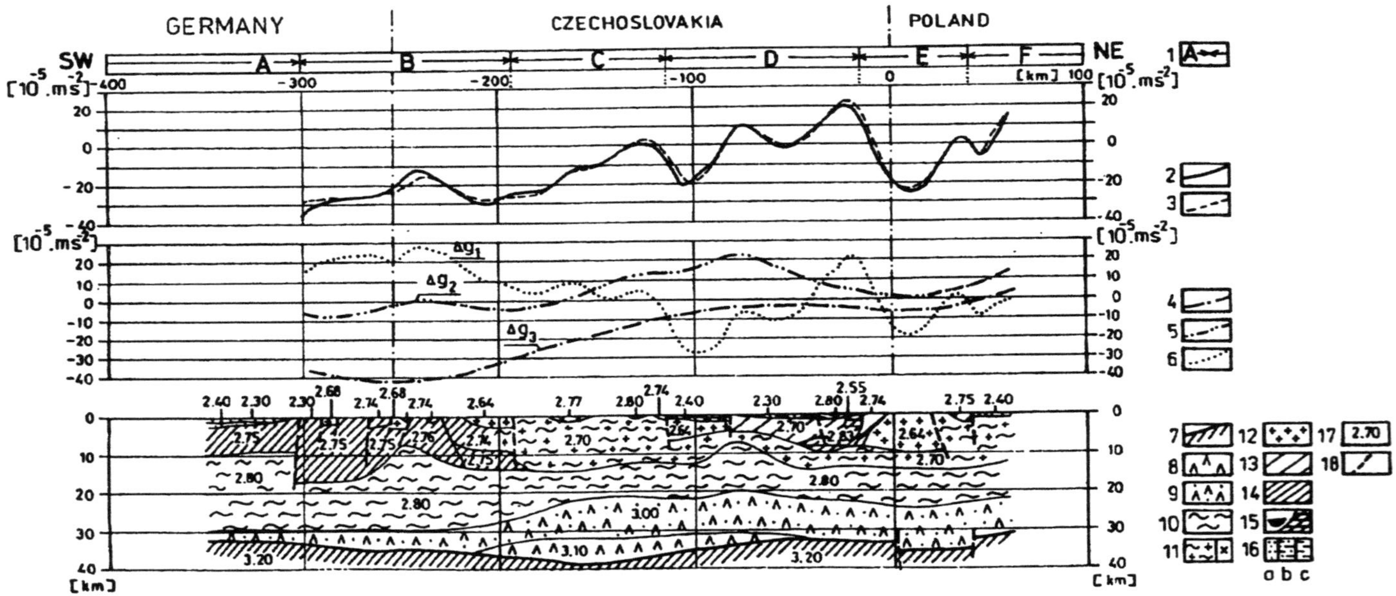

Fig. 6.14 Density model along DSS profile No. VII. 1 – boundary of fundamental geologic units: A – Alpine Foredeep, B – Moldanubicum, C – Central Bohemian Pluton and zone of islets, D – platform cover (Permo-Carboniferous and Cretaceous), E – Krkonoše – Orlické hory region, F – sedimentary rocks of the East European platform, 2 – Bouguer anomaly, 3 – computed gravity effects of the density model, 4 – computed gravity effects of masses distributed at a depth range 25 – 40 km (Dg_3), 5 – computed gravity effects of masses distributed at a depth range 10 – 25 km (Dg_2), 6 – computed gravity effects of masses distributed to a depth of 10 km (Dg_1), 7 – Moho discontinuity course (after Beránek et al. 1977), 8 – 9 – lower part of the crust with densities of 3.0 and 3.1 Mg m^{-3}, 10 – upper part of the lower crust with density of 2.8 Mg m^{-3}, 11 – granitoid rocks with densities of 2.7 Mg m^{-3}, 12 – granitoid rocks with densities of up to 2.7 Mg m^{-3}, 13 – weakly and medium metamorphosed rocks of the Late Proterozoic and Palaeozoic, 14 – intensively metamorphosed rocks of the Precambrian, 15 – volcanic rocks, 16 – sedimentary rocks (Neogene, Cretaceous, Permo-Carboniferous), 17 – densities in Mg m^{-3}, 18 – assumed tectonic boundary.

Comparing the thicknesses of interpreted layers with densities of 3.0, 2.8 Mg m^{-3}, and less, three basic types of crust can be distinguished.

In SW part of the Alpine Foredeep, the intermediate layer with density of 2.8 Mg m^{-3} is present in large thickness. Its thickness increase is balanced by thinning of layer with density of 3.0 Mg m^{-3}, which, at the sector between – 400 km and – 215 km, reaches maximum values of 3 – 5 km.

The transient zone extending from the boundary of the Moldanubicum of the Central Bohemian pluton (–190 km) to NE is characterized by an increased thickness of low-density rocks at the surface, compensated by an increased thickness of the high-density basal level rocks with densities of 3.0 – 3.1 Mg m^{-3}.

The third type of the crust, developed beneath the platform of the Permo-Carboniferous and Cretaceous cover and further extending NE to Poland, is characterized by equal thicknesses of the three basic density levels.

Density Model Along Ininternational DSS VI Profile

In Fig. 6.15 a density model along DSS VI profile is presented. Analogously as in the case of DSS VII profile, let us first discuss separately the gravity effects of the three depth levels:

1. the gravity effect of the deepest level (25 – 40 km) corresponds well to the Moho discontinuity topography. The low correlation of the Moho discontinuity topography with the Bouguer anomalies, in places of increased thickness of the crust, has often been explained by an increased density of rocks in the surficial layer. For instance, between 100 km and 250 km, the high-density crystalline rocks of the Moldanubicum and between 300 km and 390 km, high-density rocks of the Brunovistulicum occur at the Earth's surface.
2. the gravity effect of the layer 10 – 25 km corresponds well to the course of the upper boundary of the layer with an assumed density of 2.8 Mg m^{-3}. This layer is thinner in the Krušné hory – Thuringian region and in the Carpathians (5 km) than in the core region of the Bohemian Massif (7 – 10 km),
3. the gravity effect of the near-surface layer to depth of 10 km correlates best with the course of the Bouguer gravity.

From the point of view of the Earth's crustal structure the Bohemian part of DSS VI profile can be divided into three fundamental regions separated by the Ohře zone and the Peripienian lineament. These two basic dislocations can be considered as deep ones extending through the entire crust. Other features, as e.g. the Přibyslav zone, etc., probably penetrate only the surficial granitic layer of the crust.

In order to designate the three fundamental regions let us use the established terms, i.e. the Krušné hory – Thuringian region, the core region (joint regions of the Teplá – – Barrandian, the Moldanubicum, and the Brunovistulicum), and the West Carpathians.

In the Krušné hory – Thuringian region, the top part of the crust with density of 2.7 Mg m^{-3} reaches a thickness of 15 km, the boundary between the upper crust

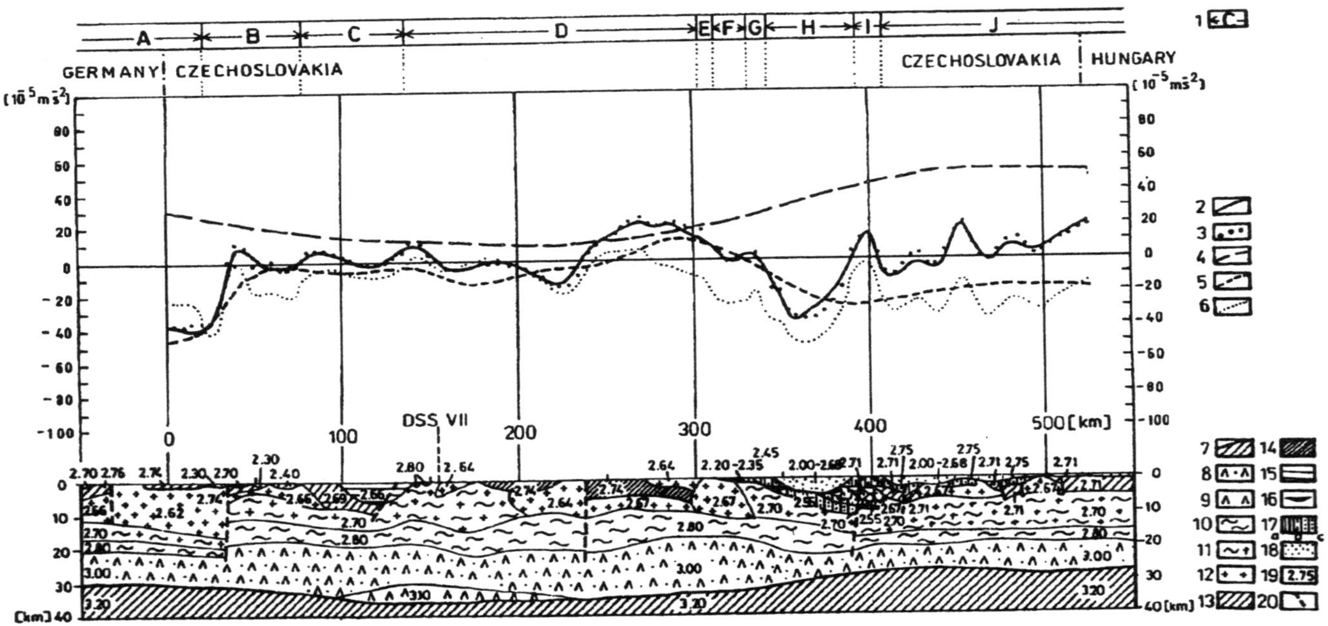

Fig. 6.15 Density model along DSS profile No. VI. 1 – boundary of fundamental geologic units: A – Krušné hory – Thuringian region, B – platform cover in the Ohře zone, C – Teplá – Barrandian region, D – Moldanubicum, E – Moravo–Silesian zone, F – Carpathian Foredeep, G – Flysch Belt of the West Carpathians, H – Vienna Basin, I – core mountains of the Malé Karpaty Mts, J – Danube Basin, 2 – Bouguer anomaly, 3 – total computed gravity effects of the density model, 4 – computed gravity effects of masses distributed to a depth of 10 km, 5 – computed gravity effects of masses distributed at a depth range 10 – 25 km, 6 – computed gravity effects of masses distributed at a depth range 25 – 40 km, 7 – Moho discontinuity course, 8–9 – lower crust (the lower part with densities of 3.0 and 3.1 Mg m^{-3}, respectively, 10 – lower crust (the upper part with density of 2.8 Mg m^{-3}, 11 – granitoid rocks with density of 2.7 Mg m^{-3} (the lower part of the upper crust), 12 – granitoid rocks with density lower than 2.7 Mg m^{-3} including orthogneisses, 13 – weakly metamorphosed Late Proterozoic and Palaeozoic rocks, 14 – intensively metamorphosed Precambrian rocks, 15 – Palaeozoic rocks, 16 – volcanites, 17 – folded sedimentary formations in the West Carpathians, a) Mesozoic of the Inner West Carpathians, b) Klippen Belt, c) Flysch Belt sediments, 18 – unfolded sediments (Neogene basins at the foot of the Krušné hory Mts., Neogene of the Carpathian Foredeep, Neogene of the Vienna Basin, Neogene of the Danube Basin, Permo-Carboniferous beds in the Central Bohemian Carboniferous Basin), 19 – rock density in Mg m^{-3}, 20 – assumed tectonic boundary.

(2.8 Mg m^{-3}) and the lower crust (3.0 Mg m^{-3}) extends approximately to a depth of 20 km, while the total crustal thickness is ca. 30 km. The negative Bouguer anomalies of this region are due to much thicker low-density rocks of the upper crust on the account of the high-density rocks of the lower crust. The deep Ohře tectonic zone in the area of Doupovské hory Mts. is manifested by a striking horizontal gravity gradient. This zone served as a vent of the basaltic Tertiary volcanism, of the older basic volcanism and the acid Permo-Carboniferous volcanism.

The second region, the so called Bohemian core, is characterized by an increased thickness of the continental crust of 30 – 39 km including its lower "basaltic" layer. The upper-lower crust boundary ranges from 9 to 15 km in depth. Majority of positive and negative gravity structures in this region are caused by rocks in the top part of the crust, which is quite heterogeneous. Granitic rocks of the Central Bohemian pluton (2.7 Mg m^{-3}) outcrop in the interval 140 – 160 km. Eastwards, these rocks are overlain by a thin layer of crystalline schists. Towards W, the granitic rocks form a basement of the Proterozoic and Paleozoic units of the Barrandian, whose thickness is up to 9 km. The high-density metamorphics (2.74 Mg m^{-3}) in the Moldanubicum outcrop in intervals 180 – 200 km and 240 – 270 km, i.e. on both sides of the assumed Přibyslav dislocation. From 310 km towards E, granites of the Brno Massif plunge under the Paleozoic and Neogene cover of the Carpathian Foredeep as well as under the nappes of the Outer Flysch Belt. The gravity effect of the greatest depression of the Moho discontinuity between 100 km and 240 km is, to a considerable extent, compensated by the effect of high-density rocks in the upper crust and by an increased thickness of high-density lower crust rocks.

For the third region of the West Carpathians, a reduced thickness of the crust (25 – – 28 km) is typical as well as a reduced thickness of its bottom layer. The boundary between the upper and lower crust is approximately at the depth of 16 km. Mesozoic rocks of the Malé Karpaty Mts., Inovec Mts., and Tríbeč Mts. cause striking positive gravity anomalies.

Density Model of the Crust Along International DSS V Profile

Profile DSS V passes from the North European platform over the West Carpathians into the Pannonian Basin (Fig. 6.16). On the basis of gravity effects at three depth levels, the following conclusion can be drawn:

1. the gravity effect of masses at deepest levels (25 – 40 km) correlates with the Moho discontinuity topography. It represents a part of the regional trend of the Bouguer anomaly curve,
2. the gravity effect of the intermediate-level rocks (10 – 25 km) maximally correlates with boundaries within the crust. It represents a substantial part of the West Carpathian gravity low with a considerable decrease in the Czechoslovak – – Hungarian border area, where a maximum compensation of this gravity effect and the one of the following layer could be observed,

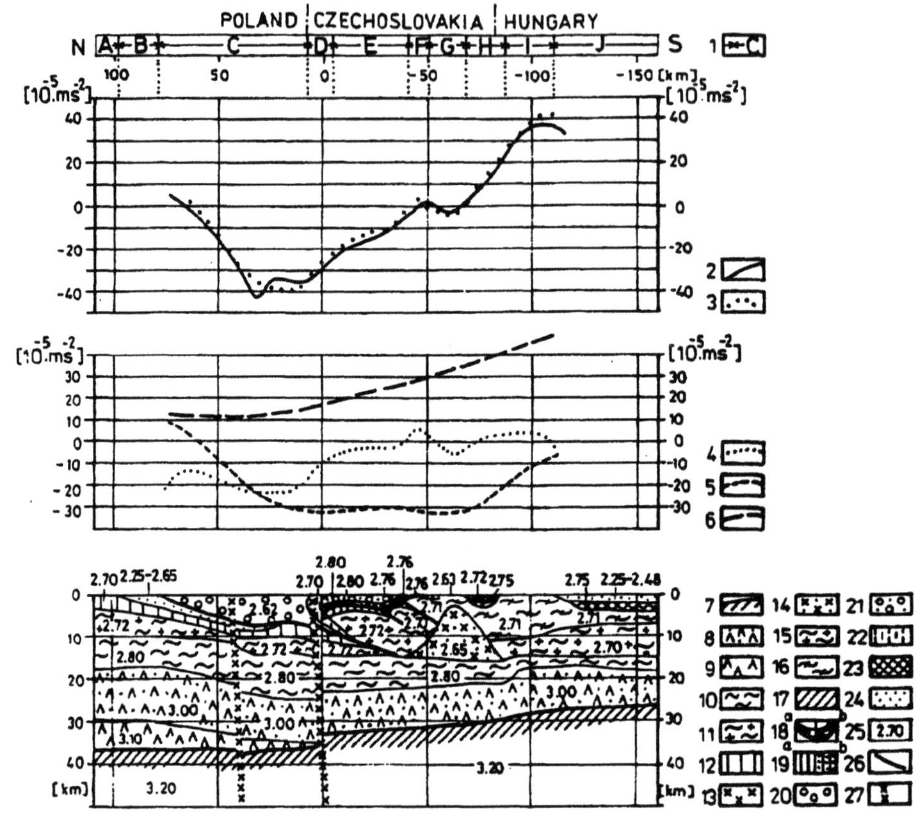

Fig. 6.16 Density model along DSS profile No. V. 1 – boundaries of fundamental geologic units: A – Palaeozoic platform cover, B – Carpathian Foredeep, C – Outer Flysch Belt, D – Klippen Belt, E – Inner Carpathian Palaeogene, F – Mesozoic rocks of Galmus, G – Palaeozoic rocks of the Volovské vrchy Mts., H – Mesozoic rocks of the Slovakian Karst, I – Palaeozoic rocks of the Szendrö Mts., J – Pannonian Basin, 2 – Bouguer anomaly, 3 – total computed gravity effects of the density model, 4 – gravity effects of anomalous masses to a depth of 10 km, 5 – gravity effects of anomalous masses between depths of 10 km and 25 km, 6 – gravity effects of anomalous masses between depths of 25 km and 40 km, 7 – Moho discontinuity course, 8–9 – lower crust (lower part with densities of 3.0 and 3.1 Mg m^{-3}, respectively, 10 – lower crust (upper part with density of 2.8 Mg m^{-3}), 11 – crystalline schists (q = 2.70 Mg m^{-3}), granitoids (q = 2.72 Mg m^{-3}) (the lower part of the upper crust). 12 – Palaeozoic platform cover, 13 – Alpine granites, 14 – granitoids, 15 – crystalline schists and granitoids of Tatricum and Veporicum with densities of 2.72 Mg m^{-3} and 2.70 Mg m^{-3}, respectively, 16 – Palaeozoic rocks of the Volovské vrchy Mts. and of the Szendrö Mts., 17 – Mesozoic Envelope unit of the Veporicum, 18 – Mesozoic rocks of the a) Galmus, b) Slovakian Karst, 19 – Mesozoic rocks of the a) Tatricum (the Envelope unit), b) Krížna nappe, 20 – Palaeogene sediments of the Outer West Carpathians, 21 – Sediments of the Inner West Carpathian Palaeogene, 22 – Klippen Belt, 23 – Mesozoic rocks of the Hungarian Midmountains, 24 – sediments of the Pannonian Basin, 25 – rock density in Mg m^{-3}, 26 – nappe boundary, 27 – deep fault.

3. the gravity effect of the top-level masses (0 – 10 km) correlates, first of all, with the thickness of the Flysch Belt sediments and of sediments underlying them.

In the northern part of the profile, the crust–mantle boundary reaches a maximum depth of 45 km. The lower crust is assumed to be formed by a thick layer of basic rocks (3.00 Mg m^{-3}) and by ultrabasic rocks (3.10 Mg m^{-3}) in its lowest part. The upper part of the crust (layers of densities 2.80 Mg m^{-3} and 2.72 Mg m^{-3}) exhibits a "normal" thickness of 12 – 14 km and is covered by a Palaeozoic platform cover (2.70 Mg m^{-3}), sediments of the Carpathian foredeep which are overthrust by nappes of the outer flysch zone.

The inner West Carpathians are characterized by the Moho discontinuity dipping northwards and by thinner lower crust (2.80 Mg m^{-3}). For this region, a complex structure of the upper crust is typical. It consists of portions of the Mesozoic and crystalline nappes (densities of 2.72, 2.70, 2.75 Mg m^{-3}), partially Palaeogene cover (2.60 Mg m^{-3}), and deep granitoid bodies (2.63 Mg m^{-3} – 2.65 Mg m^{-3}) of Palaeo-Alpine age. It is assumed that young intrusive bodies can also be present in this region. The deep fault zones which separate this region from the Carpathian Foredeep, are considered Neotectonic structure.

The typical element of the Earth's crust in the Pannonian Basin is the small depth of the Moho discontinuity (25 – 26 km), the reduction of the thickness of the lower crust, and considerable thickness of uncemented Tertiary sediments (i.e. of 3 km, and in other regions of 7 – 10 km).

Density Model Along the Trans-Carpathian 2T Profile

The density model of the section (Fig. 6.17) is based on the deep seismic reflection profiling using the CDP method with a prolonged recording time (see Chapter 2.4).

The regional seismic profiles carried out for prospecting for hydrocarbons show that the Brunovistulicum basement of the Carpathians (Dudek 1980) extends beyond the Pienniny Klippen Belt beneath the inner West Carpathians, where it occurs at depths of 10 – 13 km. Moreover, these profiles document that the Magura group of nappes (Flysch Belt) has a scaled structure, and that underneath it, a strikingly reflective medium exists which can be classified as lower nappes of the Krosno formation – Dukla nappe units of the outer Flysch Belt. The rocks there exhibit a relatively anomalous porosity and can be a source of the West-Carpathian gravity low (similarly to the East Carpathians).

Results of the drilling in deep boreholes in the Magura group of nappes resp. in the Krosno formation – Dukla nappe (inner and outer Flysch Belt respectively) indirectly confirm this hypothesis. As the analyses have shown the rocks of the Magura nappes are substantially less porous than those of the outer Flysch Belt, and their porosity decreases along the Carpathians from W to E. A decrease of differential densities of the rocks of the Magura nappes with depth was confirmed by results of deep boreholes Jarošov-1, Vizovice-1, Jablůnka-1, and Smilno-1.

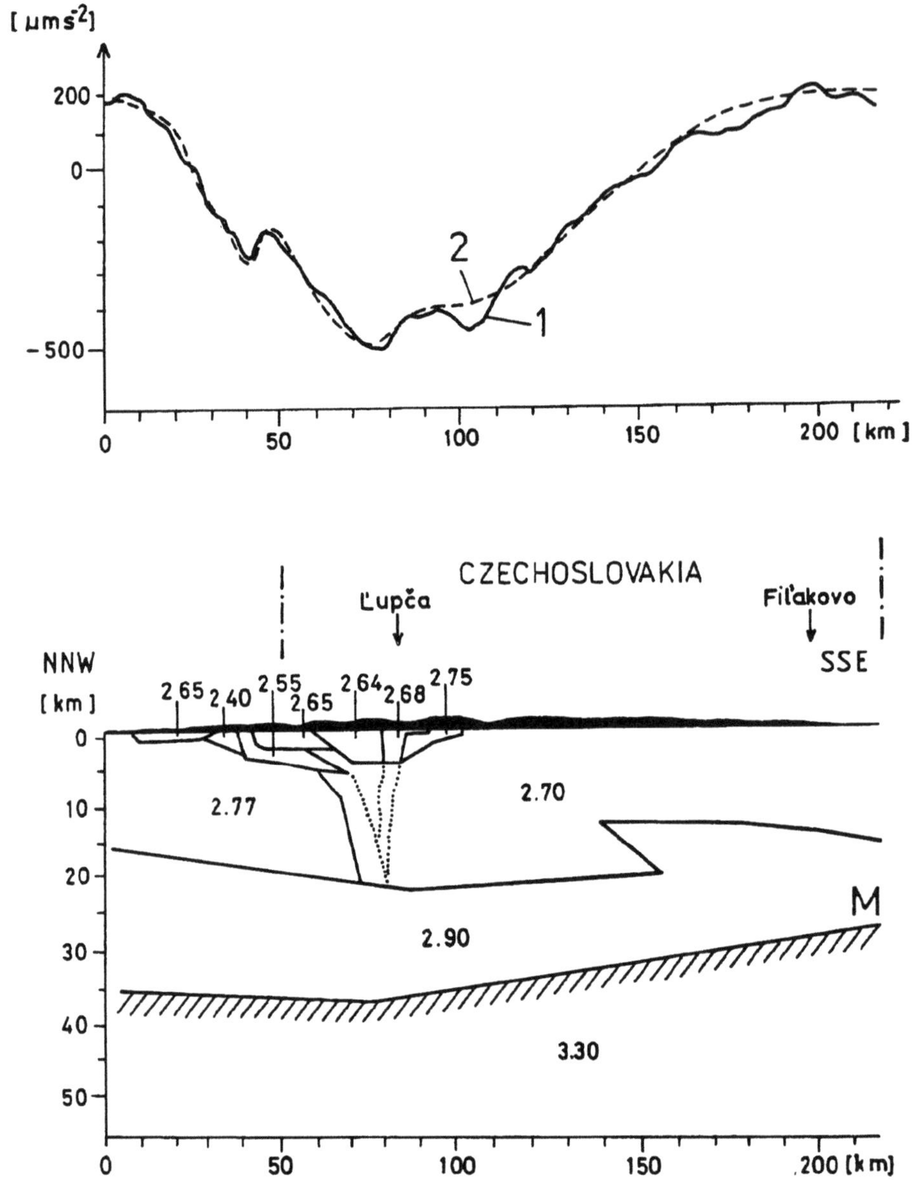

Fig. 6.17 Density model along trans-Carpathians 2T profile: 1 – Bouguer anomaly, 2 – computed gravity effects of the model: numbers 2.40 – 3.30 designate densities in Mg m^{-3}.

Rocks of the Krosno formation – Dukla nappe of the outer Flysch Belt exhibit higher porosity even at great depths (the Ševčenkovo-1 borehole in the Ukrainian Carpathians). Great differences exist in the different densities of individual nappes. The lowermost nappes, i.e. the Borislav – Pokuty nappe (classified by Soviet authors also as the Carpathian foredeep) and the Ždánice – Subsilesian nappe are generally of the highest porosity (i.e. 10 – 15%). Upper nappes, the Skole, the Silesian, and the Fore-Magura ones, consist of rocks of lower porosity (5 – 10%). Within the Silesian nappe itself, however, great differences in porosities exist, e.g. between the low-porosity Godula facies and the porous Krosno Formation. The rocks of the uppermost Dukla nappe exhibit nearly the same porosity as the rocks of the Magura group of nappes. The data on density were taken from Polish and Soviet (Bortnickaja et al. 1977) measurements.

Applying Parker's theory of ideal bodies (Tomek et al. 1979) in the Moravian–Silesian Beskydy Mts. (an outer margin of the West Carpathian gravity low) and in the region between Dukla and Bardejov in east Slovakia it was derived that the top part of the most important source of the gravity low lies at depth of maximum 4 to 5 km. Also the anomaly width, the protuberances of the minima in the Jablunkov and Ziwiec troughs, the continuity of the East Carpathian and West Carpathian anomalies, etc. document at least a partially upper-crustal origin of the gravity low (Tomek 1982).

Profile 2T begins at the margin of the Carpathian overthrust of the Outer Flysch Belt near Bialsko in Poland, extends over the Magura group of nappes, the Pienniny Klippen Belt, all the inner Carpathian units, and ends in the Gemericum at the border with Hungary. While the northern gravity high on the profile belongs to the positive block of the Precambrian Brunovistulicum, the south one belongs to the positive anomaly of the Pannonian Basin.

The gravity low is mainly due to porous rocks of the outer Flysch Belt. The other contribution comes from the Brunovistulian upper crust subducted into the lower crustal position beneath the accretionary wedge. Between 72 – 82 km, the profile crosses the Ziwiec tectonic window of the Subsilesian nappe, which causes a local negative anomaly with amplitude of about 100 m s^{-2}. The contribution of the Magura group of nappes is minimal. It follows from the Polish data (borehole Bystra IG-1), and from the Czechoslovak borehole Oščadnica-1, that the Magura group of nappes contains "denser" rocks than in east Moravia.

A palaeosubduction can be speculated on during the Badenian from the contact of blocks with different densities and from the tectonic activation of the outer Flysch Belt including *N* parts of the inner West Carpathians. The "teeth-like" inhomogeneity within the Earth's crust points at remnants of older, probably Cretaceous overthrusts. The Brunovistulicum as the basement of the outer, and probably also of the inner, West Carpathians, was not active in the overthrust faulting.

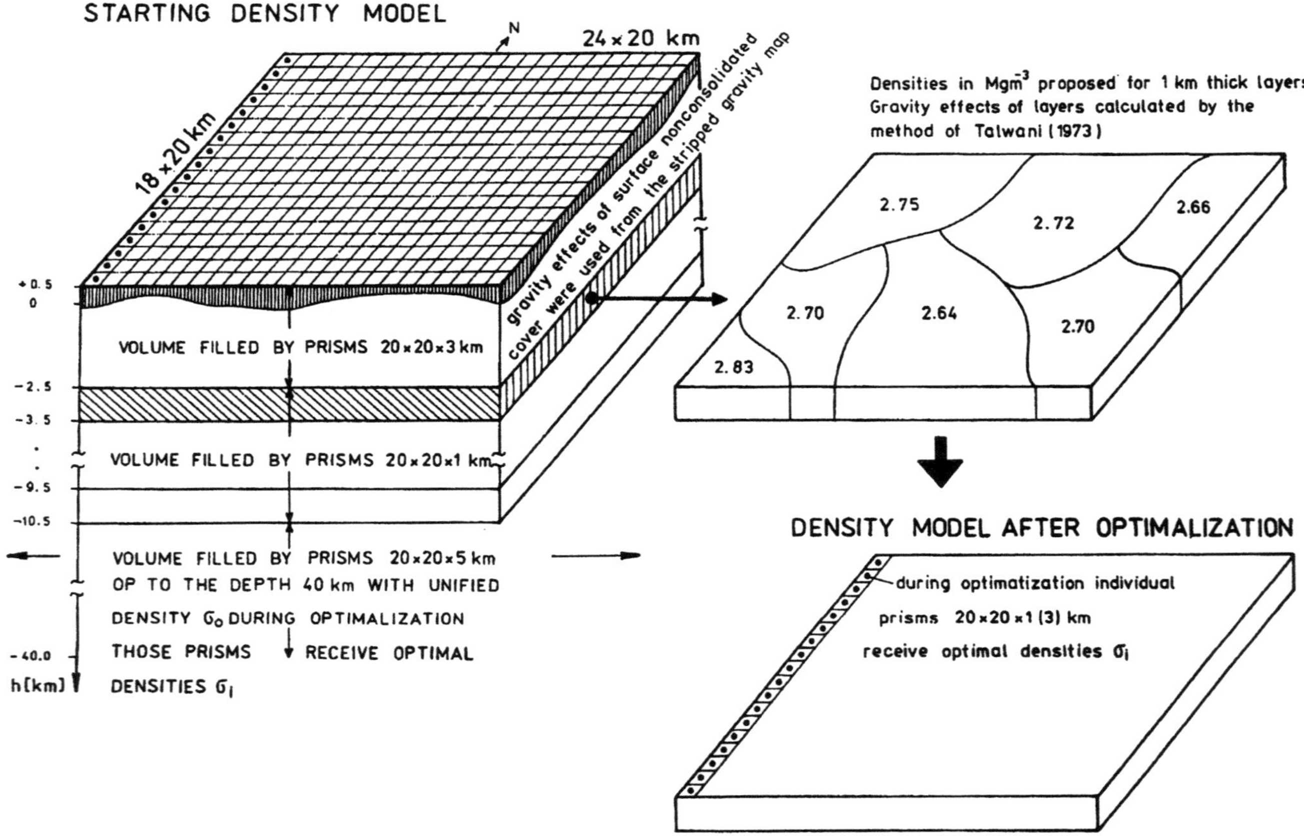

Fig. 6.18 Starting density model for 3-D modelling and its optimization.

6.5 3-D Density Models of the Earth's Crust in the Bohemian Massif

M. Pick[1], I. Charvátová-Jakubcová[1]

The starting 3-D density model of the Bohemian Massif was compiled from the following data:

1. gravity map of Czechoslovakia (Ibrmajer 1965),
2. interpretation of the gravitational field (Buday et al. 1969),
3. rock densities of the Bohemian Massif (Ondra and Hanák 1981),
4. horizontal geologic section at levels 0 km, – 3 km and – 10 km (Suk and Vacek 1984).

The 3-D density model was compiled in the computation region of 18 × 24 squares with the dimension of 20 km, which includes a volume of 360 × 480 × 10.5 km (Fig. 6.18) where the geologic sections were interpolated at an interval of 1 km. The gravity effect of the model was calculated, using the Talwani method, at the points of the square grid of 20 × 20 km.

The effect was then compared with the Bouguer gravity map and the constant shift was removed,

$$\delta g = [G(M) - G(z)] - \sum \frac{[G(M) - G(z)]}{M} \quad .$$

In this way, the field of deviations dg was obtained above the whole model.

In the course of the optimization process the methods of Pola (1967) and of Berthelmes and Kautzleben (1983) were used. They modelled the disturbing gravitational field by a set of mass points, located in the space field of deviations. Points and masses joined by them are searched; their location and magnitudes are touched up by successive approximation. These methods converge fast and always yield the same result. The method was developed by Pick and Jakubcová and has been described in Blížkovský et al. (1985b).

Fig. 6.19 illustrates resultant density models at depth levels of – 3 km, – 8 km, and – 10 km. The models significantly contribute to analysis of positively and negatively disturbed gravitational field of generally NE–SW direction in the Bohemian Massif.

It appears that in positively disturbed regions in the – 10 km section, a lower part of the crust is involved than in negatively disturbed regions, where the deeper part is masked by gravity effects of low-density rocks (sedimentary formations in east part of the Bohemian Massif, granitoid bodies at shallow depths of the Moldanubicum and of the Krušné hory Mts.).

A distinct contrast is observed between the more complex deep structure of the Bohemian Massif and the simple structure of the Brunovistulicum. These results correspond well with the 2-D density models constructed along DSS profiles No. VI and VII, mutually supporting each other.

---

1 Geophysical Institute, Boční II/1401, 141 31 Praha 4, Czech Republic

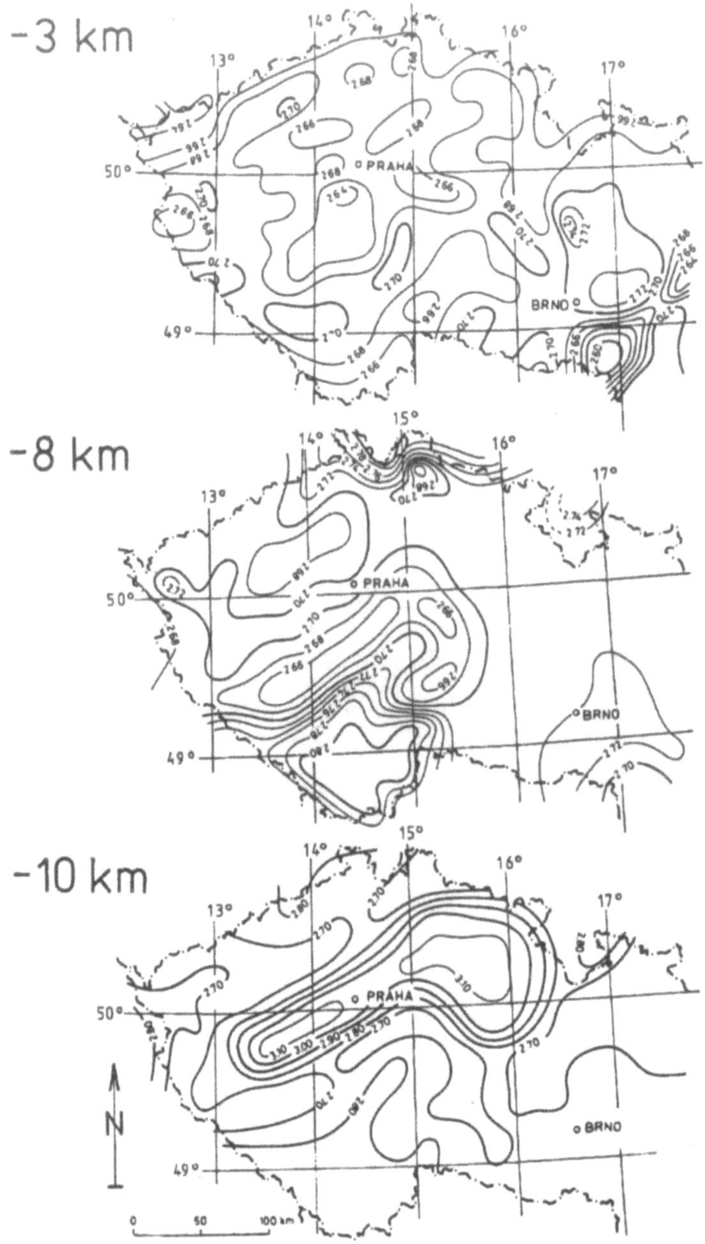

Fig. 6.19 Density models (Mg m^{-3}) at depths levels – 3 km, – 8 km, and – 10 km.

7 Geomagnetic and Palaeomagnetic Fields

7.1 Geomagnetic Field of Central Europe

V. Bucha[1]

The regional geomagnetic field provides a fairly unambiguous information on geological formations occurring on the Earth's surface or at small depths. Magnetic methods, however, have not been frequently applied to investigating deeper parts of the Earth's crust; it had been, indeed, presumed that due to the wide spectrum of the magnetic properties of rocks, the local anomalies had obscured weaker signals of sources at greater depths and thus had not allowed realistic conclusions to be drawn. But synthetically processed regional mapping and aeromagnetic maps provide abundant material which enables us to study the tectonics with regard to the distribution of regional magnetic anomalies.

In the geomagnetic map of Central Europe (Fig. 7.1), the boundary of the Russian Platform is defined by a zone of pronounced geomagnetic anomalies, extending from SE towards NW in the territories of the USSR and Poland. A comparison with geologic transverse profiles shows (Sokolowski and Znosko 1958) that the zone of positive magnetic anomalies runs in the vicinity of the axis of the Danish – Polish geosyncline, and the depth of the Permian system (most probably palaeovolcanics) at about 6 km, might indicate their appreciable thickness and strong remanent magnetization. Two arcuate zones of positive magnetic anomalies rim the entire configuration of the Carpathian – Balkan overthrust – the northern belt follows the Magura uplift, the southern zone is characterized by a series of faults that originated in the northern parts of the inner Carpathian structures; neovolcanics are characterized by an inhomogeneous geomagnetic field. The belts of anomalies in the area of the Danube Basin are confined to dominant regional faults.

A comparison of the geomagnetic map of Central Europe (Bucha 1968, 1980, 1986, Krutichovskaya et al. 1971) with tectonic features shows (Fig. 7.1) that the magnetic field is relatively quiet, and that the zone of positive anomalies follows the Carpathian arc. Two zones with an intensity of up to 400 nT can be clearly distinguished. The north zone extending from Vienna through Zlín and farther, via Krakow, towards the southeast characterizes outer Carpathian structures. The main cause of increased magnetization

[1] Geophysical Institute, Boční II/1401, 141 31 Praha 4, Czech Republic

are amphibole granodiorites. Local anomalies in the Ostrava region are caused by teschenite formations. The inner arcuate zone of positive magnetic anomalies runs across South Slovakia through Čop and farther eastward to Vinogradov following the Carpathian–Balkan arc. Middle and Lower Palaeozoic rocks covered with Palaeogene and Neogene sediments are considered as its source. Local anomalies in the zone are caused by neovolcanics.

As regards the magnetic field of the Inner Carpathian structures, a relatively quiet field in the area of the Danube Basin passes in the southern part of Hungary into an anomalous zone extending northeast. The crystalline basement of the Palaeozoic age, and the Mesozoic volcanics are characterized by positive anomalies. In the southwestern part of Hungary, anomalies corresponding to the Pliocene and Pleistocene basalt volcanics can be distinguished.

A group of positive anomalies with an intensity of up to 500 nT in the western part of Romania can be connected with the uplifted block of the Hercynian sub-platform basement formed by Precambrian schists, gneisses, amphibolites, and gneiss intrusions. Corresponding to the pattern of the magnetic field in the central part of Romania, the rocks of this block submerge below the Neogene sediments of the Transsylvanian depression (Krutichovskaya et al. 1971).

Specific geophysical features manifest themselves in the contact area of the Bohemian Massif and the Carpathian system, which is distinguished not only by a

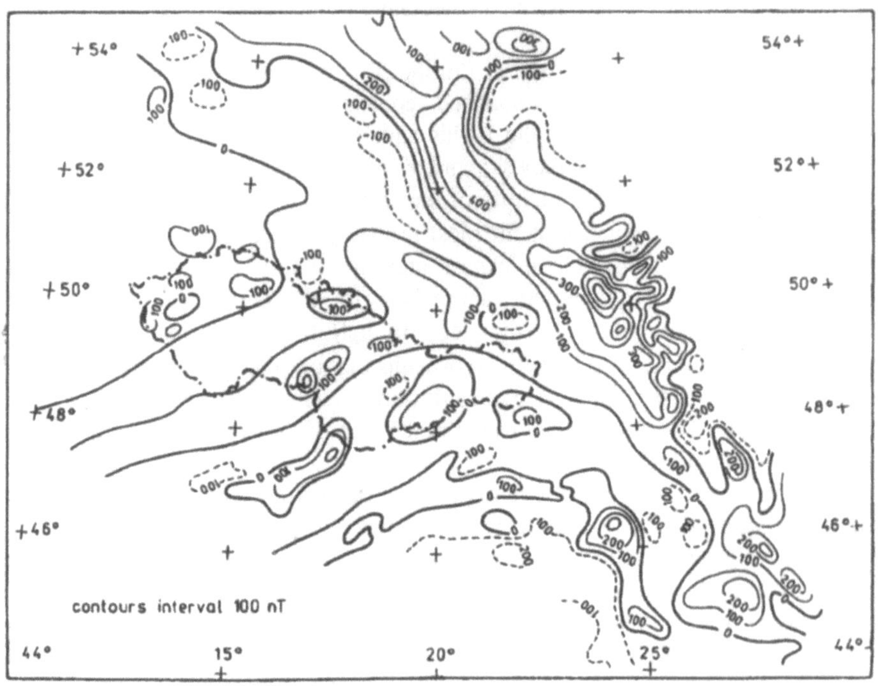

Fig. 7.1 Magnetic field pattern in Central Europe.

pronounced zone of positive geomagnetic anomalies, but also by a parallel belt of negative gravity anomalies, and by a zone of increased electro-conductivity. The pattern of the three zones makes a mutual displacement by 10 – 30 km evident (Bucha 1980, 1981, 1986). The cause may be the anomalous values of physical properties of rocks: the zone of positive magnetic anomalies situated westwards may reflect the occurrence of rocks with high magnetic susceptibility, possibly forming the subsurface elevations of the crystalline basement; the great thickness of low-density Tertiary sediments can be regarded as major cause of negative gravity anomalies. The zone of increased electric conductivity can be associated with the presence of rocks saturated at depths of 15 – 20 km by salt solutions or hydrocarbons.

The negative gravity anomalies can be logically accounted for by a considerable thickness of lower-density sediments of the Vienna Basin and the Flysch Belt, which can be connected with the tectonic processes leading to the thrusting of the Carpathian nappes over the Bohemian Massif. The zone of geomagnetic anomalies, stretching from SW to NE in the study area, is divided transversely into several blocks. The anomaly with its maximum in South Moravia reaches as far as the southwestern part of Slovakia and encompasses also the area of northeastern Austria. The consequences of this source are also manifested in southern Slovakia and northeastern Austria. The body, whose surface is identical with the crystalline basement can be considered to be the basic cause of the magnetic anomaly (see Section 7.4).

7.2 Geologic Interpretation of Main Geologic Structures in the Bohemian Massif

K. Šalanský[1]

The set of aeromagnetic maps covering the entire Bohemian Massif was compiled using the normal field for 1958.0. Absolute values of the normal field for this were in the range of 47 000 to 47 500 nT. In the Bohemian Massif, values of the anomalous magnetic field ΔT range from negative values of lower than – 1 000 nT to several thousands of nT. The extreme positive values are sporadic occurring above strongly magnetized basaltoid neovolcanics. Highest values of magnetic anomalies in the Bohemian Massif were obtained by aeromagnetic measurements over the Doupovské hory Mts. Positive values reaching 1 000 nT can be considered exceptional, even though they occur over several geologic units, associated mainly with basic complexes of palaeovolcanics and intrusives. They can sporadically indicate magnetite ore deposits. In the amplitude spectrum local and regional magnetic anomalies not exceeding 100 nT prevail. Distributions of anomalies with amplitudes exceeding 25 nT, 100 nT, and

[1] Geofyzika Praha, Geologická 2, 152 00 Praha 5, Czech Republic

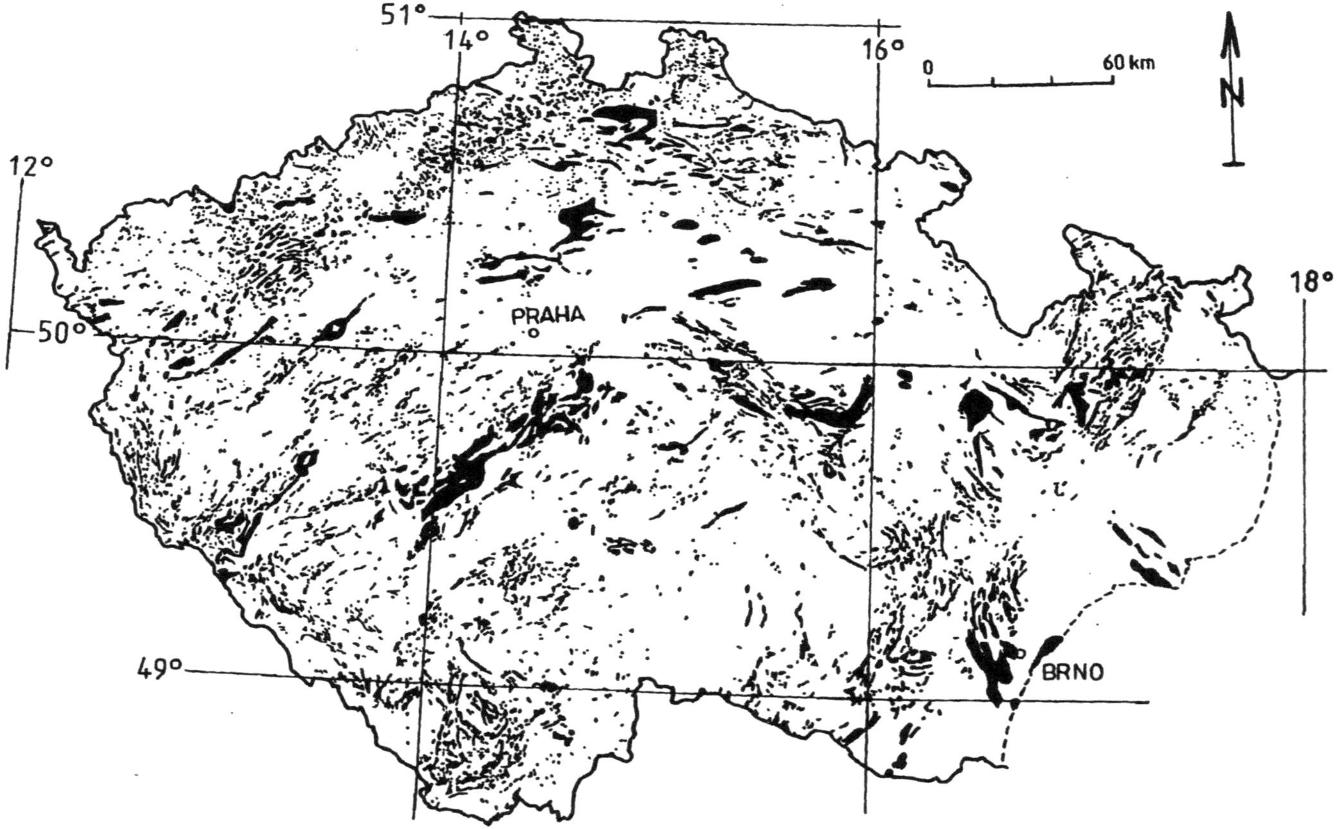

Fig. 7.2 Magnetic anomalies of ΔT in the Bohemian Massif (amplitudes over 25 nT).

500 nT are shown in Figs. 7.2, 7.3, and 7.4, respectively. Mostly anomalies with amplitudes exceeding 25 nT occur in the Bohemian Massif, except in regions of granitoids, of rocks of the platform cover, and of rocks of Moravian–Silesian culm. Considerably lower is the number of anomalies exceeding 100 nT. In the crystalline units, there are anomalies with amplitudes > 100 nT. Anomalies exceeding 500 nT are confined only to neovolcanics, to the Central Bohemian magnetic anomaly, the Železné hory massif, and the Brno Massif. In other regions the anomalies (> 500 nT) occur only sporadically as isolated features, e.g. in the whole South Bohemian Moldanubicum, there are no anomalies reaching 500 nT.

As far as the regional magnetic anomalies are concerned they are shown with various accuracy already in older magnetic maps, e.g. in the Aeromagnetic map of Czechoslovakia compiled on a scale 1:1 000 000. Recently, a map of regional anomalies of the Bohemian Massif was compiled from the data obtained by aeromagnetic mapping on a scale 1:25 000 during the period 1960 – 1972, by analytical continuation of anomalous field ΔT on the 950 m level (above terrain surface) (Pokorný et al. 1985). The map is shown in Fig. 7.5 in somewhat simplified form. In the map, the course of regional magnetic anomaly is clearly observable. In the analytical pattern, it is necessary to distinguish the regional anomalies, which originated by summing numerous local anomalies caused by near-surface rocks, from the regional ones caused by deeper sources (depth greater than 1 km). They are a part of regional magnetic structures of the Bohemian Massif (Šalanský 1983). Several anomalies exist, though, which are caused by deep sources. On the basis of the reconnaissance aeromagnetic mapping of Czechoslovakia, Man and Mašín (1967) tried to divide the anomaly field ΔT into local

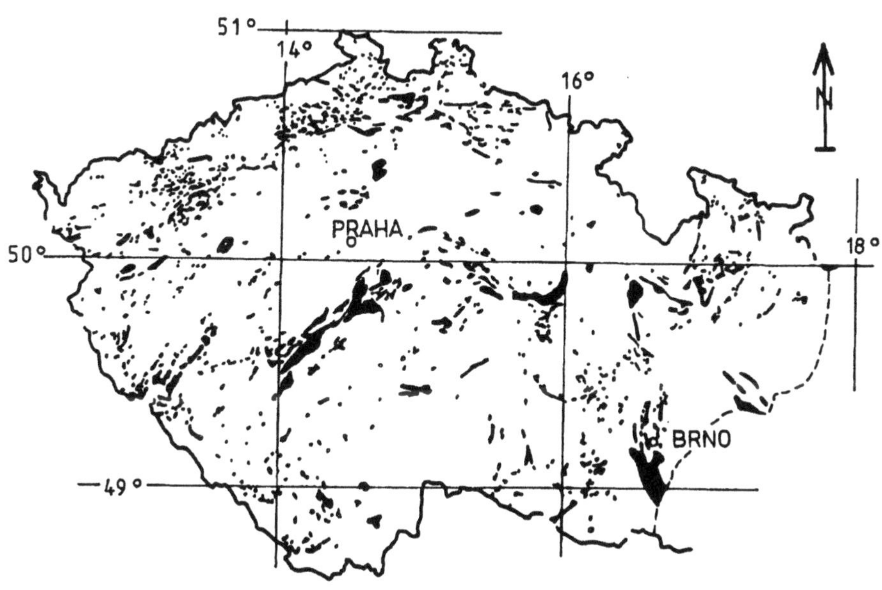

Fig. 7.3 Magnetic anomalies of ΔT in the Bohemian Massif (amplitudes over 100 nT).

and regional components. In the Moravian regional anomaly, in the Teplá crystalline unit, in the Bohemian Cretaceous Basin in the Louny area, and in the Krkonoše–Piedmont Basin, the sources are assumed at depths greater than 5 km.

While the gravitational field pattern of the Bohemian Massif is divided into positive and negative regional bands, the magnetic field pattern is considerably more complex due to occurrences of other areally extensive sources, which disturb the zonal character of the anomalous magnetic field. Moving from E to W, the positive-anomaly region of neovolcanics of the Doupovské hory Mts. and of the České středohoří Mts. with numerous intensively magnetized volcanic bodies are in the neighbourhood of the positively disturbed Krušné hory Mts. region. We may assume the presence of extensive subvolcanic to intrusive bodies in the region of neovolcanics. The Lugicum (minus the Orlické hory Mts.) and Silesicum occur as magnetically positively disturbed areas. At deeper levels, they are probably extensively matter-differentiated, prevailingly basic, intrusive complexes. Similarly, the West Moravian positive anomaly with its centre near the town of Moravský Krumlov, represents a total effect of a high-density block of hidden intrusive complex at depth of several kilometres. The Moravian magnetic anomaly represents the greatest magnetic structure at the boundary of the Bohemian Massif and West Carpathians.

Other regional magnetic anomalies are caused by combined effect of the near-surface and deep sources. Among these are the regional magnetic highs of the Teplá crystalline unit, the Přeštice magnetic high, the Central Bohemian magnetic high, the Kutná Hora crystalline unit, and the Železné hory Mts. which include another regional magnetic

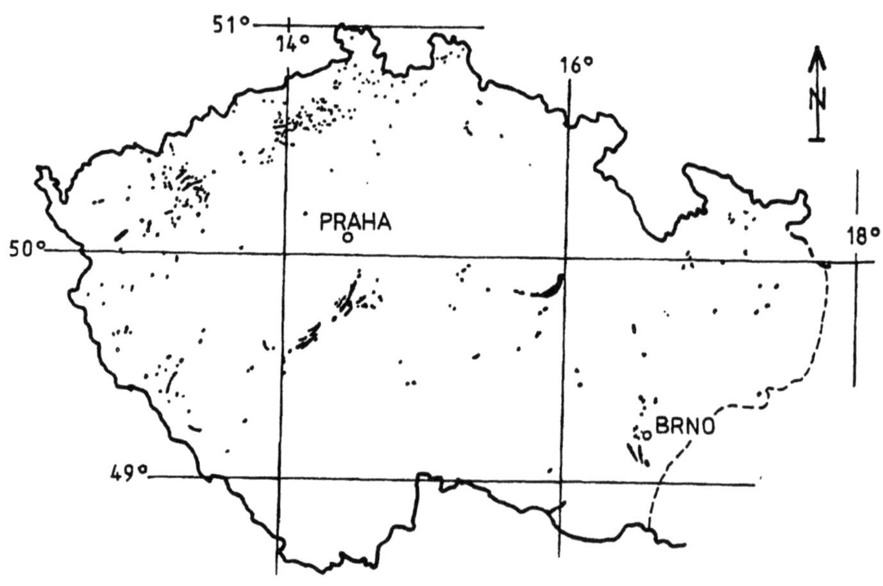

Fig. 7.4 Magnetic anomalies of ΔT in the Bohemian Massif (amplitudes over 500 nT).

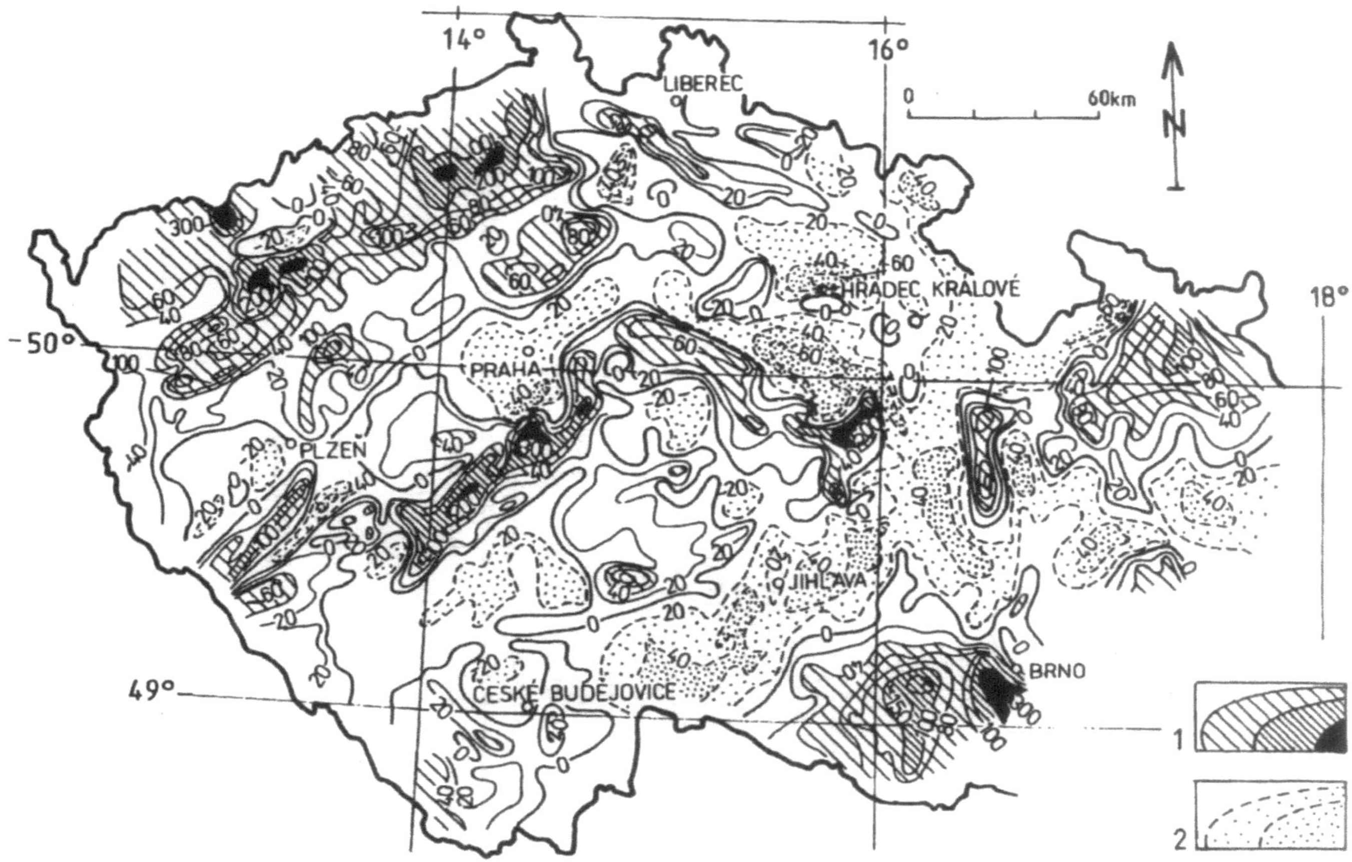

Fig. 7.5 Magnetic (ΔT) field pattern at 950 m level above the ground surface (after Pokorný et al. 1985). 1 – positive anomalies, 2 – negative anomalies.

high. Regional positive anomalies of this type are caused mostly by complexes of palaeovolcanics and of volcano-sedimentary rocks sometimes in combination with intrusive complexes (Šalanský 1983).

Extensive areas of negative magnetic anomalies occur over complexes of nonmagnetic sedimentary rocks or over Hercynian granitoid massifs (e.g. the Central Moldanubicum Pluton).

From a comparison of the regional pattern of the Bouguer gravity with the geomagnetic field pattern it is obvious that above the blocks of high-density, prevailingly basic volcanic and plutonic rocks, high ΔT values are observed. On the other hand, granitoid massifs are characterized by extensive strong gravity depressions and by a regional decrease of ΔT values of the geomagnetic field. The Krušné hory Mts. and Krkonoše Mts. blocks are exceptions, as above them, striking negative gravity lows coincide with slightly increased positive geomagnetic field intensity. In such cases, we may assume the occurrence of basic rocks beneath granitoids in those places where the crustal thickness is reduced.

In the Bohemian Massif, magnetic anomalies caused by near-surface or surficial sources prevail. It is therefore obvious that magnetometry data contain an important geologic information for basic geologic investigations but limited information for the research into a deep geologic structure of the crust. The geomagnetic field thus largely reflects shallow geologic structure, providing a great amount of information on magnetically differentiated objects. The geomagnetic field pattern can be used to classify the Bohemian Massif. This classification has to be done, first of all, on the basis of correlation of anomalies due to the same types of sources, correlation of gradients of the fields, correlation of individual, anomaly axes direction. To some extent amplitudes of the anomalies can also be used for the classification, as even minor changes in the content of ferri-magnetic accessories cause great changes of the amplitudes. They have to be used carefully, though, in order not to slip into correlating heterogeneous geological bodies of various age and genesis.

The geomagnetic field of the Bohemian Massif can be divided into regions of positive and negative regional anomalies of the first order. These can be further divided into subregions of the second order, etc. The first-order positive anomaly regions are as follows:

1. Krušné hory region with adjacent neovolcanics,
2. Central Bohemian anomaly,
3. Kutná Hora unit and Železné hory crystalline unit,
4. Silesian region,
5. West Moravian anomaly.

The Moravian magnetic anomaly is not included as it is already in the Carpathian Foredeep and Flysch Belt. Out of the less expressive positive anomaly regions, the Bohemian Moldanubicum and the Český Les Mts. with the Domažlice crystalline unit can be included here. Weakly positive regions are in the major part of the Šumava Mts. Moldanubicum.

The first-order negative anomaly regions are:

1. East part of the Bohemian Cretaceous Basin and Orlické hory Mts.,
2. Central-Moldanubicum massif.

Granitoids and parametamorphites as well as palaeovolcanic rocks in crystalline units, most likely determine the positive or negative character of the geomagnetic field. Extensive positive anomalies as the Silesian and West Moravian ones are probably due to deep intrusive complexes of gabbroid to ultrabasic rocks. In negative-anomaly regions, granitoids either crop out or occur at shallow depths. Intensive migmatization of para-gneiss substrate also causes negative anomalies (Orlické hory Mts., Kralický Sněžník Mt.).

In the Bohemian Massif, aeromagnetic mapping resulted in localization of more than 2000 anomalies. Their linear types are abundant, isometric and irregular shape are less frequent.

The vast majority of anomalies are caused by normal polarization of the geomagnetic field so that positive anomalies are accompanied by minima at their northern side. Positive anomalies without accompanying minima are less frequent. Negative anomalies caused by a reverse polarization are fully negligible. In regions of neovolcanics and sporadically also in other regions dipole anomalies occur normally as well as reversely polarized. Linear anomalies are often grouped into more extensive magnetic bands thus forming regional magnetic structures. Effects of magnetic sources located side by side or one above the other are superposed resulting in complex magnetic anomalies. Joint anomalies must be separated from each other using computational procedures. A relatively good geologic investigation coverage of the Bohemian Massif and knowledge of magnetic properties of outcropping magnetized rocks make an interpretation of local ΔT anomalies easier. Even though, there is a need to proceed to the ground magnetometry follow-up elsewhere in order to define contours of aeromagnetic anomalies more accurately. The aeromagnetic map is sufficient for use in geologic research and in regional structural and material pattern recognition, in delimiting extents of palaeovolcanic rocks, contact aureoles of hidden intrusive bodies, fault systems, etc.

7.3 Geologic Interpretation of the Geomagnetic Anomalies in the West Carpathians

M. Filo[1], P. Kubeš[1]

Introduction

Synoptic map of geomagnetic anomalies (Fig. 7.6) provides a basic idea of the geomagnetic field character in the West Carpathians. It was compiled from the results of ground and aerial magnetic mapping on scales from 1:25 000 to 1:100 000, which were completed between 1955 – 1983. A substantial part of the territory is covered by measurements of the vertical component (ΔZ) of the total vector of the geomagnetic field. As original maps of ΔZ were compiled by various authors for different geomagnetic epoches, contours do not match. The measurements, conducted at the contact of the Bohemian Massif with the West Carpathians, in the Danube Basin, in the Slovenské rudohorie Mts., and in East Slovakia, show a comprehensive character. In regions of the Tertiary volcanic rocks, aeromagnetic measurements conducted at altitudes of 80 m, 300 m, and 500 m were used. The least data on the geomagnetic field character are available for high mountains. For the above-stated reasons, the map presented can be used for a qualitative analysis of the geomagnetic field. In order to make a quantitative analysis with the aim to determine fundamental parameters of magnetized rocks we use the data from original maps and profiles of ΔZ or ΔT.

Fundamental Elements of the Geomagnetic Field of the West Carpathians

The compiled map indicates a presence of two regional magnetically active zones of NE–SW to E–W directions. Their courses correlate well with the main geologic-morphological structures of the Carpathian arc.

The zone in the north (VIII in Fig. 7.6), mean width approximately 60 km, and whose length is around 600 km, extends in SW directions far into Austrian territory and E, into the Polish territory, where it gradually disappears (Krutichovskaya et al. 1971). The zone rims the contact of the Bohemian Massif with the West Carpathians. A geologic-geophysical interpretation of this zone as well as its individual anomalies is given in Chapter 7.5.

The zone in the south shows far a more complex pattern. It extends from NW Hungary over South Slovakia and North Hungary, reaching into the USSR territory. The most intensive anomalies are confined to young volcanic mountains (I), the less intensive ones to the Slovenské rudohorie Mts. (IV, V, VI). Areal anomalies with a relatively low amplitude (to 100 nT) are observed over morphological depressions with great thickness of the Tertiary and Quaternary sediments (II, III, VII).

[1] Geofyzika Bratislava, Geologická 18, 825 52 Bratislava, Slovakia

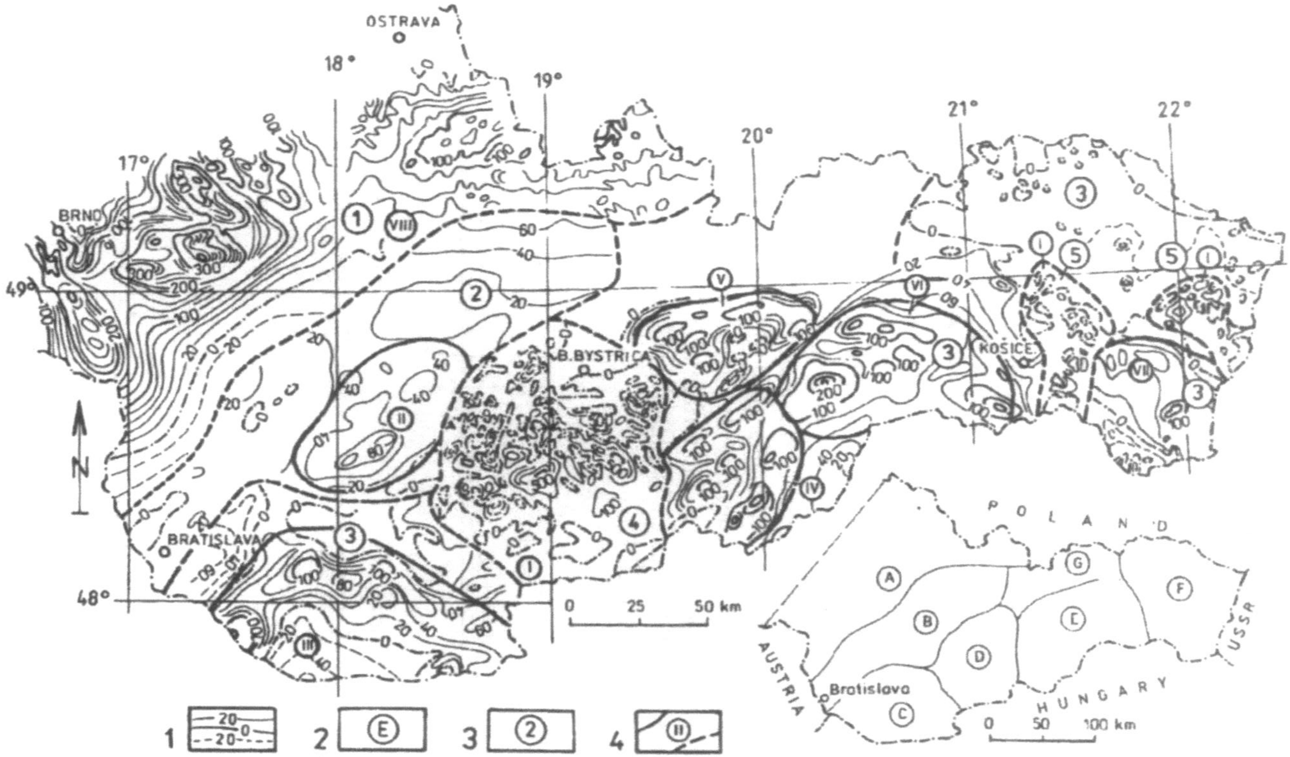

Fig. 7.6 Synoptic map of magnetic field pattern in the West Carpathians. 1 – contours of Z, ΔT (nT), 2 – authors of original maps: A – V. Čekan, J. Doležal, J. Odstrčil, A. Šutora, R. Válek; B – I. Gnojek, K. Dědáček; C – R. Běhounek, V. Čekan, O. Man, K. Müller, A. Šutora; D – K. Šalanský; E – M. Filo, D. Glova, M. Krs; F – V. Čekan, O. Man, K. Šalanský, A. Šutora; G – no data, 3 – type of original maps: 1 and 3 contour maps of Z, contour map of ΔT (80 m above the ground surface), 4 – contour map of ΔT (500 m above the ground surface), 5 – contour map of ΔT (300 m above the ground surface), 6 – Main anomalous region: I – neovolcanics mountain range, II – the Topoľčany depression, III – the Danube lowland, IV, V, VI – the Slovenské rudohorie Mts., VII the East Slovakian lowland, VIII – contact between the Bohemian Massif and the West Carpathians.

In the majority of the depressions the anomalies represent a total of rocks of variable petrographic type, different age, diversified magnetic properties, occurring at wide range of depths.

In the map, numerous boundaries of linear type exist, which can be interpreted as fault and thrust planes extending to various depths (Filo and Medo 1977). In regions of intensive magnetic activity, typically in central volcanic zones, semicircular to circular structures occur.

Results of Quantitative Interpretation of Anomalies

In the course of analysis of magnetometric investigation results obtained in individual parts of the West Carpathians, geologically significant anomalies were quantitatively interpreted along profiles to determine the following main parameters: depth of the upper side of the body, the dip of the body, and the mean magnetization of rocks or rock complexes. In order to solve the problem, Talwani's formula (1973) was applied, which solves the direct magnetic problem for two-dimensional bodies defined by susceptibility, and by coordinates of polygon points. As, in many cases, the rock bodies involved are more or less 3D, the modelling results cannot be considered unique enough to yield fully the geometry and spatial position of anomalous geologic objects. To construct the models, results of magnetic properties, study of all fundamental petrographic types of rocks of the West Carpathians occurring on the surface or in boreholes, as well as data from other geophysical methods application (seismics, gravimetry, and geothermics) and present geologic knowledge were used.

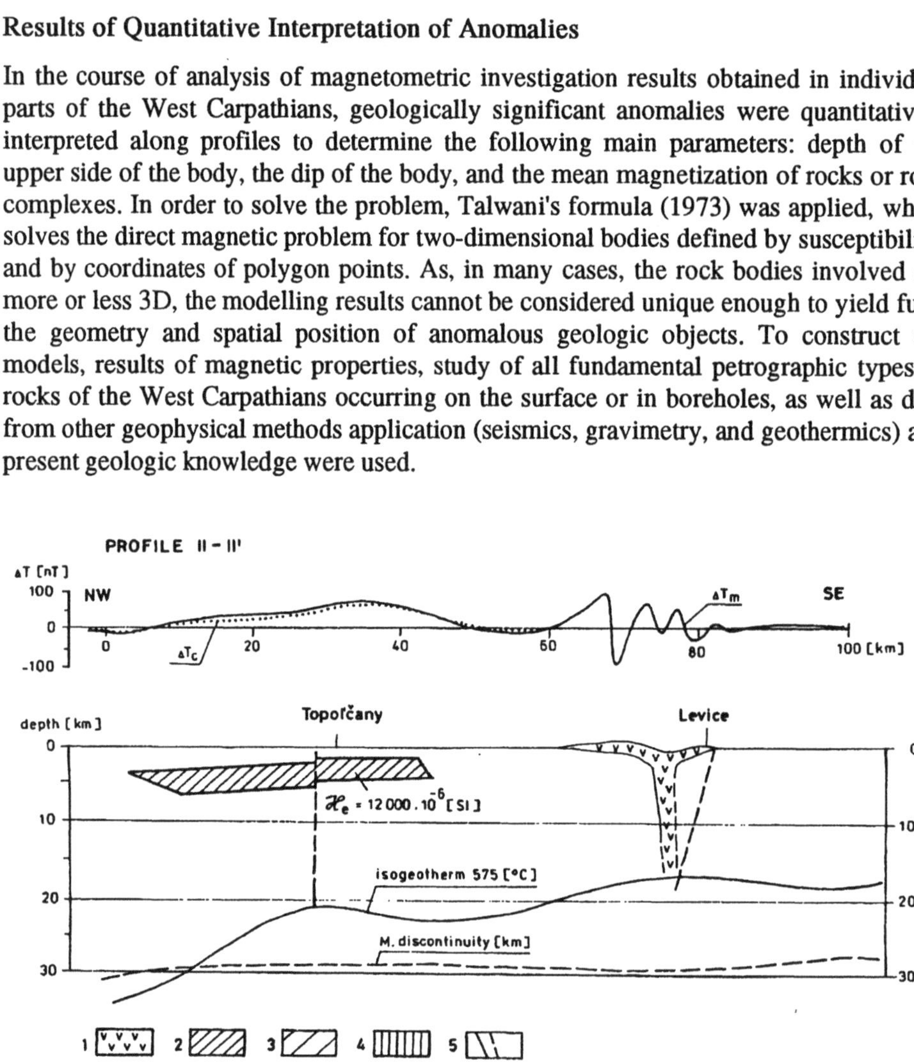

Fig. 7.7 Interpretation of the magnetometric data along profile II–II'. Region II – Topoľčany magnetic anomaly. 1 – andesites and their volcanoclastics, 2 – basic to ultrabasic rocks, 3 – geologically undetermined rocks, 4 – metamorphic rocks, 5 –interpreted faults.

Results of modelling of magnetic bodies from anomalous regions (II, III, IV, and VII in Fig. 7.6) are documented by profiles II–II' (Fig. 7.7), III–III' (Fig. 7.8), IV–IV' (Fig. 7.9), VI–VI' (Fig. 7.10), and VII–VII' (Fig. 7.11). The situation of profiles is shown in Fig. 7.12.

Output models best suit the condition of minimum differences (i.e. of the maximum accordance) between the measured (ΔZ_m, ΔT_m) and the calculated (ΔZ_c, ΔT_c) curves.

Geologic Interpretation of Magnetic Anomalies

Areal and spatial illustration of magnetized rocks in the central and inner West Carpathians (Fig. 7.12) is the result of geophysical-geologic interpretation of geomagnetic field anomalies.

Products of andesite and rhyolite volcanism of high explosiveness show the greatest areal extent. The entire inner side of the Carpathian orogene is rimmed by products of intensive volcanic activity of Central and East Slovakia forming the greatest volcanic complex. In the geomagnetic field they are represented by intensive anomalies with normal and reversed magnetization. The main source of these anomalies are unaltered andesites and their volcano-clastics concentrated around the circumferences of individual central volcanic zones. Over the central zones, a relatively quiet field of low amplitude exists, which is caused by lowering the values of total magnetization of original rocks as a consequence of intensive hydrothermal alterations. The central zones

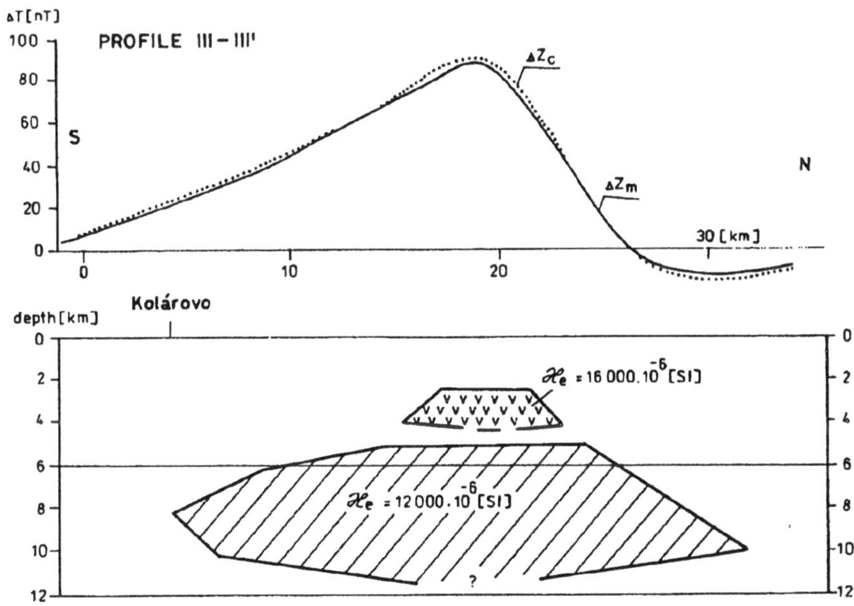

Fig. 7.8 Interpretation of the magnetometric data along profile III–III'. Region III – Kolárovo magnetic anomaly, dtto 7.7.

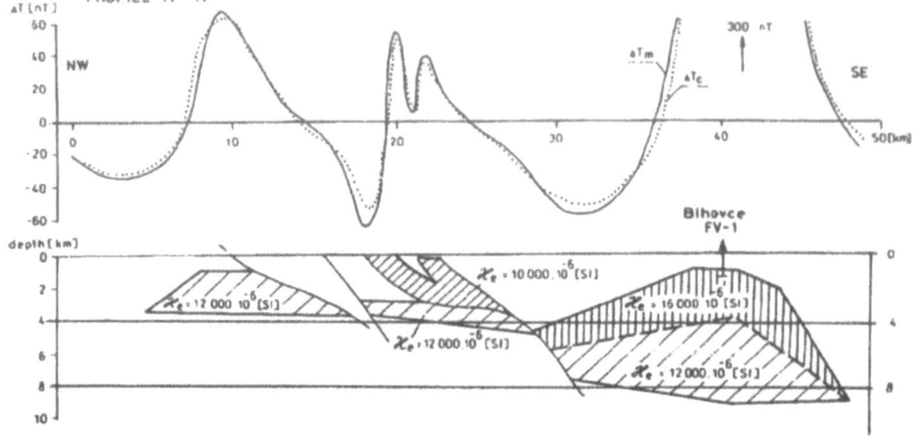

Fig. 7.9 Interpretation of the magnetometric data along profile IV–IV'. Region IV – Blhovce magnetic anomaly, dtto 7.7.

also include intrusions of various petrographic types (granodiorites, diorites, diorite and andesite porphyry, etc.). The central zones show significant metallogenic phenomena, containing polymetallic and precious metals mineralisation.

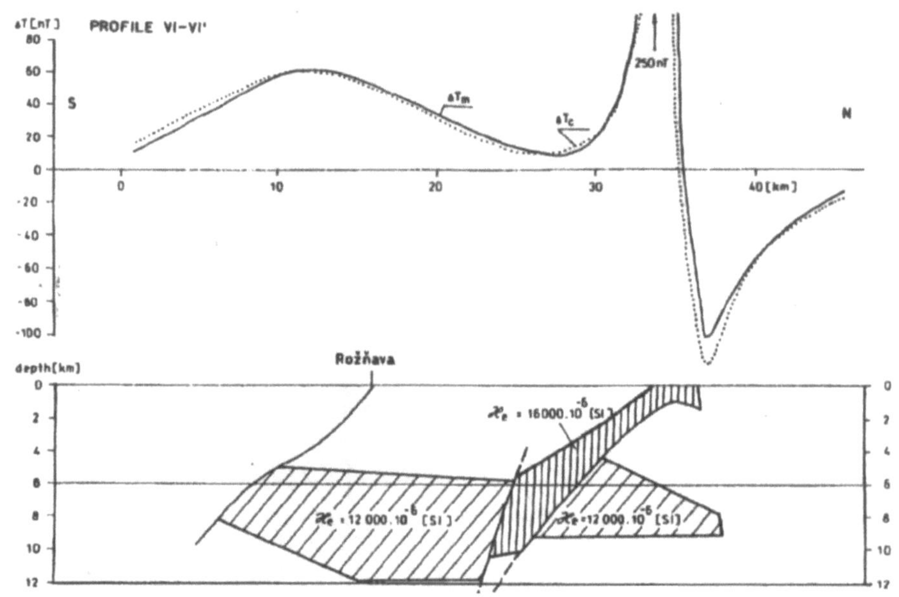

Fig. 7.10 Interpretation of the magnetometric data along profile VI–VI'. Region VI – Rožňava magnetic anomaly, dtto 7.7.

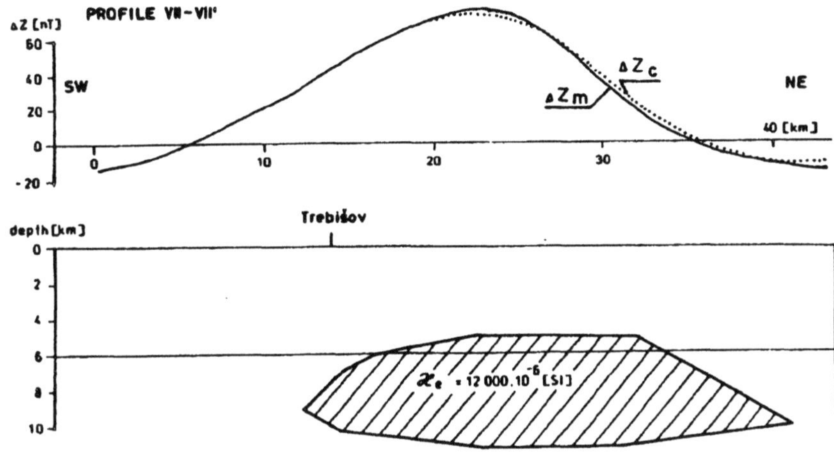

Fig. 7.11 Interpretation of the magnetometric data along profile VII–VII'. Region VII – Trebišov magnetic anomaly, dtto 7.7.

Products of andesite volcanism of great areal and vertical extent occur in the entire Danube Basin (E and SE of Bratislava). Their upper parts are at depths of 1.5 – 4.5 km. Geophysical interpretation of anomalies was confirmed by structural boreholes.

Metamorphosed basic rocks of the early Palaeozoic are another source of magnetic field anomalies of the Danube Basin. Magnetized rocks are oriented into NW–SE direction accompanying the Ludina fault system.

Geologic interpretation of the Kolárovo anomaly (Fig. 7.8) is ambiguous. The anomaly is at the northern margin of striking positive gravitational field anomaly of isometric shape, the centre of which is at the town of Kolárovo. The anomaly could be due to basic rocks at a depth of around 5 km.

NE of the town of Bratislava, there is an extensive complex of magnetized rocks (the Topoľčany anomaly). The top side of the complex is interpreted at a depth of approximately 1.5 km (Fig. 7.7). Above the complex, magnetized bodies of quartz diorites, amphibolites, and melaphyres occur.

Magnetized rocks of the Slovenské rudohorie Mts. (the region bound by towns of Banská Bystrica, Poprad, Košice, and by the border with Hungary) are geologically significant. Magnetized rocks occur at different depths (Figs. 7.9 and 7.10).

In the southern part of the region, numerous ultrabasic bodies occur, the largest one of which is south of the town of Košice. Its position has been confirmed by boreholes. All the known and presumed ultrabasic rocks bodies occur in the vicinity of striking physical boundaries and geologically detected tectonic lines of the thrust type. Based on their position in the structural-tectonic pattern of the central and inner West Carpathians it can be assumed that prior to processes of orogeny it formed relatively continuous zone of an original WSW–ENE direction. A large ultrabasic body at the town of Košice

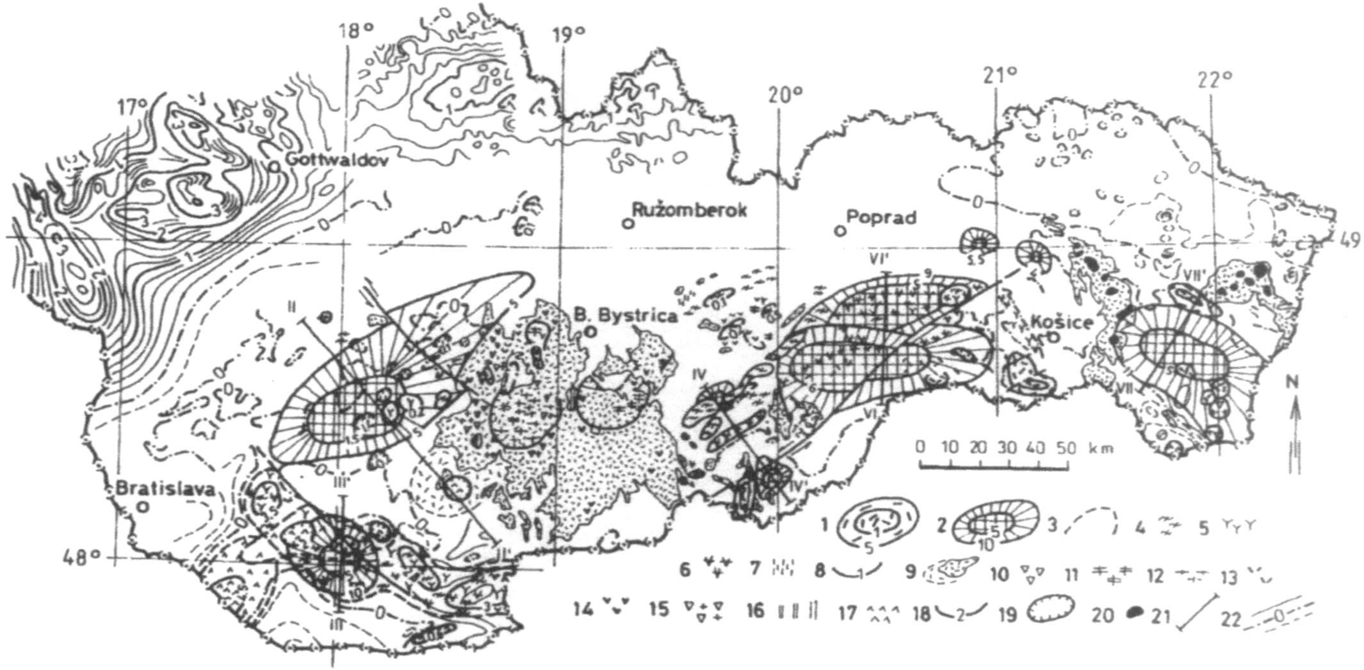

Fig. 7.12 Occurrence of magnetized rocks of Central and Inner West Carpathians. 1 – basic to ultrabasic rocks, metamorphites of basic rocks, 2 – magnetized rocks geologically undetermined (numbers designate depths to the upper and lower limitations of the rock body – in km), 3 – limits of assembly of rocks of similar petrographic type – the depth is not specified, 4 – 8 – palaeovolcanics, 4 – amphibolites, 5 – gabbros, 6 – diabases, 7 – basic rock bodies of the Mesozoic, 8 – contours of rock bodies and their depth in km, 9 – 20 – neovolcanics, 9 – volcanic unspecified, 10 – andesites, 11 – diorites, 12 – granodiorites, 13 – rhyolite extrusions, 14 – andesite extrusions, 15 – intrusive–extrusive rock bodies, 16 – basaltic lava flow, 17 – buried effusive rocks, 18 – contours of buried effusive rocks (depth in km), 19 – caldera extension limit, 20 – central volcanic zones, 21 – interpreted profiles, 22 – contours of ΔZ and ΔT (nT).

as well as bodies of originally basic and ultrabasic rocks in northern Hungary (Posgay 1967) belong to this zone. It cannot be eliminated that magnetized rocks east of the town of Poprad are ultrabasics.

Another source of geomagnetic field anomalies are basic rocks – diabases, amphibolites, gabbroamphibolites, gabbrodiorites, which prevail in the northern part of the Slovenské rudohorie Mts. Metamorphosed rocks of the early Palaeozoic of the Veporicum, i.e. phyllites and mica schists, exhibit an increased magnetization.

In the centre of the Slovenské rudohorie Mts., a striking positive geomagnetic field anomaly and coinciding negative gravitational field anomalies were detected. A structural borehole confirmed highly magnetized granite of Cretaceous age, which in its apical parts transits to aplitic granite and aplite in a depth interval of 700 – 1600 m (the final depth of borehole penetration). A relatively rich high-temperature mineralization of W-Mo type is related to this intrusion.

Other granite bodies, existing in the whole central part of the Slovenské rudohorie Mts., are considered basically low-magnetic. They overlie the Miocene complex of basic rocks, its upper boundary being at a depth of about 5 km.

A large complex of basic rocks is assumed by us also in the central part of the East Slovakian Basin (E of the town of Košice). The complex underlies the Tertiary sediments at a depth of about 5 km (Fig. 7.11).

Conclusion

The interpretation of geomagnetic anomalies confirmed the existence of a striking magnetized zone at southern margin of the central West Carpathians which represents a contact of the Carpathians and the Pannonian block. In the zone, magnetized rocks of various petrographic types and different ages occur. Most of them are of magnetic origin which documents that the contact zone of the inner West Carpathians and the Pannonian block has survived an intensive and long-lasting volcanic activity.

7.4 Geomagnetic Anomalies and the Crystalline Basement of the Vienna Basin

V. Bucha[1]

Investigations into the regional geomagnetic anomalies at the contact of the Bohemian Massif with the West Carpathians including the adjoining part of Austria have commanded attention of geologists, particularly from the point of view of explaining their causes and sources characterized by relatively strong geomagnetic manifestations. Obviously the zone of anomalies stretching from SW to NE in the study area is divided

1 Geophysical Institute, Boční II/1401, 141 31 Praha 4, Czech Republic

transversely into several blocks (Fig. 7.13). The southern regional block (unless we consider the inexpressive positive geomagnetic anomaly at Stockerau – Vienna) reaches as far as the southwestern part of Slovakia and encompasses also the northeastern part of Austria. In the northwest, this anomaly attains the highest values, apparently due to local sources SE of Brno, but a distinct maximum, obviously of regional extent, can be found in the area between Břeclav and Velké Pavlovice. Since the consequences of this source show themselves at distances of 50 – 80 km from the maximum in southern Slovakia and in northern Austria, a possible interpretation is that the source of the anomaly is a body whose surface coincides with the crystalline basement, and which rises closest to the surface at the elevation between Břeclav and Velké Pavlovice, plunging to the southeast. This interpretation is also corroborated by the results of seismological sections investigated in Geofyzika Brno. It is a case analogous to two other regional geomagnetic anomalies with maxima in the environs of the Chřiby Mts., and in the north with a maximum in the Moravian–Silesian Beskydy Mts. According to our conclusions,

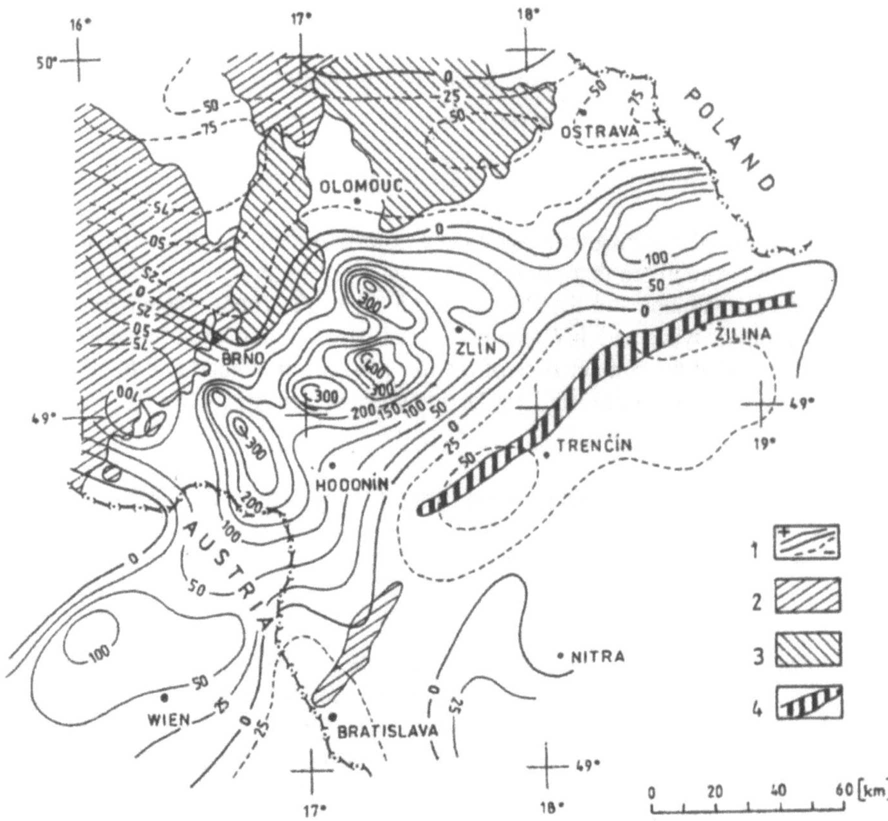

Fig. 7.13 Map of regional anomalies of vertical component of the magnetic field intensity at the contact of the Bohemian Massif with the West Carpathians (contours in nT). 1 – positive (full line) and negative (dashed line) magnetic anomalies, 2 – crystalline complex, 3 – Palaeozoic, 4 – Klippen belt.

the depths of the sources at their maxima would correspond to 3.5 – 5 km. It has been assumed so far that it is a question of either explaining the anomaly as due to regional sources located at depths greater than 7 km or due to local bodies of amphibole granodiorites whose susceptibility is substantially higher in these regions than in the other parts of the crystalline basement, and that the anomalies are not indicative of elevation structures.

But if we compare the relief of the crystalline basement, which we inferred from geologic section, with the distribution of regional geomagnetic anomalies, we can see a similarity between the distribution of the maxima of anomalies (Fig. 7.13) and elevations of the crystalline basement (Fig. 7.14). The same applies to depressions in the manifestations of both factors (Bucha 1981,1986).

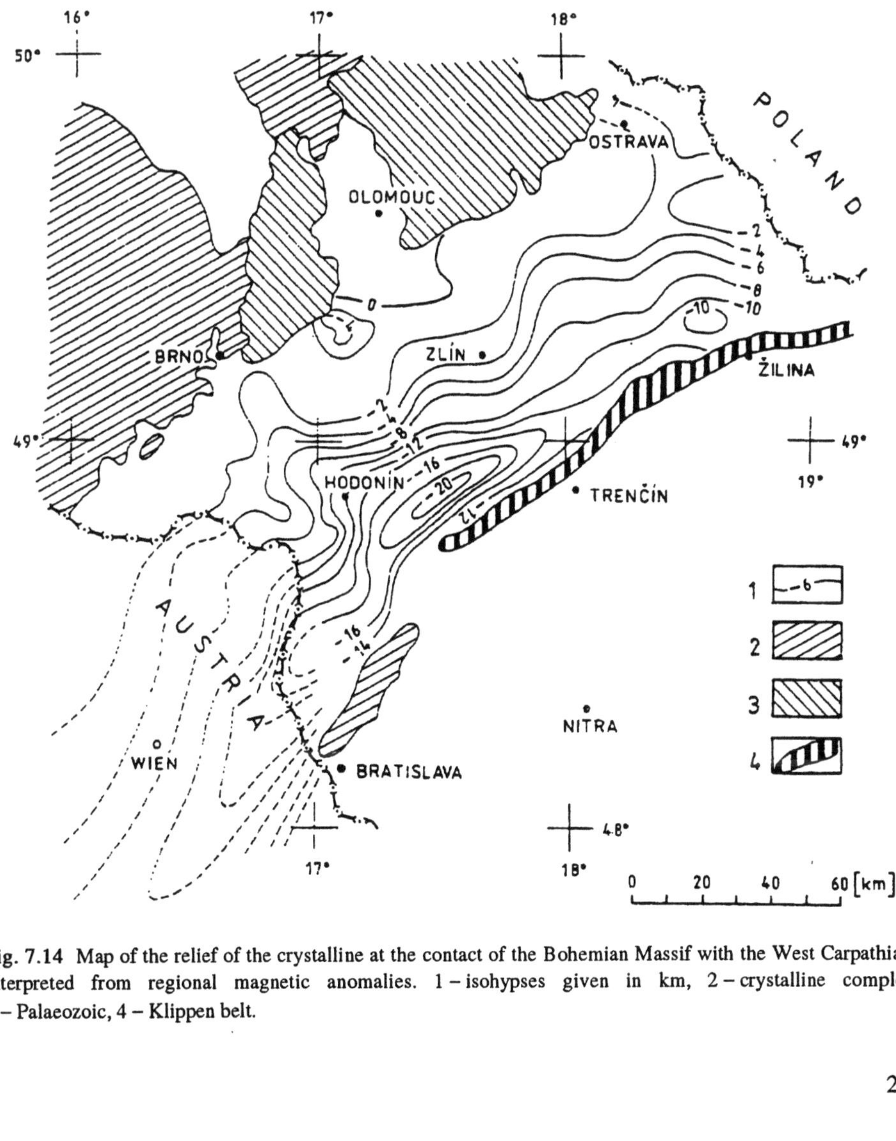

Fig. 7.14 Map of the relief of the crystalline at the contact of the Bohemian Massif with the West Carpathians interpreted from regional magnetic anomalies. 1 – isohypses given in km, 2 – crystalline complex, 3 – Palaeozoic, 4 – Klippen belt.

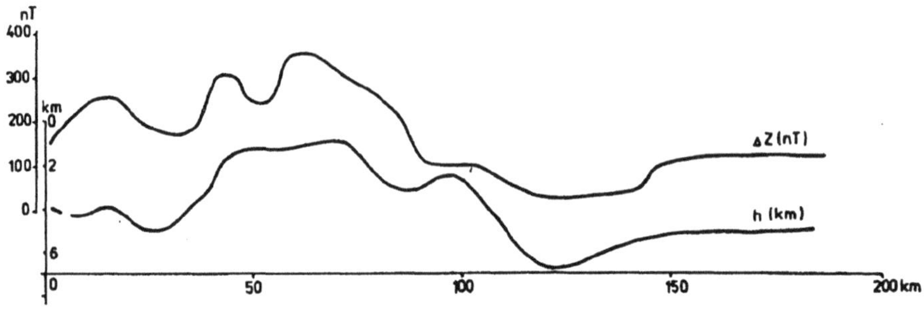

Fig. 7.15 Correlation of the course of the geomagnetic field of ΔZ with the depth to the crystalline unit along a SW–NE profile intersecting the centre of positive geomagnetic anomalies (see Fig. 7.13).

In the case of a positive correlation coefficient for the relation between the values of geomagnetic anomalies and the heights of the elevations of the submerged crystalline basement, it was possible to use geomagnetic data for investigating the subsurface structure of the crystalline basement. For this purpose we used all available data on the crystalline basement relief obtained from boreholes and seismological sections. We found that the magnetic susceptibility of rock samples of the crystalline basement from two main boreholes Jablůnka (6370 m) and Jablunkov (3200 m) also exhibited increased values. Then we evaluated the anomalous pattern of the geomagnetic field along the axis of the entire anomalous geomagnetic zone in eastern Moravia (Fig. 7.15) and compared it with the relief of the crystalline basement along the whole length of the line Břeclav – – Zlín – Jablunkov. The results obtained provide an interesting image (Fig. 7.15) with increased values of the geomagnetic anomalies corresponding to elevations of the crystalline basement and vice versa. From the pattern of the curve it follows that low-anomaly values of the geomagnetic field are typical of depressions on the crystalline basement, while high-anomaly values are due to elevations on it. As the geomagnetic field continues to the NE we observe that low-anomaly values (about 20 nT) correspond to great depths of the crystalline basement in the environs of the town of Vsetín (6370 m), and increase in the direction of Jablunkov (about 100 nT) where the crystalline basement is at a smaller depth, i.e. 3200 m. Thus we can conclude that along the axis of regional geomagnetic anomalies in eastern Moravia there is a positive correlation between the values of geomagnetic anomalies and the height of elevations of the plunged crystalline basement relief (Bucha 1986).

Considering the pattern of the submerging crystalline basement which ensues from the known geologic sections adjacent to seismological sections we can see elevations and depressions alternating there. The processing also involves Austrian sections. The northern one illustrates well the Steinberg fault (Kröll et all. 1981). At a first approximation, these sections then allowed the construction of a map of the depths of the crystalline basement in the entire contact area of the Bohemian Massif and the Carpathians (Fig. 7.14). It shows pronounced depressions where, according to geologic assumptions, the crystalline basement should reach depths of up to 15 km, namely in the area between Hodonín and Velká and in SW Slovakia. We can see here also distinct elevation structures and a considerable increase of the gradient, especially along the Steinberg fault and its continuation northward in the direction of Břeclav. A subsequent comparison of the distribution of regional geomagnetic field anomalies with the pattern of the crystalline basement relief obtained from seismological sections makes their mutual resemblance evident. Very conspicuous is the coincidence between the prominent positive geomagnetic anomaly with its centre in the area between Břeclav and Velké Pavlovice (Fig. 7.13) and the elevation of the crystalline basement (Fig. 7.14), which also follows from the interpretation of the seismological sections investigated in Geofyzika Brno.

This resemblance is shown even in Austria, where the main hydrocarbon deposits occur on the southern and southeastern slopes of the above-mentioned positive regional geomagnetic anomaly. The elevations of the crystalline basement of the Ždánice Forest, the Chřiby Mts., and their continuation towards Zlín are coincident, too, with the maxima of positive geomagnetic anomalies. The same applies to the region of the Moravian–Silesian Beskydy Mts. In the opposite way, the Nesvačilka trough and the depression near Valašské Meziříčí, as well as some other depressions, are characterized by negative geomagnetic anomalies in the same way as the minimum between Senica and Velká, and they coincide with a great depth of the crystalline basement (Fig. 7.13).

Thus, from presented results it follows that, in principle, the geomagnetic maps in this region can be used for determining the approximate pattern of the crystalline basement relief, and consequently, for verifying major elevations.

It is necessary, however, to discover what the causes are of the increased magnetization of rocks forming the plunged crystalline basement. The magnetic characteristics of pre-Variscan and Variscan plutonites greatly differ. We found Variscan tonalities, the Sázava type in particular, to be more strongly magnetized than e.g. the Benešov type. It is a case analogous to tonalites of the pre-Variscan (Cadomian) age of the Brunovistulicum, not only in its surface part but also beneath the flysch nappe.

We cannot presume, however, that this type of rocks (amphibolite) or metabasites occur everywhere in the eastern part of Moravia in the flysch base from the Austrian to the Polish borders as it would follow from the pronounced positive anomalous zone along the entire contact of the Bohemian Massif with the West Carpathians.

Experimental results proved a considerable growth of magnetic susceptibility of samples after they had been saturated with mineral oil and heated to 250 °C (Bucha 1980). These conclusions were verified by Hrouda, who used samples of the crystalline basement of the Lubná locality and found that the samples from boreholes penetrating oil-bearing formations (Lubná –1,5,7) exhibited susceptibilities of about one order of magnitude higher than the samples from boreholes penetrating oil-free rocks (Lubná – 2,3,4,6).

The findings by Uhmann, who investigated the magnetic susceptibility of samples forming the crystalline basement from borehole Ždánice-9, are of interest too. While at a depth of 932 – 940 m the values ranged from 15 000 to $20\,000 \times 10^{-6}$ [SI], susceptibilities of samples from greater depths (940 – 964 m) were lower by one order of magnitude. Usually, indeed, the upper weathered part of the crystalline basement exhibits lower magnetization than the deeper parts.

Thus, if we want to elucidate positive values of geomagnetic anomalies along the western part of the West Carpathians, we cannot avoid the premise concerning substantial effects of hydrothermal alterations of weathering or reduction processes induced by the presence of hydrocarbons and causing expressive changes in magnetic properties of the crystalline basement rocks in the course of their existence.

7.5 Interpretation of the Geomagnetic Anomalies Between the Bohemian Massif and the Inner West Carpathians

O. Man[1]

In the region bound by outcrops of the rocks of the Bohemian Massif, actually of the Brunovistulicum, in NW, and by the inner West Carpathians, in SE, the anomalous geomagnetic field is more expressive than elsewhere in Czechoslovakia.

The magnetic mapping of this whole region was accomplished in the period from 1962 to 66. It was an areal ground survey with density of 1 point per square kilometre, in the course of which Z component was measured relatively with the mean error of about 4.2 nT. Fig. 7.16 shows, in a simplified form, the detected anomalies related to the reference field presented by Bucha (1957).

In the mentioned region there is a positive anomaly which rims the Carpathian arc on the outside even in Austria and Poland (see Chapter 7.3). This regional anomaly exhibits an amplitude from several tens of nT to about 150 nT. In its background very expressive "local anomalies" are observed, namely in SE Moravia. A similar relation of the regional anomaly and local anomalies is also found on the inner side of the Carpathian arc (see Chapter 7.3). Let us thus formulate the following hypothesis: there is no need to search for a distinct source of regional anomalies as they represent only a spatial accumulation of local ones.

On the basis of a comparison of synoptic maps of the geomagnetic field, presented in Chapters 7.2 and 7.3, with the stripped gravity map (see Figs. 6.10 and 6.11) it follows that striking geomagnetic anomalies in Czechoslovakia occur prevailingly in the margins of positive gravity field anomalies. The positive gravity anomalies can be mostly characterized as regions secularly sinking, thus, on average less denuded, exhibiting a lower degree of metamorphosis, containing a large volume of basic rocks, often overlain by platform sediments, and manifesting a young volcanism. Such relations are, to a certain degree, applied also in both the mentioned zones of geomagnetic anomalies confined to the Carpathian arc. These two zones are separated by a negative gravity anomaly coinciding with an inexpressive anomalous geomagnetic field.

---

1 Geofyzika Praha, Geologická 2, 152 00 Praha 5, Czech Republic

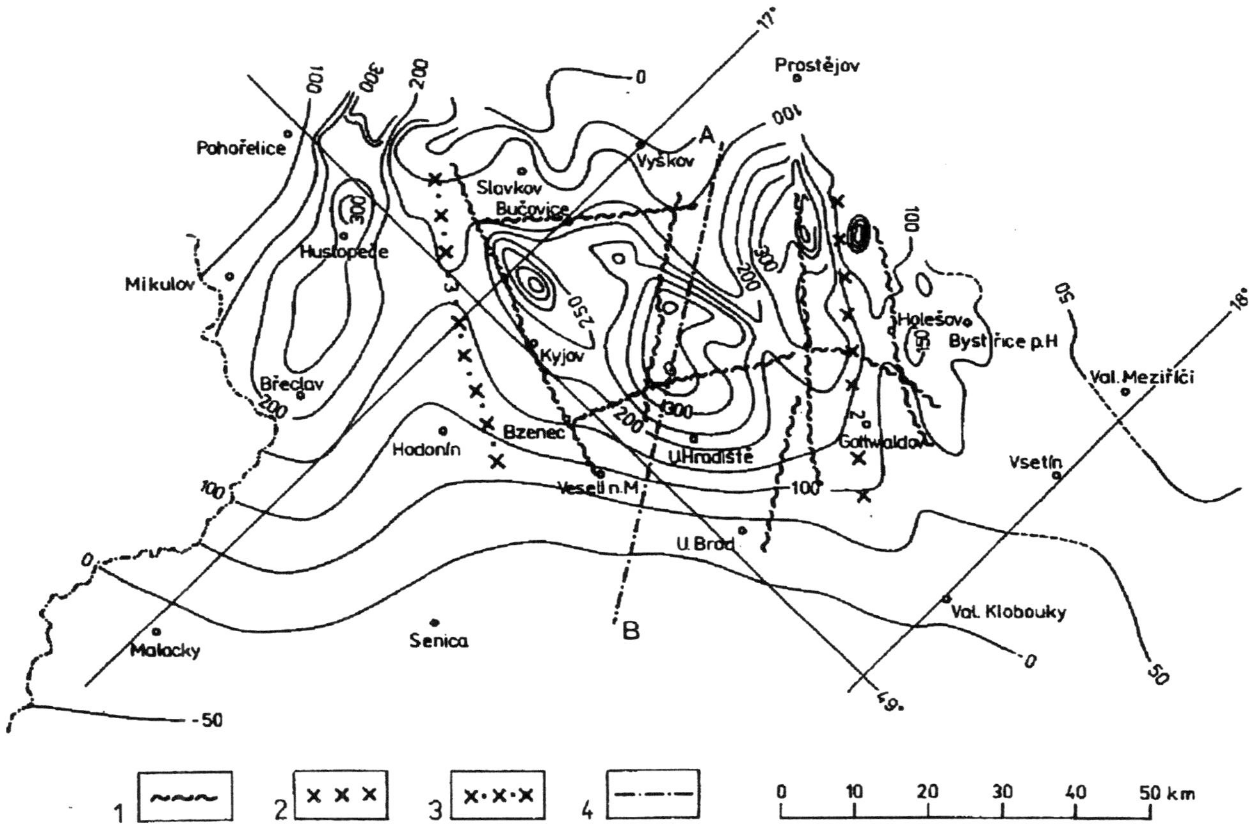

Fig. 7.16 Synoptic map of contours of the vertical component of the magnetic field intensity (in nT). 1 – some faults which are related to geomagnetic field anomalies, 2 – Haná zone, 3 – Nesvačilka depression, 4 – profile A–B selected for 2-D interpretation.

In the inner part of the outer zone of geomagnetic anomalies there is the Carpathian geoelectric anomaly (CPA) (see Chap. 5).

In the whole region interpreted heat flow cannot be uniquely characterized. In the south part of the region, low heat flow is observed, which, in places, is lower than 50 mW m^{-2} which results in plunging of the isotherm, e.g. of the Curie temperature of magnetite, i.e. 575 °C (see Chap. 4, profile EEGT 5). This fact can, however, be only of local meaning for an interpretation of geomagnetic anomalies, because further NE, around Zlín, high heat flow values exceeding 70 mW m^{-2} can be found.

The geologic structure of the interpreted region has recently been from borehole as well as seismic data. In regard to the relation of geomagnetic anomalies to this structure, the following facts can be claimed:

1. Faults which intersect sedimentary formations to reach the crystalline basement as e.g. the Haná zone, the Nesvačilka trough and others (Fig. 7.16), are striking. Some of these faults are presented also in Figs. 3.8 and 3.9. Let us note that the Haná zone is probably related to the Lužice – Jílovice dislocation, and with the Přerov – – Štiavnica fault zone (see Fig. 3.10).
2. Local positive geomagnetic anomalies often correspond to elevations of the crystalline basement.
3. Geomagnetic anomalies do not show any obvious dependence on the sedimentary formations structure (Devonian and younger ones).

These facts result in the conclusion that the substantial part of the anomalous field is due to sources in the crystalline unit or even at greater depths. Here, the crystalline unit is formed by the Brunovistulicum rocks, i.e. SW of the Haná zone prevailingly by granitoids, and NE of it, metamorphites.

A possibility of such an interpretation of geomagnetic anomalies was quantitatively confirmed by simple 2D models corresponding to A–B profile (see Fig. 7.16). The profile was chosen to extend perpendicularly to the regional anomaly, and passes through the maxima of the anomalous geomagnetic field. Because the local anomalies cannot be considered two-dimensional, the values of anomalies were averaged in the direction perpendicular to the profile to a distance of 6 km, i.e. in a zone of 12 km in width. This pattern of the field was interpreted using two different models corresponding to different ideas on crustal structure. The first model follows from the conception that granitoids forming the basement extend down only to about a depth of about 10 km, being homogeneously magnetized in their entire volume. Model II, suggested by Filo and Kubeš (Fig. 7.17) assumes, on the contrary, that rocks forming the basement extent down to the Moho discontinuity, whose generalized course can be at a depth of about 33 km (see Chap. 2.3). The composition and physical properties of ferromagnetic minerals at such a depth range must vary in dependence on P–T conditions. In the approximation two homogeneously magnetized layers of rocks were considered, the first one from the surface of crystalline unit down to the depth of 10 km, the second one from this depth down to the Moho discontinuity (see Fig. 7.17).

For both the models the procedure of the inverse problem solution was the same. The crystalline basement surface, approximately known from boreholes and other data, was taken as a starting approximation of the upper margin of the anomalous body. Using the Golub method (Golub 1965), i.e. from the condition of the minimum mean-square deviation of model effect from observed data, vectors of magnetic polarization and constant correction of the reference geomagnetic field were determined. These parameters remained preserved while the upper margin of anomalous body was modified using the iterative procedure proposed by Man (1987). The interpretative procedure minimizes the above-mentioned mean-square deviation and thus terminates the procedure as soon as its value decreased to 3 nT.

Results of the interpretation were as follows: the first model required only a minor modification of the known crystalline basement surface, and the magnetic polarization was found to approximately 2500 nT, and its direction was corresponding approximate to the induced magnetic polarization. Model II yielded poorer agreement between the inverse problem solution and the known crystalline basement surface, and the magnetic polarization values were found to be 1450 nT in the upper part of the body, and 825 nT in its lower part. Inclination of both the magnetic polarizations in the profile section was approximately 60°, while the induced magnetic polarization would correspond to 68°. The reference geomagnetic field correction was in both the cases negligible. It is worth mentioning that the maximum value of the integral of the vertical component of magnetization along the vertical coordinate was nearly identical for both the models, i.e. approximately 20 000 A.

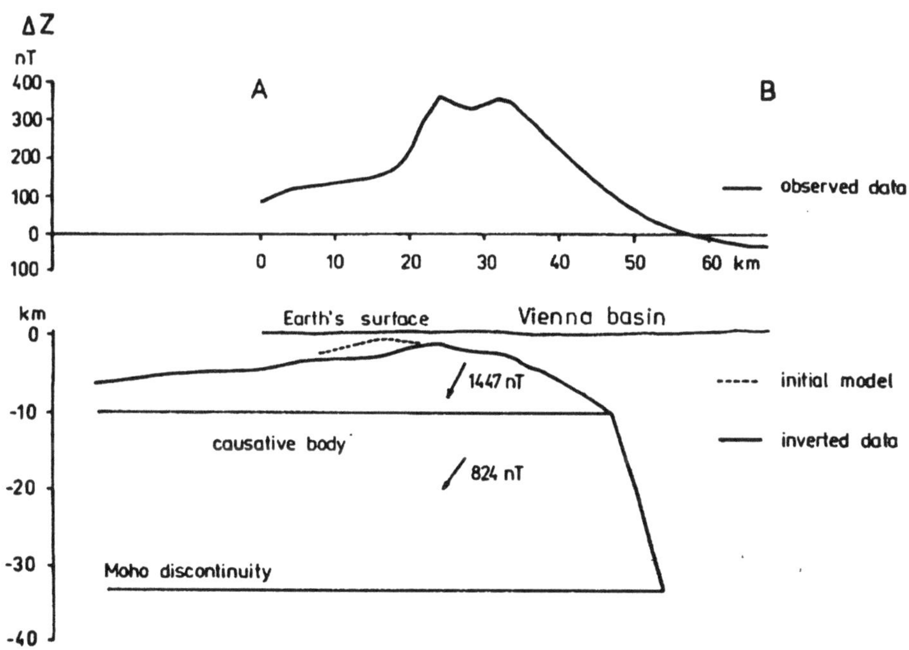

Fig. 7.17 2-D model II corresponding to profile A–B.

Let us now compare these results with the values of core sample susceptibilities from the crystalline unit (Doležal 1980). The magnetic polarization found for the first model corresponds to an equivalent susceptibility of approximately 0.05 [SI]. Such a high value has been found in the set of core samples only exceptionally. The value of 0.029 [SI], which corresponds to the magnetic polarization specified for the upper part of model II, is more acceptable.

A certain disagreement between the inverse problem solution and the known surface of the crystalline unit is understandable for the following reasons:

1. Our model was only two-dimensional while the actual geologic structures are three-dimensional.
2. The idea of homogeneously magnetized basement simplifies the actual conditions. It is known that the basement is formed by several types of rocks, whose magnetic properties vary in a wide range (Doležal 1980).

The significance of these models is thus the fact that they have supported our hypotheses quantitatively: that the sources of the regional and local geomagnetic anomalies are identical, and that these sources have to be looked for in the crystalline basement. At the same time, the hypotheses showed that a substantial part of the anomalous geomagnetic field can be explained, under the simple assumption that the basement rocks are homogeneously magnetized, only due to the relief of the crystalline basement. As this relief has actually been well known, especially in places where it is at depths less than 2 km, the share of another factor, i.e. inhomogeneities in the basement, could be found by the "susceptibility mapping" technique (Paterson and Reeves 1985).

7.6 Palaeomagnetic Research on the Bohemian Massif

M. Krs[1]

The Bohemian Massif has been investigated by palaeomagnetic methods since the beginning of the 1970s. The activity has been aimed at the palaeomagnetic dating of epigenetic ore deposits, at research into fundamental magnetic parameters of minerals depending on their structure and chemical composition, at detection of fundamental palaeomagnetic characteristics of biostratigraphically dated rocks of the Upper Carboniferous and younger, (the rocks not affected by the Variscan folding) and at development of corresponding instruments used in palaeomagnetic research; preliminary results were published in a monograph (Krs 1969). The rocks strongly affected by the Variscan folding, but also some younger rocks containing more components in the natural remanent magnetization (NRM), have required development of a new apparatus which would make use of magnetic vacuum for a thermal demagnetization. The new apparatus is called the magnetic vacuum control system (MAVACS). It was devised in the Prague Dept. of Geofyzika Brno. A highly sensitive magnetometer ROCOMA

1 Geological Institute, Rozvojová 135, 165 00 Praha 6, Czech Republic

(rotating coil magnetometer) is used as a sensor, its offset is below the value of 0.1 nT (Halousek et al. 1986). An introduction of the multicomponent analysis of magnetization made it possible to interpret better components of magnetization which originated before the folding of rocks, and to define more precisely the components of characteristic magnetization.

The palaeomagnetic dating of epigenetic ore mineralisation in Central Europe is especially significant for lithosphere investigations, as it supplies, besides other data, a palaeotectonic development of the ore deposits as well as of those rock blocks in which the deposits occur. An exclusion or substantial suppression of the effect of post-mineralization tectonics upon the mean palaeomagnetic direction was achieved by a statistical processing of palaeomagnetic directions of large sets of oriented samples collected either from various locations in the same deposit, or from several deposits of the same age in a particular metallogenic region. For numerous deposits, it has been confirmed that older vein mineralization, e.g. of the Upper Palaeozoic age, oriented samples had to be collected from a relatively large region. Young vein mineralization or older ones genetically associated with great plutons could be sampled at fewer locations than were actually done. The deposits of the latter group were affected by post-mineralization tectonics minimally. Samples collected from deep parts of deposits in mining workings brought some advantages: effects of exogenetic factors upon changes of palaeomagnetic characteristics have not been detected. The palaeomagnetic method thus besides dating also yields the possibility of studying the palaeotectonic history of epigenetic deposits (Krs 1966).

Detailed laboratory study of magnetic properties of rocks samples, taken from mineral deposits of prevailingly hydrothermal, pneumato-pegmatitic type has shown that physical properties of various minerals pigmented by hematite are suitable for a palaeomagnetic research. Fine-grained quartz–hematite aggregates, so called cinopel, jaspers, ferruginous quartz, etc., show an exceptional stability. In these aggregates, grains of high-coercitivity hematite are actually perfectly isolated by a quartz substance and thus protected from effect of chemical solutions circulating within the ore deposits. It has also been shown that hematite is a more common component of ore deposits than it has been realized and mentioned in publications. Portions of ore-mineralization finely pigmented with hematite genetically connected with overall mineralization can often be found. Rock samples collected from such parts of deposits have been weakly magnetized but are exceptionally magnetically hard. Magnetite of hydrothermal origin is magnetically very soft, thus unsuitable for a palaeomagnetic research. Pyrrhotite is rather suitable for studying the thermal history of deposits thanks to its low Curie temperature, but only in those cases when it represents statistically isotropic aggregate.

Deposits were studied on Czechoslovak, and also German territories to investigate deposits of various geneses and of diverse tectonic developments. Fig. 7.18a shows mean positions of palaeomagnetic poles derived on the basis of rocks samples collected from epigenetic ore deposits, prevailingly from the Bohemian Massif and partly from the West Carpathians. Fig. 7.18b presents mean positions of palaeomagnetic poles

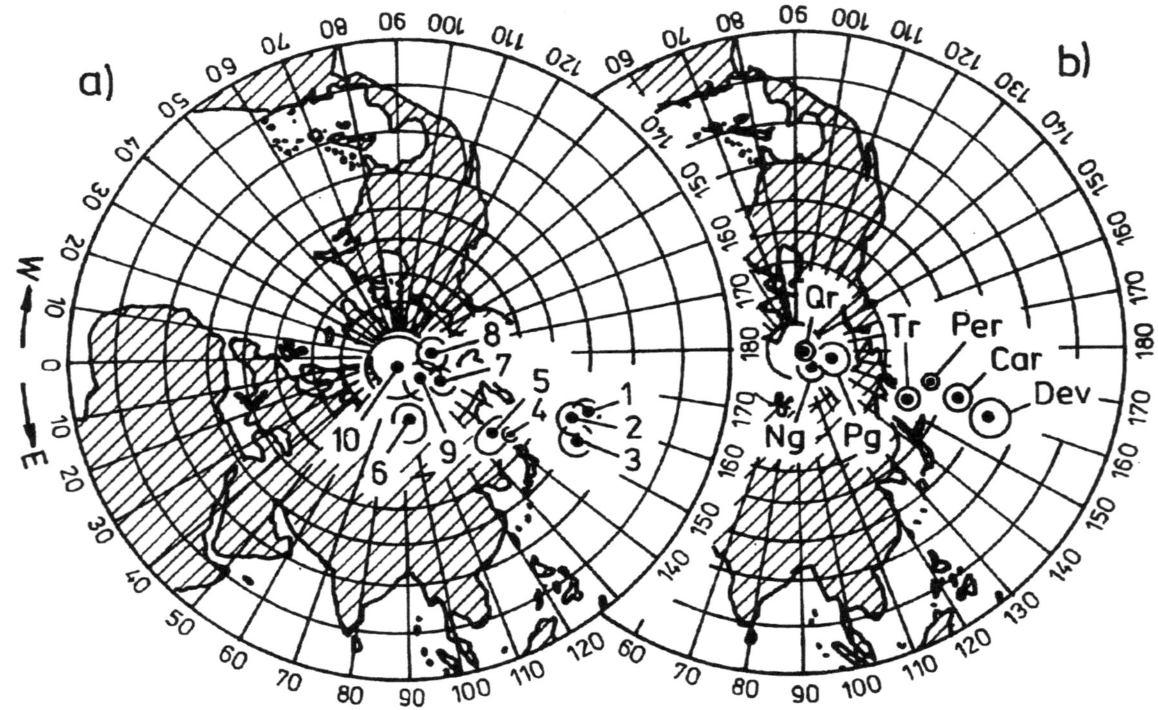

Fig. 7.18 Apparent polar-wandering paths derived from the products of epigenetic mineralization in Central Europe, and derived statistically from biostratigraphically dated rocks from the North European platform. a) Mineral deposits: 1 – Krásno, stockwork "Huber"; 2 – Freiberg, Halsbrücke, Freiberg – Brand, I. mineralization cycle; 3 – Plavno, Potůčky, Krásno, hematite-bearing veins and greisen body; 4 – Vrančice, Skalka, Bytíz, Kamenná, Brod, Ústaleč, hydrothermal veins of the Central Bohemian Pluton; 5 – Oldřichov, granodiorite with grains of hematite; 6 – Slovenské rudohorie Mts. region, Rákoš, Mlynky, Grétl, Rudňany, Poráč, Zlatník, Potok, Slovinky, Drnava, Nižný, Medzev, Bankov near Košice, siderite veins and magnesite body; 7 – Jáchymov (Joachimsthal), vein "Geschieber"; 8 – Freiberg, Halsbrücke, Brand II. mineralization cycle; 9 – Vrchoslav, fluorite - barite vein; 10 – Banská Štiavnica, cinopel (Krs 1966). b) Mean palaeomagnetic pole positions derived from biostratigraphically dated rocks on the North European platform: Qr – Quaternary, Ng – Neogene, Pg – Palaeogene, Tr – Triassic, Per – Permian, Car – Carboniferous, Dev – Devonian (Krs 1982).

statistically derived for the North European platform, using the data published by various authors. At first sight, an accordance of both the apparent polar-wandering paths is obvious, thus also the following geochronical aspects of the palaeomagnetic method.

Palaeomagnetic directions of epigenetic ore deposits occurring in tectonically stable regions were not appreciably affected by the post-mineralization tectonics, as far as the mineralization is confined to the period which had led to a final cratonization of the regions. Palaeomagnetic results obtained from the Krušné hory Mts. and the Slavkovský les Mts. have provided a background to the hypothesis that a regeneration of deposits has not taken place. The Variscan mineralization often occurs in the proximity of the Neoidic mineralization while in the Variscan mineralization the original palaeomagnetic characteristics have been preserved. This circumstance, which was confirmed at several occurrences, is strongly against assumptions of thermal or chemical re-working of the Variscan deposits by a younger mineralization. A direct evidence against the hypothesis on ore deposits regeneration in situ are the hydrothermal veins with banded structure, in which individual bands are normally and reversely magnetized (Krs and ›Šťovíčková 1966).

An apparent polar-wandering path, derived from measurements on rock samples collected from epigenetic ore deposits in the Bohemian Massif, strikingly coincides with the apparent polar-wandering path statistically derived for the North European platform from the Carboniferous to the Quaternary. This fact bears a special significance for studying movements of lithospheric plates, because the products of epigenetic mineralization are confined to the process taking place in deep parts of the lithosphere.

A great quantity of the Upper Carboniferous and the Lower Permian sedimentary rocks of the "red beds" type, often accompanied by volcano-detrital rocks deposited in basins and troughs in the Bohemian Massif, were a subject of the paleomagnetic research in the past (Krs 1968). Many of these rocks were suitable for the paleomagnetic research even in natural state or after the magnetic cleaning using an alternating and thermal field. Majority of the NRM of these rocks exhibited one-component or two-component remanent magnetizations. For this reason, the formerly used methods of magnetic cleaning were sufficient. The samples were being collected in such a way, that in each investigated region they represented a section through the rocks that had originated within a period of several millions of years.

In earlier works, the mean palaeomagnetic directions were calculated from the palaeomagnetic directions of individual layers determined on rock samples collected from several outcrops in the particular region. It was assumed that in these directions the values of palaeosecular variation were averaged out. Calculations of the mean palaeomagnetic directions from the site/locality means have recently been favoured. However, palaeomagnetic data calculated this way and the mean palaeomagnetic directions are analogous to those obtained by the first method of calculation; the data scatter is different.

An introduction of the MAVACS system to palaeomagnetic research led to an improvement of laboratory procedures. The use of the multi-component analysis of NRM has meant an improvement in interpretational procedures.

For this reason at least a part of former palaeomagnetic data was verified by new laboratory and interpretational procedures. Strongly remagnetized rocks of the Lower Permian from the Intra-Sudetic basin and from the basin at the foot of the Krušné hory Mts., were selected. Palaeomagnetic pole positions were derived from the mean palaeomagnetic directions calculated from the so called strata means as well as from the positions of virtual poles. Positions of poles calculated by both the methods are in accordance, see Table 7.1 (Krs et al., 1988). These results are also in accordance with the data derived earlier on the rocks, which had not been remagnetized at all, or only partially.

Determining the palaeomagnetic parameters of rocks of the whole Phanerozoic represents the fundamental task of the palaeomagnetic investigations in the Bohemian Massif. The pre-Variscan rocks suitable for palaeomagnetic analysis occur in the Barrandian. For older rock formations, some problems arose with a reproduction of palaeomagnetic directions determined with the use of standard laboratory procedures. An introduction of the laboratory apparatus MAVACS in the course of thermal demagnetization, and the use of the multi-component analysis of magnetization, yielded some partial results which made it possible to conduct systematic palaeomagnetic investigations of rocks from the Cambrian to the Devonian in the Barrandian.

An extensive palaeomagnetic investigation was conducted on the Ordovician rocks. Groups of layers were numbered from the Tremadocian, to the Ashgillian. Samples were collected from paleo-volcanics and their tuffs, numerous oolitic iron ores with variable Fe content, red shales and red silicites. Palaeomagnetic directions could not be

Table 7.1 Trutnov – Náchod Saxonian Basin, Trutnov Formation (Saxonian), red sandstone alternating with subordinate red-brown aleuropelites, locality between towns of Náchod and Horní Radechová. Comparative paleomagnetic studies. Mean palaeomagnetic direction calculated from strata means using *R.* Fishers's (1953) statistics, strata means are given unit weights: $D_m = 202.43°$, $I_m = -10.15°$, $k = 32.53$, $k = 32.53$, $\alpha = 16.35°$. Corresponding palaeomagnetic pole position: $\varphi_p = 40.92°$ N, $\lambda_p = 165.96°$ E, $\delta_m = 16.54°$, $\delta_p = 8.37°$. Mean pole position calculated from virtual pole position (corresponding to respective strata means) using *R.* Fisher's (1953) statistics, each pole is given unit weight: $\varphi_p = 40.15°$ N, $\lambda_p = 165.93°$ E, $\alpha_p = 18.91°$ N, $k = 24.56$

Numbering of strata	Sample Nos.	Palaeomagnetic direction		*k*	*n*	Temperature range [°C]	Notes
		D	*I*				
	3966A2					620 – 685	
2	3967A2	202.0°	– 10.0°	73.7	3	620 – 685	
	3968A2					620 – 685	
	3972A2					620 – 685	Multicomponent
4	3973A2	195.2°	– 10.4°	23.2	2	620 – 685	analysis
	3878A					620 – 685	
6	3980A1	220.1°	– 19.5°	75.0	3	620 – 685	
	3981A1					620 – 685	
	3988A1					620 – 685	
9	3991A1	193.5°	– 0.05°	49.8	2	620 – 685	

Table 7.2 (part 1) Palaeomagnetic pole positions derived for Ordovician and Cambrian rocks from Barrandian areas, MAVACS during thermal demagnetization and multi-component analyses were applied (Krs et al. 1986, 1987). *n* is number of samples; *N* is number of localities

No.	Age of formation		Lithologie	Mean geographic coordinates		Mean palaeomagnetic direction		α_{95} [°]
	Period	Epoch		latitude [°N]	longitude [°E]	declination [°]	inclination [°]	
1	Ordovician	Tremadocian	red silicate	9.75	13.63	126.65	– 41.70	3.63
2	Ordovician	Tremadocian	red silicate + + tuffite	49.75	13.63	131.84	– 47.57	3.66
3	Cambrian	Upper	andesite	49.98	13.783	89.88	– 58.16	5.77
4	Cambrian	Middle	graywacke	49.943	13.747	54.31	– 58.67	12.49
5	Cambrian	Middle	graywacke	49.943	13.747	54.37	– 64.34	3.47

reproduced for samples of basalts, andesites, and tuffs, due to total palaeomagnetic instability. Oolitic ores have also not confirmed a syngenetic magnetization. Red shale samples, collected both in the Barrandian and in the area east of Prague, exhibited magnetizations entirely affected by weathering. The samples of red silicites have been the only ones which turned out to be suitable for the palaeomagnetic research, even though also for these rocks a wide spectrum of magnetic properties was confirmed. Within the temperature intervals around 350 °C, components of the characteristic magnetization were confirmed, which originated prior to folding (see Table 7.2), in the whole sets of samples of red silicites (the Milín layers corresponding to the Tremadocian, the lower part of the Lower Ordovician). The calculated values of palaeogeographic latitudes around 28° North are somewhat lower than it was expected for the Ordovician rocks. For the discussion see Krs et al. (1986).

Table 7.2 (part 2) 1) Pole position calculated from locality means (n = 10); thermal demagnetization and multi-component analysis of all samples, positive fold test. 2) Pole position calculated from palaeomagnetic directions of samples (n = 143); thermal demagnetization and multi-component analysis of all samples; positive fold test.

No.	k	n	N	Palaeomagnetic pole position		Ovals of coincidence		Notes
				latitude [°S]	longitude [°E]	declination [°]	inclination [°]	Tests
1	24.83	64	3	41.52	91.82	2.72	4.44	thermal cleaning 350 °C, positive fold test
2	16.04	99	4	48.11	91.83	3.09	4.76	thermal cleaning 350 °C, AF cleaning, positive fold test
3	9.16	74	3	28.64	131.23	12.1	8.52	thermal cleaning, multi-component analysis of pilot samples
4	15.91	143	10	11.29	153.96	18.9	18.56	1)
5	12.56	143	10	16.99	157.67	18.9	5.55	2)

Another extensive palaeomagnetic investigation was conducted in the rock samples of the Middle and Upper Cambrian, nearly all from the Barrandian. The Upper Cambrian rocks are represented by samples of andesites altogether from five localities. Samples of andesite were progressively demagnetized with the application of both the alternating as well as the thermal fields. Results of both the procedures were identical, and thermal demagnetization turned out more effective. Out of the original set of 140 samples of andesites, which were demagnetized by the alternating and thermal field, 74 samples yielded the Zijderveld diagrams usable for an interpretation. Based on the palaeomagnetic data it was possible to interpret that the samples were collected from two lava flows, even though areally extensive. That is why the conditions for averaging out palaeosecular variations were not fulfilled, and the pole position calculated (see Table 7.2) corresponds rather to the position of the virtual pole. Samples of the Middle Cambrian greywackes and shales were collected from 16 outcrops. Some outcrops provided rocks magnetically so weak that total demagnetization to the highest blocking temperature was not possible. Samples of greywackes from 10 outcrops provided well-reproducible results. These rocks exhibited greater number of magnetization components in *NRM*, mostly four. As the layers of rocks studied had been intensively folded in the course of the Variscan orogene, the fold test of directions of corresponding components of magnetization could be applied. The pre-Variscan directions of magnetization were indicated. Rocks from 7 outcrops exhibited a normal polarity of magnetization, and rocks from three outcrops a reverse one.

Altogether 143 rock samples provided reproducible results. To all of them methods of the multi-component analysis were applied. Positions of the palaeomagnetic pole calculated from the locality means ($N = 10$) and from palaeomagnetic directions of samples ($n = 143$) are presented in Table 7.2 for a comparison.

7.7 Palaeomagnetic Investigations in the West Carpathians in Czechoslovak Territory

The Early Palaeozoic of the West Carpathians

P. Muška[1]

The crystalline unit of the Tatricum together with its surroundings represents relatively the most stable part of the West Carpathians. All other units are in an allochthonous position in relation to the Tatricum. The early Palaeozoic was studied to obtain information on the mutual tectonic position of the individual crystalline blocks. The Early Palaeozoic formation of Variscan metamorphosed sediments and volcanics and Variscan granitoids were sampled and created a base. The Považský Inovec Mts., represented by metamorphosed quartz sandstones, exhibit a characteristic direction of

1 Geofyzika Bratislava, Geologická 18, 825 52 Bratislava, Slovakia

remanent magnetization $D = 189.0°$ and $I = 25.9°$ ($\alpha_{95} = 39.9°$ characterizes local tectonics). The characteristic mean direction of gneisses, migmatites, paragneisses, and amphibolites of the Tríbeč Mts. is given by $D = 349.8°$ and $I = 51.9°$, that of the granitoid rocks of the Veporské vrchy Mts. by $D = 80.5°$ and $I = 71.2°$, that of granitoid rocks of the Malá Fatra Mts. by $D = 207.8°$ and $I = -54.9°$, and that of granitoid rocks of the Nízke Tatry Mts. by $D = 70.7°$ and $I = 83.3°$.

Samples for statistical data processing were selected on the basis of the stability tests in the alternating demagnetization field. The hard thermo-remanent and chemical magnetic polarizations are syngenetic with the Variscan orogene. The variable values of the calculated characteristic mean directions are due to Alpine tectonics, i.e. due to movement of individual blocks the basis of which is the studied crystalline unit core (Muška and Vozár 1985).

The Late Palaeozoic of the West Carpathians

P. Muška[1]

The geologic structure pattern of the West Carpathians is a result of the Alpine orogenic cycle – the nappe tectonic of the post-Middle Cretaceous and the Tertiary. The palaeomagnetic research of the main late Palaeozoic horizons of volcano-sedimentary sequences has pointed out a complexity of geologic structure of the main tectonic units such as the Tatricum, Veporicum, Hronicum, Gemericum, and Zemplinicum (Muška and Vozár 1987).

The Tatricum from the nappe structure point of view, considered as "autochthonous", is characterized, from the palaeomagnetic research aspect, mainly by differences in values of inclination caused by vertical movements of individual segments in the Alpine phase as well as in the period of consolidation during the Tertiary and Quaternary. The results of the palaeomagnetic research into the Malá Fatra Mts. (1), and the Považský Inovec Mts. (2) are presented in Table 7.3 (1 and 2).

The data from the palaeomagnetic research of the Veporicum can be interpreted as a reflection of the intensive northwards shift of the east part of the Veporicum relatively towards the west part towards the Tatricum which is indicated by its counter-clockwise rotation.

The tectonic unit of the Hronicum generally exhibits a fan-shaped character of geologic structure of complexes which form the north slopes of the Nízke Tatry Mts. Deviations of the declination are in the range of 23° to 84°. Away from older formations the values increase towards overlying layers. Fig. 7.19 illustrates the distribution of directions which are characteristic for the nappe unit of the Hronicum in the mountain ranges studied.

1 Geofyzika Bratislava, Geologická 18, 825 52 Bratislava, Slovakia

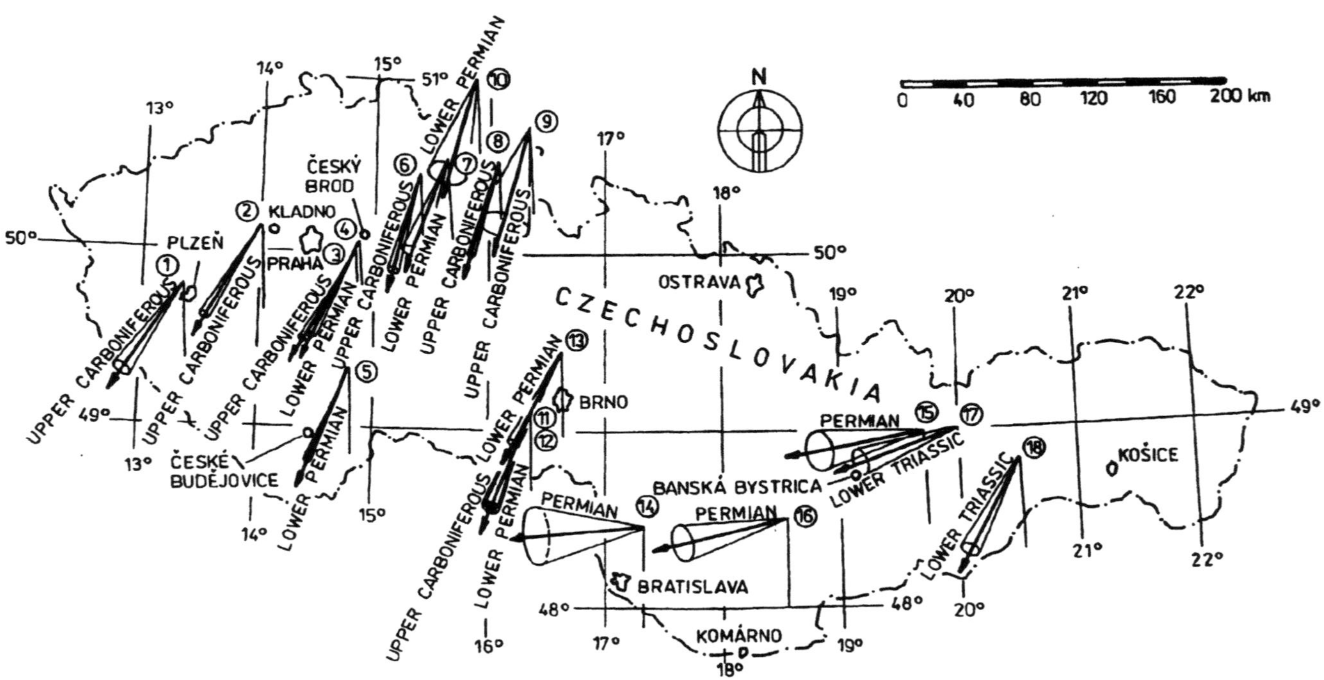

Fig. 7.19 Mean palaeomagnetic directions of Upper Carboniferous and Lower Permian rocks of the Bohemian Massif (1 – 13), and of Permian and Lower Triassic rocks of the West Carpathians of Czechoslovakia (14 – 18): 1 – Plzeň basin: Westphalian C – Stephanian, red sediments, 2 – Kladno – Rakovník basin: Westphalian C – Stephanian, red sediments, 3 – Blanice graben, northern part: Autunian, red sediments, 4 – Blanice graben, northern part, Stephanian red sediments, 5 – Blanice graben, southern part: Autunian, red sediments, 6 – Krkonoše Piedmont basin: Stephanian, red sediments, 7 – Krkonoše-Piedmont basin: Lower "Rotliegendes", red sediments, 8 – Intra-Sudetic basin: Westphalian B – Stephanian, prevailingly extrusive rocks, 10 – Polish side: Autunian, red sediments, 11 – Boskovice graben, southern part: Stephanian, red sediments, 12 – Boskovice graben, southern part: Autunian, red sediments, 13 – Boskovice graben, central part: Autunian, red sediments, 14 – Little Carpathians, Choč unit: Permian melaphyres, 15 – Low Tatras, Choč unit: Permian melaphyres, 16 – Třibeč, Choč unit: Permian, melaphyres, 17 – Choč nappe: Werfenian shales (Lower Triassic), 18 – "Autochthonous" region: Werfenian shales (Lower Triassic). Note: 1 – 8, 11 – 13, 17, and 18 – data of Krs; 9 and 10 – data of Nairn; 14 –16 – Data of Muška.

The values obtained in groups of layers of the North and South Gemericum are interpreted as having been clockwise rotated up to 80° in the Permian.

The counter-clockwise rotation of the Permian to the Lower Triassic formations corresponds to the Alpine orogene. This movement is in connection with the nappe northward overthrust of the Gemericum, especially with the thrust over the south part of the Veporicum. Tables 7.3 and 7.4 present results of the study unit in the Slovenské rudohorie Mts..

The Late Palaeozoic of the Zemplinicum (Tables 7.3, 7.4) exhibits a characteristic *RMP* direction which is coincident with the direction statistically determined for the North European platform. The values differ only in the vertical direction which corresponds to an emergence of the east part of the Zemplinicum.

Table 7.3 Palaeomagnetic directions related to the West Carpathians.

No.	Age of formation		Lithologie	Mean geographic coordinates		Mean palaeomagnetic direction		α_{95} [°]
	Period	Epoch		latitude [°N]	longitude [°E]	declination [°]	inclination [°]	
1	Permian	---	sandstone	49.00	18.60	82.3	56.1	11.7
2	Permian	---	sandstone	48.43	18.45	257.3	1.1	17.5
3	Permian	Upper	sandstone slates	48.50	20.50	335.3	41.6	9.9
4	Carboniferous	Upper	sandstone	48.38	21.67	207.2	−46.6	14.7

Table 7.4 Palaeomagnetic pole positions related to the West Carpathians.

No.	*k*	*n*	*N*	Palaeomagnetic pole position		Ovals of coincidence		Notes
				latitude [°]	longitude [°]	δ_m [°]	δ_p [°]	Tests
1	10.9	16	---	31.3 N	87.2 E	16.8	12.1	AF cleaning, dip correction
2	11.0	8	---	7.9 S	61.6 W	17.5	8.7	AF cleaning, dip correction
3	11.8	20	---	58.7 N	112.3 W	12.1	7.4	AF, TH cleaning, dip correction
4	6.9	17	---	60.7 S	34.0 W	18.9	12.1	AF, TH cleaning, dip correction

Palaeomagnetic Investigations of the Flysch Belt of the West Carpathians

M. Krs[1]

A review of palaeomagnetic investigations in the West Carpathians of Czechoslovakia was presented by Krs et al. (1982). The palaeomagnetic investigations were carried out systematically in all tectonic units as follows (after Andrusov (1973)): I – Carpathian foreland, II – Carpathian Foredeep, III – Flysch Belt, IV – Klippen Belt, V – the zone of the inner West Carpathians including the Tatricum, Veporicum, Hronicum, and Gemericum.

In the Dukla unit of the Flysch Belt, in the southern margin of which the Magura nappe occurs, the youngest rocks so far were studied. But only the samples of red pelitic sediments of the Lower to Middle Eocene age yielded paleomagnetic results. The mean palaeomagnetic direction (see Fig. 7.20) was derived on rock samples collected from outcrops located at an extreme distance of 45 km.

In the Moravian–Silesian Beskydy Mts., pelitic sediments of the Cenomanian to Lower Turonian age were studied, collected from outcrops located at an extreme distance of 24 km. In the outer arc of the Alpine–Carpathian system, teschenites of the Hauterivian–Baremian age were studied. From them, however, only some petrographic types were suitable for the investigation, such as diabase picrite, olivine fourchite, as long as they were not intensively altered, see Fig. 7.20.

Recent results from the investigations of rocks of the Flysch Belt were obtained from the Orava region. The Flysch Belt of the Orava region is represented by the Magura nappe which consists of flysch sediments of the Upper Cretaceous to the Paleogene. In the Neo-Alpine tectonic process, the sediments were gradually folded and thrusted over outer structural units towards the North European platform. In the course of the last movements, within the Neogene–Quaternary period, imbrications verging southwards were formed. The complex tectonic evolution in a relatively plastic flysch sequence resulted in the present imbricate structure including separate lithofacial-tectonic partial units.

The rock samples collected from 15 localities in the Orava region were subjected to the combined cleaning, the multi-component analysis of remanent magnetization, and a detailed petromagnetic study. The MAVACS apparatus (Příhoda et al. 1989) was used to thermally demagnetize the rock samples. Results of the thermal demagnetization were processed using the LINEFIND algorithm. The analysis of dispersion of the directions derived from individual rock samples was made. Component *A*, isolated during the initial stages of the demagnetization, corresponds in direction to the recent geomagnetic field. In higher stages of the demagnetization, component *B* was successfully defined in numerous rock samples, which corresponds to palaeomagnetization with a high degree of probability. In some cases, we used the method of great circles, in the plane of which there is the mean direction of *B* component.

---

[1] Geological Institute, Rozvojová 135, 165 00 Praha 6, Czech Republic

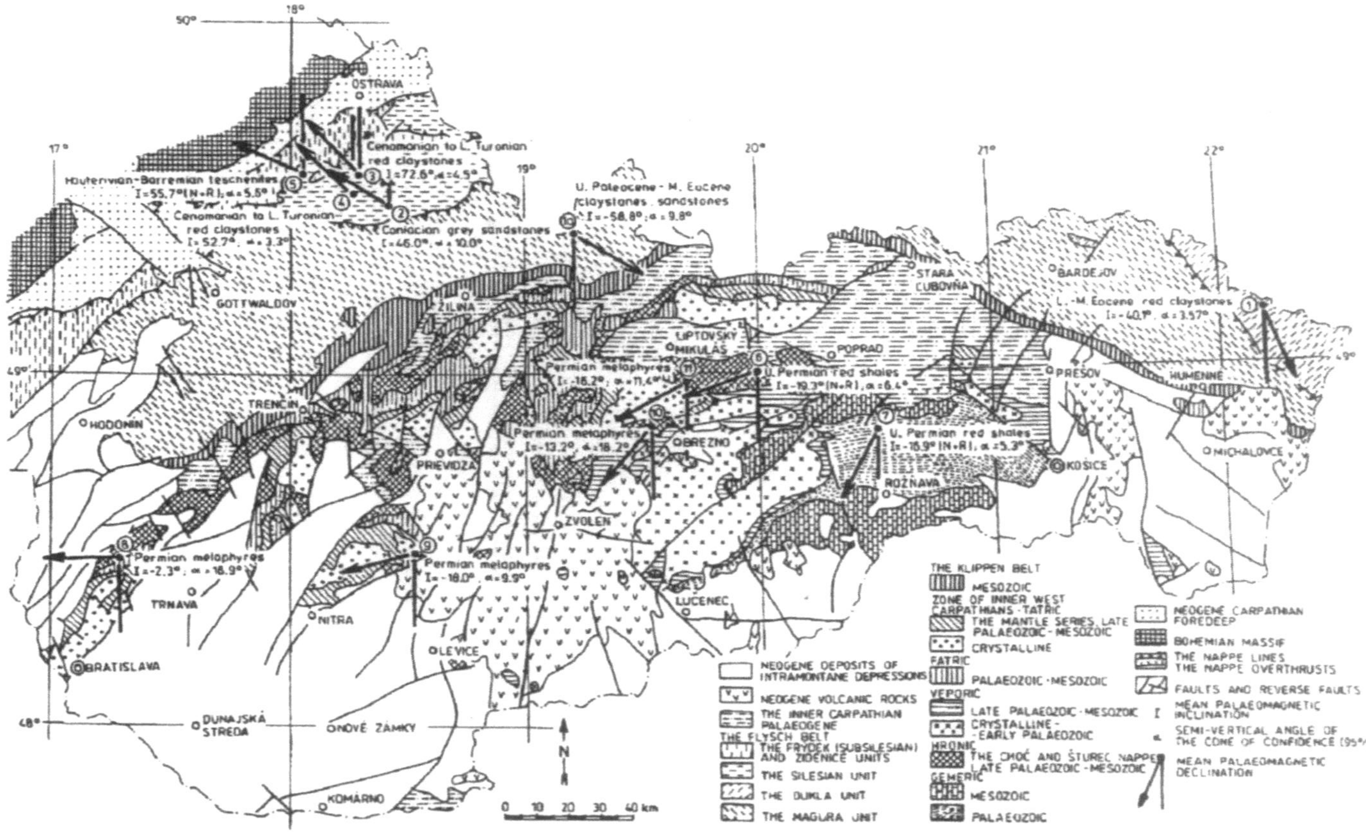

Fig. 7.20 Review tectonic scheme of the West Carpathians and the mean palaeomagnetic directions.

Fig. 7.20 shows the mean palaeomagnetic directions in the synoptic map of the West Carpathians, originally presented in Krs et al.(1982), which here includes also results from the Orava region (designated la in the map). Palaeomagnetic data for regions 1 to 11 are reviewed in the above-cited work. The results from the Orava region indicate a palaeotectonic rotation counter-clockwise. Similar palaeotectonic rotation have also been shown by red shales, sandstones, and teschenites from the Silesian unit. It seems that this rotation has no relation to curvature of the outer bands of the West Carpathians. On the contrary, the rotation of a larger block with parts are also in the West Carpathians.

Fig. 7.21 shows, in a stereographic projection, the mean positions calculated by Fisher's statistics (Fisher 1953) for the North European platform (Krs 1982). Numbers 1, 1a, 2, 3, 4, and 5 designate positions of poles derived from measurements on rocks of the West Carpathians. The pole positions do not fit the apparent polar-wandering path for the North European platform and match the region of African pole positions. The results suggest a palaeogeographic affinity of the outer Flysch Belt rocks of the West Carpathians with the African lithospheric plate.

7.8 Contribution of the Complex Interpretation of Magnetic and Palaeomagnetic Data to the Palaeogeography of Blocks of the Bohemian Massif

M. Krs[1] , P. Pruner[1]

The research into the palaeomagnetism of Palaeozoic rocks of the Bohemian Massif has not yet been completed. That is why only a partial interpretation of the data could be performed. Numerous geophysical measurements have been conducted in the Bohemian Massif, however, the complex interpretation of which yields a particular global-tectonics model.

This interpretation, though, is indirect. The model interpreted has thus to be further checked and modified in the course of the next investigations.

In the work by Bernard et al. (1983), a fossil Proterozoic deep-tectonics zone extending through the Bohemian Massif was derived, which was called the system of central deep-tectonics sutures. The zone is affected by transversal tectonics to such a degree that the zone is disintegrated into partial segments. These segments have formerly been interpreted without considering their mutual connections. Their similar features follow from global-tectonic assumptions, and from a similarity of some magnetic and gravity manifestations. Zoubek (1972) has pointed out the connection between the segments of the zone considered, and its palaeotectonic significance. The deep-tectonics structure of the Bohemian Massif has been studied in numerous papers. There, deep-tectonics lines and crustal zones, and exceptionally the subcrustal zones, were defined. Bernard et al. (1983), however, describe an interpretation of the deep

---

[1] Geological Institute, Rozvojová 135, 165 00 Praha 6, Czech Republic

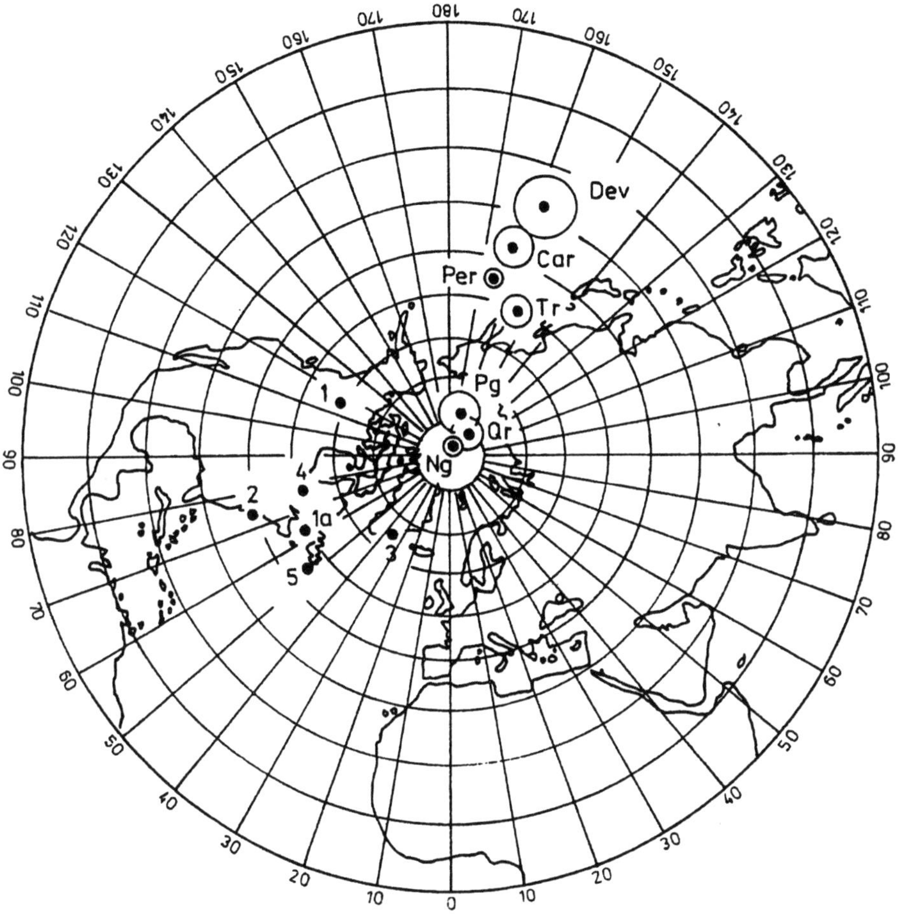

Fig. 7.21 Apparent polar-wandering path derived statistically for stable Europe (Krs 1982). Dev – Devonian, Car – Carboniferous, Per – Permian, Tr – Triassic, Pg – Palaeogene, Ng – Neogene, Qr – Quaternary. Black dots designate the mean pole positions around which circles of reliability for $P = 0.05$ are circumscribed (Fisher 1953). Numbers 1, 1a to 5 designate palaeomagnetic pole positions derived from outer zones (i.e. the Flysch Belt) of the West Carpathians. 1 – Dukla unit, Lower to Middle Eocene red shales, 1a – Magura unit, Orava region, Upper Palaeocene – Middle Eocene shales and sandstones, 2 – Silesian unit Coniacian, grey sandstones and shales, 3 – Ondřejník, Silesian unit, Cenomanian – Lower Turonian red shales, 4 – Silesian unit, Cenomanian – Lower Turonian red shales, 5 – Silesian unit, Hauterivian – Barremian teschenites.

palaeotectonic zone, which at the time of its origin and afterwards had communicated with the asthenosphere. As it resulted from the comparison with the geochemical and metallogenic data, the zone had played a role in distribution of endogenous mineralization.

Regional magnetic field in Czechoslovakia exhibits the following principal types of magnetic anomalies:

1. Anomalies caused by neovolcanics which, with regard to alternations of normal and reverse magnetization, and to variable values of Q-coefficients, are characterized by an inhomogeneous magnetic field extending over large areas.
2. In the contact zone between the Bohemian Massif and the West Carpathians, there is an areally extensive magnetic anomaly, caused by deep sources. Its interpretation is discussed in other chapters of this work. It cannot be excluded that the anomaly's existence is conditioned by the effect of deformation of the isotherm and due to a change of magnetic properties of rocks.
3. The third type of magnetic anomalies were classified into one category although anomalies are distributed in partial segments oriented prevailingly in the NE–SW direction. An affiliation of some density boundaries with these segments is obvious. The given density boundaries, however, express the tectonic structure in upper parts of the lithosphere, only.

According to the Morávek' opinion (cf. Bernard et al. 1983) the deep character of this zone and its multiphase development determine its fundamental metallogenic features, which appear to be similar to those of the Archaic greenstone belts. The main ore-mineralization belongs to the Proterozoic, Lower Palaeozoic, and early Variscan endogenous mineralization prevailingly of disseminated to stratiform sulphide character, to stockworks, and to some veins of late Variscan age. Here, metals such as Fe, Mn, Ni, Cu, Zn, Au, and Sb prevail. Authors of this study claim that the system of central deep-tectonic sutures extends both to the east beneath the Carboniferous Upper Silesian Basin, and to the west to the Vosges Mountains and the Armorican Massif (the peri-Moldanubian fault in the sense of Zoubek 1972). Palaeomagnetic investigations have partly augmented and defined blocks in the Bohemian Massif, as is specified further in the text.

Paleomagnetic investigations in the Upper Carboniferous and Lower Permian rocks of the Bohemian Massif demonstrated the first problems of palaeogeographic reconstructions (Krs 1968). The palaeomagnetic results, derived for the Permian and Lower Triassic rocks of the West Carpathians, pointed out striking rotations of partial nappes. This interpretation followed from original work (Kotásek and Krs 1965). Fig. 7.19 shows the mean palaeomagnetic directions and the confidence cones calculated according to Fisher (1953). The Figure demonstrates striking differences in palaeomagnetic declinations of both the tectonic systems. It also shows weakly expressed differences in palaeomagnetic directions in the Bohemian Massif.

Fig. 7.22 presents the palaeogeographic latitudes (palaeolatitudes) for the Autunian, calculated from the palaeomagnetism of rocks from the Blanice and Boskovice furrows, and from the palaeolatitude for "Rotliegendes" from the rocks of Intra-Sudetic and Krkonoše-piedmont Basins. An orientation of the palaeolatitudes is identical within the limits of accuracy, which shows that the blocks with marked palaeolatitudes did not undergo strong tectonic deformations in horizontal direction during and after the Lower

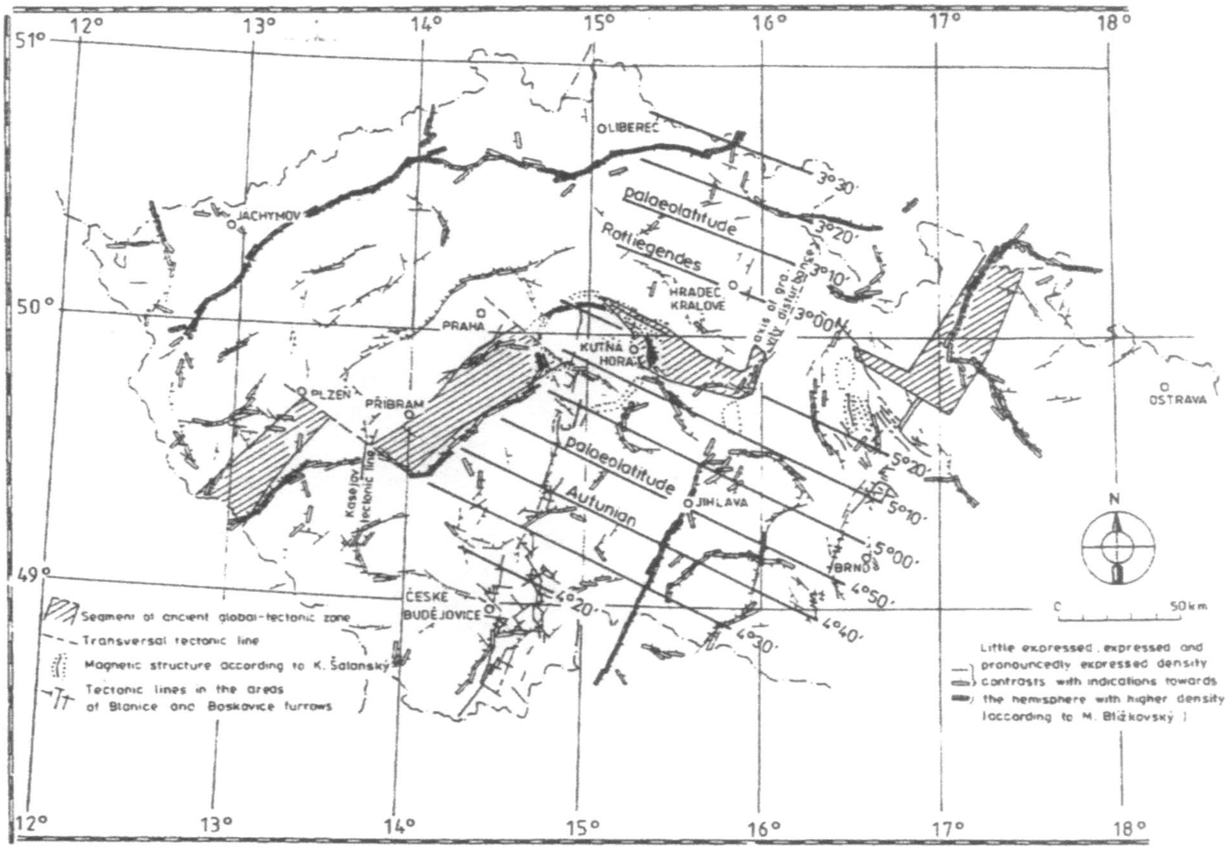

Fig. 7.22 Lower palaeolatitudes, Bohemian Massif, on the global-tectonics sketch and the map of density contrasts.

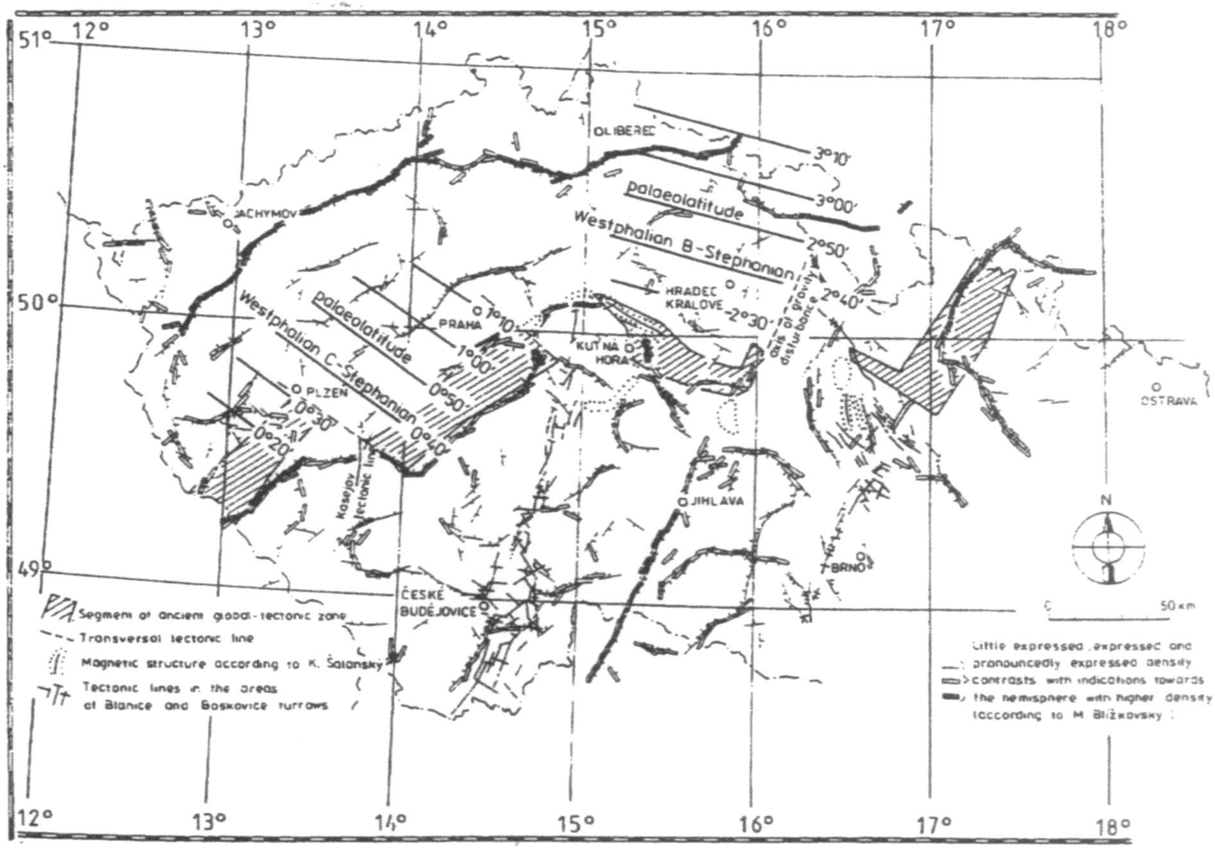

Fig. 7.23 Westphalian – Stephanian palaeolatitudes, Bohemian Massif, on the global-tectonics sketch and the map of density contrasts.

Permian. Fig. 7.23 shows palaeolatitudes for the Westphalian and Stephanian, calculated from palaeomagnetism of rocks from the Plzeň and Kladno Basins on one side, and from palaeomagnetism of rocks from the Intra-Sudetic Basin and the Krkonoše-piedmont Basin on the other side. Inaccuracy in determining the palaeolatitudes has not exceeded ± 3°. Even with regard to such an inaccuracy, there remains a striking difference in the orientation of palaeolatitudes. Analogous deviation shows itself also in the axes of individual segments of the above discussed global-tectonics zone. This deviation obviously originated during the Asturian orogenetic phase.

7.9 Magnetic Properties of Rocks and Their Interpretation

V. Kropáček[1], A. Kapička[1]

Different physical-chemical conditions, effective in originating individual types of rocks, lead to different content of ferrimagnetic minerals, to their different chemical composition, texture and crystal size. Despite the fact that these ferrimagnetic minerals (magnetite, maghemite, hematite, pyrrhotite, solid solutions such as magnetite–ulvöspinel, hematite–ilmenite, etc.) are present in the rocks only in low concentrations (0.05 – 15 % of volume), their magnetic manifestations strikingly surpass those of the remaining large majority of dia- and paramagnetic minerals (feldspars, quartz, micas, pyroxenes, amphiboles, clay minerals, etc.). That is why the vast majority of rocks behave as weak ferrimagnetic rocks, i.e. they exhibit magnetic hysteresis, saturated magnetization, Curie temperatures, remanent magnetizations of various origin (thermoremanent, chemical, thermochemical, piezoremanence, viscous, detritic, etc.).

From the point of view of interpretation of magnetic properties of rocks it is necessary (besides other physical and chemical quantities) to consider also the time-scale, i.e. either to consider the magnetic manifestation of rock to be an integral unit, or to relate the observed magnetic properties to a particular period during the rock genesis. As long as the magnetic properties of rocks for a particular time period are interpreted it is necessary to consider whether the minerals now present in the rocks were present in them also in the particular time period, and to find out which component of observed properties corresponds to them (i.e. resolving the primary and secondary mineralization).

When studying the magnetic properties of rocks the following two types of magnetic characteristics are distinguished:

1. Magnetic characteristics which depend only on the chemical composition, texture, and concentration of ferrimagnetic minerals present in the rock,
2. magnetic characteristics which depend both on the parameters of point 1. above, and on other factors, such as the grain size and shape of grains, their distribution, and

1 Geophysical Institute, Boční II/1401, 141 31 Praha 4, Czech Republic

Table 7.5 (part 1) Mean values of magnetic parameters of volcanic rocks.

rock type	M [gm^{-3}]	J_n [μT]	J_{s0} [mT]	κ × 10^{-6} [SI]	H_c [kA/m]
polzenite	3.000	10.86	8.010	45 510	7.98
melilitic nephelinite	3.190	6.57	21.60	145 080	6.40
olivinic melilinite nephelinite	3.155	14.58	5.06	34 770	7.28
nephelinite (Krušné hory)	3.176	7.11	13.17	81 980	6.62
nephelinite	3.058	3.30	21.93	110 820	6.22
analcimite	2.990	3.20	9.21	45 690	9.55
olivinic nephelinite	3.000	3.48	5.82	25 860	8.29
olivinic alkaline basalt	2.987	3.58	3.81	25 590	8.98
basanite	2.984	3.51	4.82	26 630	8.27
limburgite	2.986	5.59	3.57	20 660	14.97
basalt sensu stricto	2.850	3.79	7.60	43 405	7.55
tephrite	2.842	3.37	6.04	34 000	8.16
trachybasalt	2.780	2.93	10.34	50 440	9.96
trachyte	2.587	0.727	3.64	18 630	7.07
phonolite	2.540	0.014	0.073	177	6.12
dyke rocks	2.870	3.05	3.09	19 180	---

their actual state, effects of active physical fields (magnetic field, temperature field, pressure field).

Diversity of these individual magnetic characteristics reflects very sensitively the diversity of conditions of origin and genesis of rocks, changes in the course of existence of rocks, history and changes of the Earth's magnetic field.

Magnetic Properties of Volcanic Rocks

In order to have an idea of how values of some magnetic characteristics of various types of rocks are distributed, the mean values x_{mean} were determined (considering that vast majority of magnetic characteristics exhibit a lognormal distribution)

$$x_{mean} = \prod_{i=1}^{N} (x_i)^{\frac{1}{N}} \quad .$$

In Table 7.5, the mean values of natural remanent magnetic polarization J_n [μT] are presented, together with those of saturated magnetic polarization J_{s0} [mT], those of the magnetic susceptibility κ [SI], those of coercive forces H_c [kA/m], those of the Curie temperature Θ [°C], those of the Koenigsberger ratio Q_n, those of the ratio J_{rs}/J_{s0} (the saturated remanent magnetic polarization versus the saturated magnetic polarization), and those of the ratio H_{cr}/H_c, i.e. of the coercive force of remanence H_{cr} versus the coercive force H_c . In Table 7.5, N designates the number of studied rocks of a given type, i.e. of young alkaline volcanic rocks of the Bohemian Massif.

Table 7.5 (part 2) Mean values of magnetic parameters of volcanic rocks.

rock type	Θ [°C]	Q_n	J_{rs}/J_{s0}	H_{cr}/H_c		N
polzenite	533	3.09	0.221	1.31	P	15
melilitic nephelinite	---	0.59	0.154	2.01	MN	3
olivinic melilinite nephelinite	228	18.2	0.283	1.42	OMN	10
nephelinite (Krušné hory)	520	1.14	0.165	1.93	N (KH)	16
nephelinite	460	0.75	0.142	1.78	N	6
analcimite	390	1.32	0.292	1.81	A	8
olivinic nephelinite	315	2.22	0.286	1.50	ON	33
olivinic alkaline basalt	237	2.95	0.303	1.58	OB	36
basanite	245	2.28	0.260	1.57	BN	30
limburgite	254	5.91	0.436	1.80	Li	8
basalt sensu stricto	335	1.61	0.254	1.98	Bss	6
tephrite	330	1.88	0.301	1.71	Te	8
trachybasalt	---	0.76	0.098	1.63	TB	3
trachyte	515	0.59	0.103	1.99	Tr	26
phonolite	?	0.60	0.103	1.94	F	26
dyke rocks	515	2.10	---	---	ŽH	17

It follows from Table 7.5 that the highest values of J_n are reached by olivine types of rocks which contain pseudo-single-domain grains of the diagram J_{rs}/J_{s0} vs. H_{cr}/H_c (Fig. 7.24). The highest values of the Koenigsberger ratio Q_n also correspond to it.

The mean values of analogous magnetic parameters for alkaline-calcareous and alkaline basaltoids of the West Carpathians are presented in Table 7.6.

Figs. 7.25 and 7.26, where the determined mean values of J_n and κ, respectively, and the interval ($\log J_n - 3\,\sigma$, $\log J_n + 3\,\sigma$) are shown, were compiled to study the distribution of values of natural remanent magnetic polarization J_n, and of magnetic susceptibility κ in whole sets of rocks of the Bohemian Massif and West Carpathians. There, r is the standard deviation of log J_n (i.e. an interval within which there are

Table 7.6 (part 1) Mean values of magnetic parameters of alkaline basaltoids.

rock type	J_n [nT]	J_{s0} [mT]	κ × 10^{-6} [SI]	H_c [kA/m]	Θ [°C]
basaltic andesite	2250	1.78	8 190	12.55	270
pyroxene andesite	6090	4.04	17 460	11.18	355
dacite	2464	1.08	6 366	9.48	510
rhyodacite	31.6	0.072	201	17.40	570
rhyolite	15.7	0.163	297	8.70	510
andesitic porphyrite	2403	5.60	26 380	7.80	---
dioritic porphyrite	4000	11.15	46 900	8.25	---
granite	774	4.27	16 430	5.74	---
ignimbrite	2490	2.14	10 200	14.21	525
alkaline basaltic rock	2770	4.27	23 750	9.45	265 (495)

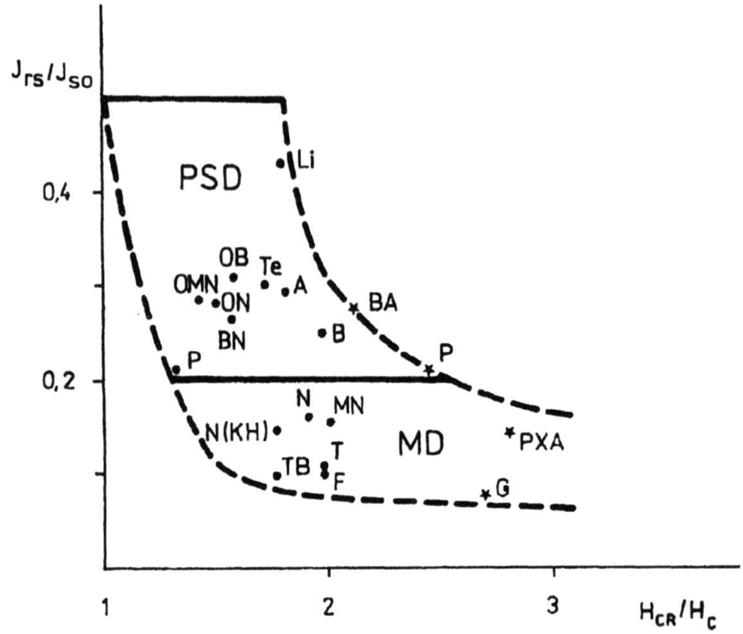

Fig. 7.24 Distribution of values of J_{rs}/J_{s0} vs H_{cr}/H_c for some types of rocks. Designation the same as in tables 7.5 (part 2) and 7.6 (part 2).

theoretically 99.7 % of all individual values and the observed maximum and minimum values of J_n and κ, respectively). The only observed exception was the value of J_n for the sample of bazanite (the Provodínské kameny locality) where the rocks were magnetized by lightning. Similar characteristics should be compiled for the majority of rocks occurring in the Bohemian Massif and West Carpathians.

Table 7.6 (part 2) Mean values of magnetic parameters of alkaline basaltoids.

rock type	Θ_n	J_{rs}/J_{s0}	H_{cr}/H_c		N
basaltic andesite	3.60	0.280	2.11	BA	12
pyroxene andesite	4.59	0.145	2.78	PXA	21
dacite	5.09	0.273	2.45	D	5
rhyodacite	2.06	0.238	---	RHD	3
rhyolite	0.70	---	---	R	6
andesitic porphyrite	1.20	---	---	AP	4
dioritic porphyrite	1.48	---	---	DP	4
granite	0.62	0.072	2.64	G	5
ignimbrite	3.21	0.210	3.68	I	8
alkaline basaltic rock	1.53	0.252	1.812	AB	17

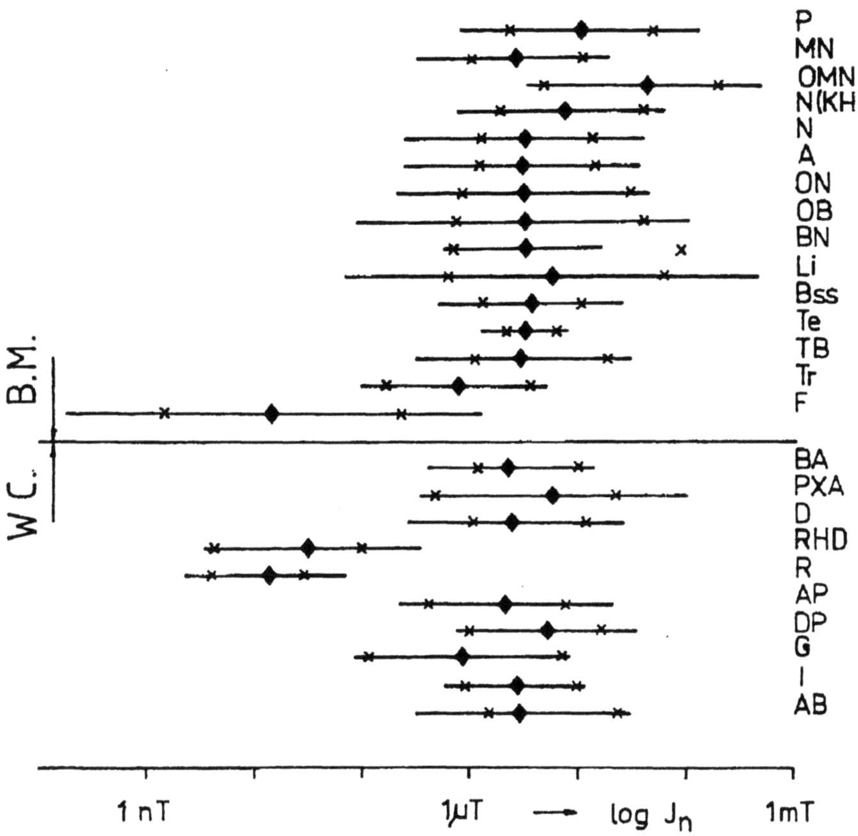

Fig. 7.25 Distribution of J_n values for individual types of rocks. Designation the same as in tables 1 and 2. ♦ – the mean value, × – maximum, minimum.

Magnetic Properties of Rocks Exposed to High Pressures

Magnetic properties of rocks depend, to a considerable extent, on external pressure. They can change substantially in deep parts of the lithosphere. Methods of measuring the magnetic parameters in situ thus allow studying some parameters under actual pressure conditions, but the measurement's accuracy is relatively low in the majority of cases. When interpreting magnetic properties of rocks in deep parts of the Earth's crust, the laboratory measurements under various pressure conditions are of great significance. The aim of these experiments is to define the fundamental physical processes which take place during an exposure of ferrimagnetic minerals to pressure, and which affect their magnetic properties on a macroscopic scale, and then, on the basis of these mechanisms, to interpret the changes of magnetic properties of rocks from various depths.

Out of the parameters describing magnetic properties of rocks, the remanent magnetic polarization (J_r) and the initial magnetic susceptibility (κ) are of decisive significance for interpretations. Pressure changes of these parameters are the direct

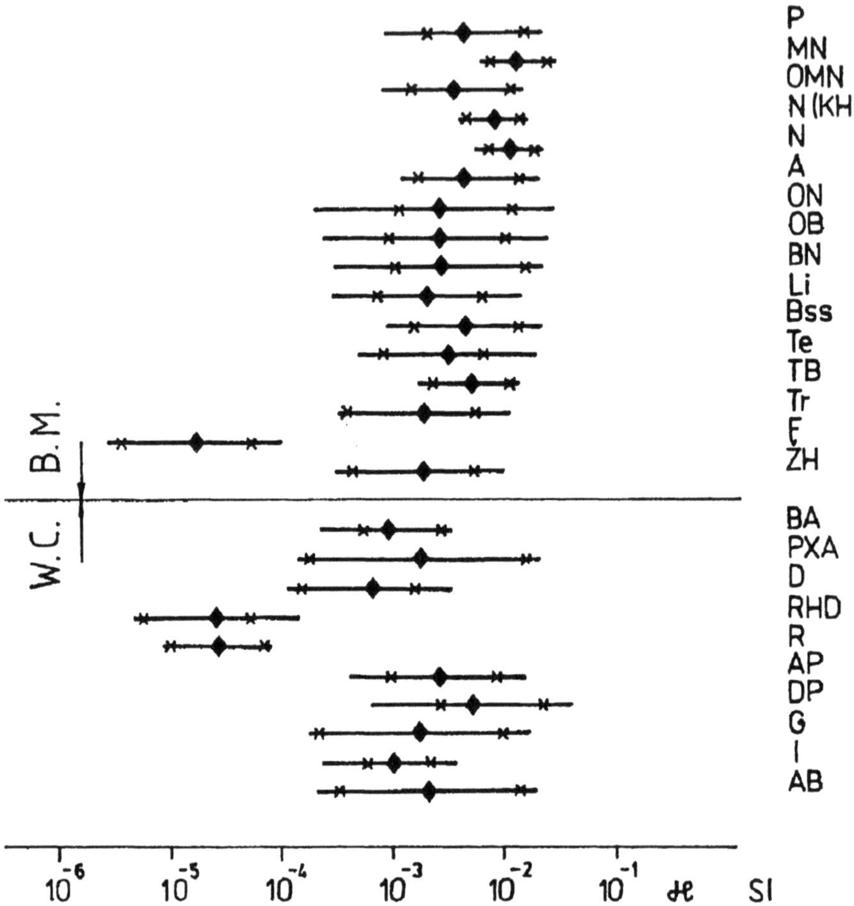

Fig. 7.26 Distribution of values of κ for individual types of rocks, designation as in tables 4 and 5. ♦ – the mean value, × – maximum, minimum

results of the following fundamental mechanisms in ferrimagnetic minerals: the rotation of spontaneous magnetization in domains, and movements of the domain walls. Magnitudes of reversible changes of these parameters for particular rocks can be expressed quantitatively by means of the coefficients of stress sensitivity (e.g. Davis 1983). An interpretation of irreversible changes which are connected with processes of irreversible movements of the domain structure, is complicated due to numerous factors affecting the domain wall stability (defects in crystals, impurities, grain interfaces, etc.). General description of dependences $J_r^{ir} = f(p)$, $\kappa^{ir} = f(p)$ has thus appeared unreal.

Reversible and irreversible changes of J_{rs} of basalts from numerous localities in the Bohemian Massif caused by effects of the uniaxial and hydrostatic pressures of up to 200 MPa were analyzed in detail. The hydrostatic pressure roughly corresponds to the conditions at a depth of 7 km. In the whole set, the reversible changes represented only 14 – 35 % (for the hydrostatic pressure) and 12 – 48 % (for the uniaxial pressure) of the total changes. The far more radical decrease of J_{rs} is caused by the effect of non-hydrostatic pressure (i.e. uniaxial) pressure. The changes of J_r for pressures in the range of $\tau \leq 40$ MPa are described by the expression

$$J_r(\tau) = J_{r0}(1 - \beta\tau) \quad ,$$

where J_{r0} is the initial value of polarization. The constant of stress sensitivity β for basalts is in the range of $0.4 - 2.1 \times 10^{-3}$ MPa^{-1} (Bezuglaja and Kapička 1981). These values correspond to lower values of β for intrusive rocks, which are presented in literature. For hydrostatic pressures, the reversible changes are observable only at high pressures. Here also, the mechanisms of rotation of spontaneous magnetization has probably taken place, as a result of inhomogeneous transfer of the pressure through a nonmagnetic matrix (Kapička and Savěnko 1979).

Courses of irreversible changes of J_{rs} under uniaxial pressure are analogous for all basalts, because for high pressures, they approach the limiting value. In order to describe their stability quantitatively, coefficient $K = J_r^{\sigma}/J_{rs}$ was defined, where J_r^{σ} is the remanent magnetic polarization for the stress at the breaking strength representing pressure σ. Values of K for basalts are in the range of 0.35 – 0.74. With regard to diverse mechanisms of reversible and irreversible changes of J_{rs}, no correlation has been found between coefficients β and K.

On the contrary to the pressure demagnetization J_r, the pressure remanent magnetic polarization (*PRM*) depends substantially on the magnitude of the external field. *PRM* is a significant kind of remanent magnetic polarization as it originates under the natural conditions of a particular rock's existence. *PRM* is always greater than the isothermal remanent magnetic polarization (*TRM*) in the same magnetic field. As the stress increases, *PRM* approaches the limiting value. The *PRM* of basalts exhibits a relatively low stability. In all rocks of the set it was demagnetized by the field $H_{max} \approx 1.2 \times$ $\times\, 10^{-4}$ Am^{-1}.

The initial susceptibility and its anisotropy under uniaxial as well as hydrostatic pressures is another parameter studied in detail. The results show that the changes of susceptibilities of basalts are caused by two diverse mechanisms: either by a change of intrinsic susceptibility of ferrimagnetic mineral κ_i or by a change of the demagnetization factor. For elastic deformations of rocks, only the first process is characteristic, while deformations above the yield point can be caused by both the processes.

As the uniaxial pressure increases, the parallel direction susceptibility κ of basalts decreases. For low pressures ($\tau \leq 60$ MPa) the decrease is approximately defined by the following linear expression:

$$\frac{\kappa(\tau)}{\kappa_0} = 1 - \alpha\tau \quad ,$$

where κ_0 is the susceptibility of undeformed rock. Values of α for basalts are in the range of $0.5 - 8.1 \times 10^{-3}$ MPa^{-1} (Kapička 1984). The ratio of coefficients of stress sensitivity for two perpendicular directions $\alpha^{\perp}/\alpha^{||}$ is a characteristic quantity determining the susceptibility stability with regard to the uniaxial pressure. This ratio for basalts of the Bohemia Massif varies between 0.1 – 0.31, being smaller of the assumed theoretical value. The change of directional susceptibilities is accompanied by a radical change of the susceptibility anisotropy. From the measurements of reversible changes of anisotropy under the uniaxial pressure two conclusions follow (Kapička 1988):

1. If the anisotropy of basalts in undeformed state is determined by the shape effect, then already under a relatively low pressure ($\tau \approx 50 - 60$ MPa), the magneto-elastic energy is decisive for the anisotropy first of all,
2. the main cause of the pressure change of anisotropy is the radical change of the parallel directional susceptibility κ. Pressure changes of the mean susceptibility κ_{mean} of basalts under the uniaxial and hydrostatic pressures are closely connected with the pressure change of anisotropy (Kapička 1990). In both the cases, κ_{mean} decreases as the pressure increases, but for hydrostatic pressure the changes are very small.

Characteristic changes of the susceptibility anisotropy of basalts result from deformations above the yield point (Kapička 1984). The main cause of small plastic deformations ($\varepsilon \approx 0.2$ %) are irreversible changes of intrinsic susceptibility of κ_i, which are associated with the growth of concentration of dislocations in ferrimagnetic minerals. Changes of demagnetization factors connected with deformations of ferrimagnetic grains take place only in cases of high plastic deformations. It has been shown that the initial susceptibility especially, and its anisotropy, can be a sensitive indicator of the start of plastic deformation of rocks.

Experimental results of the study of pressure changes of both the remanent magnetic polarization and the magnetic susceptibility show that the decisive cause of these changes in deep parts of the Earth's crust can be, first of all, local concentrations of stress. Assuming only hydrostatic pressure, the changes of magnetic parameters are then relatively small as they are caused only by an inhomogeneity of the pressure transfer via nonmagnetic minerals to ferrimagnetic ones.

8 Geological and Petrological Models of the Earth's Crust in Central Europe

8.1 Reviews of Geologic Investigations by Drilling

Z. Mísař[1], D. Ďurica[2], B. Leško[3], J. Weiss[4]

Information on geologic structure of the lithosphere can be obtained, apart from geophysical methods, mostly by deep drilling. In the Bohemian Massif and the West Carpathians boreholes are distributed irregularly; the majority of boreholes has been situated in areas of present or prospective coal mining, and of oil and gas exploitation. That is why the boreholes do not provide adequate information on geologic structure of Czechoslovak territory (Fig. 8.1).

Investigation of the basement of the North Bohemian Cretaceous Basin yielded valuable results. There, several boreholes in the section were drilled to a maximum depth of 1934.5 m (Všeň 1). In this way the general structure of the basement was verified. In east part, metamorphic rocks of greenschist facies mainly occur, that can be considered the continuation of the Letovice and Polička crystalline unit towards the north to the Zábřeh unit. All these units are of late Proterozoic age. An ultrabasic rock body has recently been verified at a depth of 620 m in a place of intensive magnetic and gravity field anomalies near the town of Svitavy. NE of the Železné hory Mts. the Přibyslav deep fault extends further into the basement of the Cretaceous. The Železné hory pluton projections extend far towards the north. In the Železné hory Mts. rim, further to NW, prevailingly metamorphosed early Palaeozoic rocks form the basement.

Weakly metamorphosed late Proterozoic and early Palaeozoic underlie sediments in the western part of the basin. In part of the basin, which is adjacent to the Krušné hory Mts., granulites and ultrabasic rocks of the Moldanubian type also occur. The connection of the Central Bohemian Carboniferous and the Lužice Permo-Carboniferous was confirmed as well as the existence of a vast granitoid massif of Louny.

1 Faculty of Natural Sciences, Charles University, Albertov 6, 128 43 Praha 2, Czech Republic
2 Central Institute of Geology, Malostranské náměstí 19, 118 21 Praha 1, Czech Republic
3 Geofyzika Brno, Ječná 29a, 612 46 Brno, Czech Republic
4 Geological Institute, Masaryk University, Kotlářská 2, 611 28 Brno, Czech Republic

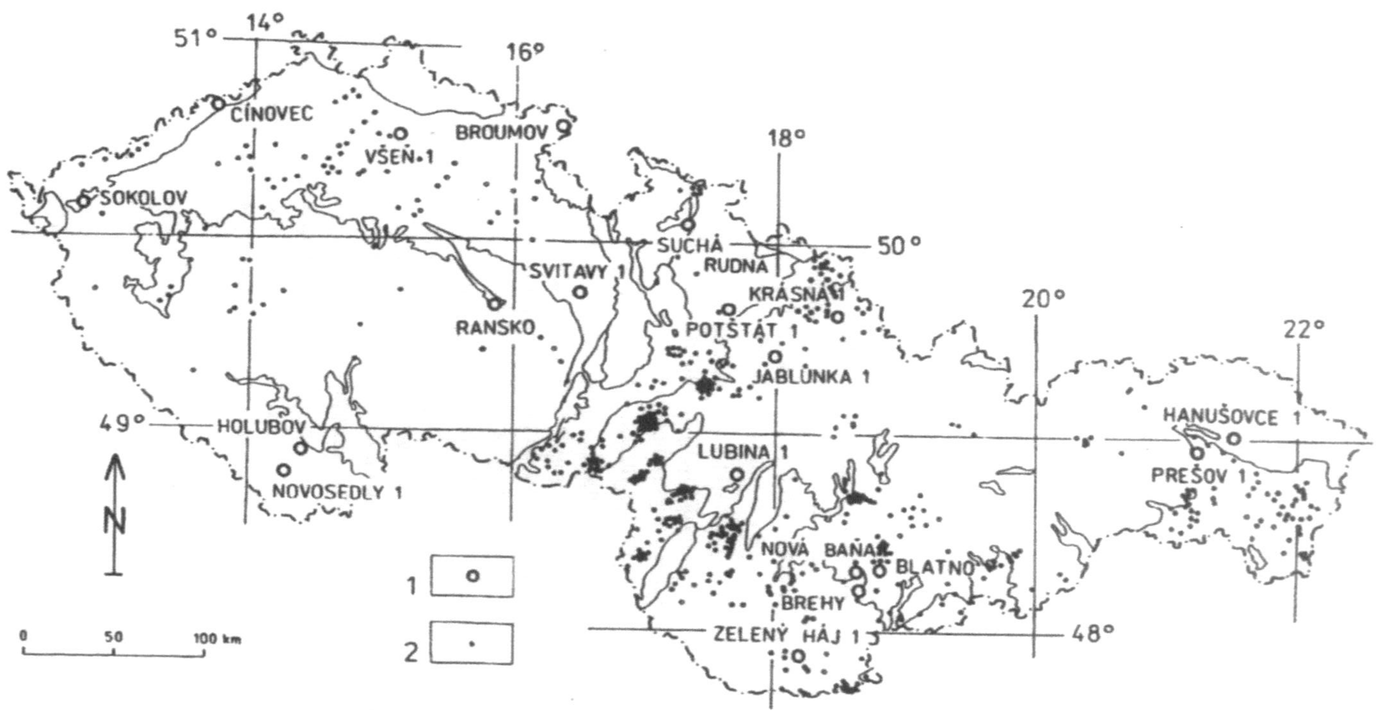

Fig. 8.1 Distribution of deep boreholes in Czechoslovakia. 1 – basic boreholes, 2 – prospect boreholes.

The structure of Moldanubicum and its composition have so far been confirmed by deep boreholes only sporadically. A borehole in the granulite body of Kleť (SW of the town of České Budějovice) was terminated at a depth of 1409 m, and showed various types of granulites with garnet peridotites. Another deep borehole (Novosedly-1) in this region penetrated a group of the Moldanubicum containing crystalline limestone and graphites down to depth of 1576 m. Although numerous boreholes of maximum depth of 1000 m confirmed the composition of individual parts of the Moldanubicum (migmatites, gneisses, etc.), they did not contribute knowledge of its fundamental structure. The shallow boreholes situated at the eastern margin of the Moldanubicum, that penetrated its dislocation contact of a thrust character with the Moravicum, are more significant in this respect. The boreholes also confirmed the northwestern convergence of the eastern margin of the Třebíč Massif. A borehole situated at the northern margin of the Moldanubicum in the Ransko massif penetrated peridotites down to a depth of 1320 m and thus enabled interpretation of the structure of this complex gabbro-peridotite conical body.

In other geological units of the Bohemian Massif, deep boreholes have been drilled only exceptionally. In the Saxothuringian region, borehole Cínovec penetrated young tin-bearing granite down to a depth of 1596 m, and thus confirmed the existence of albitization, greisenization, and Sn-mineralization also at this depth. Several boreholes of 1000 m depth confirmed the structures of thermal springs. A borehole in the central part of the Sokolov Basin penetrated the basin's fill (Tertiary), crystalline schists of Proterozoic age, and was terminated in granite of the Karlovy Vary massif at a depth of 950 m.

Shallow boreholes in the Silesicum, drilled to depths of about 500 – 880 m, confirmed the contact of the Devonian of the Vrbno group of layers with the Culm of the Nízký Jeseník Mts. In the Culm of the Šternberk – Horní Benešov belt, drilling activity remains. More boreholes are to be drilled which should strike the basement of the culm formation and, if possible, also of the early Paleozoic (Devonian).

The Carpathian Foredeep and the Flysch Belt in Moravia and Slovakia represent the largest area where a great number of deep boreholes were drilled (the deepest one is Jablůnka-1 – 6505 m; mean depths from about 2 to 4 km). Information on the geologic structure of the area has been provided by structural exploration, and production boreholes. It was found that the basement, reached by more than 100 boreholes, is formed by the Cadomian crystalline unit – metamorphites in northern part, and extensive Brno pluton in southern part. There is a platform cover constituted of sediments of the Devonian and Carboniferous, Jurassic, Oligocene, and Miocene. All these units are overlaid by overthrust nappes of the Flysch Belt. The overthrust length documented by boreholes reaches nearly 40 km. The nappe structure exists also in the Fore-Carpathian basement including the Devonian in the vicinity of Hranice. The isotopic age of the basement rocks detected for the rock samples collected from numerous boreholes ranges from 550 Ma to 660 Ma. Borehole Broumov-1 (depth 3 000 m) drilled to the basement of the Intra-Sudetic basin, and boreholes Dětřichov and Potštát (4 100 m) which both confirmed folded structure and lithostratigraphy of the Lower Carboniferous of the Nízký Jeseník Mts. represent the newest boreholes.

Horizontal stratification of the crust of the West Carpathians in its entire thickness represents the most significant information obtained from the existing, gravimetric, and magnetometric measurements. In order to decipher the geologic structure of the crust beneath the West Carpathians, as well as to derive geologic information of all stages of its orogenic development it is vital to explain the genesis of this stratification. In this respect boreholes are contributing a lot, because on the basis of the data from them, numerous geophysical data can be identified, and thus their geologic meaning be interpreted. It needs to be done, however, after a brief evaluation of stratigraphic-palaeontological, sedimentary, volcanic, petrographic-metamorphic, and kinetic criteria.

In the Klippen Belt foreland (i.e. in the Magura, Silesian, Subsilesian, and in east Slovakia, Dukla nappes) several deep boreholes were drilled. Some of them have proved an Alpine involvement of southern margins of the North European platform (Krásná-1) and its far reaching overriding by the Flysch Belt nappes in the course of the Neogene orogenes (boreholes Berndorf-1, Zistersdorf-1A on Austrian territory, and numerous boreholes in the Vienna Basin). Geophysical investigations signal, however, an existence of numerous elevated parts of the platform and of its geologic units in the neighbourhood of the Klippen Belt along the entire Carpathian arc.

The Pieniny Klippen Belt as a specific phenomenon of the West Carpathians is a nappe unit which is a result of Eocene and Neogene orogenes, during which the inner West Carpathian units of the crystalline complex, Palaeozoic, and Mesozoic formations were progressively and gradually thrust over the North European platform margins as over a plastic-like and incompetent lithologic substance. This function of the Klippen Belt is documented by both its relation to the Flysch Belt and to inner units, to seismic velocity pattern as well as the gravity field distribution pattern along its entire course in the West Carpathians. In some places this function of the Klippen Belt has already been confirmed by deep boreholes.

Borehole Hanušovce-1 (depth 6003 m) situated in a outcropping Klippen Belt in east Slovakia has proved its overriding the Magura nappe towards NE (at an angle of 22° – – 25°). Borehole Lubina-1 (depth 3300 m) situated in the vicinity of Stará Turá in west Slovakia, about 4 km SE of an outcropping Klippen Belt, has confirmed its complete tectonic "rolling out", and its immediate overfault contact of the Cretaceous group of layers of the Manín nappe with the Eocene sediments of the flysch rocks of the Biele Karpaty unit. Analogous geologic situation existing in the east Calcareous Alps has been confirmed by borehole Berndorf-1 (depth 5000 m). In this borehole, situated about 16 km SE of the northern margin of Mesozoic nappes of the Austro-Alpine unit, it was discovered that a group of these nappes had overthrust an imperceptible thickness of flysch sediments the basement of which is formed by sediments of the Oligocene–Miocene molasse deposited on the crystalline complex of the Bohemian Massif.

The Klippen Belt function in the genesis of the West Carpathians is also documented by a deposition environment in the Považský Inovec horst. Based on results of geophysical measurements the horst is elevated by 9 – 12 km relatively to the present level of the surrounding Pliocene sediments, so that in its present morphological section, those geologic features are exposed which characterize the Earth's crust at a level of 8 - 9 km below the ground surface, i.e. the lithospheric-stratigraphic sequence affected by the Alpine dynamic metamorphosis in the course of the Paleogene orogenes. In the top of the horst elevation, Upper Cretaceous sediments of the Kláp group of the Klippen Belt occur in tectonic windows in the crystalline complex and in sediments of the late Palaeozoic of the Inovec unit of the Pieniny provenance, which are together with them, dynamically metamorphosed as a result of the Alpine orogenic phase.

It points out the facts that the Palaeogene movements had completed structure of the nappe system of the Austro-Alpine-southcontinental provenance, and that their transport towards the North European continent caused a destruction and dynamic metamorphosis under load in the South Pieninian nappes underlying it. Only the marginal part of the original sedimentation of the Klippen Belt could have escaped the dynamic metamorphosis because as plastic and incompetent masses were pushed away by the progressing Austro-Alpine (Tatricum, "Sub-Tatricum", Silicicum, etc.) nappes, as well as the south Pieniny (Iňačovce – Pozdišovce, Kričov), nappes of the Veporicum and Gemericum, and other units, towards the outer margin of the West Carpathians, and it thus did not interfer with the whole mass of the nappe system.

The nappe structure is well observable in outer belts of the West Carpathians, but in inner and south parts of the West Carpathians, it is hidden and poorly visible. This situation is caused by two factors. One of them is the Miocene to Pliocene orogenic movements that destroyed lithologic masses of original nappe structures from the Upper Cretaceous and Palaeogene epochs, and often distributed them in space contrary to their original transport movement and to their tectonic style. The second factor, which "masks" the nappe structure of the West Carpathians, are the Tertiary sediments and volcanic products, which, especially in South Slovakia and in adjacent regions of North Hungary, hide a substantial part of the Carpathian nappes. The negative effect of the second factor has been reduced by those boreholes which penetrated the Tertiary sediments and volcanics and which entered their basement, as well as by the results of geophysical measurements. The geologic units derived from the "oceanic" part of Tethys can be delimited with regard to those units which originated from its continental margin, thus from the units of the Austro-Alpine and Euxinic-Pienides provenance. In the West Carpathians, units of Tatricum, "sub-Tatricum" (Krížna and Choč nappes), Silicicum, and Bükkovicum, belong to the Austro-Alpine group of nappes. The first three form, due to the Miocene–Pliocene movements, more or less isolated blocks-horsts which extend in an arc-like fashion from the Malé Karpaty Mts. over the Považský Inovec Mts., Velká Fatra Mts., Malá Fatra Mts., Vysoké Tatry Mts., Nízke Tatry Mts., to East Slovakia. The maximum east extent of the Tatricum was detected by borehole Šariš-1 in the Torysa river basin, where the Tatricum was verified in the basement of the inner Carpathian Palaeogene and the Krížna nappe at a depth of 3850 m. The Krížna nappe, however, continues along the southern margin of the Klippen Belt further SE into the Ukraina territory, where it was verified by boreholes near Užhorod in the basement of neovolcanic extrusions and of the Pliocene sediments.

A spatial distribution of the Choč group nappes in the Carpathian arc towards the south has not yet been adequately investigated because in many boreholes, the nappe presence in the basement of the Tertiary sediments and neovolcanics has often been assumed or stated only on the basis of the late Palaeozoic sediments (the "melaphyre series") occurrence. These sediments, though, have not been adequately evaluated from the palaeontology, petrography, and isotopic studies point of view.

It was detected, for instance, in the East Slovakian Neogene Basin and in the Košice depression, that only the Middle Triassic carbonates and dolomites (boreholes Ďarkov-1, 2, 3, and Prešov-1) can be attributed to the Choč unit, and that the pelitic, psammitic, and carbonate rocks, originally classified as "Permian–Carboniferous", represent lithospheric sequences of the Mesozoic (Middle Triassic–Cretaceous) of the "oceanic" crust.

Old evaluations of some core samples from the surroundings of Banská Štiavnica, Nová Baňa, and Levice have left a doubt concerning the spatial distribution of the Choč unit in the basement of neovolcanics and Tertiary sediments. The lithologic-petrographic character of rocks, which were penetrated by boreholes, and which have, so far, been considered elements of the Krížna and Choč units (e.g. boreholes Zlatno-1, 2; Brehy Gk-14, Nová Baňa Gk-13), exhibits lithologic features of the Mesozoic formations of the Pieniny of the West and East Alps. Moreover, here, these rocks were differently affected by dynamic metamorphosis, and were intensively reworked by tectonics; these facts, in the West Carpathians, are not characteristic features of upper units of the Austro-Alpine group because, in the course of the Palaeogene and Miocene orogenes, they behaved more or less passively in the tectonic movement. Tectonic movements of this epoch had prevailingly taken place in deep parts of the crust affecting Pieniny and the units in their basement, thus at the level of platform units of the North European platform.

The Silica nappe forms karst plains from Moldava to Šafárikovo in South Slovakia. Its maximum west extent is to the Danube Basin, where the nappe was detected in borehole Zelený Háj-1 near Hurbanovo.

The spatial distribution pattern of Mesozoic and Palaeozoic lithologic units enable us to assume that the units of the Carpathian Pieniny form a wide and extensive tectonic window (semi-fenster), which is tied to the tectonic window of the Hohe Tauern Mts. through Rechnitz in the East Alps. The West Carpathian tectonic window of the Pieniny is limited by the Silicicum and Bükkovicum in the south, and by nappes of Tatricum and "sub-Tatricum" in the north. The Pieniny Klippen Belt and the Magura Flysch Belt, as parts of the Carpathian Pieniny, delimit, in the Miocene structure of the West Carpathians, the outer extent of Pieniny nappes against units with the lithologic-stratigraphic content, which had originated on the North European platform.

Effective verification of fundamental elements of the structure of the Bohemian Massif units and their mutual relations appears to be possible in future. It is more significant than drilling to greater depths to verify the lithologic development. A possibility of making a prognosis of mineral resources occurrences is closely connected to the verification of each unit's structure. From the theoretical problems and practical views it is necessary to resolve, first of all, the existence of either nappe or non-nappe structure of the Moldanubicum, especially in its central parts (the position of granulites and ultrabasic rocks of the mantle). A verification of the relationship of the Moldanubicum with the surrounding units appears important also. These marginal zones are often associated with uranium mineralization. The detection of the contact character of the late Proterozoic with the Moldanubicum in the zone extending from Klatovy over the Jílové zone to the Železné hory Mts. (the centre of acid volcanism) is extremely significant. Boreholes in the Hrubý Jeseník Mts. and in the Nízký Jeseník Mts., where occurrences of polymetallic mineral ore deposits of the Devonian to Lower Carboniferous age can be assumed, will investigate ore mineralization mainly.

Deep boreholes at the eastern margin of the European platform will help in verifying the interpretation of seismic reflection sections, according to which a complex of nappe structures has been assumed here. Geophysical investigations suggest numerous elevation regions of the platform and of its geologic formations mainly in the vicinity of the Klippen Belt along the whole Carpathian arc. These are, first of all, the elevation zone extending from Radošovce and Holíč over Starý Hrozenkov, Drietoma, Klokočov – – Turzovka, Oravská Polhora to East Slovakia, to the Čergovské pohorie Mts. and Vihorlat Mts. In order to investigate these prospective zones to discover oil and gas it is necessary to carry out an extensive research and exploration by means of deep boreholes which should be drilled down to depths of 5 000 – 8 000 m.

With regard to the positive results in prospecting for oil and gas in the Vienna Basin it is necessary to continue research and exploration program and make a complex geologic evaluation based on geophysical and drilling data with the aim to prospect for oil and gas deposits also in non-anticline structures.

8.2 Peridotites and Eclogites as Indicators of the Structural and Lithologic Evolution of the Earth's Crust: Bohemian Hercynides (Variscides), Czechoslovakia

Z. Mísař[1]

Introduction

Serpentinites, garnet peridotites, spinel peridotites, and eclogites are important crustal constituents in the European Hercynides. The structural position, geochemical character, and plate tectonic setting of this lithologic association in Cadomian, Caledonian, and Hercynian fold belts has been extensively discussed (Burg and Matte 1978; Gebauer and Grüenenfelder 1979; Mísař 1979; Fuchs and Matura 1980; Kornprobst 1980; Bonhomme and Fluck 1981; Dewey 1982; Godard 1982; Lasnier 1982a; Peucat et al. 1982; Vogel et al. 1982; Gogné 1983; Weber and Behr 1983; Behr et al. 1984; Mísař et al. 1984 a, c; Neuman 1974; Gogard and Cabanis 1985; Kiesl et al. 1985; Tollman 1985; Chaloupský 1986; Matte 1986; Pin and Peucat 1986; Suk 1986).

The ages of protoliths for high-pressure ultrabasic and basic associations have been distinguished from subsequent structural and metamorphic events by geochronological investigations. The ages of peridotite and eclogite protoliths range from middle to late

1 Faculty of Natural Sciences, Charles University, Albertov 6, 128 43 Praha 2, Czech Republic

Proterozoic (1100 – 600 Ma) in the Caledonides (Medaris and Wang 1986 – Almklovdalen, Norway; Sanders 1985 – Glenelg Scotland), the Hercynides (Peucat et al. 1982; Godard 1982 – Baie d'Audirene and Vendee regions, France; Gebauer and Gruenenfelder 1979 – Münchberg gneiss, Germany; Vogel et al. 1982 – Cabo Ortegal and Sobrado regions, Iberia), and the Alps (Ivrea zone).

The isotopic dates on metamorphism and deformation are, however, younger than the protolith ages. The age of eclogite metamorphism in the Scandinavian Caledonides has been dated at 490 and 420 Ma. In the French Hercynides blueschist metamorphism of ultrabasic and eclogitic rocks has been reported at 440 – 400 Ma and metamorphism of eclogites and peridotites associated with granulites at 380 Ma (Peucat et al. 1982; Pin and Peucat 1986). Similar "early" Palaeozoic K/Ar and Rb/Sr ages have been given for the metamorphism of lherzolites, eclogites, and granulites of Cabo Ortegal (den Tex 1979; Vogel et al. 1982), and for Austrian granulites (Arnold and Scharbert 1973).

"Late" Palaeozoic ages (380 – 320 Ma) are related to younger granulite metamorphism and distinct retrogression of eclogite mineral assemblages (Bonhomme and Fluck 1981; Schellhorn et al. 1982; Van Breemen et al. 1982; Weber and Behr 1983). Hercynian metasomatism is also reflected in the 266 ± 41 Ma of phlogopite from the Finero peridotite in the Alps.

Two approaches must be adopted in order to understand more fully the origin and emplacement of ultrabasic and eclogitic rocks in the crust of the Hercynides. The first approach consists of the recognition of older relicts of continental or oceanic crust. Older continental crust occurs in Schwarzwald gneiss (2000 Ma; Todt and Busch 1981), the Limousian region (2400 Ma; Schellborn et al. 1982) and the Armorican massif (2400 – 1300 Ma; Peucat et al. 1982). Older oceanic crust was found in the Armorican and Central massifs (1300; Godard 1982; Lardeaux and Santalier 1985).

In the Bohemian Massif older continental crust is indicated by zircon ages from the Gföhl and Keprník gneisses (Van Breemen et al. 1982). In the Barrandian basin, late Proterozoic deep-water sediments intercalated with basic volcanics and some ophiolite centres (deformed and metamorphosed to greenschist facies) are overlain by transgressive Lower Cambrian sediments (Mísař et al. 1984b).

The second approach deals with the recognition and distinction of various segments of the Hercynian (Palaeozoic) crust that have had different structural and lithologic histories. In any geodynamic model of the Hercynides, the interaction of the Euroasian (Laurasia) and African (Gondwana) continents must be considered (Matte 1986) as well as the development of rift and fold belt structures related to the closing of the Iapetus ocean (Dewey 1982), and the growth and destruction of the Rheic ocean.

Structures of the Rheic ocean that were in the early Palaeozoic controlled by NW–SE and E–W transform faults perpendicular to the Iapetus suture became more active in Devonian and Carboniferous times. Rejuvenated shear zones (e.g. the two Armorican zones, the Pyrenean fault, the Coimbra – Cordoba and Pays de Bray shear zones, the Elbe fault zone, and others) separated older Hercynian crustal segments, or microplates, from each other (Vai and Cocozza 1986) and formed, in toto, a distinct strike-slip orogene (Badham 1982). Deep faults, secondary shear zones, and overthrusts in different crustal segments of the European Hercynides have different orientations due to an anti-clockwise rotation of Rheic segments, combined with a general oblique movement along the margins of African and European plates. In central and eastern parts of the Moldanubian zone and in the Moravo-Silesian zone of the Bohemian Massif, the most significant structures, including strike-slip faults, underthrusts, and overthrusts, are oriented NE–SW and NNE–SSW.

The position of eclogites and ultrabasic rocks (some related to ophiolites, and others not) in the crust of the Hercynides of the Bohemian Massif can be considered from the point of view of the two approaches discussed above, as well as from those of specific geologic histories and crustal lithologies. This article aims at considering this problem from all these viewpoints.

Lithology and Geochronology

The ultrabasic and eclogitic rocks of the Bohemian Massif belong to several genetic lithologic groups that differ in age and in associated country rocks.

1. Peridotites associated with garnet amphibolite and granulite ultrabasic rocks (with or without pyrope) occur as allochthonous bodies in metamorphic rocks of garnet amphibolite and granulite facies. Serpentinized peridotites (lherzolites and harzburgites) are more abundant than serpentinized dunites. Metasomatic reaction zones between ultramafic rocks and adjacent acidic country rocks are common, having been formed under amphibolite facies conditions at the end of the main Hercynian tectonic event (Machart 1984, Mísař and Jelínek 1981).
   The ultrabasic rocks of the first group may be further divided into two subgroups, i.e. 1a and 1b. Ultrabasic rocks of subgroup 1a usually contain pyrope, have metamorphic reaction zones (locally tectonically modified), and are genetically connected with retrograded eclogites. Typical localities (Fig. 8.2) are Bečváry, Borek, Nové Dvory, and Biskupice (locs. 1, 2, 3, and 21).
   Peridotites of subgroup 1b consist of spinel and garnet peridotites which are associated with granulite; common eclogites are mostly absent from this group. Peridotites of this type occur at Mohelno (loc. 4) where a distinctive anatectic and bimetasomatic contact zone has developed between peridotite and granulite (Mísař 1980, Mísař et al. 1985), in southern Bohemia at Kleť (loc. 18), and in the deep borehole in the basement of the České Středohoří Mts. (loc. 25), where transitional eclogites also occur (Fiala and Paděra 1984).
   The age of group 1's rock family is uncertain. Protoliths of the ultrabasic and eclogitic rocks are probably Precambrian. Emplacement took place during older (440 – 400 Ma) and younger (380 – 320 Ma) Hercynian tectonic-metamorphic events, but definitely before the intrusion of Hercynian granitic plutons (ca 320 – – 300 Ma).
2. Peridotites associated with amphibolites and greenschists. Lherzolites, harzburgites, and dunites of the second rock family are also allochthonous but were emplaced in metamorphic rocks of amphibolite or greenschist facies (Mísař 1984; Tonika 1971;

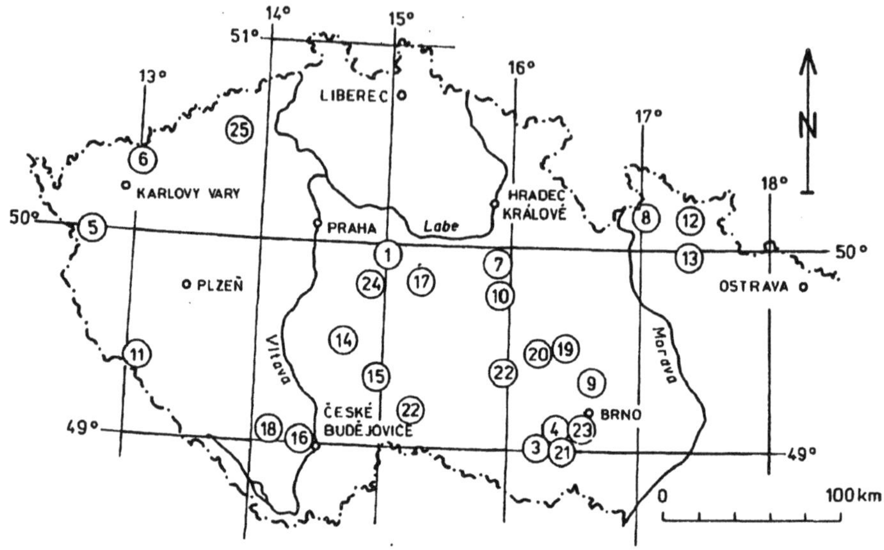

Fig. 8.2 Distribution of ultrabasic rocks localities and eclogites in the Bohemian Massif. Numerals correspond to the references in the text.

Klápová 1985). Typical localities are at Mariánské Lázně, Krušné hory Mts., and Železné hory Mts. (locs. 5, 6, and 7). Some of the peridotites have been interpreted as ultrabasic tectonites or ultrabasic cumulates of Late Proterozoic ophiolite complexes (Mísař 1966; Mísař et al. 1984a, b, c; Kastl and Tonika 1984; Jelínek et al. 1984). Country rocks of these peridotites are amphibolites, metabasites s.l., metagabbros, and, in some places, prograde eclogites (Mísař et al. 1984b, c; Tonika 1971). Typical localities are Letovice, Staré Město, and Mariánské Lázně (locs. 9, 8, and 5). Serpentinites without any relation to basic metavolcanics occur as isolated lenses in micaschists of the Železné hory Mts. (Mísař 1979).

3. Concentric peridotite-gabbro complexes. Peridotite, plagioclasse peridotite, and melanotroctolite in the Ransko gabbro-peridotite massif (Mísař et al. 1984) belong to the concentric type of gabbro-peridotite intrusions with which sulphide mineralization is associated. Layered structures are typical of this type of intrusion, characterized by a triad of plagioclasse peridotite + troctolite + pyroxene gabbro, and a tholeiitic differentiation trend. A typical locality is the Staré Ransko intrusion (loc 10), which is probably of late Proterozoic – Lower Cambrian age.
4. Ultrabasic rocks related to basic volcanics and gabbros. Ultrabasic rocks, some of which may be of cumulate origin, occur in relation to late Proterozoic basic volcanics and gabbros, as in the hortonolites of the Kdyně massif (loc. 11) and to Devonian volcanics and hypabyssal rocks of the Jeseník and Sobotín massifs (locs. 12, 13). In the Brno Massif, lenses of ultrabasic rocks associated with

Devonian and pre-Devonian metadiorites and metagabbros are of oceanic or volcanic arc affinity.

5. Eclogites differ in mineral composition, geochemistry, and occurrences in various country rocks. Typically, they exhibit a complex polyphase metamorphic history. Moldanubian eclogites were classified by Dudek and Fediuková (1974) into three groups, based on the composition of constituent garnets: 1) pyrope-almandine-grossular eclogite (common eclogite), 2) pyrope-almandine eclogite (transitional eclogite), and 3) pyrope eclogite (griquaite of Paděra 1972). Group 1 was subdivided according to geologic occurrences in the following manner: types 1a and 1a' occur as layers in peridotite, type 1b is constituted by separate masses but it is associated with nearby peridotite (disrupted peridotite-eclogite bodies), and type 1c occurs in gneiss and migmatite, not associated with peridotite. Dudek and Fediuková (1974) pointed out that all three groups of eclogite occur as layers in peridotite, but only group 1, common eclogite, occurs in gneiss and migmatite that is devoid of peridotite. New temperature – pressure estimates for these rocks were provided by Jelínek (1985), Fiala (1985), Fediuk and Fediuková (1985), Jelínek et al. (1985), and Mísař et al. (1984a).

Eclogites (griquaites) of the third group defined by Dudek and Fediuková (1974) occur only as layers in garnet peridotites. In such "eclogites", garnet contains more than 65% of pyrope component, diopside has a low content of Na, and olivine and/or enstatite-bronzite may be present. These rocks are mineralogically and chemically related to the enclosing garnet peridotites, and a garnet pyroxenite would be better a term for them, rather than eclogite or griquaite, in order to emphasize their magmatic-metamorphic origin in ultrabasic mantle protolith. Garnet pyroxenites occur at Bečváry, Borek, Nové Dvory, and České Budějovice (locs. 1, 2, 3, and 16). Common eclogite and transitional eclogite also occur at the localities where they have been strongly retrograded under amphibolite facies conditions.

Many eclogites in Moldanubian gneisses and migmatites are common eclogites after Smulikowski (1972) and Dudek and Fediuková (1974). Garnet in these eclogites has a distinctly lower pyrope content, clinopyroxene is omphacitic, and primary (?) brown hornblende locally occurs. Common eclogite clearly indicates crustal metamorphic conditions and may be considered as the metamorphic product of tholeiitic basalt.

Retrograde recrystallization, including symplectitization and amphibolitization, is a common later process. According to Klápová (1985) most eclogites in two mica gneisses, orthogneisses, and migmatites in the Krušné hory Mts. (loc. 6) have a polyphase history. They experienced a pre-eclogite stage in the amphibolite and greenschist facies, were then subjected to a high-pressure eclogite stage in which kyanite was stabilized, and were finally recrystallized to symplectite and amphibolite during post-tectonic retrogression. In this respect the kyanite eclogite of the Krušné hory Mts. is similar to eclogites in alpine domains (Maggetti 1985; Spear et al. 1985) in the Saxothuringicum in Germany (Schmidt 1982), in Corsica (Pequignot and Caron 1985), in Schwarzwald (Wimmenauer and Ádám 1985), in Cabo Ortegal (Vogel et al. 1982, Godard 1982), in Spitzbergen (Hirajima et al. 1985), and in many other regions. All these alpine-type eclogites probably originated from basic protoliths derived from oceanic crust of various ages. They were subjected to polyphase metamorphism which involved hydrous phases, as was the case at the classic Sambagawa locality (Takasu 1985).

In the Bohemian Massif another good example of a prograde kyanite eclogite that was derived from gabbro is found in the Mariánské Lázně complex (Tonika 1971, Kastl and Tonika 1984). Similar eclogitization of gabbros and diorites related to subduction processes has been described from the Armorican Massif (Godard 1982, Godard and Cabanis 1984) and the Massif Central (Piboule and Briand 1980).

Ultrabasic and Eclogitic Rocks as Products of Geotectonic Cycles

The crust of the Bohemian Massif was, of course, reworked by Hercynian tectonic, metamorphic, and magmatic events. However, in the crust of the Hercynides it is necessary to distinguish elements of Palaeozoic fold belts from reworked elements of older basement. The degree of reworking varies in individual tectonic units (zones) of the European Hercynides in relation to the amount of Palaeozoic and Precambrian elements. Ultrabasic and eclogitic rocks of different families (see rock families 1 – 5 above) also characterize the main tectonic and stratigraphic units of the Bohemian Massif. The units are as follows: Moldanubian, Moravian-Silesian, and Saxothuringian with a specific inner Bohemian region. All units can be correlated with corresponding zones in the European Hercynides, including the Central German Crystalline Rise. A specific crustal segment referred to as the Brunovistulikum (Dudek 1980, Suk 1986) occupies a specific position between the Moravian-Silesian unit and the West Carpathians.

Rhenohercynian Zone

The Rhenohercynian zone of European Hercynides most probably occurs in the NE part of the Bohemian Massif in the Nízký Jeseník Mts., and in the Drahany Plateau. Ultrabasic rocks are scarce in this zone, which is normal for the Rhenohercynian zone, except for the Lizard complex in England. Ultrabasites are present in the Sobotín amphibolite massif (loc. 13), where they have been completely transformed to serpentinites, hornblendites, and soapstones (Jelínek and Souček 1981).

Central German Crystalline Rise

The Central German Crystalline Rise may or may not be seen in the basement of the Rhenohercynian zone in the Silesian region (the Hrubý Jeseník Mts.). Ultrabasic rocks are absent in the pre-Devonian nappe slices of this region. The border calc-alkaline volcanics of Devonian and Lower Carboniferous ages alternate with shallow water sediments. Although the Devonian volcanics are tholeiitic in character, there is no evidence for the presence of ophiolite complexes (Hercynian ocean floor) in the Devonian and Lower Carboniferous section.

Brunovistulikum

The Brunovistulikum of Dudek (1980) is overthrust by the Moravian and Moldanubian zones in the west and forms the basement of the West Carpathians. There are very few occurrences of ultrabasic rocks (Dudek 1980). Metamorphosed basic volcanics, diorites, and gabbros in the Brno Massif (Weiss 1978, Štelcl et al. 1986) may represent "oceanic" elements emplaced in the margin of a continental crustal segment.

Saxothuringian Zone

The Saxothuringian zone is limited by the Central German Crystalline Rise in the north and contains segments of the pre-Hercynian ocean floor and continental crust and of the Hercynian continental crust.

Segments with Ocean Floor Rocks

Late Proterozoic zones that contain deep clastic sediments, basic volcanics with tholeiitic and calc-alkaline trends, and ophiolite complexes are concentrated in the inner Bohemian region of Mísař et al. (1983). The contact zone between the Moldanubian and central Bohemian regions is marked by the late Proterozoic Jílové volcanic belt (loc. 24), which consists of older OF and IAB tholeiites and younger andesites and dacites, and is considered to have been erupted along the continental margin (Waldhauserová 1984).

The position of ultrabasic rocks surrounded by metavolcanics and metagabbros in the Mariánské Lázně amphibolite massif is largely controlled by a deep fault oriented NE-SW. Complementary ophiolite members were metamorphosed under conditions of the amphibolite and eclogite facies, some of which were transformed to kyanite eclogites.

The ophiolite complex at Letovice (Jelínek et al. 1984; Mísař et al. 1984 b) is tectonically dismembered and contains ultrabasic tectonites in isolated allochthonous bodies and ultrabasic cumulates at the base of a layered gabbro zone. Equivalents of the Letovice ophiolite complex in the form of ultrabasites, gabbros, and metavolcanics occur in the Staré Město mica schist zone and beneath the Sowie Góry Mts. allochthonous block in Poland (Mísař 1979). A similar ophiolite association occurs at Krems in Austria, where it is overthrust by the Gföhl crystalline nappe.

The Ransko gabbro-peridotite massif, which contains plagioclasse-bearing peridotites, was emplaced at the intersection of several deep faults at the southern margin of the Central Bohemian region at the end of the Cadomian geotectonic cycle (Mísař 1974 a).

Ophiolite complexes also occur in other ocean floor segments of the Saxothuringian zone in Germany. Many allochthonous ultrabasic rock bodies and ophiolite complexes are located along the central Saxonian lineament (Werner 1981).

The general age of ocean floor in the Saxothuringian zone is late Proterozoic, and the metamorphic grade of the volcanic rocks of the ocean floor and country rocks does not exceed that of amphibolite facies. Alpine type eclogite with kyanite and hydrous phases formed under relatively low temperature and high pressure conditions. The protoliths of eclogites from the Mariánské Lázně complex (Kastl and Tonika 1984), Krušné hory Mts. (Klápová 1985), Munchberger gneiss massif (Gebauer and Grüenenfelder 1979), Sněžník Mts. in Poland (Smulikowski K. and Smulikowski W. 1982) are probably of the same age and origin. The high-pressure eclogite in the Schwarzwald (Wimmenauer and Ádám 1985) is also of some interest. The tectonic environment of the Saxothuringian zone was not favourable for the ocean floor formation during the Hercynian geotectonic cycle.

Segment of Continental Crust with Ultrabasic and Eclogitic Rocks

Segments of this character represent counterparts to the ocean floor segments of the Saxothuringian zone. Continental crust segments are usually composed of tectonically interlayered Hercynian and pre-Hercynian structural levels.

Granulites and granulite gneisses (e.g. in the Saxonian granulite Mts., Sowie Góry Mts., and Rychlebské hory Mts. are typical ultrabasic country rocks. Pyroxenites and transitional eclogites are commonly intercalated with the ultrabasites. The protolith age of the ultrabasic and pyroxenitic rocks is uncertain, but the late Proterozoic or older age is probable. Better known is the age of the transformation of protoliths into eclogites at 440 – 400 Ma and 380 – 300 Ma, corresponding to two fundamental events in the Hercynian orogeny.

Ultrabasic bodies with pyroxenites may represent allochthonous sheets of subcontinental upper mantle that have been emplaced at several levels in the continental crust. Optimum sites for such tectonic mixing are long deep faults, major shear zones, overthrusts, and zones of subduction. All ultrabasic rocks and their associated pyroxenites in continental segments of the Saxothuringian zone are comparable to those of the Moldanubian zone, as described in the following section.

Moldanubian Zone and Kutná Hora – Svratka Complex

Peridotites, garnet peridotites, and eclogites are concentrated along some well-defined deep faults (Machart 1984; Mísař 1974) and are interpreted as tectonic slices of the upper mantle within continental crust. The main orientation of faults containing allochthonous bodies of ultrabasic and eclogitic rocks is NE–SW in the central Moldanubian zone at the localities of Bechyně, Choustník, and České Budějovice and Včelnice (locs. 14, 15, 16, 22). In the northern parts of the Moldanubian zone and in the adjoining Kutná Hora unit, the most important deep fault with ultrabasic bodies is the NW–SE Sázava fault (loc. 17). Inclusion of pyroxenites, peridotites, and garnet amphibolites in E–W striking biotite gneisses and pyroxene hornblende gneisses at Plaňany (Fišera 1981) may delineate a tectonic melange. Ultrabasites and eclogites in the Jihlava region are distributed along NNE–SSW deep fault and associated mylonites.

Not all ultrabasic and eclogitic rocks in the Moldanubian zone, however, are distributed along linear structures. Mísař (1974) described a type of central regional structure in which ultrabasic rocks and garnet amphibolites are surrounded by granulite and granulite gneisses. In this case peridotites with or without pyroxenites occur as tectonic slices in granulite, which together constitute a tectonic unit of large dimensions. Good examples of this structure are found at Mohelno, Kleť, and Náměšť (locs. 4, 18, 23). Near St. Leonard in Austria, there is a serpentinite-garnet peridotite-granulite nappe, which is a portion of the larger Gföhl nappe (Fuchs and Matura 1980).

Contacts between peridotites and country rocks are strongly tectonized at most localities, for example, at Nové Dvory, Bečváry, Borek, and Níhov (locs. 3, 1, 2, 19). In contrast, metamorphic reactions zones formed under amphibolite facies conditions have been preserved at the contacts between smaller ultrabasic bodies and surrounding granulite, as at Kleť and Bory (locs. 18, 20).

Pressure – temperature (*P–T*) calculations have yielded temperatures of 850 – – 1100 °C and pressures of 15 to 30 kb for most Moldanubian peridotites (Fiala 1966, 1985; Fediuk and Fediuková 1985). Exceptional temperatures of 1200 – 1350 °C have been obtained for spinel and garnet peridotites at Mohelno. These data and unusual high-temperature contacts between the peridotites and granulites may have been emplaced as a hot tabular diapir into granulite at a lower crustal level and subsequently uplifted with granulite.

Inclusions of peridotite, pyroxenite, eclogite, and of other rocks at the Bory quarry (loc. 20) document the composition of the lower and upper continental crust in this part of the Bohemian Hercynides (Mísař and Jelínek 1981). The compositions of coexisting minerals in banded peridotites yield "frozen" mantle conditions of 1 200 °C ± 100 °C and 30 kb (Mísař et al. 1984a). Lower temperatures have been obtained for garnet pyroxenite (1000 °C), and symplectitic eclogite (750 °C). At the Bory locality the interdigitation of upper mantle material and Hercynian continental crust occurred along deep fault zones (ductile shear zones) extending from the Moho to upper levels of the continental crust.

Such an emplacement mechanism can be valid for all occurrences of peridotites surrounded by granulite and granulite gneisses. The emplacement of the hot peridotite mass at Mohelno remains a unique example of an upper mantle diapir in the Hercynian continental crust.

Shear zones and diapir tectonics in the Moldanubian area must be related to Hercynian (Devonian and Lower Carboniferous) tectonic events that pre-dated deposition of the Visean culm clastic sediments in the Rhenohercynian zone and intrusions of granite in the Saxothuringian and Moldanubian zones.

Conclusions

The crust of the Bohemian Hercynides is a product of several geotectonic cycles, including the pre-Cadomian, the Cadomian, possibly the Caledonian, and the Hercynian ones, among which the effect of the Cadomian and Hercynian cycles are best documented.

In the Saxothuringian zone, the late Proterozoic ocean floor and associated ophiolitic ultrabasites occur along the Saxonian lineament, in the Mariánské Lázně amphibolite complex, in the Letovice meta-ophiolite complex, and the Staré Město mica schist zone. All these ophiolite sequences originated in deep sedimentary basins, probably as marginal seas. The continental segments of the Saxothuringian zone also contain many ultrabasic and eclogitic rock bodies that are generally interpreted as tectonic slices of upper mantle material within continental crust.

In the Moldanubian zone, the association of peridotites with granulites and granulite gneisses reflects complex tectonic and metamorphic events during the Hercynian geotectonic cycle. Tectonic mixing of subcontinental upper mantle material (peridotite and pyroxenite), lower crustal elements (granulite), and even some upper crustal material occurred along deep shear zones. At Mohelno the interdigitation was accompanied by the emplacement of a hot peridotite diapir.

The Bohemian Massif is devoid of classical ophiolite sequences of Hercynian age; ocean floor was probably not produced in this region during the Hercynian cycle. Although the Devonian and Lower Carboniferous basic volcanics in the Jeseníky Mts. at the margin of the Rhenohercynian zone are tholeiitic in composition, they are not associated with other lithologies characteristic of ophiolite suites.

Eclogites in the Bohemian Massif have diverse origins. Kyanite eclogites in Cadomian metamorphites (Krušné hory, Mariánské Lázně) are of alpine type and have been produced by polyphase prograde metamorphism of basic volcanics and gabbros. In continental segments of the Saxothuringian and Moldanubian zones, pyroxenite layers in peridotite were formed by upper mantle magmatic and metamorphic processes, as were most likely the common eclogites within peridotite. Other occurrences of Moldanubian eclogites, surrounded by gneiss and not associated with peridotite, may possibly have originated from low-pressure protoliths of basaltic dikes, sills, or lava flows. A characteristic feature of all types of eclogites is their later amphibolitization.

Ophiolite complexes and protoliths of alpine-type eclogite are very likely of late Proterozoic age. The later process of eclogitization was probably related to one of the main Hercynian metamorphic and tectonic events (440 – 400 Ma and 380 – 330 Ma). The protolith ages of upper mantle peridotite and pyroxenite are uncertain, although the tectonic mixing of upper mantle and lower crust along deep shear zones in the Bohemian Massif is related to Hercynian events.

8.3 Major and Trace Elements in "Eclogites" and Their Host Garnet Peridotites, Data from the Bohemian Massif

P. Jakeš[1], E. Jelínek[2], J. Fiala[3], S. Ross Taylor[4]

Introduction

Garnet peridotites have attracted the attention of earth scientists for a number of years since the rocks were thought to be representative of upper mantle materials. The occurrence of eclogites within garnet peridotite bodies have attracted attention since they could be derived by melting of the upper mantle garnet peridotite. The emplacement of large bodies of garnet peridotite in crystalline schist areas such as high grade metamorphic terrains of the Moldanubicum in Central Europe remains a matter of controversy, (see e.g. Massif Central France occurrences – Bernard at al. 1981; Norwegian Caledonides – Brueckner 1977; Medaris 1980; Medaris and Wang 1986; or Ronda Massif in Betic Cordillera, Spain – Obata 1980).

We present major and trace element data on garnet pyroxene rocks ("eclogites") and their host peridotites from five localities which represent the principal geologic occurrences in the Bohemian Massif. In this section the term "eclogite" is used for garnet pyroxene rocks (not necessarily omphacitic pyroxene bearing) in order to emphasize the metamorphic nature of the ga-cpx paragenesis and also to stress the continuous transition from Mg-rich (Na-poor compositions) to Fe-rich "basaltic rocks" compositions. Eclogites occur in major regional units of the Bohemian Massif, usually related to the ultrabasic rocks. This is demonstrated by the thin lenses of eclogite occurring within the ultrabasic rock bodies; in large bodies, however, the relationship is more obscure. Isolated bodies of eclogites and bodies of ultrabasic rocks do occur in the same geologic horizon of the area. Eclogites closely related to the ultrabasic rocks are usually rich in Mg. The Mg-rich eclogites (with 1 wt % Na_2O) were named griquaites by Smulikowski (1968). Thus an analogy with the eclogite inclusions in kimberlitic pipes is implied. In fact there is a close relationship in major element chemistry. The common eclogites, i.e. those with basaltic chemistry, with nearly equal proportions of Fe-Mg and Na_2O (about 2 wt %) were considered by Smulikowski (1972) to have no relationship to the ultrabasic rock bodies, and to be characteristic of the gneissic and migmatitic terrains. Since Jelínek (1985) observed numerous transitional compositions between the Mg-rich and common eclogites we dispute Smulikowski's point and argue in favour of a continuous series of eclogite compositions related to the ultrabasic parent in this particular (Moldanubian) case.

---

[1] Central Institute of Geology, Malostranské náměstí 19, 118 21 Praha 1, Czech Republic
[2] Faculty of Natural Sciences, Charles University, Albertov 6, 128 43 Praha 2, Czech Republic
[3] Geological Institute, V Holešovičkách 41, 180 00 Praha 8, Czech Republic
[4] Research School of Earth Sciences, The Australian National University, Canberra, A.C.T., Australia

Mg-rich eclogites form concordant layers of a few centimetres in thickness most frequently within the garnet peridotites. The common eclogites, on the other hand, form relatively large lens-like bodies tens to hundreds of metres long and several metres thick. These bodies occur in varied geologic environments (e.g. garnet bearing ultrabasic rocks, ultrabasic rocks without garnet granulites or gneisses) although in this study we present data on the eclogites geologically clearly related to garnet-peridotite. Both common eclogites and Mg-rich eclogites are mutually related by numerous geologic and compositional transitions. Transitional members occur in ultrabasic bodies as irregularly-shaped lenses of 0.1 – 0.5 m in size. They also occur as thin layers within the common eclogites.

The AFM ((Na + K) – Fe – Mg) diagram may be used (Fig. 8.3) to distinguish among the "eclogite" types. If complemented by empirical lines representing molar Fe/Mg ratios 0.16 and 0.32 the eclogites are divided into common eclogites, transitional eclogites, and Mg-rich eclogites. There is a relationship between the size of eclogite bodies and their chemical composition (Jelínek, unpubl.): large bodies are common eclogites, whereas small ones tend to have high Mg contents. The studied localities represent all three types: common eclogites (Nové Dvory-5), transitional eclogites (Zöblitz-6), Mg-rich eclogites (Mstěnice-1 and Bečváry-2).

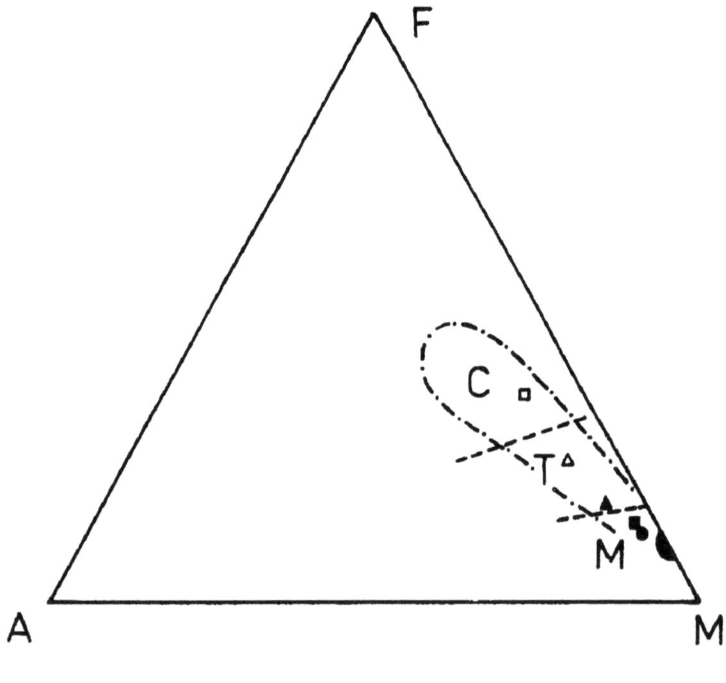

Fig. 8.3 The AFM of the studied eclogite/garnet peridotite association. Dashed line divides the field of Moldanubian eclogites into the magnesian (Mg-rich) (M), transitional (T), and common (C) types. Data by Paděra (1972). Full circle – Mstěnice No. 1, full square – Bečváry No. 2, full triangle – Zöblitz No. 3, open triangle – Hranice No. 4, open square – Nové Dvory No. 5.

Description of the Localities

Bečváry (No. 2). This locality is found in the vicinity of Kolín (Central Bohemia), geologically it is a part of the Kutná Hora crystalline complex, a subunit of the Moldanubian complex. The outcrop consists of a garnet-bearing peridotite (lherzolite) lenticular body of 700 × 1100 m in size surrounded by orthogneisses and muscovite-biotite gneisses. The serpentinized garnet lherzolite has a distinct planar foliation and up to 5 cm thick layers of Mg-rich eclogite are developed along the foliation.

Hranice (near Dobešovice- 4). The transitional eclogite occurrence is situated in the same regional unit as in Bečváry, though the serpentinized garnet peridotite body is of small size (100 × 300 m) and situated within muscovite-biotite gneisses. The eclogite lens is of an irregular shape with maximum thickness of 40 cm and follows the border between the dunitic and lherzolitic part of the garnet peridotite.

Mstěnice (No. 1) and Nové Dvory (No. 5). Both the localities are in the SE part of the Moldanubicum (near the town of Znojmo, in the south of Moravia) within the body of serpentinized garnet peridotite of 2.5 × 1.0 km in size situated within the Gföhl gneisses (felsic biotite gneiss probably formed by the retrogression of granulites, e.g. Jakeš 1969, Dudek et al. 1974). The eclogite layers follow the distinctly banded structure of serpentinized peridotite. According to Jelínek (unpublished), there types of eclogites could be distinguished three: a) layers of up to 10 cm in thickness of Mg-rich eclogite in serpentinite (sample Mstěnice is of this particular type), b) layers of 10 – 15 m thick with a parallel structure within the common eclogite following the structural pattern of the surrounding serpentinite (sample Nové Dvory is of this type), c) layers of up to 10 cm thick of transitional eclogite enclosed in the "b" type (no sample of this type has been studied).

Zöblitz near Ansprung in Germany (No. 6). This locality is a part of the Krušné hory Mts. (Erzgebirge) crystalline complex. The sample is from a 15 cm thick layer of transitional eclogite which is folded together with the surrounding serpentinized garnet lherzolite. The lherzolite body forms a relatively large lens (3.0 × 0.5 km) within a magmatic gneiss with scattered granulite occurrences (Behr et al. 1965).

Data and Methods

The major and trace element data discussed here are presented in Tables 8.1, 8.3, and 8.4. Major element analyses were carried out in the chemical laboratories of the Geological Survey in Prague (M. Huka). Trace element analyses were carried out by S. Ross Taylor at the Australian National University, Canberra, using spark source mass spectrographic methods. The limitations, detection limits, accuracy, and precision of the data is discussed by Taylor and Gorton (1977).

The transition elements were determined using optical spectrography by Jelínek (unpubl.).

Table 8.1 Major element analyses of garnet peridotites and eclogites

	Mstěnice (1)		Bečváry (2)		Zöblitz (3)		Hranice (4)		Nové Dvory (5)	
	GP	E	GP	E	GP	E	GP	E	GP	E
SiO_2	40.40	45.72	40.90	44.82	39.75	46.37	40.58	45.40	40.65	42.29
TiO_2	0.08	0.14	0.13	0.16	0.10	0.40	0.08	0.15	0.08	2.20
Al_2O_3	1.64	11.96	3.16	14.50	2.42	13.83	2.39	16.16	2.53	16.06
Fe_2O_3	4.75	1.82	5.22	1.90	5.33	2.38	4.98	1.70	4.96	2.68
FeO	2.70	3.70	2.81	4.02	2.84	4.94	2.49	6.97	2.56	9.98
MnO	0.12	0.14	0.11	0.125	0.11	0.19	0.10	0.18	0.10	0.21
MgO	38.00	21.45	33.90	20.02	35.90	19.05	35.50	13.50	35.55	11.00
CaO	1.63	9.48	2.00	10.80	1.12	9.32	1.25	13.67	1.40	12.30
Na_2O	0.18	0.60	0.24	0.66	0.14	1.40	0.12	1.29	0.11	1.38
K_2O	0.012	0.01	0.016	0.008	0.003	0.029	0.008	0.004	0.006	0.014
P_2O_5	tr	tr	tr	tr	0.05	0.04	tr	0.04	tr	0.04
CO_2	0.46	0.10	0.17	0.044	0.20	0.053	0.23	0.063	0.28	0.027
Cr_2O_3	0.30	0.38	0.29	nd	0.29	nd	0.307	nd	0.32	nd
H_2O^+	9.35	3.88	10.12	2.17	11.09	1.48	10.90	0.36	10.13	0.67
H_2O^-	0.55	0.52	0.66	0.28	0.54	0.22	0.69	0.13	0.66	0.13
total	100.17	99.38	99.06	99.50	99.88	99.70	99.63	99.52	98.66	99.18

Major Element Data

All garnet peridotites studied exhibit similar major element composition (Table 8.1). Because of the high water content due to serpentinization of peridotites, the analyses were recalculated using water-free basis (Table 8.2).

Garnet peridotites have typical low SiO_2 contents, high Mg contents, high Mg/Fe ratios, low Ti contents. There are no substantial differences among garnet peridotites from different localities, although conspicuous differences in the associated eclogites exist. The garnet peridotites from the Bohemian Massif cannot be distinguished from garnet peridotite inclusions in diamond-bearing pipes or from the other garnet peridotite bodies (e.g. Norwegian see Medaris and Carswell 1990) using major element chemistry (e.g. Carswell and Dawson 1970) The abundances of Al and Ca are characteristically "chondritic" low, though differences in Al content exist as a result of garnet fractionation. The oxidation state of iron is relatively high, due to serpentinization of the original paragenesis. From the extremely low (probably pristine) K and Na values it

Table 8.2 Recalculated analyses of garnet peridotites and eclogites

	Mstěnice (1)		Bečváry (2)		Zöblitz (3)		Hranice (4)		Nové Dvory (5)	
	GP	E	GP	E	GP	E	GP	E	GP	E
SiO_2	44.90	47.87	45.97	46.18	45.04	41.30	46.10	45.85	45.93	43.19
TiO_2	0.09	0.15	0.15	0.16	0.11	0.41	0.09	0.15	0.09	2.24
Al_2O_3	1.82	12.52	3.55	14.94	2.74	14.11	2.72	16.32	2.86	16.32
Fe_2O_3	4.95	1.91	5.87	1.96	6.04	2.43	5.66	1.72	5.60	2.72
FeO	3.00	3.87	3.16	4.14	3.22	5.04	2.83	7.04	2.89	10.14
MnO	0.13	0.15	0.12	0.29	0.12	0.20	0.11	0.18	0.11	0.21
MgO	42.23	22.46	38.10	20.63	40.67	19.43	40.33	13.64	40.17	11.18
CaO	1.81	9.93	2.25	11.13	1.27	9.51	1.42	13.81	1.58	12.50
Na_2O	0.20	0.63	0.27	0.68	0.16	1.43	0.14	1.30	0.12	1.40
K_2O	0.013	0.01	0.018	0.008	0.003	0.03	0.01	0.004	0.007	0.014
P_2O_5	tr	tr	tr	tr	0.06	0.04	tr	0.04	tr	0.04
CO_2	0.51	0.10	0.19	0.045	0.23	0.054	0.26	0.063	0.32	0.027
Cr_2O_3	0.34	0.40	0.33	nd	0.33	nd	0.35	nd	0.36	nd

appears that serpentinization did not influence the major element chemical composition of peridotites to a notable degree.

The major element chemistry of common eclogites resembles that of basaltic composition, though in Mg-rich eclogites the composition is picritic. The abundance of Al corresponds to that of tholeitic basalt, but the Ca-content values are relatively high. If the Al–Ca–Fe–Mg or Fe–Ca–Mg relationships are examined, though, there appears to be much similarity between picritic eclogites from kimberlites as well as some similarity with low potassium basalts or komatiites. The contents of alkali elements in the eclogites are slightly higher than those in peridotites though much lower than in basaltic rocks of similar compositions. The major element abundances of eclogites appear to be controlled by the changing proportions of two "eclogite" minerals, i.e. garnet (high Mg and Al) and clinopyroxene (high Mg and Ca values).

Literature and unpublished data on the composition of garnets and clinopyroxenes show high temperatures and pressures of ga-cpx (Fiala 1966; Jelínek 1985; Fediuková 1988; Medaris et al. 1990) equilibrium as well as an interesting relationship (Fig. 8.4) between rock and mineral compositions. A very close relationship between the garnets of Mg-rich eclogites and associated lherzolites is obvious. In common eclogites, however, garnets contain a larger proportion of the almandine component and clinopyroxenes are richer in hedenbergite and omphacite than those of Mg-rich

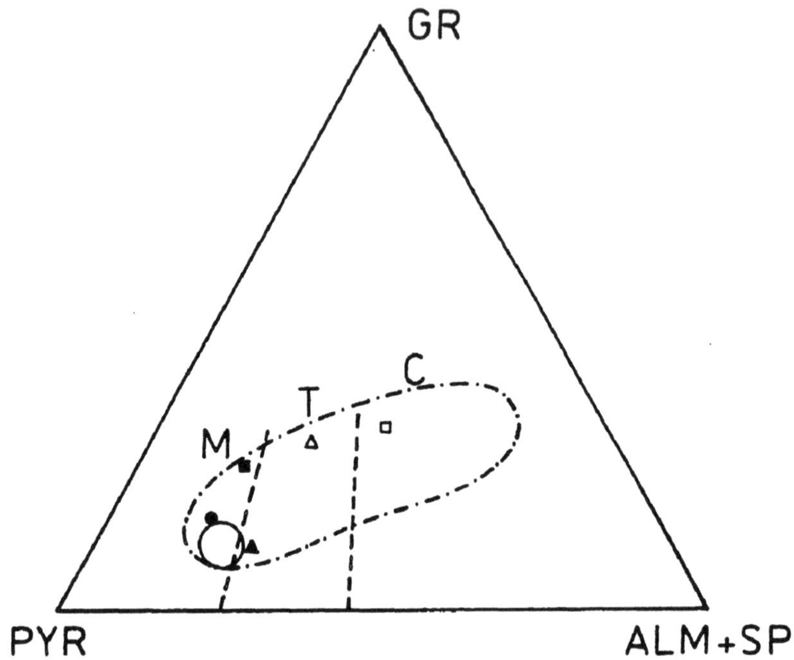

Fig. 8.4 Garnet compositions of eclogites and their peridotites in the diagram: pyrope – almandine + + spessartine – grossular. Data courtesy Mathé (1969), Dudek (1971), Jelínek (1985), Dudek and Fediuková (1974), and Fiala (unpublished). Boundary lines are drawn according to Fiala and Paděra (unpublished). Symbols are the same as in Fig. 8.3. The large open circle represents garnet compositions of garnet peridotites.

eclogites. This is explained by the bulk composition and later equilibrating the cpx-ga paragenesis at deep crustal or upper mantle conditions.

Trace Element Data

Garnet peridotites are characterized by extremely low contents of large cations (Cs, Rb, Pb, and Sr) – see Table 8.3. The contents of the high field strength cations (Th, U, Zr, Hf, Nb, Y) are also extremely low. This fact precludes any major contamination of the original rocks either due to serpentinization processes or due to the crustal emplacement. We believe that the abundance is characteristic of the original rocks. The associated eclogites show extremely low contents of large lithofile ions, though there is an increase from Mg-rich eclogites to common eclogites (e.g. Ba increases from 3.4 ppm (in Mg-rich eclogite) to 107 ppm in common eclogite). There is also an increase in Zr, whereas the other LIL and HPS elements do not show systematic variations. The comparison of LIL contents of the garnet peridotites and associated eclogites indicates higher concentrations of LIL elements in eclogites. These levels, however, are not as high as would be expected if the eclogites represented basaltic (partial) melts of peridotite. Even the lowest partition coefficients for liquid/solid equilibria do not allow the coexistence

Table 8.3 Data on trace elements in garnet peridotites

	Mstěnice (1)		Bečváry (2)		Zöblitz (3)		Hranice (4)		Nové Dvory (5)	
	GP	E	GP	E	GP	E	GP	E	GP	E
Cs	0.10	nd	nd	nd	nd	nd	nd	nd	nd	nd
Rb	0.02	0.11	0.08	0.19	0.26	1.12	0.19	0.31	0.21	0.63
Ba	0.25	3.43	6.83	5.95	6.58	9.26	2.76	7.30	4.40	107
Pb	nd	11.70	1.07	2.61	0.83	1.12	0.66	1.909	0.28	5.21
Sr	4.72	45.34	10.92	36.17	5.23	28.34	4.81	50.33	7.42	63.05
Th	nd	0.44	nd	nd	0.08	0.17	nd	nd	nd	0.63
U	nd	0.57	nd	nd	0.03	0.06	0.36	nd	nd	1.34
Zr	1.43	6.83	7.18	12.20	3.47	14.41	1.75	9.43	4.40	34.18
Hf	nd	0.39	nd	nd	0.14	0.36	nd	nd	nd	nd
Nb	nd	4.02	0.78	1.40	0.38	2.96	0.24	0.89	nd	6.21
Y	0.83	6.98	3.36	14.98	1.35	8.61	1.52	28.66	1.68	18.08
Ti	480	840	780	960	600	2400	480	3600	480	13200
V	45	170	7	155	40	180	35	230	70	330
Cr	2700	2500	240	1300	1800	620	2200	450	2100	120
Co	140	40	12	70	95	35	180	45	110	65
Ni	2200	380	160	350	2100	320	2300	170	1800	140

of garnet peridotite and basalt (eclogite) containing such low LIL contents. Thus, we suggest that the LIL element contents of eclogite parageneses were derived from a depleted source. Mg-rich eclogites represent the extreme end of LIL depletion, whereas common eclogites (e.g. sample Nové Dvory) have a small LIL depletion, showing close correspondence to MORB tholeiites.

Rare Earth Elements

The REE abundances are presented in Table 8.4 and the chondrite normalized patterns are shown in Fig. 8.5. Both rock types, i.e. garnet peridotites and eclogites show LREE depleted patterns. The LREE depletion for the ultramafic rocks recalls the analogies with depleted alpine-type peridotites and ultramafic rocks of depleted upper mantle origin. In this respect they are similar to the other ultramafic rocks, e.g. Norwegian or Lac de Lherz peridotites, or ultramafic rocks of Troodos (Garmann et al. 1975; Frey 1970; Kay and Senechal 1976, respectively), and could represent "pristine" mantle minus a small portion of basaltic liquid.

The LREE depleted patterns for the eclogite, however, are surprising since several authors have advocated ideas that the eclogites represent "frozen in" basaltic liquids within the peridotites. The analogy of LREE depleted eclogites with the LREE depleted rocks similar to MORB (e.g. Garmann et al. 1975) is confusing since MORB rocks are depleted in La, Ce, Pr though heavier REE are chondritic and pattern from Nd to Lu is usually not deflected. In these eclogites, the Norwegian eclogites and the eclogite studied by Philipotts et al. (1972) from the Kakanui breccia, there is a systematic La enrichment on account of Lu for the chondrite normalized pattern. This difference is considered crucial.

The analogy with MORB depletion (Garmann et al. 1975) seems difficult to advocate since the depleted mantle source in the MORB source area is mineralogically different (i.e. does not contain garnet). Thus the chemical composition of basaltic liquid and REE pattern is controlled by minerals such as olivine or pyroxene. These have little effect on REE fractionation (i.e. from LREE to HREE ratios). In regions of garnet stability REE compositions are controlled by this mineral. The depleted nature of mantle in the MORB source region in the absence of garnet is reflected in the partial melts. The LREE depleted pattern of MORB is characterized by the depletion of La–Nd. This is apparently a feature of the present day MORB source region. The REE pattern of the eclogites thus exclude the concept that they represent a "frozen in basaltic liquid". Basaltic liquid in equilibrium with garnet peridotite should be either LREE enriched or "chondritic" (Mysen and Holloway 1977) although their REE abundances are by an order magnitude higher than in garnet peridotites. The eclogites appear to be more

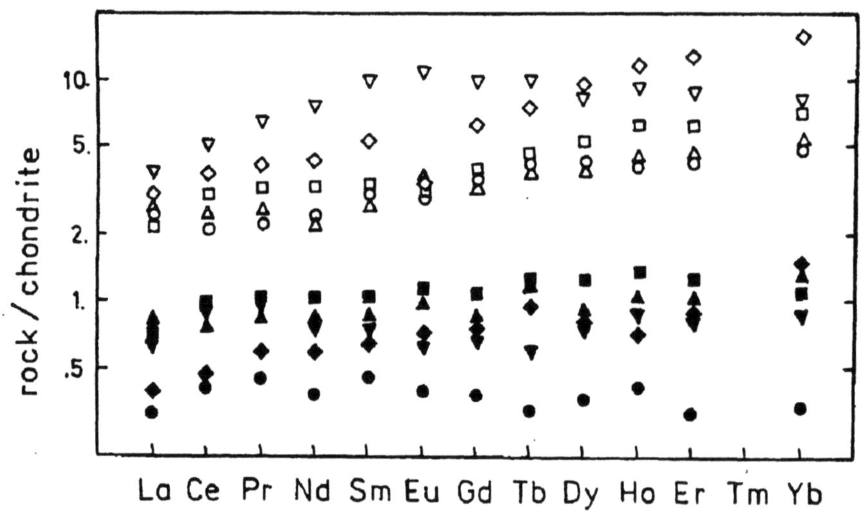

Fig. 8.5 The REE contents of eclogites (open symbols) and of their host peridotites (full symbols), normalized to chondritic values (localities Mstěnice No. 1 (circles), Bečváry No. 4 (squares), Zöblitz No. 3 (triangles), Hranice No. 4 (diamonds), and Nové Dvory No. 5 (inverted triangles).

Table 8.4 Abundances of rare Earth elements

	Mstěnice (1)		Bečváry (2)		Zöblitz (3)		Hranice (4)		Nové Dvory (5)	
	GP	E	GP	E	GP	E	GP	E	GP	E
La	0.09	0.72	0.23	0.65	0.23	0.79	0.12	0.90	0.20	1.13
Ce	0.36	1.71	0.88	2.49	0.66	1.99	0.38	2.84	0.78	4.16
Pr	0.06	0.27	0.13	0.38	0.11	0.29	0.06	0.50	0.11	0.74
Nd	0.23	1.33	0.58	1.91	0.48	1.25	0.37	2.51	0.44	4.30
Sm	0.09	0.67	0.24	0.72	0.17	0.55	0.15	1.10	0.15	2.10
Eu	0.03	0.24	0.09	0.22	0.08	0.28	0.05	0.25	0.04	0.81
Gd	0.10	0.86	0.30	1.04	0.22	0.85	0.19	1.65	0.17	2.52
Tb	0.02	0.20	0.06	0.23	nd	0.18	nd	0.38	0.03	0.47
Dy	0.12	1.25	0.39	1.63	0.27	1.22	0.22	2.84	0.215	2.63
Ho	0.03	0.31	0.09	0.45	0.08	0.32	0.06	0.85	0.06	0.63
Er	0.065	0.93	0.26	1.375	0.21	0.98	0.17	2.76	0.175	1.71
Tm	nd	nd	nd	nd	nd	nd	nd	nd	nd	0.27
Yb	0.065	0.97	0.24	1.39	0.30	1.02	0.28	3.26	0.17	0.27

LREE depleted than garnet peridotites. This can be seen in Fig. 8.3 where the eclogite abundances were normalized to their respective parent garnet peridotite REE abundances. The abundances of REE generally increase from Mg-rich eclogites towards the common eclogites, though the pattern remains in general unchanged (Fig. 8.6). In common eclogites Eu anomalies appear, suggesting plagioclase fractionation.

Transition Metals

The content of transition elements in the eclogites are normalized by the respective contents in the garnet peridotites (Fig. 8.7). The contents of V and Ti increase with increasing Fe/Mg ratios (Fig. 8.8). The reverse applies to Cr, Ni, and partly to Co. Mg-rich eclogites (samples M1 and B2) have a transitional metal pattern more similar to the garnet peridotites, whereas the composition of common eclogites is close to the "basaltic" distribution. The transition metal element abundances show a pattern similar to those observed in the LIL element and also in the REE patterns. The amounts of Ti and V, i.e. those elements favouring partial melts rather than solid residua increase from Mg-rich eclogites to common eclogites.

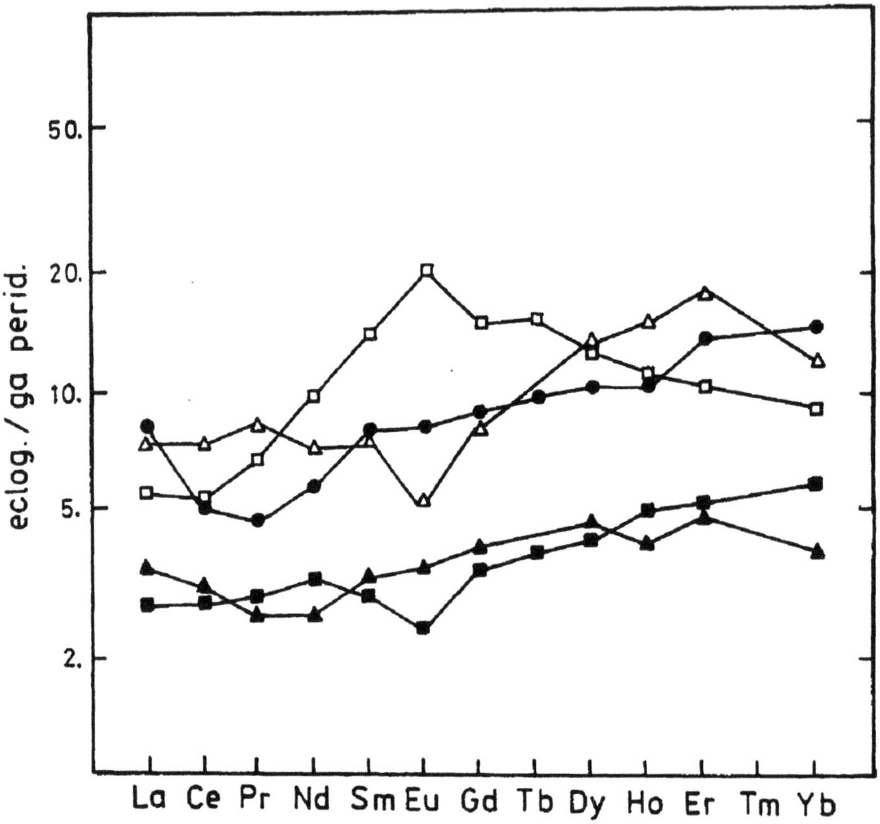

Fig. 8.6 The REE contents of eclogites from garnet peridotites normalized to their respective host rocks (garnet peridotites), indicating the LREE are depleted in the "eclogites" relative to peridotites. This suggests the garnet precipitation control over the initial REE distribution. (Symbols same as in figs. 8.4 and 8.5).

Discussion

A wide variety of opinions regarding the origin of eclogites within ultramafic bodies have been published. Thus Dudek (1971) is of the opinion that eclogites and their ultrabasic host rock represent the primary upper mantle material. Paděra (unpublished) thought that eclogites represent a "felsic" fraction of peridotites, which is similar to the concept that the eclogites represent a variety of metamorphosed basaltic rocks. Similar opinions have been expressed by Austrian authors and summarized, e.g. by Dempsey and Scharbert (1975). Ringwood (1975) considered the alkali-poor eclogites in the kimberlite pipes to result from a partial melting of subducted oceanic crust converted to quartz eclogites. The pattern of kimberlite inclusions, however, is LREE enriched. It thus seems that LREE depleted eclogites are more likely to be candidates for "eclogite fractionation" in a subduction-controlled environment. Jelínek (1985) advocated a close

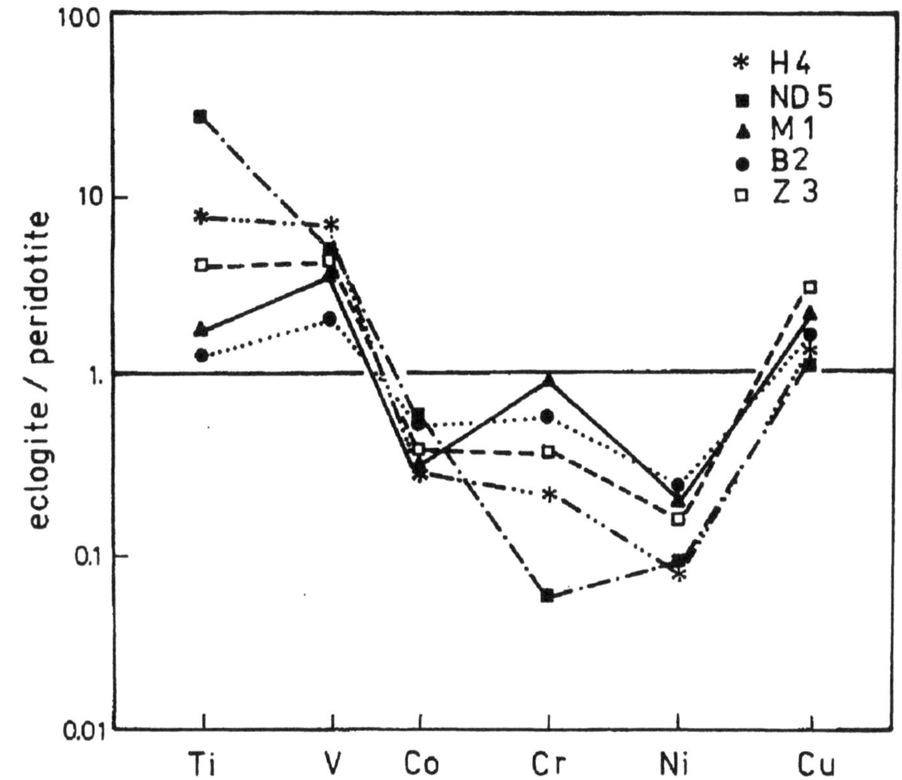

Fig. 8.7 Transition elements behaviour in the studied eclogites and garnet peridotites. Eclogite abundances are normalized by the mean values in garnet peridotites. M1 – Mstěnice, B2 – Bečváry, Z3 – Zöblitz, H4 – Hranice, ND5 – Nové Dvory.

compositional relationship of garnet peridotites and Mg-rich eclogites. We would stress the observations on the existence of eclogites with "crustal" and also "LREE depleted" patterns. Garmann et al. (1975), e.g. reported LREE depleted pattern for Norwegian eclogites and advocated the similarity between them and MORB basalts.

Eclogites from kimberlite inclusions are considered to be derived from basaltic and picritic magmas. It appears likely that a significant portion of such magmas may not reach the surface (for density reasons) and, therefore, could crystallize at depths (Ringwood 1975) or underlie the continental crust. It should be noted that major element similarities between the garnet peridotites of kimberlite inclusions as well as similarities of Mg-rich eclogites are not reflected by the REE abundances and patterns. It is thus difficult to reconcile the formation of peridotite and eclogite patterns by a single stage process. While the LREE depletion of garnet peridotites could result from the extraction of a low-melting point silicate melt, this hardly applies to eclogite layers. Garnet

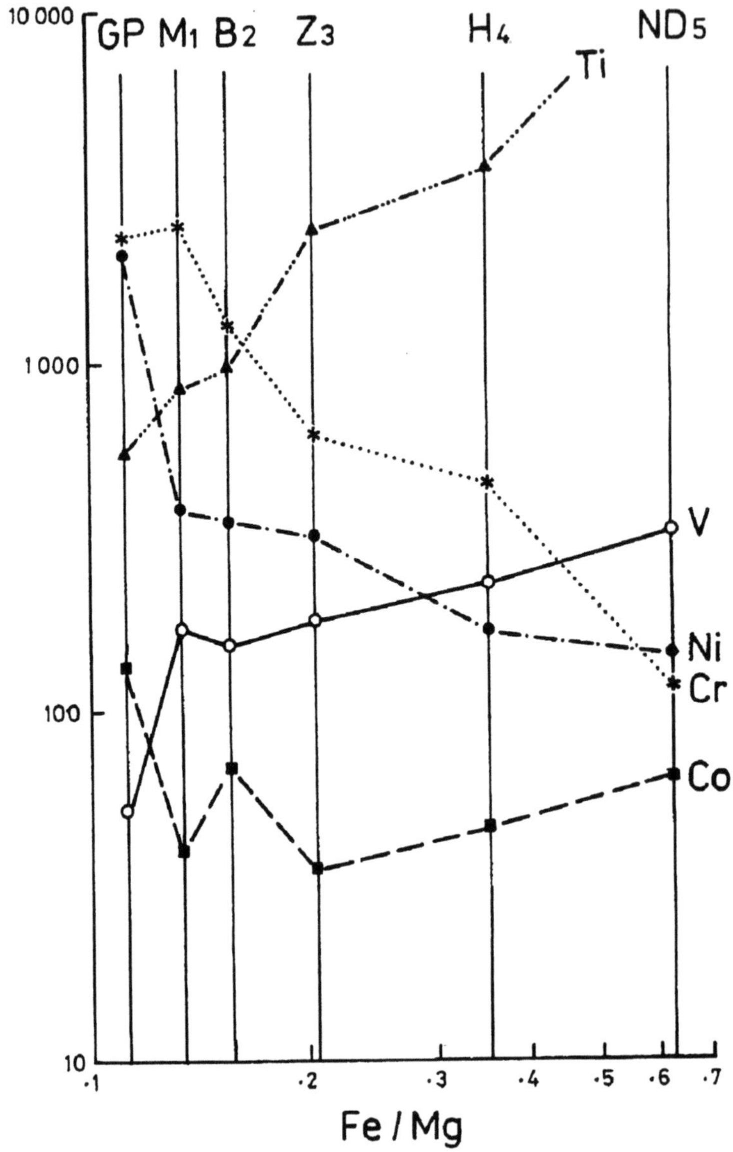

Fig. 8.8 The transition element content (in ppm) as compared to the atomic Fe/Mg ratio (log scale) in eclogites and garnet peridotites (GP). M1 – Mstěnice, B2 – Bečváry, Z3 – Zöblitz, H4 – Hranice, ND5 – Nové Dvory. Note the increase of Ti and V, and the decrease of NI and Cr with increasing Fe/Mg.

precipitation from the melt may have led to the observed REE patterns in eclogites. Such a melt could have had an LREE enriched pattern corresponding to melt-garnet peridotite equilibrium. Thus the phases controlling the major element composition and the fraction of REE are similar, i.e. garnet and clinopyroxene, and magmatic garnet pyroxenites become "metamorphic eclogites". There is an analogy with Mg-rich eclogites and mafic (ga-cpx) layers in the Ronda peridotite (Obata 1980). Garnet pyroxenite fractionation from a basaltic liquid could have played an important role in the information of major element chemistry of eclogites which are apparently accumulative in major and trace element chemical compositions. The fact is supported by experimental evidence (Green and Ringwood 1972; Obata and Dickey 1976). The REE data from the eclogites suggest also that plagioclase fractionation is involved. Since the plagioclase fractionation is not possible in garnet peridotites (plagioclase is not stable) the fractionation must have taken place in the field of garnet and feldspar stability, i.e. approximately under a pressure of 16 to 29 kb. Transition element abundance (especially Ni) provides, however, another insight into the mechanism of formation of Mg-rich eclogites (Fig. 8.8). The Ni abundances are relatively low in respect to high Mg values. This fact could be accounted for not by olivine fractionation but by the precipitation of another Mg-rich (though relatively Ni-poor) phase such as garnet.

We advocate a two-stage model in order to explain the origin of eclogite layers within the garnet peridotites. We assume that garnet peridotite produced basaltic melts (in high pressure environments); this melt separated from the residual garnet peridotite and balanced with it (i.e. ga-opx-cpx mineralogy). The melts precipitated cumulates of garnet and clinopyroxene. Later, these melts were squeezed out of the source space leaving only thin layers of ga-cpx within the garnet peridotites (these are the Mg-rich eclogites). If, however, the crystals of early precipitated garnet were remixed with the melt and were not completely separated (or completely squeezed out of the parent garnet peridotite) the transitional or common eclogites would form. We believe we can observe "eclogite fractionation" although via magmatic garnet-pyroxene precipitation. The estimated proportion of garnet peridotites to eclogites varies from 100:1 to 10:1, and we believe this represents the proportion of garnet peridotite to eclogite in the subcontinental depleted mantle.

8.4 Petrologic Indications of Deep Structure of the West Carpathians

D. Hovorka[1]

Introduction

The West Carpathian segment of Neo-European Tethys, similarly to the entire global structure, is characterized by an extremely complex tectonic structure. The structure originated in the Cenozoic, and partly, already in the Hercynian period. It is characterized by the existence of numerous allochthonous units (nappes) consisting of Mesozoic rocks and, in some tectonic units, of pre-Mesozoic complexes.

The mosaic of blocks in the present erosional section of the mountain range provides only limited possibilities for a reconstruction of rock content of the deep zones of the crust of the West Carpathians. Their geographical position and the areal extent of fundamental geologic units is shown in several maps, here. In the last few years remnants of intensively, Hercynian retromorphic recrystallized (equilibated) rocks of high-grade origin (granulite and eclogite facies; Hovorka and Méres 1989a, b) have been found within the gneiss-migmatite complexes. Hercynian metamorphic recrystallization took place under conditions of the green-schist to amphibolite facies of low/intermediate baric type.

When trying to approximate the structure and especially the rock content of deep zones of the West Carpathians, we start from the present state of knowledge on: the type, magnitude, form, and genesis of bodies of Variscan magmatites of granite series, volcanism of the Palaeozoic and Mesozoic of ophiolite complexes, and xenoliths of spinel peridotites in late Cenozoic alkaline basalts/basanites. The late Cenozoic volcanism in the whole is also significant for learning about deep zones.

A review of the subject provides, at the same time, basic data for solving the palaeodynamics and palaeogeography or palaeoreconstruction of units of the present space of the West Carpathians.

Role of Magmatites of the Granite Series in the Composition of the Crust

Bodies of magmatites of the granite series form backbones of the mountain ranges oriented conformably to the course of the Carpathian segment of Tethys. In the present relief, the bodies occur on the surface mainly in the central zone in form of minor massifs also in the inner zone (when the West Carpathians are divided into the outer, central, and inner) on Czechoslovak territory. The massifs are prevailingly of Variscan age, locally smaller bodies are of Alpine age.

According to present knowledge, Variscan magmatites originated by anatexis of supercrustal rock sequences (Hovorka 1979; Cambel et al. 1980; Hovorka and Spišiak 1983; Cambel et al. 1985). The low initial ratios $^{87}Sr/^{86}Sr$ (0.705 – 0.707) detected in

1 Geological Institute, Comenius University, Mlynská dolina, 842 15 Bratislava, Slovakia

massifs of Tatras (Burchart 1968) and Malé Karpaty Mts. (Bagdasarjan et al. 1982) can be explained as due to the presence of large quantity of mantle material in the zone of anatexis. The high isotopic Sr ratios in massifs of other rock types suggest that melt was generated in a crust of continental type.

A basic rock type of Variscan magmatites is represented by granodiorites/tonalites, which for some massifs in marginal parts, are transitions to porphyritic facies of granodiorites to granites. In some massifs, aplitic-pegmatitic zones are present, forming their marginal parts (the Tatras). The basic aspects of genesis of the massifs are as follows:

1. There is no developed sequence of rocks that could suggest appreciable processes of fraction crystallization and differentiation. Here, biotite is fundamental and abundant dark mineral. Amphibole belongs to the category of accessory minerals. It can be assumed, on the basis of composition, of basic microscopic features, as well as of REE content distribution pattern (Hovorka and Spišiak 1983), that the massifs represent products of a postkinematic cooling of an anatectic melt generated in a crust of continental character. Metamorphites of supercrustal sequence (metasediments originated from greywackes and metavolcanics) were probably protoliths.
2. Existing data on REE content in the province of Variscan anatectic granites of the West Carpathians provide controversial information. According to Hovorka and Spišiak (1983), the magmatites exhibit striking Eu anomaly which is also characteristic for all distinguished fundamental types of plutonites, while according to Cambel et al. (1985), the same types (massifs) of magmatites are characterized by a lack of Eu anomaly. Such distribution patterns of normalized REE contents could indicate a cumulate origin of some massifs, e.g. the Modra massif in the Malé Karpaty Mts. (Cambel and Vilinovič 1987). There is no microscopic evidence available in the entire mountain range, however, in order to apply the process.
3. The peraluminous types found (Hovorka and Fejdi 1983), containing modal garnet, sillimanite, etc., represent evolution facies of the fundamental type of anatectic melts.
4. For some massifs (folds of Široká and Červené vrchy in the Tatras, the Vepor massif especially its north part) their allochthonous position was proved. An absence of "roots" of the massifs complicates attempts to reconstruct the structure of deep zones from the point of view of place and way of origin of anatectic melts of Variscan age.
5. The majority of Variscan magmatites of the central zone of the West Carpathians is characterized by the presence of schlieren (skialites) of various size and diverse degree of anatexis influence. The schlieren represent relicts in imperfectly homogenized melt. They are always associated with minerals of metasediments and of metaeruptive rocks of the amphibolite facies provenance.
6. The lateral (areal) extent of anatectic melting in the Earth's crust has not yet been solved. The problem whether the anatexis was a regional process or an areally limited one has remained. Despite the fact that the author of this work tends to favour an allochthonous position of most parts of Variscan magmatic massifs with regard to the place of generation of anatectic melts, the observed zonality of some

massifs, which is documented by more leucocratic facies in their marginal parts, still provides the possibility of considering them mutually unrelated bodies in the course of their ascent and thus also of their cooling. This way the anatexis in the crust would have occurred only in places of higher heat flow (Hovorka 1979; Hovorka and Spišiak 1983). The geological causes of the phenomenon are not known, however. This idea follows from commonly accepted considerations that the normal geothermal environment is not sufficient for anatectic processes to take place in the Earth's crust of normal thickness. The heat could have been supplied by different sources. One of the most significant of them could be intrusions of basic melts from the mantle.

The problem of the ascent of Variscan magmatites of the West Carpathians into their present position in the crust has not been sufficiently studied. Several models of ascending melts in the continental crust have recently been published. The theory of diapiric ascent of crystal melt is based on isoenthalpic ("self-heated") phenomena of the ascending crystal melt. In the course of each ascent, the system is actually decompressed while the melt volume increases (on account of solid phases). If the hypothesis concerning the existence of block mosaic with the continental and oceanic and suboceanic crusts in the Variscides of Central Europe (Mahel' 1980; Hovorka 1986) is considered, then the model by Brown (1977, 1979), suggesting that the anatectic melting takes place mainly in a collision environment – in our case a collision of blocks with the continental lithosphere – is applicable also to the region discussed.

Diapirism is a slow process in the course of which the crystal melt ascent takes place in the vertical direction (Marsh 1982). A basic condition for its existence is a density contrast between the ascending melt and the surrounding environment. If the crystal melt cools down and stiffens, the process stops. The late Cenozoic ascent of cooled massifs of magmatites of the central zone which follows is, due to tectonic processes asynchronous. According to Kráľ (1982), first the Slovenské rudohorie Mts. (75 Ma) ascended, then the Nízke Tatry Mts. (28 – 35 Ma). Ascents of external massifs took place 10 – 22 Ma.

Role of the Early Palaeozoic Volcanism in Evolution of the Crust

The discontinuous occurrences of complexes which are considered late Palaeozoic in age (only partly confirmed) in the central zone rest prevailingly on migmatite and orthogneiss complexes presumably of Proterozoic age. If the prevailingly type of metasediments – paragneisses – is also included in the early Palaeozoic sequences of Central zone, then the rocks of greywacke formation with intercalations of rocks of various composition – greywackes of lower grade to quartz sandstones, shales, sediments with increased Fe and Al contents, sediments with organic substance, and locally, also carbonates (Hovorka 1974) represent the fundamental facies of protolite. From the study it follows, that this sedimento-genetic material was derived from blocks with continental lithosphere, and transported a short distance. In forming the rock content of the Palaeozoic level the synchronous basalt and rhyolite volcanism has also played a substantional role (Hovorka 1986; Grecula and Hovorka 1987).

On the basis of the present knowledge of metamorphosed complexes lithology it can be assumed that in the early Palaeozoic zones existed with mainly basalt volcanism and zones (blocks) with lithosphere of continental section put together in the Precambrian (Maheľ 1980; Hovorka 1986). According to Hovorka (1986) the whole environment can be analogous to pack ice with megablocks of the Euroasian and African shields and with individual blocks of continental section situated among them. The space in between the blocks, i.e. the "water channels", were filled with formations which had originated on the oceanic/suboceanic crust. Metamorphosis-recrystallization Variscan processes together with the origin and consolidation of magmatic masses had caused that from the mosaic of blocks and "rift or riftoid" zones between them (filled mostly with early Palaeozoic volcanosedimentary sequences) a "megabreccia" originated. Within the megabreccia, the Proterozoic facies (maybe even older) of blocks were pushed close together with the newly originated "increments" of the early Palaeozoic age.

The units recently forming the fill of the inner zone of the West Carpathians consist of different rocks. According to Grecula (1982), sequences of sediments and volcanics originated in a rift zone. They form an early Palaeozoic complex with several facies developments, differences of which were accentuated also by the origin of partial tectonic units (nappes) thrust over their foreland (i.e. over the complexes of the central zone of the West Carpathians). Also in this partial segment of the Carpathians, granitic melts (in comparison with the central zone being strikingly more acid in type) had generated in the crust and subsequently penetrated the higher level of crustal structure up to the level of greenschist complexes as in the case of Central zone.

The presence of basic, acid, and very probably also intermediary volcanics of tholeiitic series in the Palaeozoic of the central and inner West Carpathians, indicates their generation both in the mantle and in active continental margin tectogene. It is obvious from this that the early Palaeozoic volcanism had conditioned the transport of mantle material and had caused a redistribution of the lower crust material.

Variability of Products of Volcanic Activity in the Mesozoic – Reflection of Diverse Geodynamic Conditions

Volcanic activity had taken place in various partial sedimentary basins of the Mesozoic. Rocks of teschenite association (teschenites, picrites, alkaline basalts, monchiquites, ouachites, etc.) are characteristic for flyschoid sequences (Hauterivian – Barremian) of the Silesian unit of the outer West Carpathians. It is the classical region of teschenites (Hohenegger 1881) and picrites (Tschermak 1886) occurrences. On the basis of synchronous course of volcanic activity with sedimentation in a basin of a particular tectonic unit, we consider the volcanism of the given type, in accordance the with ideas of Hovorka and Spišiak (1987), to be a consequence of initial rifting of the slightly thinned, but still fairly thick continental substrate of the given unit. These processes probably reflected the initial movements at the contact of the Epivariscan platform with complexes of the mobile zone of the Tethys. The original alkaline character of partial melts derived from the mantle was strongly affected by processes of fractional crystallization which took place mainly in secondary magma chambers in the crust and in thick bodies of volcanics/subvolcanics. The partial mantle melts were also modified by processes of assimilation of rock sequences underlying the sedimentary basin (prevailingly carbonate rocks of the Mesozoic, Hovorka and Spišiak 1987). The absence of huge bodies at the subvolcanic to plutonic level is also indicated by distribution pattern of geophysical fields.

An association of alkaline basalts/basanites is characteristic for the Mesozoic tectonic units of the present central zone (the Krížna nappe and "mantle" units). Its upper-mantle origin and character had not been affected subsequently by processes of fractional crystallization, although the detected processes of assimilation of mostly carbonates affected, in several places, their subsequent geochemical composition. Probably the only indicator of conditions in the adjacent deep basement are the picrite occurrences in the wider surroundings of Banská Bystrica. The picritres occur in two tectonic units (Slavkay 1979), into which they penetrated along N–S lineament. In the course of this penetration, crustal material was assimilated (increased K, etc.; Hovorka and Spišiak 1987).

Mesozoic rocks of the recent inner West Carpathians are of a strikingly different character. They belong to both the effusives of the tholeiite sequence (as a member of incomplete ophiolite formation) and the calc-alkaline effusives. The latter form a sequence rhyolite–dacite–andesite to andesitoid basalt, locally with shoshonite trend of differentiation. They have been found mainly in the Bükk Mts. in Hungary and in the Drienok nappe in Central Slovakia.

From the known character of volcanic sequences of individual tectonic zones of the West Carpathians of the first order it follows, that volcanics of the present outer and central zone had been generated uniquely in the upper mantle. Consequently, their occurrences in the central zone represent primitive upper-mantle melts. On the contrary to it, the volcanics of the inner West Carpathians originated within a complicated tectonic background. Unfortunately, the substantial space reduction and allochthonous accumulation of Mesozoic units containing volcanics make it impossible to characterize the role of the Mesozoic magmatic melts in deep zones beneath the present extent of the West Carpathians more specifically.

Significance of the Ophiolite Complex in a Structure

In the central zone of the West Carpathians, a few bodies of peridotites (actually of their metamorphosed equivalents) are known. With regard to their outcrops as well as to their buried bodies, the inner zone of the West Carpathians, actually southern partial unit of the Centralides (the crystalline core of the Kohút zone) adjacent to it, is most significant.

Despite various intensity of metamorphic-recrystallization processes (under conditions of greenschists to low-grade amphibolite facies), the original character of ultrabasites of the spinel peridotite type (lherzolites, harzburgites, seldom also other types) is evident (Hovorka et al. 1985).

The problem of the presence of peridotite bodies at two stratigraphic levels in this zone, i.e. in the early Palaeozoic (the micaschist complex in the Kohút crystalline complex, and in the sequences of the Gemericum near Vyšný Klátov), and in some tectonic units of the Mesozoic (the Meliata group of Gemericum) has not been solved uniquely. From the West Carpathians point of view the bodies occur in spatially close, actually overlaid environments.

Bodies of ultramafic rocks occur (actually they are indicated) within a zone of several tens of kilometres wide, and nearly 150 km long, extending in ENE–WSW direction. The zone coincides with the zone of assumed basic rocks masses (Fusán 1985). In the direction parallel with the zone axis, also late Cenozoic alkaline basalts containing xenoliths of spinel peridotites occur (Hovorka and Fejdi 1980). It means that this had been a zone (even though its spatial position had fluctuated during the specified period) of penetration of ascending mantle material into formations of supercrustal sequences of various ages from the Mesozoic (actually even the Palaeozoic) to the late Cenozoic. A lack of direct data e.g. from deep boreholes, etc., in this zone but generally also in other parts of the West Carpathians make quantitative specification of the position of the upper-mantle ophiolite complexes impossible.

Late Cenozoic Volcanism – the Youngest Vertical Movements of Masses in the Earth's Crust

The inner (southern) side of the Carpathian arc is characterized by the presence of the late Cenozoic volcanic mountain ranges. Rock associations belong to two fundamental sequences: a) calc-alkaline with the range of rock types from rhyolites, over dacites, andesites, up to basaltic andesites; b) alkaline (alkaline basalts/basanites).

Places and ways of generation of the Late Cenozoic magmatic melts, which penetrated to the ground surface or cooled down at subvolcanic levels, have been, during last few decades of years, the subject of several hypotheses (Kuthan 1967; Fusán et al. 1971; Konečný and Slávik 1974; Lexa and Konečný 1974; Hovorka 1980; Bielik et al. 1986a).

From the point of view of deep zone compositions of the West Carpathians, the late Cenozoic volcanism is exceptional. This follows from the fact that it was active in the period when:

1. the arcuate course of the West Carpathians was already formed,
2. the thrusting of extensive allochthonous units or rock blocks was already completed,
3. the crust of the continental section was already generated,
4. the last volcanic activity phases (alkaline series of effusives) were already confined to the period of commencing cratonization, specifically in the Pannonian block.

Factual information on the upper mantle composition beneath the West Carpathians and the adjacent Pannonian Basin has been provided by xenoliths of spinel peridotites in alkaline basalts/basanites occurring at the southern margin of volcanic mountain ranges and in the Pannonian Basin (Hovorka and Fejdi 1980). It follows from the results of the above-mentioned authors' work (l. c.) as well as from the work by Embey and Isztin

(1976), that identical xenoliths (spinel peridotites) occur in alkaline basalts in areas separated by hundreds of kilometres, while the surrounding rocks are basically identical (alkaline basalts/basanites).

Based on these facts, the following preliminary conclusions can be formulated: a) the melting out of alkaline basalts/basanites in the south periphery of the West Carpathians and in the Pannonian Basin happened under identical conditions of the upper mantle, b) the rocks from the environment which had yielded partial melts were the same in composition, and had corresponded to spinel peridotite or to spinel pyrolite. Types of peridotites containing garnets or plagioclases have not been found.

The xenoliths which are known from volcanics of the calc-alkaline series do not, unfortunately, provide any information on deep parts of the Earth's crust. Fiala (1954) and Hovorka and Lukáčik (1972) discovered that the xenolith types occurring there (partly resorbed and recrystallized) yield information on rock types forming the flat basement of the volcanic complexes. They document the existence of those complexes in the basement of effusives which also occur in adjacent areas on the ground surface (the Neogene sediments and rocks of the gneiss–amphibolite complex of Palaeozoic or Proterozoic).

There are no adequate geochemical data (REE, isotopes, data on the silicates phases composition, etc.) available for finding the origin of magmas of volcanic of the Late Cenozoic, and especially, for investigating their subsequent development.

Conclusion

Petrologic or petrogenetic aspects of selected rocks complexes have provided particular thought simplified to a considerable degree and ambiguous information on the processes which shared formation of the rock content of deep crust beneath the West Carpathians. Deep crustal zones had formed to their present rock content from the Proterozoic to the sub-Recent period. Magmatic processes within the crust as well as the volcanic activity of zones of tectogenes, i.e. regions which could correspond to the zones of active margins of the present continents in the West Pacific region, and up to zones of island arcs, represent the fundamental processes. All these processes have caused the "cementing" of blocks of the pre-Variscan substrate.

Assuming that the normal thermal gradient in the continental crust does not suffice to evoke anatectic processes, we consider the anatectic processes as the consequence of an existence of zones of increased heat flow. That is why the anatexis during the Palaeozoic period (which caused intensive granitization of the Earth's crust) but also in the Cenozoic did not have any regional extent.

Despite the fact that data on the presence of metamorphosed complexes of granulite or eclogite facies at least in part of the blocks of the megabreccia of Cenozoic rocks sequences is sporadic, the share of these metamorphosed complexes on the lower crust zones composition is quite probable.

8.5 Deep-Origin Xenoliths in Volcanics of Czechoslovakia

F. Fediuk[1]

Introduction

Czechoslovak geologists have been interested in xenoliths since the very beginning. The work by Farský (1987) dealt with the "olivine spheres" occurring in basaltic rock of the late Cenozoic volcano Kozákov in the north of Bohemia, and represents world's first serious information on four-phase lherzolites, i.e. on the rocks which later became a backbone of present hypotheses on the petrographic character of the upper mantle.

Czechoslovakia has an exceptionally rich geologic history of volcanic activity and, therefore quite diverse products of volcanism. During the period from 600 Ma to 0.3 Ma the following volcanic activities can be differentiated: Proterozoic, early Palaeozoic, late Palaeozoic, Mesozoic, Tertiary to Quaternary. There is no volcanic activity in the late Recent period, however. Continental, within-plate, island-arc, and oceanic tectonomagmatic settings, alkaline, shoshonitic, tholeiitic, and calc-alkaline series, acid, intermediate, basic, and ultrabasic volcanic rocks as well as oversaturated, saturated, and undersaturated volcanic rocks occur here. Terrestrial as well as underwater volcanics are present here, and their index of explosivity varies widely. All these types of volcanics contain xenoliths, which are in an absolute majority of rock fragments of supercrustal complexes: hornfelses and tactites that originated from sediments, various types of crystalline schists, volcanics of older formations, granitoids, etc. Although definitely valuable for learning the geologic structures of a particular region, they do not provide information on deep structure. The Tertiary to Quaternary basaltoids are the only Czechoslovak volcanics which can provide such information thanks to the character of their xenoliths.

Young Basaltic Volcanics of Czechoslovakia

In Czechoslovakia the Tertiary to Quaternary volcanics are set in two strikingly diverse tectono-magmatic positions. One of them, which is in the Bohemian Massif, is represented by rift, within-plate unsaturated volcanics of alkaline sequence. The second group, being in the West Carpathians, is characterized by saturated and oversaturated volcanics of the calc-alkaline sequence, among which unsaturated alkaline rocks occur only sporadically.

Alkaline volcanics of the Bohemian Massif are known in two regions i.e. northern part of Bohemia and north of Moravia (see Fig. 8.9). The first one is far more significant from both the quantitative (total areal extent of volcanics is estimated to 1 220 km^2) and degree of differentiation aspects. Melilitic, nephelinic, noseanitic, and other rocks with foids are present here, as well as basanites, tephrites, alkaline trachyandesites, trachytes,

[1] Geohelp, Na Petřinách 1987/29, 162 00 Praha 6, Czech Republic

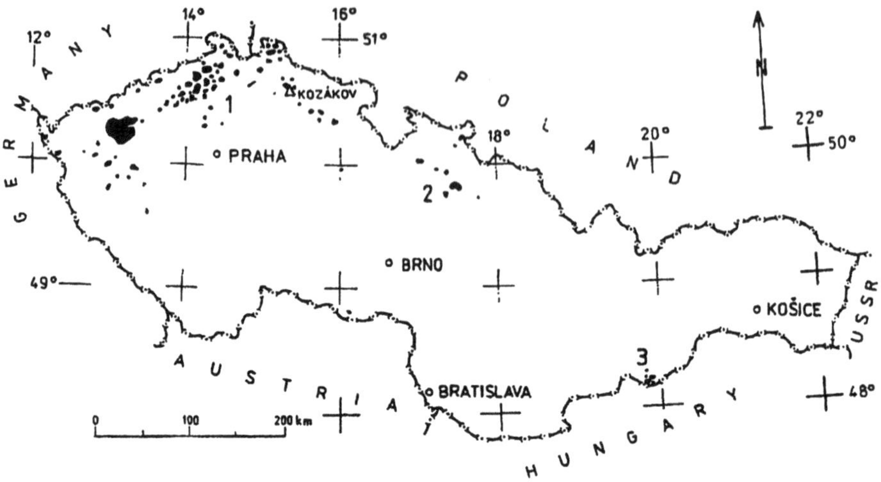

Fig. 8.9 Synoptic map of occurrences of basaltic rocks containing ultramafic nodules in Czechoslovakia. 1 – North-Bohemian, 2 – North-Moravian, 3 – South-Slovakian regions. The most significant locality – the Kozákov volcano of the Lower Pliocene age is designated by a triangle.

and phonolites. Only basaltic rocks, however, contain xenoliths of deep origin. The North Bohemian volcanic region includes the surroundings of Cheb, the Doupovské hory and the České Středohoří Mountains with adjacent parts of the Krušné hory Mts., the areas between the towns of Frýdlant and Semily, Mladá Boleslav and Pardubice.

The second region, in the extreme north of Moravia, includes volcanics of the Nízký Jeseník Mts., of Ostrava, Opava, and Javorník surroundings. In this region, there are far fewer volcanic rock occurrences than in the northern part of Bohemia. Total areal extent of volcanics in the north of Moravia is estimated at 15.5 km^2. Altogether 30 occurrences of volcanics are known here. All of them consist of basaltoid rocks, the composition of which varies from olivine nephelinite (in places containing melilite, leucite or nosean), over the prevailingly nepheline basanite to alkaline olivine basalt. Phonolites and trachytes as well as trachyandesites and tephrites are totally absent.

In the West Carpathians, there is a third region of alkaline basaltoids in Czechoslovakia, which contain deep-origin xenoliths. The volcanics are distributed in the inner part of the Carpathian arc at the border with Hungary, near the towns of Lučenec and Fiľakovo. Its areal extent is comparable with that of the region in north of Moravia, i.e. approximately 20 km^2 in total. Nepheline basanite represents the main rock, while alkaline olivine basalt is in minority. Their undersaturation does not reach the grade of foids.

The ratio of areal extent of volcanics in the three above mentioned regions can approximately be expressed as 97.5:1:0.5. Numbers of deep-origin xenoliths in them have been estimated at the same ratio. The contours in Fig. 8.10 express, in the divariant field of alkali versus silica, the frequency of variation range of volcanics from the whole

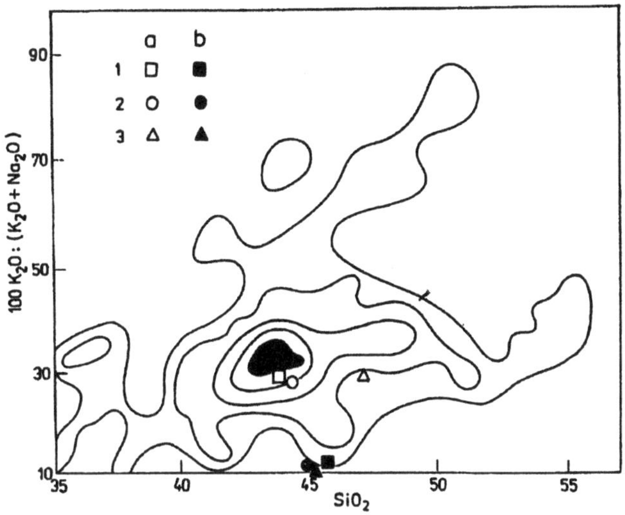

Fig. 8.10 Chemical composition of the world volcanites containing lherzolite nodules in the diagram $100\,K_2/K_2O + Na_2O$ vs. SiO_2. According to Forbes and Kuno (1967) modified by Fediuk (1973). Contours 1 – 2.5 – 5 – 10 – 15 %. The mean points a) of surrounding basaltoids of 1 – North-Bohemian, 2 – North--Moravian, 3 – South-Slovakian regions (compare with Fig. 8.9), b) of lherzolite nodules of 1 – North--Bohemian, 2 – North-Moravian, and 3 – South - Slovakian regions.

world, which are known for being host rocks of the most frequent types of deep-origin xenoliths, i.e. the four-phase lherzolites. In this diagram, points of projections of arithmetical means of chemical composition of basaltoids of all three regions in Czechoslovakia are plotted, as well as those of the mean compositions of corresponding lherzolite xenoliths. It follows from the diagram that the xenolith-bearing basaltoids of the north Bohemian and north Moravian regions are very close to the frequency centre of world volcanics containing lherzolite xenoliths, while the volcanics of the tectono-magmatic region of south Slovakia exhibit higher acidity, their alkalinity being roughly the same. Contrary to it, the variance of macrochemical composition of lherzolite nodules, characterized by the diagram, is very small for all three regions.

Field Petrography of Xenoliths

On its way to the ground surface from its deep source, a magma of volcanic materia continuously "sampled" the whole geologic section by xenoliths. From the north Bohemian volcanic region such cases are known where the rock filling the volcanic chimney, at its present denudation level, encloses xenoliths of sedimentary cover which had been deposited higher than at this level: the magma ascended through the chimney, produced frappe-fragments of rocks on the ground surface, and then again descended. In this connection, xenoliths of supercrustal origin are not of any interest to us.

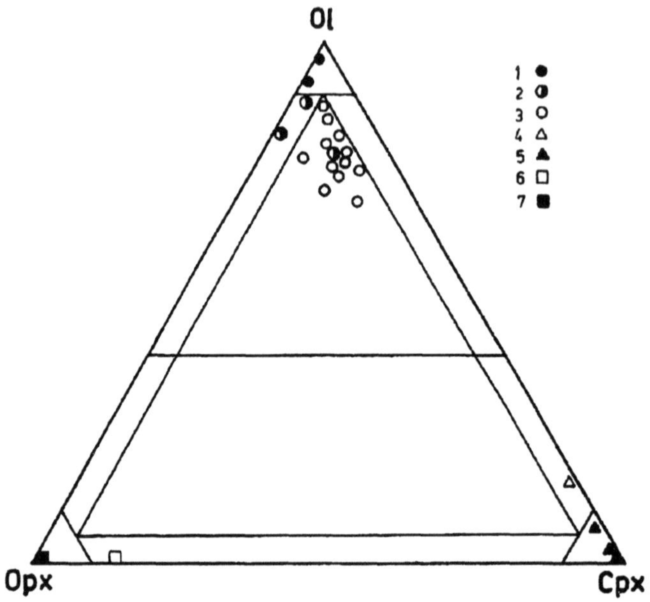

Fig. 8.11 Model composition of ultramafic xenoliths in nephelinic basanite of the Kozákov volcano. Peridotites: 1 – dunite, 2 – harzburgite, 3 – lherzolite; Pyroxenites: 4 – olivine clinopyroxenite, 5 – clinopyroxenite, 6 – websterite, 7 – orthopyroxenite.

Proceeding downwards, the first rock occurring in xenoliths that attracts our attention is charnockite – the representative rock of the deep, highly dehydrated part of the continental crust. Xenoliths of this petrographic group were discovered in several occurrences of infrequent volcanics in the north of Bohemia, specifically from the central part of the České Středohoří Mts. between the towns of Ústí nad Labem and Litoměřice (Opletal 1967; Bořický 1988). They typically contain the mineral association of andesine + perthitic K-feldspar + quartz + orthopyroxene (bronzite–hypersthene). The xenoliths are of up to 30 cm in size at the Myštice locality. Contrary to granulites, which occur in the xenoliths together with charnockites but are known also from the bodies presently outcropping, the charnockites, in situ, in the present denudation level of the Bohemian Massif have not yet been found. Xenoliths of two-pyroxene gabbroids (olivine gabbronorites), but of course without K-feldspars, were found in nepheline basanite of the Kozákov volcano (Fediuk 1973). Also in this case, it is likely that a member of the lower-crust charnockite level in the Bohemian Massif has not been discovered on the basis of other methods.

Ultramafic rocks represent a far more frequent and even the deepest originating group of xenoliths. Fediuk (1981) and Jakeš, Vokurka (1987) provided a review of them. Pyroxenites (clinopyroxenites and orthopyroxenites) are few but peridotites are quite common. Modal composition of these ultramafic xenoliths is expressed by the example of 23 nodules from basanites of lava flows of the Kozákov volcano in the classification diagram of olivine–orthopyroxene–clinopyroxene (Fig. 8.11). Clinopyroxenites as well as orthopyroxenites have, as a rule, the character of monomineral rocks. Clinopyroxenites are not only more frequent, but their xenoliths are usually greater (up to 20 cm in length). Among peridotite xenoliths, as is obvious in Fig. 8.11, both dunites and harzburgites occasionally occur, but lherzolites prevail. 17 modally analyzed peridotites exhibit the following quantitative ratios of their minerals:

volume %	range	arithmetical mean
olivine	69 – 97	79
orthopyroxene	4 – 17	8
clinopyroxene	1 – 21	11
spinel	0 – 4	2

Of all Czechoslovak occurrences of peridotites nodules in basaltic rocks, the Kozákov volcano (743 m a.s.l.) near the town of Semily and Železný Brod, is clearly the most significant. The eruption age, which took place in four or five phases, has been determined by radioactive dating as 6.3 Ma. Following a vague explosion phase, several lava flows began to flow out of the central crater. The most powerful lava flow after 2.5 km reached the stream bed of a Lower Pliocene river, which it followed for another 6.5 km. At present, numerous quarries have been opened in this lava flow, which have frequently been visited by mineral collectors. Some olivines have the appearance of precious stones. The largest-cut crystal is 15.67 carats. In the lava flow rock, 35 peridotite nodules larger than 1 cm can be found per 1 m^2, on average. This corresponds to 2 % of total volume of the volcanic material, which means, that for the areal extent of 8 km^2 of flows and for the mean thickness of about 20 m, the volcanics contain roughly 11 million metric tons of peridotite. The largest continuous nodule found was 70 cm in length. Some proofs exist, though, that in the course of the lava flow, nodules cracked and were crushed, which means that they were originally larger. Fresh-state olivine is bright, light-green in colour. Orthopyroxene is macroscopical black or brown, while clinopyroxene is usually bright green, the colour intensity, depending predominantly on Cr content, is quite variable. Even more brilliantly green are spinelides, grains of which, however, are mainly microscopic.

The localities Brtníky near Rumburk (the Plesná hill) and the Provodínské kameny near Česká Lípa belong to significant occurrences of peridotite nodules in the north of Bohemia. Altogether about 50 localities are known in this area where abundant nodules occur, and more than 100 of others with sporadic nodules. The quantity of substantial occurrences increases in east direction. Localities in the north of Moravia and in the south of Slovakia are not as famous for their occurrences as those in the north of Bohemia, although peridotite xenoliths, olivine and pyroxene xenocrysts can be commonly found there. Occurrences in the north of Moravia were studied in detail by Fediuk and Fediuková (1988), those in the south of Slovakia were studied by Hovorka and Fejdi (1980).

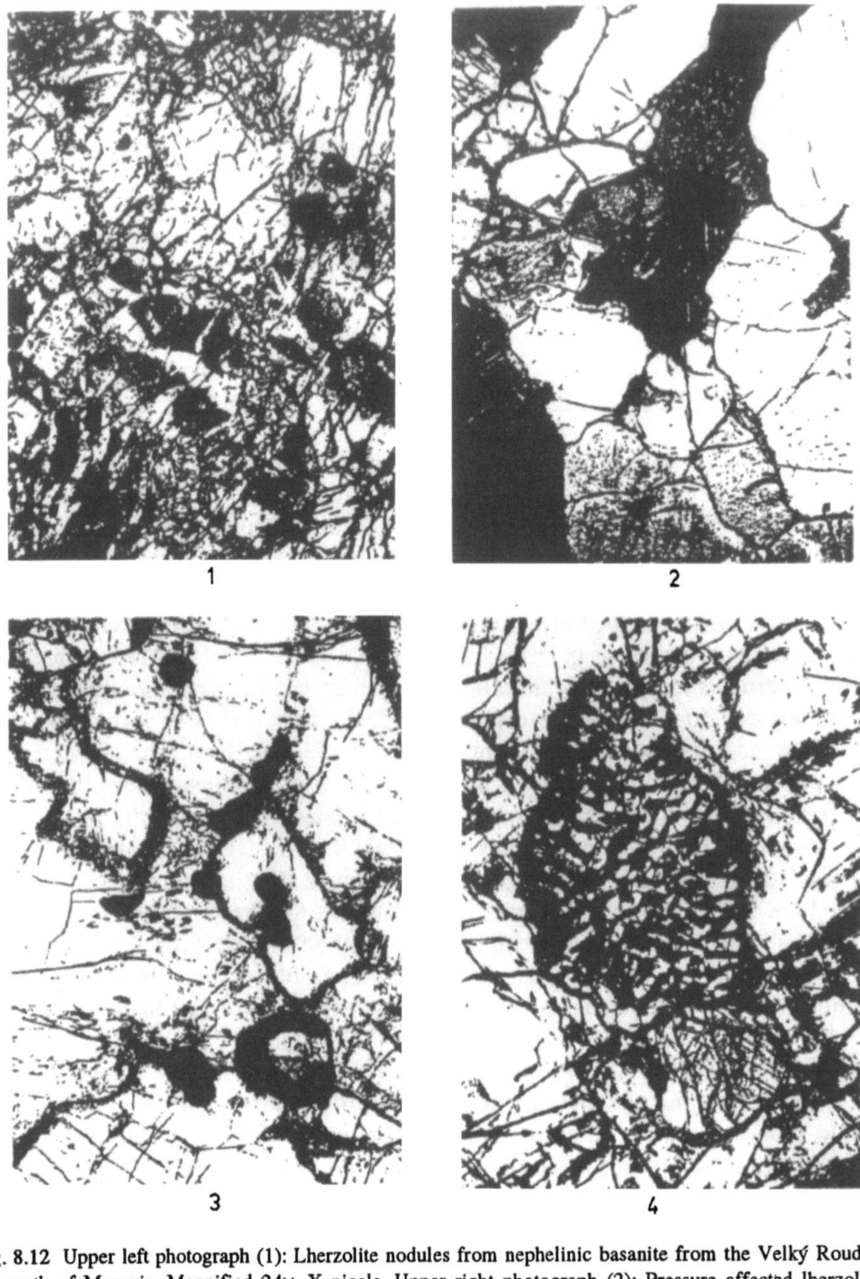

Fig. 8.12 Upper left photograph (1): Lherzolite nodules from nephelinic basanite from the Velký Roudný in the north of Moravia. Magnified 24×, X nicols. Upper right photograph (2): Pressure affected lherzolite of olivine nephelinite from the locality Zálesí u Javorníka in the north of Moravia. Magnified 24×, X nicols. Lower left photograph (3): granulose spinel in lherzolite nodule from the locality Uhlířský vrch in the north of Moravia. Magnified 24×, no analyser. Lower right photograph (4): symplectite spinel in lherzolite nodule from the locality Velký Roudný in the North Moravia. Magnified 24×, analyser.

Microscopic Character and Mineralogy of Lherzolite Nodules

From the point of view of qualitative mineralogy, the composition of lherzolite nodules is quite uniform as its variability is projected only to certain fluctuations of mutual quantitative proportions. Substantially greater differences are observed in textural properties of lherzolites. The grain-size varies quite considerably, so that besides the fine-grained to finely granular varieties not only medium but also coarse-grained varieties occur side by side. Epigranular grain-size prevails, but porphyric varieties can often be found. Exceptionally, porphyric phenocrysts of olivine of 5 cm in size were found. For the most part lherzolite exhibits directionless structure, but banded structure also exists, even though it can be only slightly indicated or striking. Besides practically undeformed samples, those samples also occur, in which protoclastic deformation of grains, especially of olivine, reaches various degrees of intensity. Examples of two diverse characteristic patterns of microscopic appearance of lherzolites are shown in photographs 1 and 2 (Fig. 8.12).

Olivine

This most abundant mineral of lherzolite xenoliths is characterized by its exceptionally small scattering of chemical composition, as is known nearly from all analogous xenoliths in the whole world.

In Table 8.5, mean values of chemical composition of olivines from all the three Czechoslovak regions are presented, calculated and adjusted to 100 from the data

Table 8.5 Mean values of chemical composition of olivines

	1	2	3
SiO_2	41.08	41.19	43.14
Al_2O_3	0.49	0.36	0.03
FeO^{tot}	8.07	9.40	10.31
MgO	50.14	48.99	46.46
CaO	0.22	0.06	0.06
	cation contents (4 oxygens)		
Si	0.997	1.004	1.050
Al	0.014	0.010	0.000
Fe	0.163	0.191	0.209
Mg	1.814	1.781	1.686
Ca	0.005	0.001	0.001
100 Mg / Mg + Fe	91.76	90.31	88.97

published in literature: the north of Bohemia (1) determined from the four analyses presented by Vokurka and Povondra (1983), the north of Moravia (2) determined from 14 analyses of olivines presented by Fediuk and Fediuková (1988), the Slovakian region (3) determined from analyses of olivine published by Hovorka and Fejdi (1980):

Orthopyroxene

It is also valid to say for this mineral that its chemical composition has a very narrow spread, both from comparison between individual regions and the comparison within a particular region. As Fediuk and Fediuková (1988) showed, it is also valid within a particular locality. The mean values presented in Table 8.6 are taken from the same publications as in the case of olivine. Only the number of starting analyses differs for north Moravia (the mean was calculated from 10 analyses) and for south Slovakian region (the mean was calculated from two analyses).

Clinopyroxene

Out of all four lherzolite minerals, clinopyroxene exhibits the largest chemical composition scatter. Its Cr_2O_3 content fluctuates wildly. The mean values of clinopyroxenes of the three Czechoslovak regions were obtained by computations

Table 8.6 Mean values of chemical composition of orthopyroxenes

	1	2	3
SiO_2	54.93	55.61	57.59
TiO_2	0.08	0.09	0.06
Al_2O_3	4.64	2.79	4.43
FeO	4.91	6.16	5.91
MgO	33.32	43.28	31.44
CaO	2.12	1.07	0.57
	cation contents (6 oxygens)		
Si	1.891	1.921	1.968
Ti	0.002	0.002	0.001
Al	0.188	0.113	0.178
Fe	0.141	0.177	0.168
Mg	1.710	1.765	1.601
Ca	0.078	0.039	0.020
100 Mg / Mg + Fe	92.38	90.97	90.50

analogous to those described in the paragraph on olivine. The mean of the north Bohemian clinopyroxenes and the south Slovakian clinopyroxene were calculated from four starting values, and that of the north Moravian clinopyroxene was calculated from 10 starting values, Table 8.7.

Spinel

Under the microscope, spinel is most frequently brown and sometimes greenish. It follows from the existing data on chemical composition of spinel (Schovánek 1977; Hovorka and Fejdi 1980; Fediuk and Fediuková 1988), the quantity and quality of which have so far not been satisfactory, that it is a common spinel containing Cr, chrome spinel, picotite, and pleonaste. From the morphology and genesis aspects two types of

Table 8.7 . Mean values of chemical composition of clinopyroxenes

	1	2	3
SiO_2	51.50	51.90	55.31
TiO_2	0.25	0.42	0.35
Al_2O_3	5.30	4.39	5.84
Cr_2O_3	0.94	1.20	0.81
FeO	3.31	3.26	2.91
MgO	16.90	16.99	15.00
CaO	21.80	20.48	18.14
Na_2O	n.d.	1.33	1.62
K_2O	n.d.	0.03	0.02
	cation contents (6 oxygens)		
Si	1.870	1.888	1.972
Ti	0.006	0.011	0.009
Al	0.227	0.188	0.245
Cr	0.026	0.034	0.022
Fe	0.100	0.099	0.086
Mg	0.914	0.921	0.797
Ca	0.848	0.798	0.693
Na	---	0.093	0.112
K	---	0.001	0.000
100 Mg / Mg + Fe	90.13	90.29	90.26

spinel occur, i.e. the anhedrally isometric granular, individualized type, and the symplectite type. Both these types, which can occur together in one sample, are documented by photographs no. 3 and 4.

Figures 8.13 and 8.14 courtesy of Fediuk and Fediuková (1988), show all the four mineral phases of lherzolite, based on 36 microprobe analyses of xenoliths from the north of Moravia, compared according to the ratios of 100 Mg/Mg + Fe. This diagram documents the above-mentioned large scatter of clinopyroxenes' chemistry as well as their highest mean Mg content. The Mg content decreases in orthopyroxenes, olivines, and spinels, respectively. The actual absence of a zonal structure of grains can serve as a characteristic feature of all the four minerals which compose Czechoslovak lherzolite nodules. As long as the porphyric or the porphyroblastic texture is developed, significant differences in the chemistry of populations of the same mineral differing in grain size do not occur.

The investigations have shown that clinopyroxene and spinel are the main carrier of trace elements in lherzolites. There is a huge difference in the minority elements content of the surrounding basaltic rocks and of the lherzolite xenolith. As a rule, the difference is in one order of magnitude. Apart from Cr, Co, and Ni, which occur in lherzolites in substantially higher concentrations than in basaltoids, the vast majority of other elements exhibit counter concentrations. The rare earths exhibit marked differences, when not only absolute values but also the ratio LREE/HREE of normalized contents are exceptionally high in basaltoids, while in lherzolites, they are low and, at the same time, balanced.

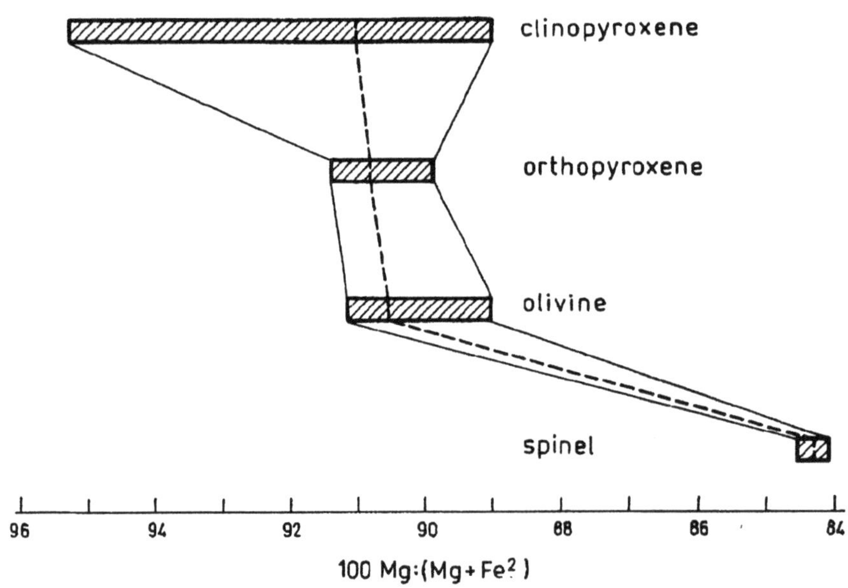

Fig. 8.13 Range of Mg-content 100 MgO/MgO + FeO in coexisting minerals of lherzolite nodules of the North-Moravian region (according to Fediuk and Fediuková).

Differences in chemical compositions of lherzolite xenolith and the surrounding volcanic material are striking. Similarly, there is an absolute lack of any relationship in mineralogical composition of both rocks. Olivines occur in both the rocks but they differ both in their morphology (anhedral grains in xenoliths while euhedral crystals in the volcanic) and especially in their composition. Olivine in volcanics contains 10 – 15 % more of the fayalite component than that in lherzolite xenoliths. In basaltoids, orthopyroxenes and spinels are entirely absent, and clinopyroxenes are strikingly diverse and not of interchangeable character.

Lherzolite Xenolith Properties Implications at Depth

First of all, it is necessary to stress that an indubitable garnet rock has not been found anywhere in ultramafic xenoliths in volcanic rocks of Czechoslovakia. This is quite the opposite to xenoliths in metamorphic rocks. In this respect, we do not consider the occurrence of xenoliths of pyrope-bearing serpentinized peridotites known from the Tertiary diatreme breccia, namely on the Linhorka Hill near Měrunice in the České Středohoří Mts., for these mantle-provenance peridotites have not been emplaced to their present supercrustal position by a volcanic rock but evidently inserted into the crustal rock context tectonically because they were struck in situ in diatremes by boreholes at a depth of several hundreds of metres. Their volcanic transport in the form of xenoliths and pyrope xenocrysts was thus only very short.

The fact, that the xenoliths in volcanics are formed by spinel and not garnet peridotites, clearly documents their upper-mantle origin at no great depth from the levels being designated as the "uppermost mantle".

Fediuk (1973), Schovánek (1977), Vokurka and Povondra (1983), and Vokurka (1983, 1988) tried to evaluate the chemical composition of minerals in lherzolite xenoliths of the north of Bohemia from the more accurate geothermometric and geobarometric viewpoint. Fediuk and Fediuková (1988) tried the same for that of the north of Moravia, and Riedel (1983) and others did the same for the south of Slovakia. Various methods were used for this purpose, which were published in literature (lists in reference of the above-cited papers) as well as original methods (elaborated mainly by Vokurka). The results obtained by majority of the methods are basically consistent with some minor exceptions. It is, first of all, the method of determining the temperature balance which is considerably easier than determining the pressure parameter. The review values summarized in the Table 8.8 have so far been considered representative (1 – the north of Bohemia, 2 – the north of Moravia, 3 – the south of Slovakia):

Table 8.8 Temperature balance and pressure parameter values of lherzolite xenoliths

	1	2	3
T [°C]	1200 – 1270	1005 – 1231	1010 – 1290
p [kbar]	1.6 – 1.9	2.2 – 2.6	1.8 – 2.5

The assumption appears highly probable that the above-presented pressure conditions represent the maximum possible values or rather slightly exaggerated values, the accuracy of which is affected rather by insufficient accuracy of analytical data or by not completing the solutions of some problems, e.g. supplementary rebalancing caused by enclosing magma, than by the methodology of the computation. From thus evaluated results of thermodynamic calculations, a realistic assumption can then be derived that balancing of lherzolite nodules took place at depths of about 50 – 80 km i.e. under conditions of the uppermost mantle.

It follows from the investigations accomplished by various authors that ultramafic nodules in volcanics of the three quite distant regions, i.e. the north of Bohemia, the north of Moravia, and the south of Slovakia, are basically identical with partial diversities; it can be deduced that a particular layer of the uppermost mantle is, in the Czechoslovak part of Central Europe, homogeneous in the first approximation. In detail, however, it is characterized by inhomogeneities which are mainly quantitatively mineralogical, but derived from that, also of chemical nature.

8.6 Geological and Petrologic Models of the Earth's Crust Beneath Czechoslovakia

O. Fusán[1], M. Suk[2]

The upper part of continental crust is, in vertical and horizontal directions, geologically and lithologically immensely articulated but in no way chaotic, for it corresponds to laws of the geologic evolution of a particular portion. This has to be accounted for in constructing geologic and petrologic models, which are compiled from borehole data, extrapolation of geologic structure to depth, evaluation of xenoliths in volcanics and metamorphites, as well as from a distribution of occurrences of exotic pebbles in sedimentary rocks.

The existence of two types of the crust in the Bohemian Massif was noted already by Zoubek (1967) who distinguished the crust of the Moldanubian elevation with an evolved thick sialic (granitic) layer, and the crust of the Teplá – Barrandian region with the crust prevailingly affected by a basaltic layer. Differences between the two crusts he explains by an increasing thickness of the granitic layer during orogenic processes. The Moldanubian elevation is older. That is why it had a consolidated granitic layer already in the Cadomian orogeny.

Zeman (Zeman in Mahel' 1984 b) extended the distinction of two types of continental crust, i.e. sialic and simatic, to the whole Bohemian Massif. The simatic crust has been active for a long time with enzymatic, generally folded basins. The following regions belong to it: Teplá – Barrandian (Bohemicum), Moravian Moldanubicum and the Hrubý Jeseník Mts. with occurrences of basic and ultrabasic rocks, and the northern part of the

1 Geophysical Institute, Dúbravská cesta, 842 28 Bratislava, Slovakia
2 Faculty of Natural Sciences, Masaryk University, Kotlářská 2, 600 00 Brno, Czech Republic

culm of the Nízký Jeseník Mts. The sialic crust, according to Zeman a reworked granitic layer with migmatites and diapirs, is in the Brno Massif, the Krušné hory Mts., and the Bohemian Moldanubicum. Zoubek attributes to the simatic type of crust in the Bohemian Massif the oceanic character. Zeman (in Mahel' 1984b) defines it as a suboceanic one, originating as a weakly sialized residuum of oceanic crust or intensively basified continental crust (Fig. 8.14). The term sub-oceanic crust has of course commonly been used in an entirely different meaning (Zverev and Tulina in Mueller 1974) so that the term para-oceanic type of crust (which was also used by Mahel' 1984) is better.

According to interpretation by Finlayson (1982) the differences between the two types of continental crust are caused by the presence of oceanic crust in the lower structure of young units and by a complementary larger extent of eclogites and mafic granulites in the upper mantle beneath the Proterozoic units. These differences extend

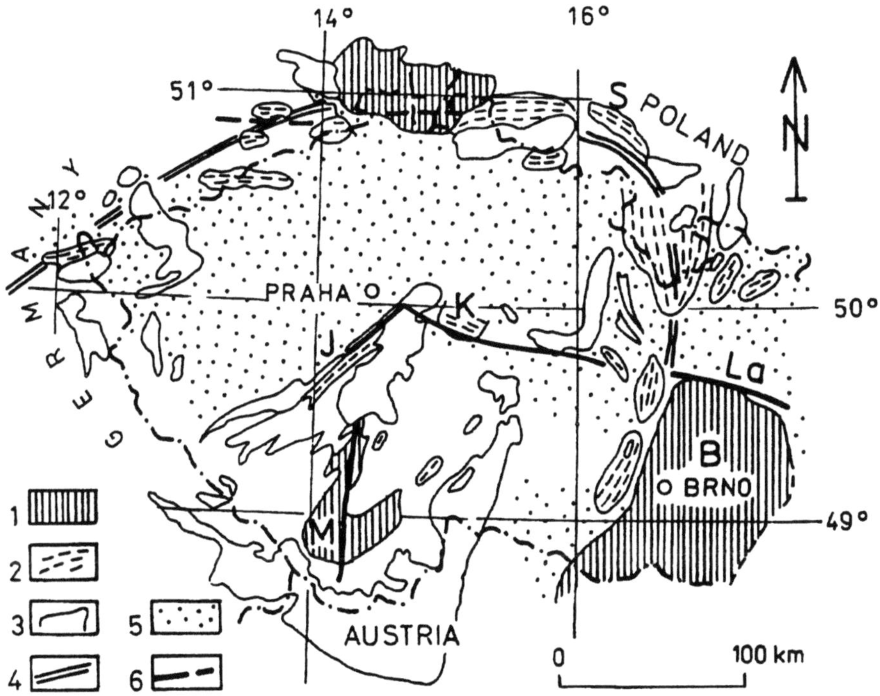

Fig. 8.14 Types of the Earth's crust in the Bohemian Massif (according to Zeman in Mahel' 1984 b). Sialic crust: 1 – cores of the Precambrian sialization with acid metavolcanites and Pre-Caledonian granitoids, 2 – appreciable bodies of orthogneisses and orthogneisses domes, original rocks being mostly of Precambrian origin, 3 – main Palaeozoic granitoids, 4 – course of the sialization ring. Simatic crust: 5 – regions of restites of the Precambrian simatic crust, 6 – main Precambrian deep faults (La – an embryo of the Labe lineament) affecting the differentiation and development of the crust. The Precambrian nucleus of the continental crust of the: B – Brunnia, M – Moldanubicum, L – Lugicum; sialization Precambrian zones: J – zone of Jílové, S – zone of the Sowie Góry Mts.

down to depths of about 60 km, where spinel and garnet peridotites (lherzolites) are identically present. According to geophysical data, the sialic and simatic crust in the Bohemian Massif differs, first of all, in the ratio of thicknesses of the upper and lower crust. The difference is contoured by striking gravity gradients (Blížkovský et al. 1986a).

In the West Carpathians, two analogous types of the crust were distinguished by Buday et al. (1969), Fusán (1985), and Čech (1982). Central outer West Carpathians exhibit a sialic type of crust, while the covered regions of the south part of the inner West Carpathians exhibit a simatic type (Fig. 8.14). From these interpretations Grecula and Roth (1978), and Roth (1980), consider the border between the two types as the boundary the crystalline unit of the Fenosarmatian platform, (which is manifested by an increase of granitic part of the crust) and of its sedimentary cover. The southern part corresponds to the Pannonian Basin with a smaller thickness of sedimentary and granitic parts of the crust, and with significant effects of the lower ("dry", basaltic) parts of the crust (Fig. 8.14), their significant part are mantle elevations. In the sialic part, the total crustal thickness is twice that in simatic part. This model could have easily been transformed in accordance with geophysical data (Ádám and Pospíšil 1984, Fig. 8.18).

From other vertical inhomogeneities of the crust in Central Europe, diapir structures and deep faults occur.

In the lithosphere, diapir structures exist at all levels. In the supercrustal part, there are e.g. salt diapirs in the Neogene of the East Slovakian Basin and of the Carpathian Foredeep, thermal domes in the crystalline regions of the Bohemian Massif (Moldanubian, Krušné hory Mts.). In deeper part of the lower crust, there are migmatitic, granitic, and probably also granulite diapirs (the diapir form of the South of the Bohemian granulites assumed already Kodym sen. 1963). It follows from the model calculations (e.g. Ádám and Pospíšil 1984; Blížkovský et al. 1985 a) that the diapir wedging-out of the lower level of the crust which has not been confirmed by deep boreholes, and which has also been explained by incorporating the lower crust and upper mantle to the overthrust tectonics in the origin of the nappe structure (Giese 1980). Diapir structures are a characteristic part of the Hercynides of the Bohemian Massif, where they have acted mainly in areas of regional migmatitization (Kumpera and Suk 1985).

Deep faults are shown by anomalous phenomena in seismic section and velocity models. Majority of them terminates at the upper crust – lower crust interface (crustal faults), or at the Moho discontinuity (deep faults). Exceptionally, they extend down to the upper mantle. On the contrary, some fault structures in the lower crust do not continue to the Earth's surface. In some cases, also boundaries of units (e.g. Rhenohercynicum and Saxothuringicum, Saxothuringicum and Moldanubicum – Weber in Giese 1985) indicated by changes in conductivity (Haack in Giese 1985), appear as analogous tilted discontinuities. Similar phenomena are known at the boundary between the outer and inner West Carpathians (Beránek et al. 1972; Praus et al. 1984).

On the contrary to the sedimentary crustal layer, in which an existence of dislocations can be reliably detected by seismic methods, their interpretation at depth is far more complicated (Beránek 1971). The possibility of Moho depth changes caused by tectonics still remains. Generally, there is no fault on the Czechoslovak territory whose depth extent below the Moho discontinuity was confirmed. Most indications exist in the contact zone of the Bohemian Massif with the West Carpathians, e.g. the aseismic zone (Beránek and Zátopek in Zátopek 1981, the Lednice zone (Weiss 1983), and the Peripienian lineament of Máška in Buday et al.(1961).

The Odra line appears the most significant lineament in the version of Mayerová et al. (1985). The line is very clearly expressed in the Moho discontinuity relief. The relationship between the Moho indications at various depths suggests the possibility that in NW as well as SE of the marginal part of the Bohemian Massif, the tectonic processes in the course of the Hercynian nappe structure origin had affected also the lower part of the crust and uppermost mantle as it was assumed by Giese (1980) and Beránek et al. (1981).

Basic rock and granulite occurrences are very often considered as being indications of lineaments of sutures and subduction zones of various ages ranging from the Precambrian to the Cenozoic.

Bard et al. (1980), for instance, associate the basic rock occurrences in West Bohemia and at the east margin of the Bohemian Massif (the Moravo-Silesian ophiolite zone) with the boundary of three separate continents in the Hercynian orogene. Zeman (1981b) characterizes the Moravian granulite belt as a sialic structure with a material transport along the deep fault "granulite Klippen Belt". Neumann (1978) characterizes the granulites of the Saxothuringicum as manifestations of a subduction zone in the European Variscides.

Basic volcanics (spilites) in the Proterozoic of the Bohemian Massif are considered by Pouba (1970), Bernard and Klomínský (1975), and Bernard et al. (1983) as indications of a significant suture analogous to the Perimoldanubian lineament of Zoubek and Vyskočil (1977). This lineament was assumed, in the entire European Variscides, to divide the Moldanubian zone since the late Proterozoic. These geologically significant interpretations, however, have not been confirmed by geophysical indications.

A clear association with the spatial distribution of neovolcanics follows from the map of the Moho discontinuity relief. The majority of occurrences of neovolcanics in the Bohemian Massif and in the West Carpathians is concentrated in zones of steep transition between deep and shallow segments of the Moho discontinuity. The vertical division of lithosphere beneath Czechoslovakia offers several possible interpretations:

The geologic model assumed, according to the law of superposition, that with increasing depth older formations occur, so that the Moho discontinuity relief reflected structural conditions at the time of the Earth's crust origin.

Beneath Czechoslovak territory the following structural levels can be distinguished (from the youngest to the older ones): Alpine-Carpathian (13 – 170 Ma), Hercynian and Caledonian (250 – 450 Ma), and Cadomian (470 – 670 Ma) which as part of the North European shield forms the basement of all the younger levels.

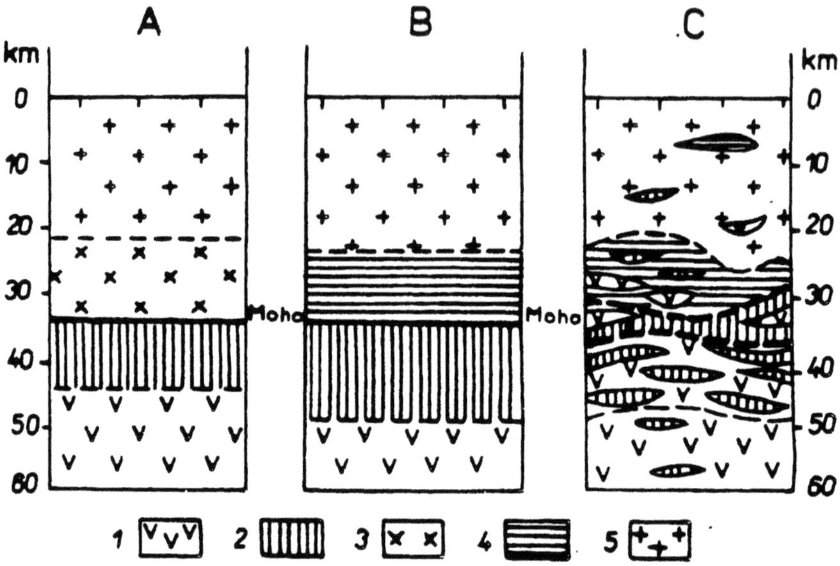

Fig. 8.15 Petrologic models of the contact of the continental type of crust with the upper mantle according to Mísař (1979 A – a model in which the Moho discontinuity corresponds to the phase change of gabbro to eclogite, B – model in which the Moho is a sharp boundary between granulites and eclogites, C – the model by Mísař for the northern margin of the Moldanubicum in which the Moho forms a 5 km layer consisting of eclogites prevailing over peridotites, which are present in lesser quantity also in the granulite level. The layer is complicated both tectonically and by intrusions of mantle material. 1 – peridotite, 2 – eclogite, 3 – gabbro, 4 – granulite, 5 – granitoids and metamorphosed rocks.

A petrological-geochemical model assumed composition changes with depth. Either the mineral composition of rocks (metamorphic facies) or the total chemical composition of rocks changes with depth.

According to Bott (1971), the upper mantle beneath the continental crust contains dunites and peridotites with blocks of eclogite, and beneath the oceanic crust it consists of ampholite (theoretically derived rocks consisting of olivine and hornblende). The deeper part of the upper mantle beneath both types of the Earth's crust is formed by pyroxenite pyrolite. Petrologic models of the Earth's crust structure in the Bohemian Massif, compiled by Mísař and his co-workers (Mísař 1979, Mísař et al. 1984a), correspond to these statements, that are also based on the data from the study of ultrabasic xenoliths in the Moldanubian granulites (Fig. 8.15).

For the deep structure of the Bohemian Massif, Mísař assumes a transition from pyroxene granulites of the lower crust over the transient layer, about 5 km thick, ("petrologic Moho" with eclogites and peridotites) into peridotites of the upper mantle.

Petrologic Data on the Upper Mantle Composition Beneath Central Europe Yield Controversial Conclusions

On the basis of composition study of eclogites, granulites, and xenoliths in granulites, Dudek (1971) and Jakeš et al. (1985) derived a hypothesis on a great differentiation of the upper mantle composition both in the horizontal and vertical directions.

Contrary to it, the study of ultrabasic xenoliths in neovolcanics of the Bohemian Massif and the West Carpathians shows (Fediuk, Chap. 8.5) that the composition of the upper mantle beneath Central Europe is homogeneous.

The geophysical model of the basic structure of lithosphere beneath the Czechoslovak territory was elaborated by Beránek et al. (1975, 1977, 1979, 1981). Mean velocities were calculated from the travel-time curves of reflected waves. The mean velocities then served to calculate the layer velocities. They bear errors following from the simplified initial hypotheses on the structure and thickness of the Earth's crust, and also from the fact that waves propagate to greater distances along their trajectories through more geologic units. From the global point of view the West Carpathians belong to the East European zone of strikingly lower velocities of *P* waves, which includes the entire lithosphere down to depths of 300 km. Contrary to it, the Bohemian Massif belongs to the zone of anomalously high velocities (Hovland and Husebye 1982).

In Czechoslovakia, several blocks have been distinguished, and each block has separate characteristics. The authors differentiate the following four parts:

1. A surface layer with velocity of up to 6.0 km s^{-1} which corresponds to the sedimentary layer
2. The depth interval with rocks of velocities of 6.0 – 6.6 km s^{-1} corresponding to granitic layer.
3. The depth interval with rocks of velocities of 6.6 – 7.6 km s^{-1} which corresponds to a basalt layer, is divided into the upper part with rocks of velocity of 6.8 km s^{-1} which could consist of high-pressure metamorphites. This interface corresponds perhaps to the Conrad discontinuity. The lower part (6.8 – 7.6 km s^{-1}) is certainly more basic in composition and extends down to the Moho discontinuity with the velocity of 8.2 km s^{-1}.
4. The boundary of the upper mantle with velocities exceeding 8.2 km s^{-1}.

A small thickness of the first (sedimentary) layer is characteristic of the Bohemian Massif. In the Palaeozoic and late Proterozoic rocks region (the Krušné hory Mts., the Bohemian Cretaceous Basin, the Krkonoše Mts.) the boundary (6.0 km s^{-1}) is at a depth of 5 km, only in the Moldanubicum it is at a depth of 17 km (Beránek et al. 1977). The unobtrusive Conrad discontinuity is at a depth of about 20 km. The thickness of the lower part of the third layer (velocities of *P* waves exceeding 6.8 km s^{-1}) changes in relation to the Moho discontinuity dept. The eastern margin of Moldanubicum is anomalous, where, in the interval of 14 – 20 km, there are rocks of high velocities overlying rocks with velocities lower than 6.6 km s^{-1} which could, according to Beránek et al. (1975), correspond to a dragging of the basalt layer in tectonic zones as is known from the Ivrea zone in the Alps (thrusting the Moldanubian block over the Moravian one ?). The change of the crustal thickness at the boundary between the Bohemian Massif and West Carpathians takes place because of the basalt layer.

The West Carpathians contain thick sedimentary complexes with low velocities at the surface and with the high vertical gradient of velocity. Masses of higher velocities are close to the surface (at a depth of 19 or 6 km in the Danube Basin). The lower part of the third layer (basaltic, velocities exceeding 6 km s^{-1}) is similar to those in the Moldanubicum, not only by its thickness, but also by the character of rocks which form it (the intensively metamorphosed blocks containing magmatites). The low-velocity channel at depths from 10 to 20 km is caused by physical changes connected with the increased temperature gradient of the region. New information, provided by reflection seismic data, added to this model in its uppermost part the discovery that in all parts of the Bohemian Massif as well as in the West Carpathians striking subhorizontal boundaries exist in the superficial ("sedimentary") layer of the crust, which correspond to:

1. thrust planes of nappes and thrust faults, e.g. in the Teplá – Barrandian Proterozoic (Cidlinský and Ibrmajer 1986), in the Lower Carboniferous of the Nízký Jeseník Mts. and at the boundary of the outer and inner West Carpathians. Analogous results have

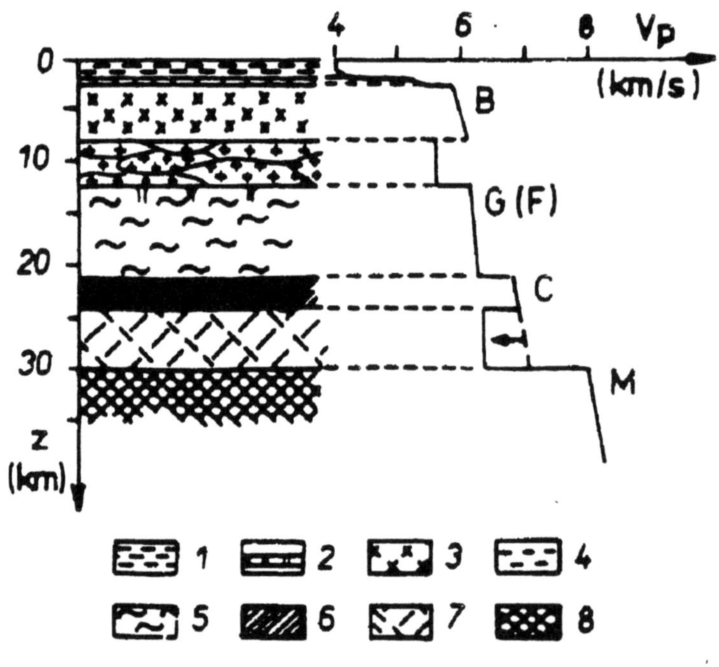

Fig. 8.16 Schematic structure of the continental crust of the European Hercynides with a petrologic interpretation by Mueller (1974), Finckh et al. (1984): 1 – platform sediments (subsurface low-velocity zone), 2 – Mesozoic and Palaeozoic sediments, 3 – crystalline basement of schists and gneisses (the positive velocity-gradient zone). 4 – zone of gradient intrusions (sialic low-velocity zone), 5 – migmatites (intermediate crustal level), 6 – amphibolites, 7 – granulites (lower crust), 8 – ultramafic rocks (the uppermost mantle).

provided an interpretation of sections Lipůvka – Tišnov (Štelcl et al. 1981). According to them, the Brunnia extends west of the Boskovice Furrow far to the west to the basement of the Moldanubicum.
2. striking stratification which could be primary or also of a tectonic origin. In the section near Mohelno (Blížkovský et al. 1986 a) (Fig. 8.17) several boundaries were detected which corresponds most likely to the stratification of the Moldanubicum and Moravicum and to their thrusting over the Brno unit (Brunovistulicum).

According to the seismic Moho discontinuity zone in Czechoslovakia, the crustal thickness ranges from 26 to 40 km. The zone of maximum depths (38 – 40 km), which extends beneath the units of the Bohemian Massif and West Carpathians (Blížkovský et al. 1985) in the W–E direction, represents the most striking phenomenon.

Investigations of rocks of a deep origin contributed to indications of the composition of lower parts of the crust. The investigations were aimed at:

1. relicts of oceanic crust, mostly incomplete and metamorphosed remnants of an ophiolite series (metabasites, eclogitic rocks, and ultrabasites),
2. complexes of ultrabasic rocks and eclogites confined to them, forming allogene bodies in the surrounding metamorphites (namely ultrabasic xenoliths in granulites of the South Bohemian and West Moravian granulite belt),
3. granulites, to which a supracrustal origin (metavolcanics and their tuffs and tuffites) is mostly attributed, are by some authors considered as tectonically transported parts of the lower crust. In the first case, the conformity of metamorphic development, rock composition (chemical composition and stage elements) and a stratiform position are all arguments, as well as, in the second case, an interpretation of the structure as deep tectonic zones around granulites bodies, and the mineral composition,
4. eclogites, in the Bohemian Massif considered both as metamorphosed palaeovolcanics (spilitized basalts) and the transported parts of the lower crust and the upper mantle.

Ernst (1965) notified the significance of xenoliths as a direct indication of the lower crust composition and of the upper mantle composition.

With regard to the structure and composition of the lower crust and of the upper mantle, ultrabasic xenoliths in basalts and granitoids and eclogites in kimberlites are discussed most frequently. Xenoliths of granulites are also discovered belonging to the rocks of deep origin.

An interpretation of the origin and nature of xenoliths is, however, far from being unique. Information derived from the interpretations can be controversial. Xenoliths in intrusive rocks can be:

1. fragments of rocks which surround the original magma chamber in the crust or upper mantle,
2. poorly meltable remnants of the upper mantle material or of the lower crust material,
3. xenoliths of rocks in upper horizons penetrated by effusive or intrusive rocks,

4. consanguineous autoliths (e.g. olivine nodules in basalts), which are the product of differentiation, and which have absolutely no relation with the upper mantle composition.

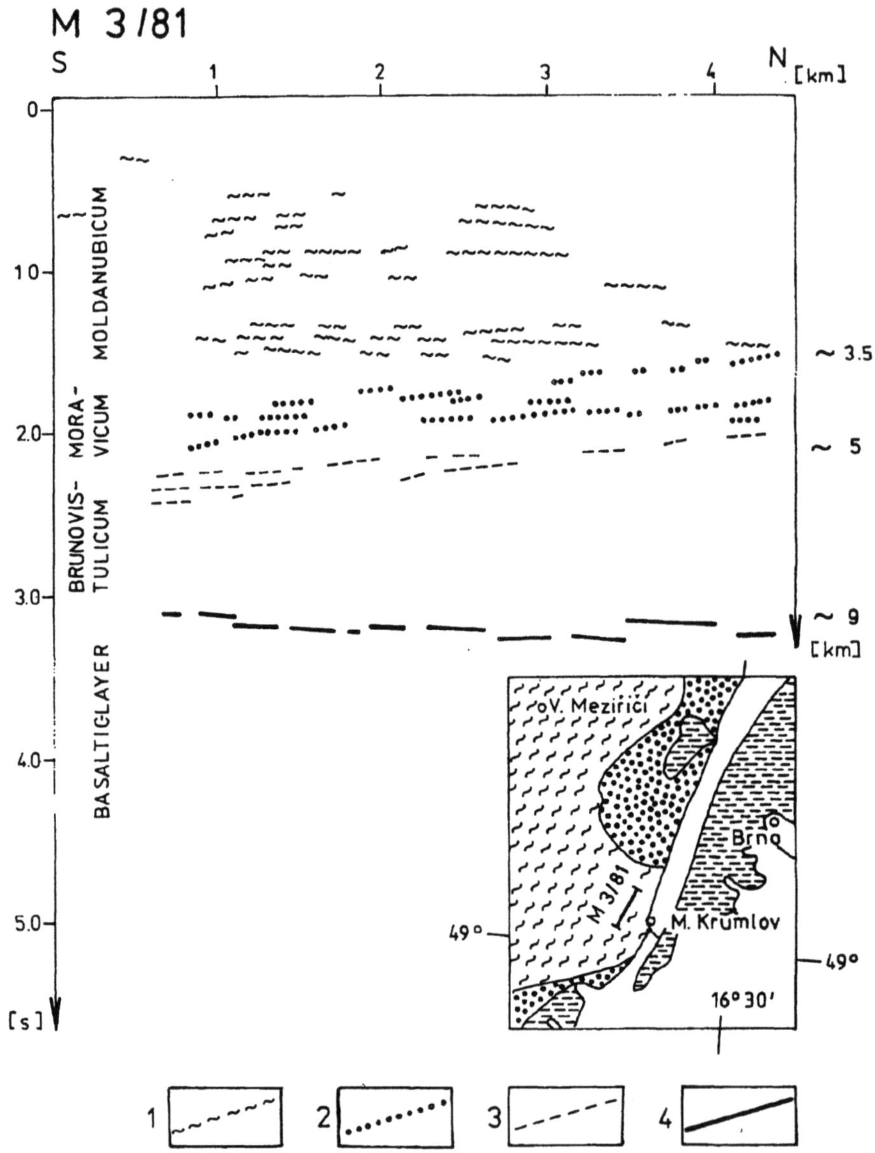

Fig. 8.17 Interpretation of the reflection seismic section M 3/81 in the Moldanubicum of the west of Moravia: 1 – reflections corresponding to the stratification of the Moldanubian granulite belt, 2 – reflections of the variegated group (Moravicum), 3 – less striking reflections corresponding to granitoid of the Brno Massif, 4 – reflections corresponding to the Moho discontinuity.

Detailed studies showed that the majority of allogene xenoliths in intrusives and effusives in the Bohemian Massif comes from the upper crust (type c) and their transport had not achieved more than tens to thousands of metres. The following xenoliths belong to them: first of all xenoliths of granites and migmatites in plutons (Krkonše Mts., Central Bohemian, and Brno) but also in basalts (the Krkonše-Piedmont Basin, Bohemian Cretaceous Basin), but also charnockites (Bohemian Cretaceous Basin) and ultrabasites (the České Středohoří Mts.). The most confirmatory indications of the deep structure are served by ultrabasic xenoliths (type a) in neovolcanics, the compositions of which, as well as the stable chemical composition independent of deep structure of individual units, and the petrographic character, correspond to the upper mantle.

The gathered data on boreholes in Czechoslovakia together with the results of investigations of the deep-origin rocks and xenoliths and with geophysical indications of the deep structure made it possible to construct geologic sections in the Bohemian Massif (Suk 1983, Suk and Vacek 1984) and in the West Carpathians (Ádám and Pospíšil 1984; Fusán et al. 1987) at depth levels of – 3 000 m, – 10 000 m, and – 15 000 m (Figs. 8.18, 8.19). All the above mentioned data show, first of all, the overall tendency to simplify the geologic structure and to homogenize geologically the deep parts of the crust.

The main property of the section – 3 000 m (Fig. 8.19a) is an entire absence of the platform cover, a significant reduction of the Hercynian units and a change of configurations of the Hercynian granitoid plutons. Among deep faults the NW–SE lines parallel to the Tornquist-Teisseyre line are the most striking ones. Contrary to them the significance of fault zones of NNE–SSW (the Blanice and Boskovice furrows, and the Přibyslav zone) is surprisingly lower. In the West Carpathians, the Miocene fill of the basins extends to depths greater than 3 km. The Klippen Belt bears the same significance as on the Earth's surface.

Section – 10 000 m (Fig. 8.19b) illustrates the boundaries of the upper part of the crust with the rocks containing volatile components and the low "dry" parts of the upper crust, in which advanced dehydrated rocks of metamorphic origin prevail. The transition zone extends down to depths of about 15 km in the Bohemian Massif and to greater depths in the West Carpathians.

In the Bohemian Massif, no Hercynian unit extends down to a depth of 10 km. The boundary between the Bohemian Massif and the Brunovistulicum becomes the most significant interface. It is strikingly shifted towards the West (Dudek 1980). Its course at depth is indicated also by the change of Wiese vectors (Praus et al. 1984).

In the Jeseníky Mts., this boundary is less expressive due to the origin of dome structures in which the rocks of the Brunovistulicum extend into the upper Hercynian structure (the Desná and Keprník domes).

Rocks of the Brno unit extend also far to the east forming the Precambrian basement of western and northern parts of the West Carpathians. It possibly extends to the line which is identical with the Slovenské rudohorie Mts. axis and with occurrences of glaucophanites and granites of the Cretaceous age. North of the line crystalline schists and granitoids prevail at depth. Here, Alpine volcanics and basic massifs, representing likely diapir elevations of the lower part of the crust, are of great extent.

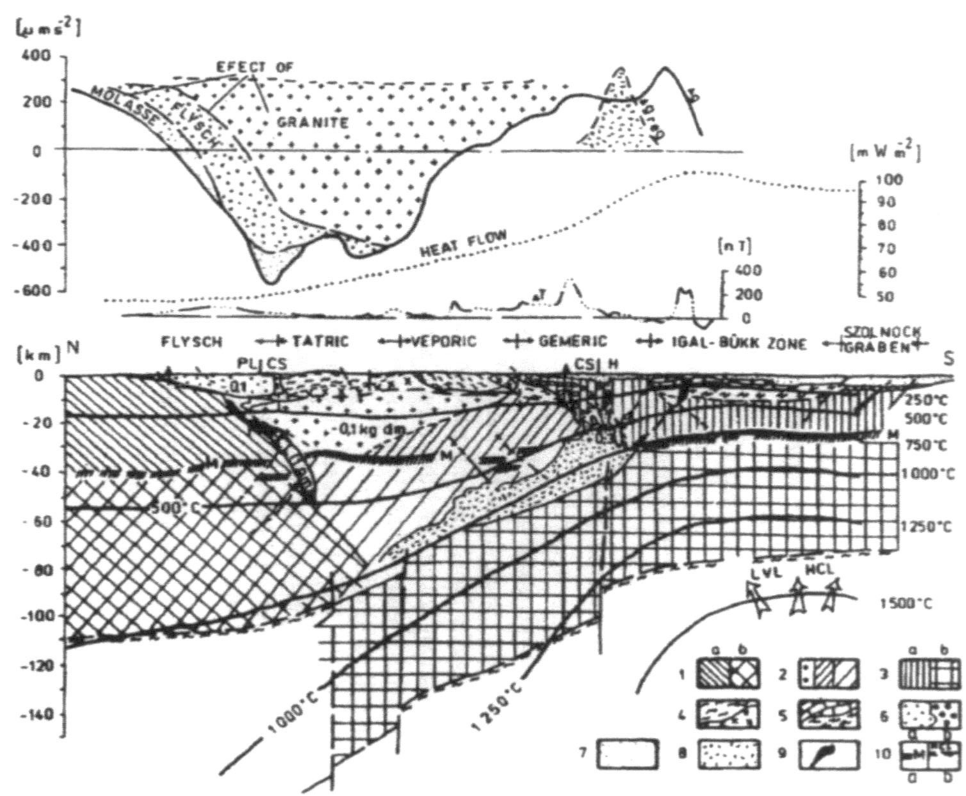

Fig. 8.18 Geophysical model of the lithosphere in the West Carpathians (Ádám and Pospíšil 1984). 1 – epi-Variscan platform, 2 – "island-arc type" of the lithosphere with granites, 3 – lithosphere of the Pannonian Basin (a – crust), 4 – Tatricum, Veporicum, and sub-Tatric nappes, 5 – Igal – Bükk zone, 6 – Flysch and the Klippen Belt, 7 – molasse, 8 – the zone of intermediary volcanism mobilization, 9 – basites actually ultrabasites, 10 – geophysical boundary (a – Moho discontinuity, b – lithosphere–asthenosphere boundary).

Section – 15 000 m (Fig. 8.19c) shows the boundary of the upper and lower crust. The upper part of the crust is composed mainly of anatectic migmatites with spinel (the Wernstein type and the Boubín type), which outcrop in diapir elevations of the southern margin of the Moldanubicum, and perhaps also of granulite rocks, which are in a similar position. In the West Carpathians, the margin of the Podolian platform is active. South of it in the highly metamorphosed crystalline schists and granitoids, elevations of the lower crust occur in larger quantity, as in the Bohemian Massif. These models were verified by continuing the gravity field to the appropriate level (Pick in Blížkovský et al. 1986 a).

Interesting results have been obtained from construction of the three-layer model of the Earth's crust structure in the Bohemian Massif (and the West Carpathians) corresponding to the theoretical one by Weber (1986).

The lower layer (lower crust) with prevailing basic rocks is built of anatectic migmatites of quartz-diorite composition and of the other dry associations of granulite facies. The presence of leucocratic granulites, frequently occurring in the brittle crust (leptynites) is often assumed but is unlikely. Consequences of horizontal movements are not so important contrary to those of vertical mass movements, and horizontal or subhorizontal boundaries of different lithology occur. The undulations of the Moho coincide with the lower boundary of the layer, showing the crustal thickness in Central Europe in general. There is a remarkable agreement between the morphology of the Moho in the central part of European Hercynides (Massif Central France, Vosges, Schwarzwald, the Bohemian Massif) and the morphology of the Moho in the Ukrainian Massif (Sollogub et al. 1980). This fact is in accordance with our concept of the Earth's crust structure in Central Europe.

The middle crustal layer in Central Europe is composed of a consolidated crust with dominating Precambrian Middle to Upper Proterozoic units. The principle rocks are dry granitoids and migmatites with mineral associations of a higher amphibolite or lower granulite facies (characteristic features are the stability of cordierite with K-feldspar and substitution of spinel for cordierite). The extent of this layer was investigated by refraction and reflection seismic measurements. They were studied on velocity models (see Chapter 2), on gravity and magnetic models (Chapters 6 and 7), and in places of occurrences of migmatite diapirs (e.g. in South Bohemian Moldanubicum), or tectonically stretched-out rocks (e.g. in the Pienniny Klippen belt separating the Outer and Inner Carpathians) and of xenoliths in granitoids and basalts. In the middle layer, most dislocations fade out and dome structures appear. The alternating positive and negative gravitational field probably reflect the layer thickness undulation, compensating the undulation of the Moho.

In some parts of the Bohemian Massif, the middle layer reaches up to the surface. It most probably includes the migmatite diapirs and other domal structures in Moldanubicum, the Brno unit and the Lužice Massif whose stratigraphy and composition could be correlated with the crystaline units of the Ukrainian Massif and the Scandinavian shield.

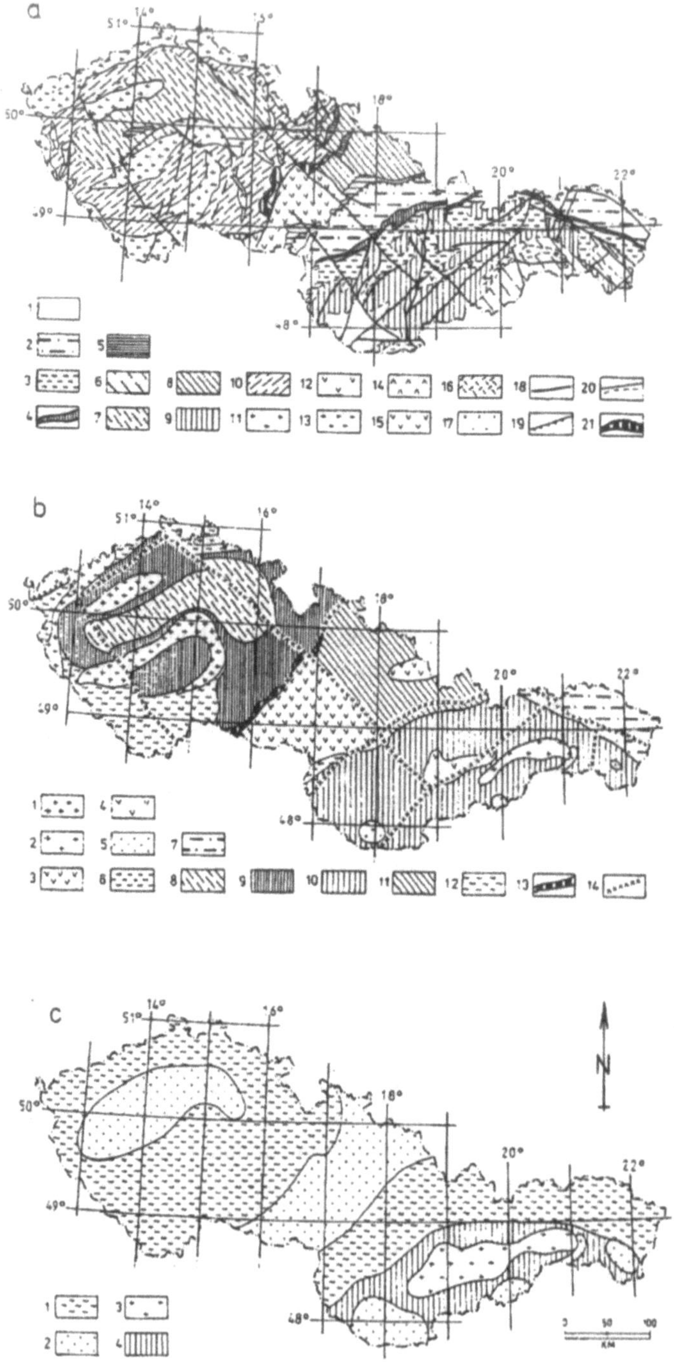
a
51°
14°
50°
18°
20°
22°
49°
48°
b
51°
14°
16°
50°
18°
20°
22°
49°
48°
c
51°
14°
16°
50°
18°
20°
22°
49°
48°
N
0 50 100
KM

Fig. 8.19

According to our data, the presence of Hercynian units (Saxothuringicum, Rhenohercynicum) at the site of borehole KTB (Bohemian Massif, Bavaria) assumed by Behr (1983, 1987) is unlikely and the shield-type lithology will be struck deeper in the borehole. The boundary of the shield-type layers is more distinct in the Western Carpathians where it is indicated by a gravity gradient delimiting a negative field. On reflection seismic profiles and on the surface it is indicated by occurrences of Alpine granites and by Tertiary neovolcanics mainly of andesitic composition. Therefore it is advisable to verify the real range of shields under younger orogenes and sediments over all continents. Then such data would enable us to understand the deep structure of the continental crust. The upper mantle layer (brittle crust) in the Bohemian Massif is built by non-metamorphic geological units of platform cover (Permian and younger in age), by folded and low- to medium- grade metamorphic (amphibolite facies) crystalline complexes, with rock indicating origin in the mature continental crust, wedgings of mobilized rocks of the deeper part of crust, and probably of the upper mantle (basaltic effusions), too. The tectonic structure of the brittle layer was affected by vertical, and mainly horizontal movements (folds, nappes, overthrusts, shear ductile zone) and by a polycyclic development since the Upper Proterozoic in Cadomian, Hercynian, and Alpine orogenies. The tectonic structure shows a notable reduction of space in the top parts of the Earth's crust as a consequence of the principal process of the geological development in Central Europe, the secular collision of Gondwana and Laurasia. The thickness of the brittle crust in the Bohemian Massif ranges mostly from 12 to 15 km.

Fig. 8.19 (left)

a) Horizontal section at a depth of –3 km. 1 – Miocene basins, 2 – Flysch of the West-Carpathian Foredeep, 3 – Inner Carpathian Palaeogene, 4 – Klippen Belt, 5 – Late Palaeozoic, 6 – crystalline units of low and medium grades of metamorphosis, 7 – Precambrian rocks of low and medium grade of metamorphosis, 8 – gneisses and migmatites of the Brunovistulicum, 9 – metamorphosed Palaeozoic of the West Carpathians, 10 – Precambrian rocks of high grades of metamorphosis, 11 – granitic rocks of Cretaceous age in the West Carpathians, 12 – basic rocks of the Cretaceous age, 13 – Hercynian magmatites, 14 – durbachites, 15 – granitoids of the Brunovistulicum, mainly of Precambrian age, 16 – Precambrian basic eruptive rocks, 17 – basic eruptive rocks in deep structure of the West Carpathians, 18 – major faults, 19 – the front of Carpathian nappes, 20 – geologic and petrographic contacts of rocks, 21 – the Moravo-Silesian zone.

b) Horizontal section at a depth of –10 km. 1 – Hercynian granitoid rocks, 2 – granitoids of the Cretaceous age, 3 – Precambrian intrusives of the Brunovistulicum, 4 – Cretaceous basic eruptive rocks in the West Carpathians, 5 – basic rocks in deep structure of the West Carpathians, 6 – Miocene basins, 7 – Flysch in the West-Carpathian foredeep, 8 – Precambrian rocks of medium to high metamorphic grades, 9 – gneisses and migmatites, 10 – Palaeozoic and Precambrian rocks of medium to high metamorphic grades, 11 – Precambrian gneisses and migmatites of the Brunovistulicum, 12 – anatectites of high densities, 13 – Moravo-Silesian zone, 14 – deep faults (lineaments).

c) Horizontal section at a depth of –15 km. 1 – anatectites of high densities, 2 – basaltic layer rocks, 3 – granitic rocks of the Cretaceous age, 4 – Precambrian and Palaeozoic rocks of high metamorphic grades.

The structure of the upper part of the Earth's crust appears to be far less chaotic than is often presumed. Regularities derived from physical and chemical properties of rocks and from geological processes must be considered. The crustal structure appears to simplify with the depth. Beneath the depth level of about 15 km in the Bohemian Massif and 20 km in the West Carpathians horizontal boundaries prevail as the composition of rocks gradually becomes more uniform.

9 Dynamic Manifestations of the Lithosphere

9.1 Recent Movements of the Earth's Surface

P. Vyskočil[1]

In the mid 1970s and at the beginning of 1980s repeated levelling measurements had been carried out in the Bohemian Massif and at its contact with the West Carpathians. The measurements were partly tied to the international system. In the study area, more detailed measurements were then carried out. Results of the measurements in the international system were used to construct a new map of vertical movements of the Earth's surface in Central European and East European countries, which was edited on a scale 1:2 500 000 in 1986 (Meschersky and Kusnetsov 1986). The fundamental data were also used for constructing the map of vertical crustal movements of the Carpathian–Balkan region, compiled on a scale 1:1 000 000 in Hungary and edited in 1985 (Joó et al. 1985, 1987). Results obtained from the repeated levelling, supplemented by results of measurements of second-order levelling lines on Czechoslovak territory, became basis for a complete adjustment of annual velocities in vertical direction and for a compilation of synoptic map of vertical crustal movements related to the reference point Želešice in the Bohemian Massif, south of Brno (Vanko et al. 1988). Results of this adjustment also became a basis for compiling the detailed map of the presently studied region on the original scale 1:200 000. The regional drawing was gradually reduced to operating scales 1:500 000 and 1:1 000 000. In reduced scale and simplified contours is the map edited in this work, see Fig. 9.1. The map on a scale 1:500 000 has recently been digitized in a square network 5 × 5 km, in order to serve in further detailed analyses.

Following the aim to estimate the main tendencies of horizontal movements in the territory under study, the original Austrian – Hungarian triangulation initiated in 1863 was analysed. Then the results of triangulation carried out from 1920 to 1945 were used as repeated measurements. Both the triangulation measurements were independently adjusted and the results of the last triangulation measurements were transformed to the original Austrian – Hungarian coordinate system.

---

1 Research Institute of Geodesy, Topography and Cartography, 250 66 Zdiby 98, Czech Republic

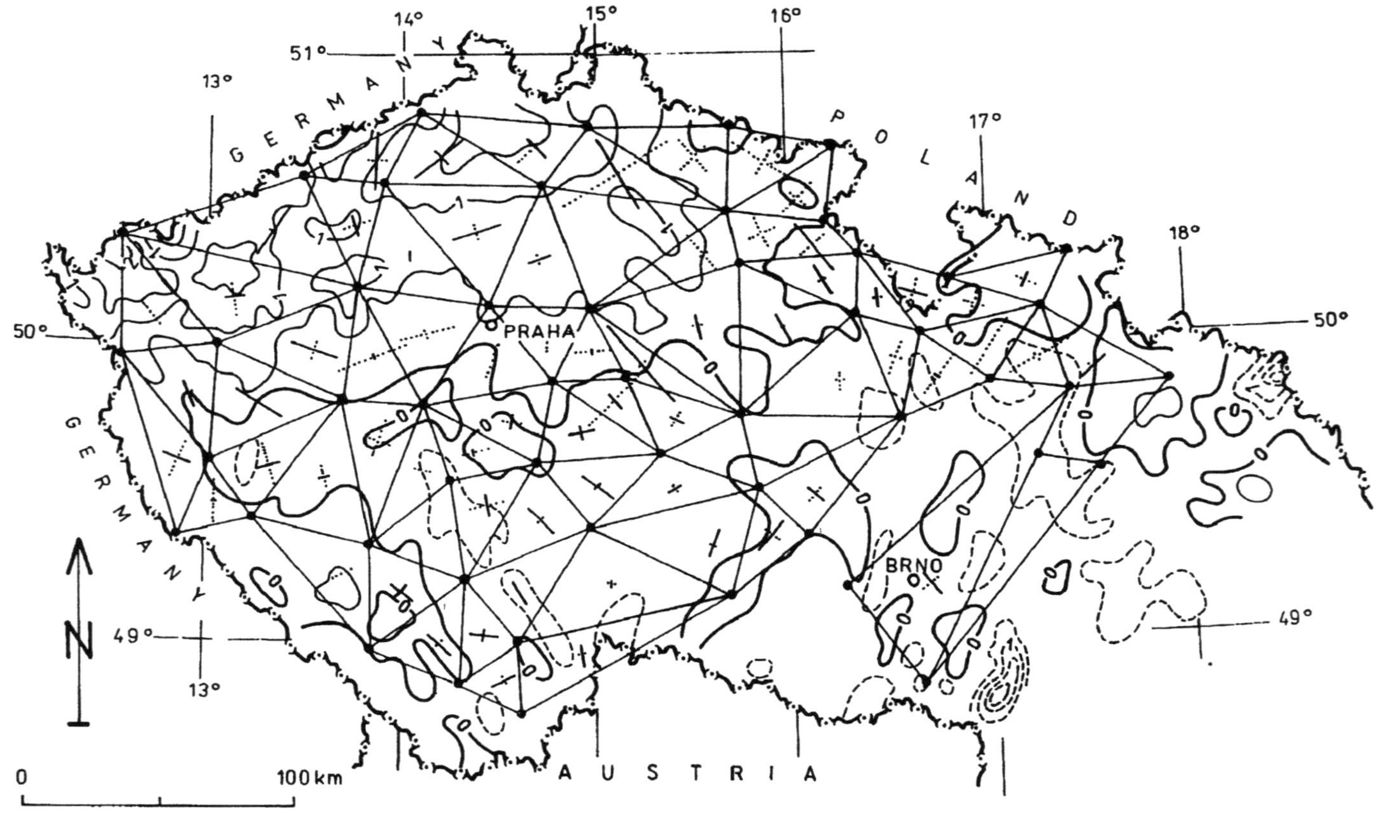

Fig. 9.1 The map of vertical movements at the territory of the Czech Republic in reduced scale and simplified contours in the interval 0.5 mm/year uplifts by full lines, subsidences by dashed lines. In addition, information on horizontal deformation are given within triangles by main axes e_1 , e_2 (Vyskočil P., 1984) of extension (full lines) or compression (dashed lines), in *u* strain.

Results of this data processing, given by the difference in coordinates at individual identical points and transformed to annual velocity of crustal movement in mm/year, became a basis for deriving the parameters of the deformation field in individual triangles of the network used (Vyskočil P. 1984). Results of these computations are plotted in the map of vertical crustal movements (Fig. 9.1). Contrary to the data on the vertical component of crustal movement defined by the value of annual velocity of movement, the data on its horizontal component are characterized by principal axes of the field of deformation in the expansion or compression regime. Such an expression is advantageous as it presents the data on horizontal deformations independently of the reference frame of coordinates. It is necessary to add, however, that the characteristics of vertical and horizontal crustal movements or deformations do not coincide in time. The data on vertical component of the crustal movement refer to the mean time interval 1950–1979, and the data on horizontal deformations refer to the mean time interval 1870–1930. As the original triangulation in Moravia is inaccurate, the region is covered by the basic rectangle which runs across the most interesting parts of the Carpathian Foredeep only partly.

Main Dynamic Tendencies

In order to contemplate the following discussions of the basic properties of the study area dynamics a schematic map of main geographic and tectonic units and dislocation lines is shown in Fig. 9.2. As far as the vertical crustal movement component is concerned, the relative zero contour divides the relatively subsiding regions of the core of the Bohemian Massif from the differentially uplifting regions of the Šumava, Český les, Krušné hory, and Krkonoše Mts. Maximum uplifts are found along the NW Bohemian–German frontier (SE of the Krušné hory fault), which reach, in a relative sense, values of up to 2 mm per year. This region is related to the belt of neovolcanics. Here, the uplifts are analogous to those in the same tectonic structures in Slovakia.

For these reasons the relative uplifts of neovolcanic regions can be considered a general feature of the tectonics of Czechoslovak territory. Intensive uplifts also take place in the Cheb area in NW Bohemia, where the registered uplifts are being associated with pre-seismic deformations which precede the earthquake swarm which occurred in this region in winter 1985/86 (Vyskočil P. 1987). A considerable differentiation of vertical crustal movements reflects also in the relative tilt of the SW part of the Krušné hory Mts. to their NW part, where the maximum changes exceeding 2 mm/year occur over a relatively short distance of about 60 km. Similar differentiation has also been observed in the Krkonoše Mts., where NW part is uplifting strikingly and the SE part subsiding. Contrary to the slightly uplifting region of the Český les Mts., the Šumava Mts. is characterized rather by subsidences which, towards the north, are associated with considerably different tendencies of vertical movements in the area of cretaceous basins in Southern Bohemia (not identified in Fig. 9.2). The core of the Bohemian Massif, including the Bohemian Cretaceous Basin are rather complicated and their discussion exceeds the aim of present work. The striking gradient of the movement along the Železné hory Mts. shows the recent mobility of this region.

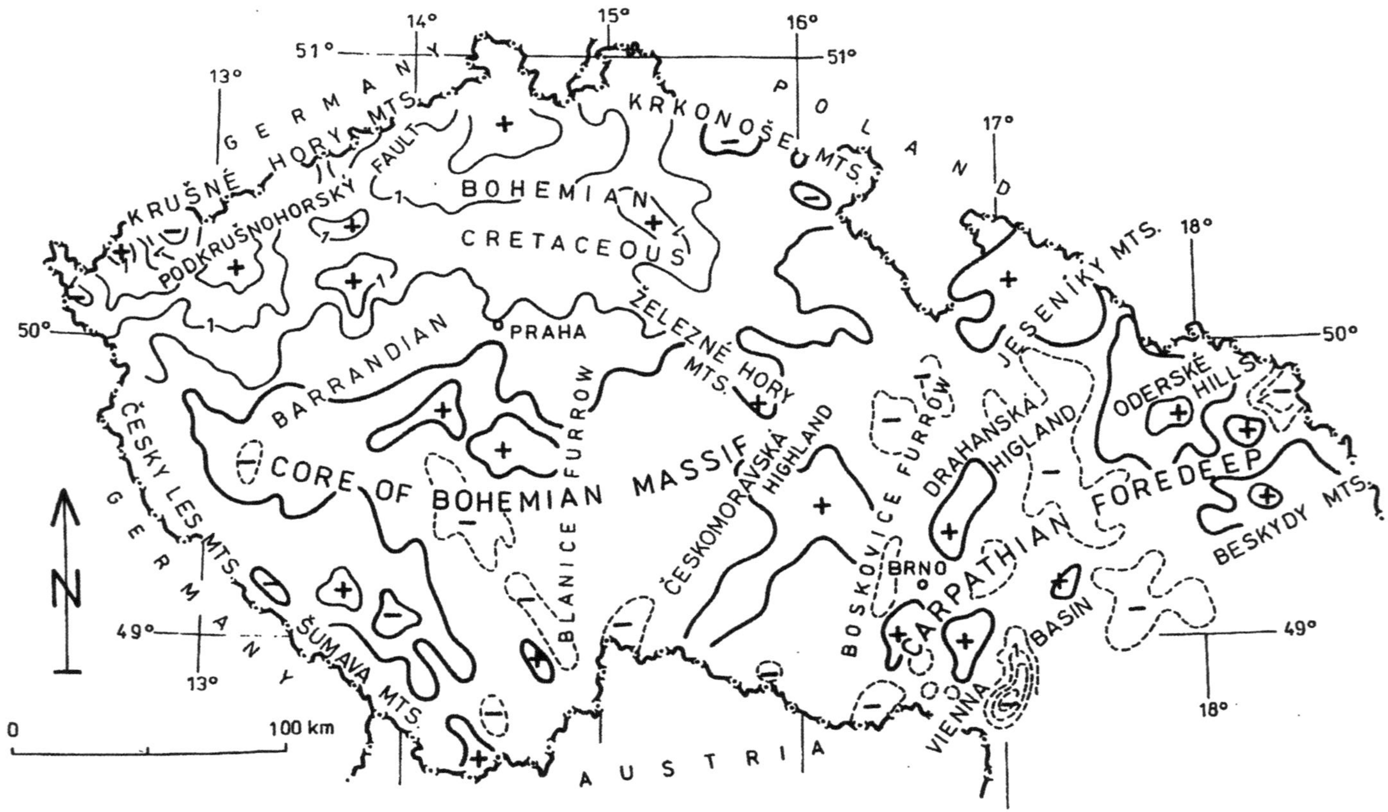

Fig. 9.2 The sketch map of the main geographical and tectonic units at the territory of the Czech Republic.

In a continuation towards Moravia, in the southern and central parts of the Českomoravská vysočina upland, as well as the Jeseníky Mts., Oderské vrchy Mts., and Drahanská vrchovina upland, uplifts take place. In the West Carpathians only the Beskydy Mts. show an uplifting tendency. The Carpathian Foredeep and north part of the Vienna Basin show a complex mobility. In the Vienna Basin striking subsidences occur which are partly due to oil, gas, and coal exploitation. The subsidences at the NE margin of the Carpathian Foredeep (the Ostrava–Karviná coal district – the Czechoslovak part of the Upper Silesian Basin) are similar in character. It cannot be excluded, however, that subsidences in both the above mentioned localities have a mixed character, i.e. that their development is conditioned by both the mining activity but also tectonic movements at the contact of the Bohemian Massif with the West Carpathians.

Comparing the map of vertical crustal movement of the same region, compiled by P. Vyskočil and Kopecký (1973) in 1971, it is necessary to stress that the basic tendencies, mentioned in the above text, are generally preserved. Only in the Cheb area, is there a striking difference which has been explained by seismic activity of the area. The main contribution of the map is in its greater detail. Here, partial elements of movement are accentuated or expressed in greater detail, as compared to the map edited in 1971.

The basic course of contours of annual velocities of vertical crustal movements is studied from the point of view of basic tectonic structure of the area under study, orientations of both the compared features exhibit a striking accordance. They are namely the Krušné hory fault, margins of the Bohemian Cretaceous Basin, but also the Blanice furrow, and other partial faults which could not have been plotted in Fig. 9.2. The SW margin of the Železné hory Mts. (the Železné hory fault) is also marked by contours as well as the Jihlava dislocation which follows the NW margin of the Českomoravská vysočina upland. The Boskovice furrow is also a striking feature as well as the Carpathian Foredeep, despite being very complex. These mutual relations between the recorded movement tendencies and the structure of the study area follow from a comparison with some geophysical indication maps in the Bohemian Massif (Blížkovský et al. 1984).

A more detailed discussion, however, must be supported by detailed analyses of geodetic, geophysical, and geologic data, which represents the main way of utilizing the present map of vertical crustal movements.

In the map of vertical crustal movements in Fig. 9.1, there are plotted the data on characteristics of the field deformation in a horizontal direction, derived for individual triangles of the original trigonometric network. Compared with the data on the vertical component of the crustal movement, this information is far more schematic for the network of identical points of the first and repeated measurements is not very dense. Beside this, the analysis of the field of deformation is based on the assumption that the medium, in which the basic characteristics, i.e. principal axes of compression or extension, are being determined, is homogeneous. This assumption, in case of larger triangles, is not satisfied a priori. Then the elements of the field of deformation need to be understood as a certain mean value, valid for the given triangle for which they were derived. Despite these primary reservations, the basic orientation of axes of compression is observable in SW Bohemia in the direction approximately perpendicular to margins of the Alpine arc which lies outside the study area. At the northern margin of the core of Bohemian Massif, these tendencies are more differentiated, and in central parts of the Bohemian Cretaceous Basin, the principal axes of deformation are relatively small. Compression increase in NE–SW direction takes place also at NE margin of the Bohemian Cretaceous Basin, and it is especially striking in the area of the Hronov–Poříčí seismo-active dislocation.

From the centre of the Bohemian Massif towards the NW, the orientation of principal axes of the field of deformation is substantially changed characterizing obviously different deformation relations across the Krušné hory fault zone. In the west foothills of the Českomoravská vysočina upland, the horizontal deformations are again smaller, while the data from the actual centre of the region are absent. Partial data from the contact area of the Bohemian Massif with the West Carpathians suggest original information on the horizontal dynamics of this area. The present as well as other analyses of horizontal repeated measurements (e.g. Vyskočil P. et al. 1987) suggest that in the south part of the foredeep the expansion tendencies of both the geologic systems of Czechoslovakia prevail, while in the north part, compression takes place. These tendencies are obviously associated with the wide range of subsidences of South Moravia, and with the striking uplift of the Beskydy Mts., and of the opposing Oderské vrchy Mts., in the north.

In the concluding discussion of the presented data on vertical and horizontal components of crustal movement, it is necessary to stress the mutual connection between the movements and structural properties of the Bohemian Massif and its margin and contact with the West Carpathians. Additional data on elements of the field of deformation in horizontal directions points at the prevailing effect of pressure forces of the alpine arc on the dynamics of the Bohemian Massif which is marked mainly by compressional deformations in NE direction. As orientations of the principal axes of deformation in horizontal direction as well as the orientation of contours of horizontal movements suggest the forces of this character are active in central zone of the Bohemian Massif in NE direction. At the NW and SE margins of this zone, a distribution of horizontal pressures takes place into directions perpendicular to the original direction of their activity. Such characteristics of the dynamics have obviously been given by a block structure of the Bohemian Massif, in accordance with the main structural elements of this geologic unit. Orientations of contours of vertical crustal movement together with partial information on horizontal deformations at the contact of the Bohemian Massif with the West Carpathians, both designate the significance of this geologic boundary as well as the dynamic properties of both the units. In the south part of the contact of these two geologic units, an extension takes place accompanied by subsidences.

Partial uplifts of the Ždánický les Mts. and the Pavlovské vrchy Hills suggest only minor block structure of this part of the foredeep. In the north part of the contact, the subsidence area narrows as the margins of the West Carpathians (Beskydy Mts.) and of the Bohemian Massif (Oderské vrchy Mts.) are uplifted, which is due to horizontal

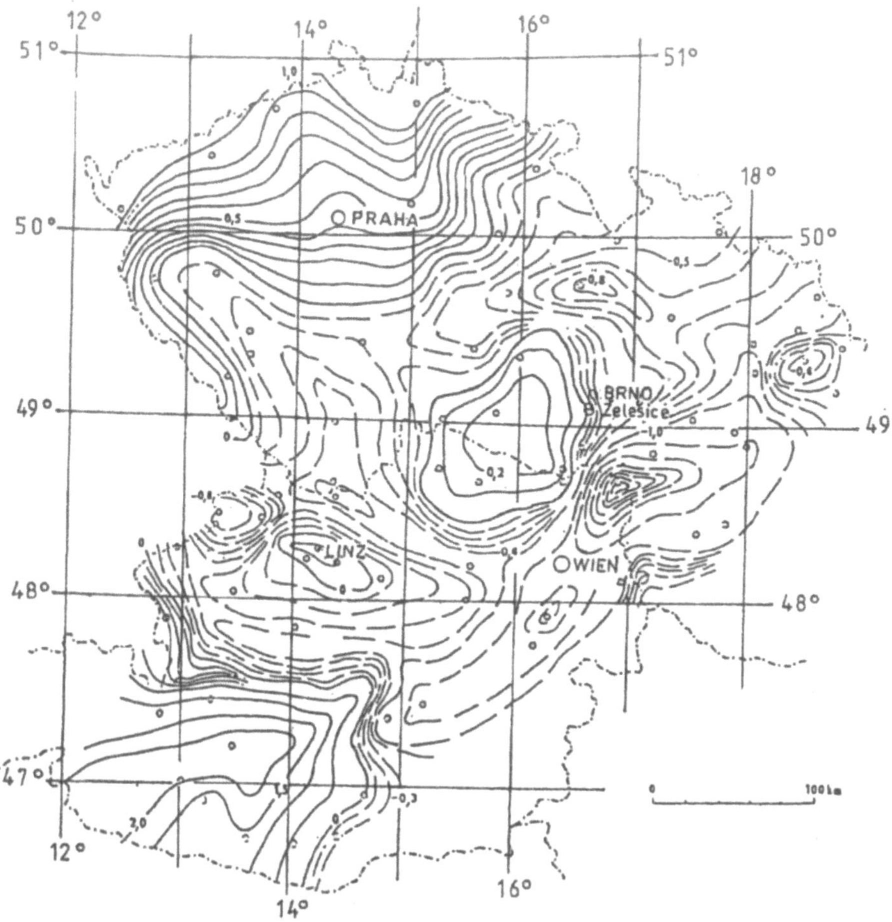

Fig. 9.3 The map of vertical movements on a part of Austrian and the Czech Republic, compiled by Höggerl N. (Austria) and Vyskočil P. (Czechoslovakia). Contours in mm/year – uplift by full lines, subsidences by dashed lines.

compression of these two fundamental units. The main compression forces of the Carpathian arc probably act on Slovakia, in a northward direction. From this point of view, the mutual dynamic relation of the Bohemian Massif and the West Carpathians can be controlled by such a movement of the margin of the West Carpathians along the margin of the Bohemian Massif, under a partial compression in the northern part and an extension in the southern part. The actual dynamics of the core of Bohemian Massif is then prevailingly determined by horizontal movements of the Alps towards NE.

The results given above and conclusions of their analysis can be completed by results achieved in 1990. The first one is the map of vertical movements in the Austrian and Czech republics, shown in Fig. 9.3. The map was compiled after joint adjustment of

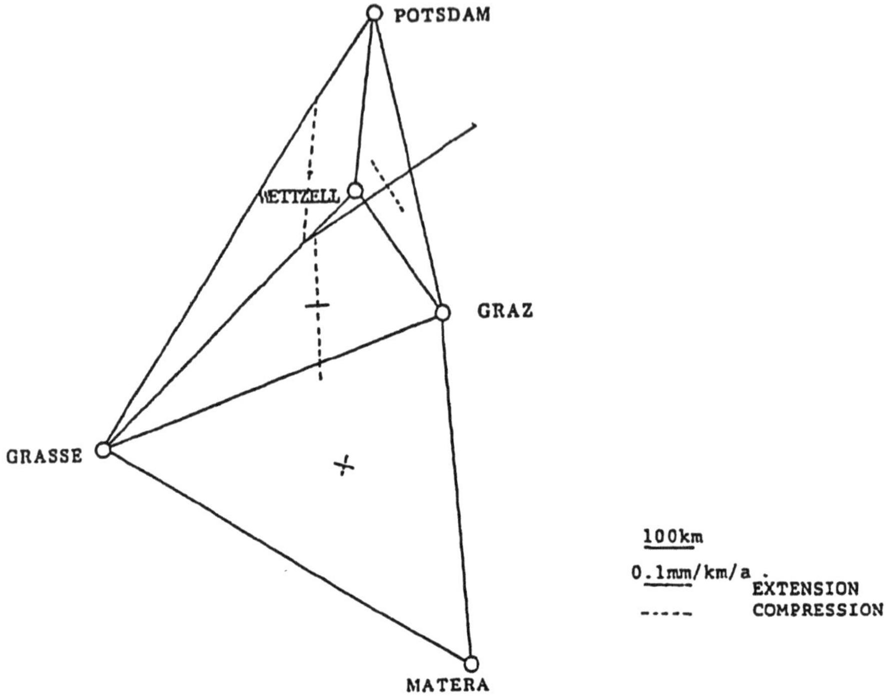

Fig. 9.4 Main axes of horizontal deformations within triangles among some of European observatories of Satellite laser measurements. Extensions given by full lines, compressions by dashed lines.

repeated levelling networks at the territories of both countries at the original scale 1:2 000 000. The contour interval 0.1 mm/year reflects the regional tendencies of vertical movements and demonstrates uplift in the NW part of the Bohemian Massif and part of the Alps. The properties of movements could be discussed in more detail but here it should be emphasized that in the western part the contours follow the "Alpine" direction, i.e. parallel to the Alpine Mts. ridge, and in the eastern part the "Carpathians" direction. The most impressive line is of subsidences towards SW and NE of Vienna. This line coincides with so called Peripieniny lineament with striking gravity as well as geoelectric anomalies.

The second result is based on the analysis of repeated satellite laser measurements from the observatories given in Fig. 9.4. The observatories Grasse, Graz and Matera (southern Italy) are in the orogenic area of the Alps and Apennine Mts., the Wetzell is in SW part of the Bohemian Massif and Potsdam is on the European platform. The main axes of horizontal deformations within triangles are given the evidence on compression as the probable pressure of the Alps towards the Bohemian Massif.

The results presented here represent only a preliminary analysis of the rich surveying material which needs to be analysed further in detail, with the use of all available geophysical and geologic data. A work of this type is relatively time consuming; it will be done in the near future. Results of such analyses will then be published. From the point of view of more detailed understanding of the mechanism of the mutual movement of the Bohemian Massif and the West Carpathians, it will be contributing to the former and present positional measurements along the contact area of both the main geologic units in Czechoslovakia. The recent results have already suggested that the relation of the Bohemian Massif to the West Carpathians is more positive, and that its dynamics has certainly been affected by the dynamic properties of the Alps.

9.2 Isostatic State of Czechoslovak Territory

V. Vyskočil[1], M. Burda[1], J. Trešl[1]

A study of isostatic state of a particular region includes, first of all, an evaluation of the method used for compensating topographic masses (i.e. masses confined to irregularities of the Earth's surface), density inhomogeneities in lithosphere, the isostatic equilibrium level, and a rate of isostatic movements of lithosphere. Isostasy is discussed in nearly all works which deal with the relation of gravity anomalies to the structure and vertical movements of the Earth's crust. The isostatic state of Czechoslovak territory is discussed e.g. in V. Vyskočil (1971, 1977), that in the West Carpathians in Popelář (1968), Kvitkovič et al. (1967), and that of the east part of the Bohemian Massif in Blížkovský (1973). Maps of isostatic anomalies were used together with mutual correlation functions between gravity anomalies, elevations above a sea level, depths to the Mohorovičic discontinuity, and neotectonic movements.

For studying isostasy more generally it is necessary to take into account, beside the distribution of masses, other sources of stress in the lithosphere.

Mutual Correlation of Gravity Anomalies, Elevations Above Sea Level, and Depths to the Moho Discontinuity

An investigation of the correlation dependence between elevations h above sea level and free-air anomalies (Faye anomalies) Δg_F, Bouguer anomalies Δg_B, and isostatic anomalies Δg_{is} will serve to get a firm idea of how the topographic relief masses are isostatically compensated. In Czechoslovakia, isostatic anomalies are calculated mainly with the use of the Airy–Heiskanen system with depths of compensation $T_0 = 20$ km, 30 km, 40 km, and 60 km.

Values of correlation and regression coefficients, expressing relations between the above-mentioned parameters that were calculated by various authors for the Bohemian Massif, West Carpathians, East Alps, the territory of Czechoslovakia, and that of Central Europe are presented in Vyskočil V. et al. (1985). The coefficients are generally in a good agreement with the assumption of isostasy. The greatest positive values of

[1] Geophysical Institute, Boční II/1401, 141 31 Praha 4, Czech Republic

correlation coefficient r were detected for free-air anomalies Δg_F that essentially correspond to isostatic anomalies Δg_{is} for $T_0 = 0$, and the most expressive negative values of correlation coefficient were found for Bouguer anomalies Δg_B that can be considered as being isostatic anomalies for $T_0 = \infty$. In all regions, a gradual transition of positive values of correlation coefficients to negative ones with increasing depth T_0 is observed. The following depths of isostatic compensation T_0 correspond to value of $r = 0$: $T_0 \approx 21$ km in the West Carpathians, $T_0 = 24$ km in the East Alps, and $T_0 > 40$ km in the Bohemian Massif.

A more detailed investigation of variability of the correlation and regression dependences in various regions is enabled by the moving window method which was applied for the Czechoslovak territory (Vyskočil V. et al. 1985). The choice of window size affects the calculated values of correlation coefficients and other characteristics. As their dimensions increase, local variabilities are smoothed out and hidden.

The most marked correlation between Δg_B and h on Czechoslovak territory (i.e. from $r = -0.7$ to $r = -0.9$) was obtained for Central and East Slovakia. In the Vysoké Tatry Mts. in the north of Slovakia, values of correlation coefficients vary in the range from -0.3 to -0.1. Positive values of correlation coefficients $r < 0.4$ are obtained only in SW Slovakia and in South Moravia. In the Bohemian Massif, the correlation coefficients values r vary from negative values ($r \approx -0.6$) in the west, north, and central part of Bohemia to prevailingly positive ones ($r \approx 0.7$) in Moravia, while in the south and east part of Bohemia, they fluctuate in the range from -0.4 to 0.4. In case of correlation between Δg_B and h, the positive values of r are in contradiction with the assumption of isostatic compensation of topographic masses. Using greater dimensions of the convenient window (Vyskočil V. et al. 1985), the variability of correlation coefficients values r will be greatly reduced and, on Czechoslovak territory, positive values of r will nearly disappear.

According to the Airy hypothesis, the compensation of topographic masses is being accomplished by corresponding undulations of the crust–mantle interface which has been taken as identical with the Moho discontinuity. The compensation testing enables us to study the correlation between Δg, h, and depths H to the Moho discontinuity. Such testing has been accomplished for Czechoslovakia and for the adjacent regions (Vyskočil V. et al. 1983; Vyskočil V. et al. 1985).

The results obtained attest to a very good correlation of h and Δg_B with depths H to the Moho discontinuity in the East Alps. Contrary to it, in the Bohemian Massif, the correlation dependences between Δg_B and H ($r = -0.15$) are very inexpressive, which also follows from a relatively great variability of values r for individual windows.

The most detailed correlation analysis was accomplished in the West Carpathians and in the adjacent regions, where beside anomalies Δg_B also Bouguer anomalies Δg_{BC}, corrected for the effects of surficial density inhomogeneities, were used. As in the east part of the study area a transition zone between the Earth's crust and upper mantle was seismically indicated, depths H and H_0 of its lower and upper boundary, respectively, were considered. Correlation dependences between h and H ($r = 0.46$), or h and H_0

($r = 0.61$) are worse than in the East Alps ($r = 0.88$), while those between H and Δg_B ($r = -0.70$), H_0 and Δg_B ($r = -0.83$), H and Δg_{BC} ($r = -0.73$), H_0 and Δg_{BC} ($r = -0.86$) are generally of the same order of magnitude as those between H and Δg_B ($r = -0.76$) in the East Alps. The dependence between the Bouguer anomalies and the crust–mantle boundary is actually accomplished by means of gravity effect ΔG_H of anomalous masses confined to the boundary. In the West Carpathians (Vyskočil V. et al. 1983), better correlation between Δg_B and ΔG_H ($r = 0.79$) was obtained than between Δg_B and H, as was actually expected. When evaluating the existing results of correlation analyses it is necessary to pay attention to the fact that the data on depths to the crust–mantle boundary have continuously been supplemented and made more precise.

Maps of Isostatic Anomalies and Their Properties

Isostatic anomalies generally show deviations of distributions of masses from the chosen isostatic system. In the Geophysical Institute of the Czechoslovak Academy of Sciences, the maps of isostatic anomalies were constructed for the Czechoslovak territory using the Airy–Heiskanen system for $T_0 = 20$ km, 30 km, 40 km, and 60 km. The map for $T_0 = 30$ km is presented in Fig. 9.5. Even though parameters of this model (i.e. constant density of the Earth's crust of 2.67 Mg m^{-3}, and the density contrast of 0.6 Mg m^{-3} at the crust–mantle boundary) substantially differ from assumed reality, the maps of isostatic anomalies are specifically significant for evaluating isostatic conditions. Positive isostatic anomalies indicate an excess of masses in the upper part of the Earth's body, while negative ones indicate a mass deficit. Properties of isostatic anomalies on Czechoslovak territory are discussed in V. Vyskočil (1977). Comparing isostatic anomalies with Bouguer anomalies (Ibrmajer 1981), the anomalies have a strong tendency to positive values but their variability decreases only slightly. The distribution pattern of the anomalous gravity field contours remains also basically unchanged. The compensation depth T_0 increased by 10 km affects the values of isostatic anomalies most in the Vysoké Tatry Mts. (by about 10×10^{-5} m s^{-2}), while in the rest of Czechoslovak territory, these effects are smaller.

The highest values of isostatic anomalies occur in SW Bohemia, in the Moravian part of the Bohemian Massif, and in the inner West Carpathians. Relatively the lowest values are in the Krušné hory Mts., in the Lužické hory and Jizerské hory Mts., and in the entire zone on the outer side of the Carpathian arc. It follows from the isostatic anomalies that over the majority of Czechoslovak territory excess masses are revealed, which could be due to density inhomogeneities at various depths. When analysing isostatic conditions it is necessary to take into account that the values of isostatic anomalies in Czechoslovakia can be lower by ca 17×10^{-5} m s^{-2} by introducing a new reference gravity system (Vyskočil V. 1978).

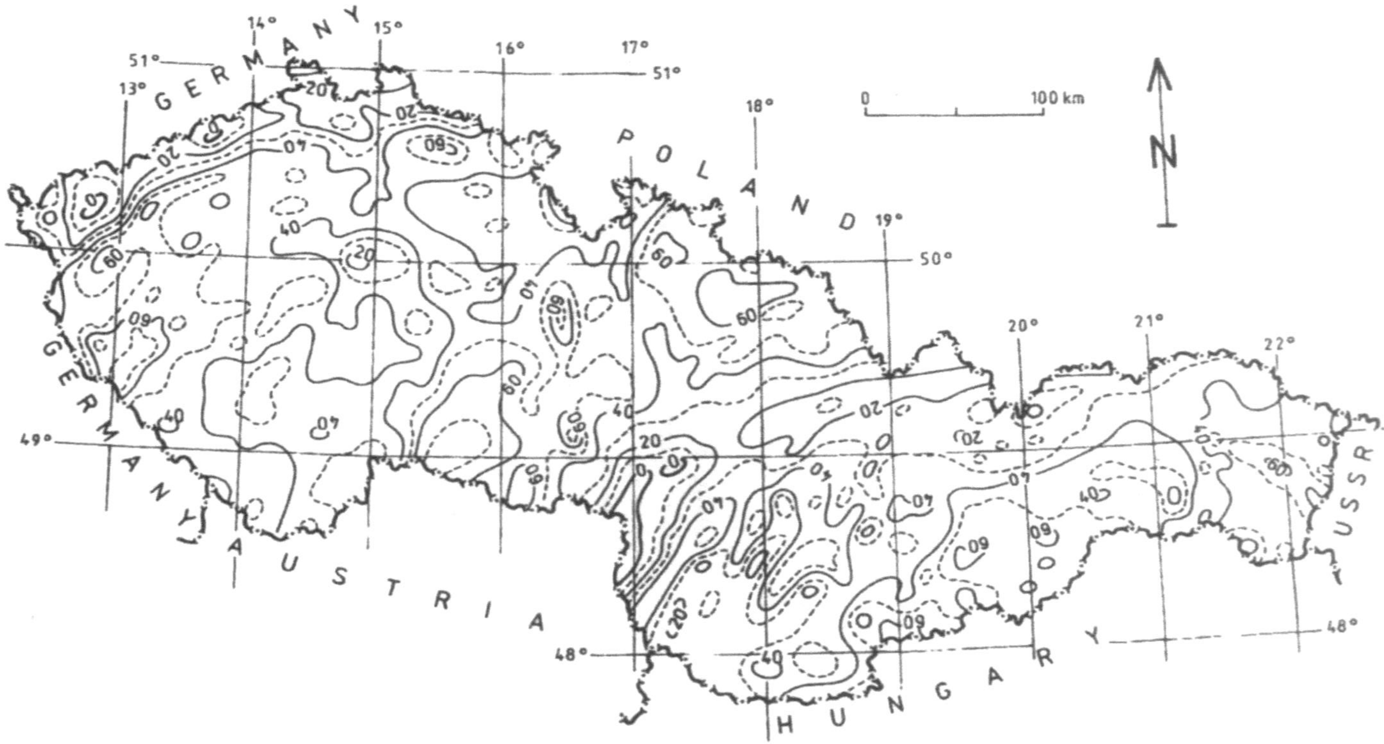

Fig. 9.5 Isostatic gravity anomalies on Czechoslovak territory. Airy – Heiskanen system $T_0 = 30$ km, contour interval in 10×10^{-5} m s^{-2}.

Masses of Vertical Columns of Lithosphere

Isostatic equilibrium can also be tested by comparing the values of static pressure p_H at the assumed level of isostatic compensation at a depth H. Values of p_H can then be calculated from the expression:

$$p_H = \int_0^{h+H} \rho(z) g(z)\, \mathrm{d}z \quad ,$$

where h is an elevation of the Earth's surface above sea level, $\rho(z)$ is density, and $g(z)$ is gravity acceleration. In the case of ideal isostatic equilibrium $p_H = \text{const.}$

With regard to the fact that changes of gravity acceleration g with the considered depth are relatively small and we do not know their accurate values, the computation of p_H is substituted by a simpler calculation of masses:

$$M_H = DS \int_0^{h+H} \rho(z)\, \mathrm{d}z$$

of vertical columns with their base DS, or of areal densities $\mu_H = M_H/DS$.

For calculating values of M_H or μ_H, knowledge of density distribution from the Earth's surface to the chosen depth H is needed. It is therefore effective to apply this calculation only to regions in which density models of lithosphere are available. In our case, density models are used which were constructed along Czechoslovak parts of international DSS profiles Nos. V (Fig. 6.17), VI (Fig. 6.16), and VII (Fig. 6.15). Areal densities μ_H were determined for depths H of 0, 10, 25, and 40 km, and in case of profile No. V, also for $H = 50$ km. Calculations of values μ_H were performed along all the profiles at 5 km intervals. The course of μ_H values is shown in Figs. 9.6. – 9.8 Because from the point of view of isostasy the local terrain irregularities of a small areal extent are not significant, during calculations of μ_H, a considerably smoothed out shape of the Earth's surface was taken into account. Table 9.1 shows the mean values of μ_H, and standard deviations m_μ.

It follows from Figs. 9.6 – 9.8 and from Table 9.1 that the undulation of curve μ_H for $H = 0$ is about the same or rather smaller than in case of curves μ_H for greater depths. It is obvious that crustal density inhomogeneities are as significant for the isostasy as the topographic masses.

For the Carpathian profile DSS V, standard deviation m_μ is the smallest for $H = 25$ km and the largest for $H = 50$ km. The fact suggests that a partial compensation of topographic masses and of superficial density inhomogeneities is involved already in the upper and intermediate parts of the crust. Density inhomogeneities at greater depths increase the disturbance of isostatic equilibrium. Curves μ_H for $H = 40$ km and

Table 9.1 Mean values of areal densities μ_H and standard deviations m_μ.

H [km]	DSS Profile V [10^6kgm^{-2}]		DSS profile VI [10^6kgm^{-2}]		DSS Profile VII [10^6kgm^{-2}]	
	μ_H	m_μ	μ_H	m_μ	μ_H	m_μ
0	1.265	0.565	0.939	0.538	1.291	0.592
10	28.031	0.673	27.555	0.773	28.377	0.800
25	70.274	0.457	69.826	0.869	70.443	0.692
40	116.535	0.745	116.378	0.720	116.422	0.613
50	148.453	0.879				

$H = 50$ km exhibit a clear difference between values of μ_H in outer and inner West Carpathians, which likely reflects strikingly different thicknesses of the Earth's crust.

On profile DSS VI, on the contrary, the greatest standard deviation m_μ was encountered for $H = 25$ km, which fact is due to an occurrence of remarkable density inhomogeneities in the upper and intermediate parts of the crust, and due to a diverse structure of the crust in the Bohemian Massif and that of the West Carpathians.

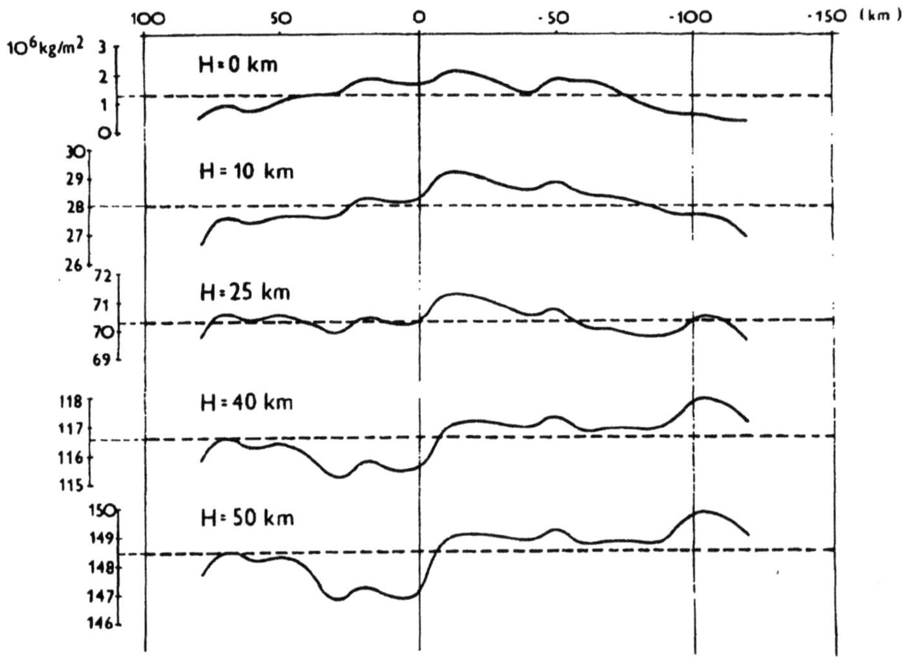

Fig. 9.6 Distribution of areal densities μ_H at different depths along DSS profile V.

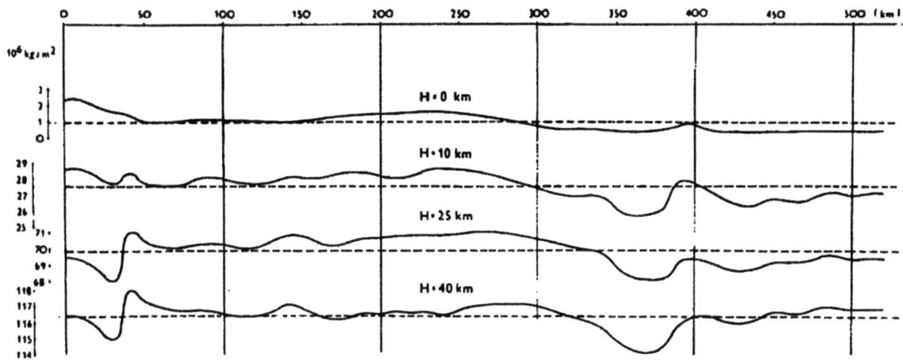

Fig. 9.7 Distribution of areal densities μ_H at different depths along DSS profile VI.

Profile DSS VII exhibits the largest standard deviation m_μ for a depth of $H = 10$ km, and the smallest one for $H = 0$ and $H = 40$ km.

Although values of μ_H, calculated for depths of $H = 40$ km and $H = 50$ km are strongly affected by the choice of the lower crust and of the subcrustal densities, their variability, expressed by the form of curves μ_H can be considered fairly reliable because it is in accordance with the course of the anomalous field. It follows from the comparison of mean values of μ_H for individual profiles that their best agreement is at a depth of $H = 40$ km. This confirms that masses of lithospheric blocks of greater areal extent are approximately the same.

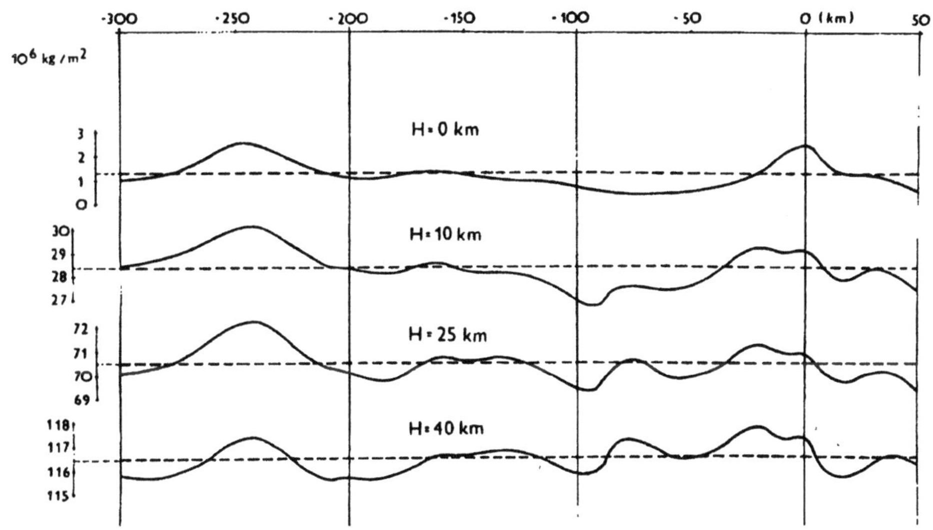

Fig. 9.8 Distribution of areal densities μ_H at different depths along DSS profile VII.

The Stress Field Caused by the Terrain Topography and by Density Inhomogeneities in the Earth's Crust

The problem of isostatic equilibrium of the Earth's crust represents only a part of the more general problem of determining the fields of stress and deformation in the lithosphere. During a long period (in the geologic time scale), elastic stresses can be preserved and transferred in the uppermost part of lithosphere only, which is approximately delimited by the isotherm of $T \approx 600$ °°C. In high temperature environments, some creep occurs and elastic stresses cannot persist.

Stresses in the lithosphere can be generated and transferred by different sources (Bott and Kusznir 1984). One of their causes is the existence of the Earth's surface topography and of density inhomogeneities in the Earth's crust. A corresponding field of stress was computed for the Czechoslovak part of international DSS VII profile (Trešl et al. 1987) with the use of new density models (Blížkovský et al. 1985 a). Components of the stress tensor were computed by means of the elastic half-plane model solving the biharmonic equation

$$\Delta\Delta\Psi = 0\,, \quad \sigma_{xx} = \frac{\partial^2\Psi}{\partial z^2}\,, \quad \sigma_{zz} = \frac{\partial^2\Psi}{\partial x^2}\,, \quad \sigma_{xz} = -\frac{\partial^2\Psi}{\partial x\,\partial z}\,,$$

with the boundary conditions:

$$\sigma_{zz}(x,0) = f(x)\,, \quad \sigma_{xz}(x,0) = 0\,, \quad (\sigma_{xx},\sigma_{zz},\sigma_{xz}) \to 0 \text{ for } z \to 0$$

for -300 km $\le x \le 0$ km (interval 5 km), $0 \le z \le 40$ km (interval 10 km). Moreover, magnitudes of disturbing isotropic stresses p and those of deviatoric stresses d were computed from the following expressions:

$$p = \frac{(\sigma_{xx} + \sigma_{xz})}{2}\,, \quad d = \left[\frac{(\sigma_{xx} - \sigma_{zz})^2}{4} + \sigma_{xz}^2\right]^{\frac{1}{2}} \,.$$

The computation was performed by means of the known formulas for stresses caused by rectangle load $\xi_1 \le \xi \le \xi_2$, $\zeta_1 \le \zeta \le \zeta_2$ with a constant disturbing density ρ.

$$p(x,z) = \rho g(\zeta_2 - \zeta_1)\frac{\varphi}{\pi}\,,$$

$$d(x,z) = \rho g(\zeta_2 - \zeta_1)\frac{\sin\varphi}{\pi}\,,$$

where g is gravity acceleration, and ζ is the angle under which the lower edge of the rectangle is seen from the point (x,z). The computation was performed in two variants (Trešl et al. 1987):

Table 9.2. Differences between maximum and minimum values of the isotropic and deviation stresses (in MPa) for different depths z (in km); A – topography, B – topography and crustal density inhomogeneities.

Depth z [km]	$p_{max} - p_{min}$ A	$p_{max} - p_{min}$ B	$d_{max} - d_{min}$ A	$d_{max} - d_{min}$ B
0	21.1	21.1	0	0
10	14.2	20.7	5.0	4.4
20	10.1	16.7	4.9	6.1
30	7.4	11.5	4.0	6.4
40	5.6	8.3	4.4	4.8

1. Independent effect of the terrain topography,
2. Total effect of the terrain topography and of density inhomogeneities (Table 9.2 A and B).

In Table 9.2, differences between the maximum and minimum values of p and d are presented for individual depths z. Horizontal variations of isotropic stress $p_{max} - p_{min}$ reach typical values of 10 MPa at depths of 30–40 km, which represents roughly 50 % of all surficial variations. Crustal density inhomogeneities do not contribute, however, to more rapid decrease of differences $p_{max} - p_{min}$ with depth, but on the contrary, they make it slightly less rapid.

Horizontal variations of deviatoric stress $d_{max} - d_{min}$ attain values of 5 MPa in the whole range of depths from 10 km to 40 km along the particular profile. Even in this case the effect of crustal density inhomogeneities does not manifest itself by more rapid decrease of differences $d_{max} - d_{min}$ with depth.

It is obvious that the computation model used represents only a basic approximation. In order to define the actual state better it is necessary to use a three-dimensional model and consider the effect of numerous other factors (i.e. variable thickness of elastic sheet and buoyancy forces at its lower boundary, faults, porosity of the medium, etc.).

Recently the more realistic model of an elastic plate floating on a liquid basement has been used (Trešl and Marvanová 1988). It was shown that large horizontal stresses may be generated by certain load wavelengths. Using such a model, the stress field along the West Carpathians profile Zakopane – Salgotarjan has been computed (Trešl 1989).

9.3 Modification of Runcorn's Equations to Convection Flows

M. Pick[1]

A hypothesis on the existence of the convection flows was presented by Holmes as early as 1931. The convection flows were supposed to work as a possible mechanism for mountain building. Long-distance undulation of the geoid could be explained by irregularities in the mass distribution of the upper parts of the Earth's body. Jeffreys estimated that these undulations might cause internal stresses of up to 5 MPa.

The problem can be solved by means of density models of the Earth. The Navier-Stokes equation could be applied if the distributions of the density, pressure, viscosity, and gravity in the lithosphere and asthenosphere were known.

Runcorn (1964, 1967) substantially simplified this problem. He supposed that the upper part of the Earth's body consisted of two layers: the outer layer being rigid and still, and the inner layer being viscous, satisfying the following conditions:

1. Its size is sufficiently large so that its lower boundary gravitational effect can be neglected.
2. The shape of the upper boundary of the viscous layer, i.e. the interface between the lower viscous and the upper rigid layers, causes the hydrostatic equilibrium.
3. The coefficient of viscosity is constant and equals to 10^{21} Pa s.

It is evident that this is an estimate rather than a computation. It can, however, be used as the first approximation.

According to Runcorn (1967), the stress between the two layers in the south–north and west–east directions can be derived by solving the simplified Navier-Stokes equation

$$\mu \frac{\mathrm{d}}{\mathrm{d}\Theta} \frac{\mathrm{d}^2 W}{\mathrm{d}r^2}\bigg|_{r=a} = \sum_{n=2}^{\infty} \frac{Mg}{4\pi a^2} \left(\frac{a'}{a}\right)^{n+1} \frac{2n+1}{n+1} \frac{\mathrm{d}S_n(\Theta,\Phi)}{\mathrm{d}\Theta}\bigg|_{r=a} ,$$

$$\frac{\mu}{\sin\Theta} \frac{\mathrm{d}}{\mathrm{d}\Phi} \frac{\mathrm{d}^2 W}{\mathrm{d}r^2}\bigg|_{r=a} = \sum_{n=2}^{\infty} \frac{Mg}{4\pi a^2} \left(\frac{a'}{a}\right)^{n+1} \frac{2n+1}{n+1} \frac{1}{\sin\Theta} \frac{\mathrm{d}S_n(\Theta,\Phi)}{\mathrm{d}\Phi}\bigg|_{r=a} , \qquad (9.1)$$

where μ is coefficient of viscosity, M is Earth's mass, Θ , Φ are geographical coordinates, W is scalar function, (poloidal function of convection flows), r is the distance from the Earth's centre, g is the gravity acceleration, $S_n(\Theta, \Phi)$ is defined lower, a', a are the mean radii of the Earth and of the interface, respectively, n is a degree of spherical harmonics.

---

[1] Geophysical Institute, Boční II/1401, 141 31 Praha 4, Czech Republic

At the same time, the well-known equation is valid for the gravity potential

$$V = U + T \quad ,$$

where U and T are normal and the disturbing potential, respectively.

$$T = \frac{GM}{a'} \sum_{n=2}^{\infty} \left(\frac{a}{r}\right)^n \sum_{m=0}^{n} (C_{mn} \cos m\Phi + S_{mn} \sin m\Phi)\, P_{mn}(\cos\Theta) \quad ,$$

with

$$S_n = \sum_{m=0}^{n} (C_{mn} \cos m\Phi + S_{mn} \sin m\Phi)\, P_{mn}(\cos\Theta) \quad ,$$

where G is the gravitational constant.

Runcorn (1967) tested these equations using a satellite model of the Earth's gravity field according to Guier and Newton up to the 8th degree.

The equations in Runcorn's modification were used by other authors too, e.g. by Liu (1980), Pěč and Martinec (not published).

However, other data such as the results of surface gravity measurements, information about geoid heights (the results from altimeter), etc., could be employed as well.

A disadvantage of the original Runcorn method is in its low resolution of results. In computations, a development of outer gravity potential into series is used. The resolution, of course, depends on number of members of the series. For instance, for a required resolution of 40×40 km, altogether 320 000 coefficients would have to be used. That is why we modified the Runcorn equations.

The first modification is clearly formal. Infinite sequences in equations (9.1) were summed and maps of gravitational anomalies and maps of geoids were used as input data. The necessary theory is presented in Pick et al. (1973) and Pick and Charvátová-Jakubcová (1988).

The second modification is substantial. In an agreement with the derivation of Runcorn formulae, the Earth was regularized, i.e. Bouguer anomalies were used instead of Faye gravity anomalies, and heights of the Bouguer co-geoid instead of heights of geoid. Corrections of geoid heights for co-geoid heights reach large values in the mountains (e.g. in the Alps it is over 20 m).

The derived relations by us agree up to members of second order with equations of Burša (1971), but the physical meaning is different, of course.

The map of stresses between the lithosphere and the asthenosphere, caused by convection currents in Europe, serves as an example of the presented method (Fig. 9.9).

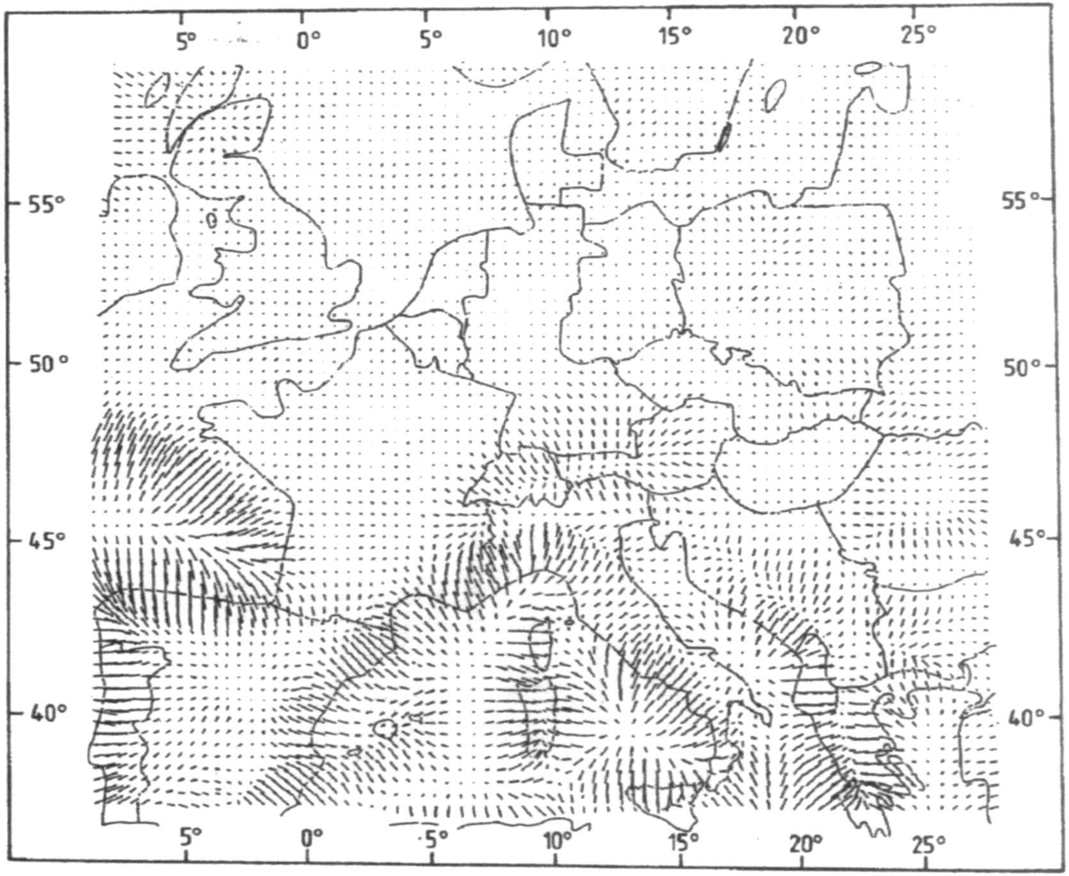

Fig. 9.9 Stresses between lithosphere and asthenosphere caused by convection flows.

9.4 Linear Structures in the Czechoslovak Part of the Bohemian Massif Derived from Geophysical Data

M. Blížkovský[1], J. Mašín[2], E. Mátlová[3], P. Mitrenga[1], A. Novotný[1], L. Pokorný[2], L. Rejl[1], K. Šalanský[2]

Introduction and Methodology

In the period 1983–86 Geofyzika Brno carried out a complex interpretation of the linear structures of the Bohemian Massif indicated by geophysical data. The primary aim was to find favourable localities for uranium prospecting, often related to fault zones in the upper part of the Earth's crust.

In the course of the work, the general distribution pattern of the linear structures was analyzed as well as the areas delimited by them. Most information was provided by gravity and magnetic maps, in basins also by seismic and geo-electric data. Radiometric data provided only additional information. In the following we shall describe briefly the methods used for constructing the final maps.

Maps of gravimetric indications were compiled on the scale of 1:200 000. They show the basic linear gravity structures (density boundaries, changes of gravity contours) and distribution of positive and negative gravity areas. They were based on maps of Bouguer anomalies on the scales of 1:25 000 and 1:200 000, and on derived maps (regional and residual, maximum absolute values of horizontal gradients, indications of vertical density boundaries).

Maps of magnetic indications were based on aerial geophysical mapping of the Bohemian Massif (1960–71) and partly on the measurements conducted after 1976. Maps of *DT* on the scale of 1:200 000 and of analytical upward continuation of *DT* at 150 m level were used to indicate shallow and deep linear magnetic structures. The final map of magnetic indications contains the axes of linear *DT* anomalies, bends and dislocations of magnetized horizons, isometric sources and interpreted linear magnetic discontinuities.

Maps of radiometric indications also resulted from aerial geophysical mapping (1960–71). They indicate areas of high and low radioactivity, axes of linear positive and negative anomalies, and local radioactivity anomalies often linked with fault lines.

Seismic indications of linear structures were based on earlier reflection and refraction measurements as well as on more recent CDP Vibroseis measurements carried out in sedimentary basins. Structural-tectonic schemes of the Cretaceous and Permo-carboniferous basement relief, of the surface of Palaeozoic and crystalline formations were used to follow faults and zones of abrupt depth changes.

1 Geofyzika Brno, Ječná 29a, 612 46 Brno, Czech Republic
2 Geofyzika Praha, Geologická 2, 152 00 Praha 5, Czech Republic
3 Uranium Survey, Tř. 1. máje 108, 461 71 Liberec, Czech Republic

Maps of electric indications were based on electric (mainly VES) measurements conducted for hydrogeological and engineering-geological purposes, and in the course of coal, oil, and gas prospecting. Striking tectonic lines from earliest works were used. Where only atectonic maps of the basement relief were available the data was reinterpreted.

All the above maps of geophysical indications were correlated, compared with available structural-geologic data, and used for construction of the maps of linear structures of the Bohemian Massif on the scales of 1:200 000 and 1:500 000 (Fig. 9.10). The linear structures are presented in three categories according to length and continuity. Most distinctive are those which bound gravity areas and subareas. Alphabetic designation of indications defines the directional systems: A = SW–NE, B = NW–SE, C = N–S, D = W–E, E = WSW–ENE, F = WNW–ESE, G = WSW–ENE.

Characteristics of the Defined Areas and Principal Linear Structures

Buday et al. (1969) divided the map of Bouguer anomalies of the Bohemian Massif into positive and negative gravity areas. This basic division was gradually specified (Pokorný et al. 1970; Blížkovský et al. 1975; Pokorný et al. 1985 a). The model in Fig. 9.11 roughly corresponds to the idea forwarded by Blížkovský et al. (1985 a).

The Krušné hory – Krkonoše Mts. negative gravity area (I) reflected the influence of extensive light metamorphic complexes and Hercynian granitoids in the upper crust. Its northern boundary is in Germany, with the exception of the Šluknov projection where the discontinuity A01 borders the gravity area above the Lužice massif. In the south, the area is bordered by line A04 corresponding to the deep Litoměřice fault. Its southwestern continuation from the crossing with C01 and B06 has not been finally specified yet. It presumably extends along line A05 across the northern part of the Bor massif. The area of the Český les Mts. can be regarded as a transition subarea (I–l/l). In the Labe valley the southern boundary turns to the W–E direction along D05, F03, and D03 at the southern border of the Krkonoše – Jizerské hory pluton.

In the Barrandian – Železné hory – Broumov area (II) positive gravity structures prevail. They are presumably associated with a thick layer of metamorphic late Proterozoic sediments and volcanites attaining the depth of 7 km (Beránek et al. 1975). They may also be associated with upwarded basic lower crust. In the Teplá – Barrandian subarea (II–l) parallel elevation and depression zones reflect the distribution of basic rocks in the upper crust. In the W, area II contacts the transition subarea of the Český les Mts. along discontinuities C01 and B02 (the Tachov deep fault – Röhlich and Šťovíčková 1968). The southeastern border with the Moldanubicum – Klodzko area (III) is formed by structures A11 and A12 (the Klatovy deep fault).

The Jáchymov deep zone B09 shifts the border by 25 km to the SE to A14 (the Benešov fault). To the NW the border is shifted along B16 (the Sázava deep fault) to discontinuity A13. Structures A12, A13, and A14 from the system of the Central Bohemian suture, in the environs of Poděbrady transversely adjoined by the Železné

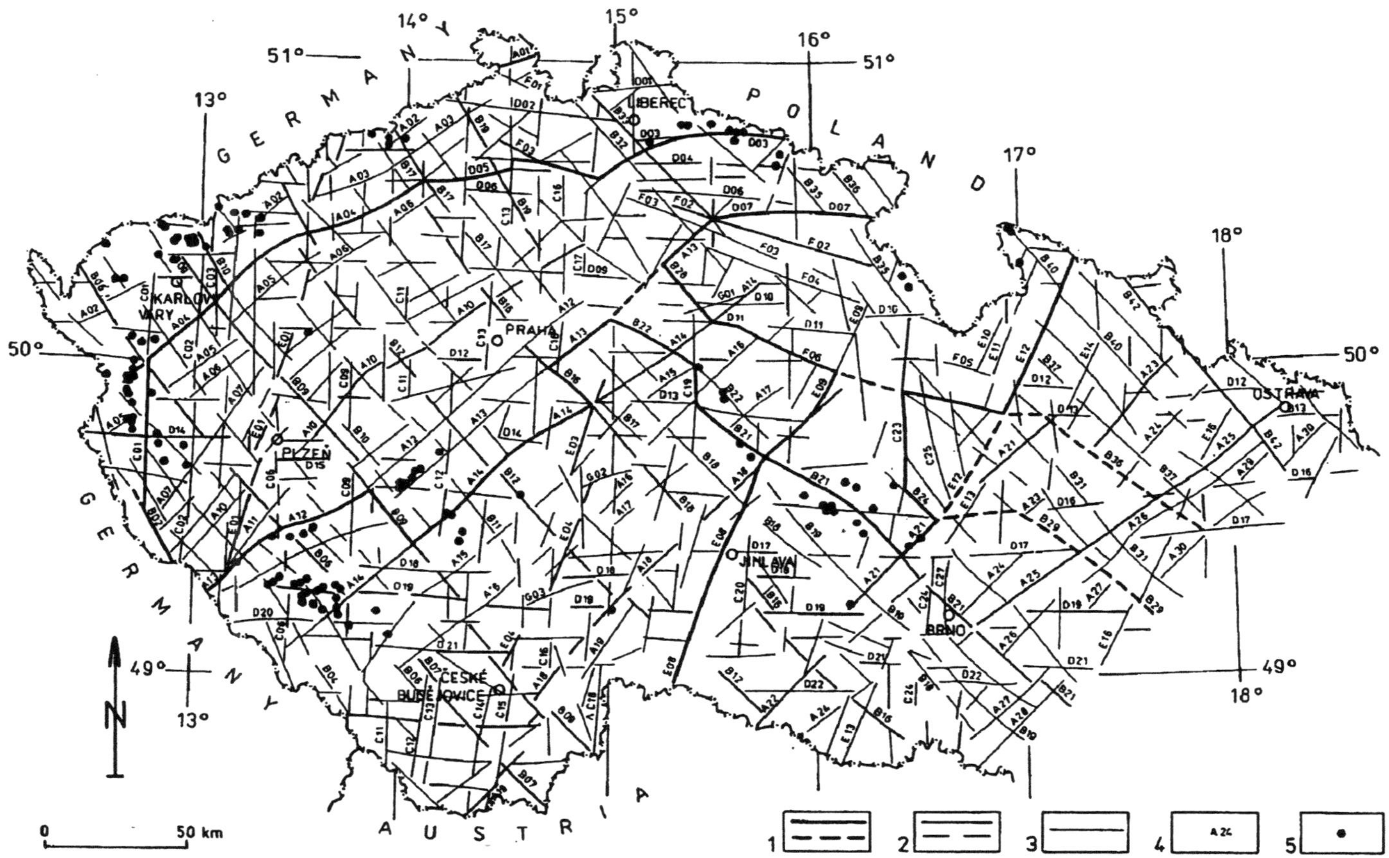

Fig. 9.10 Linear structures in the Bohemian Massif on Czechoslovak territory derived from geophysical investigations 1 – boundaries of areas and subareas according to the gravity data, 2 – distinct linear structures, 3 – other linear structures, 4 – designation of linear structures, 5 – uranian deposits and indications.

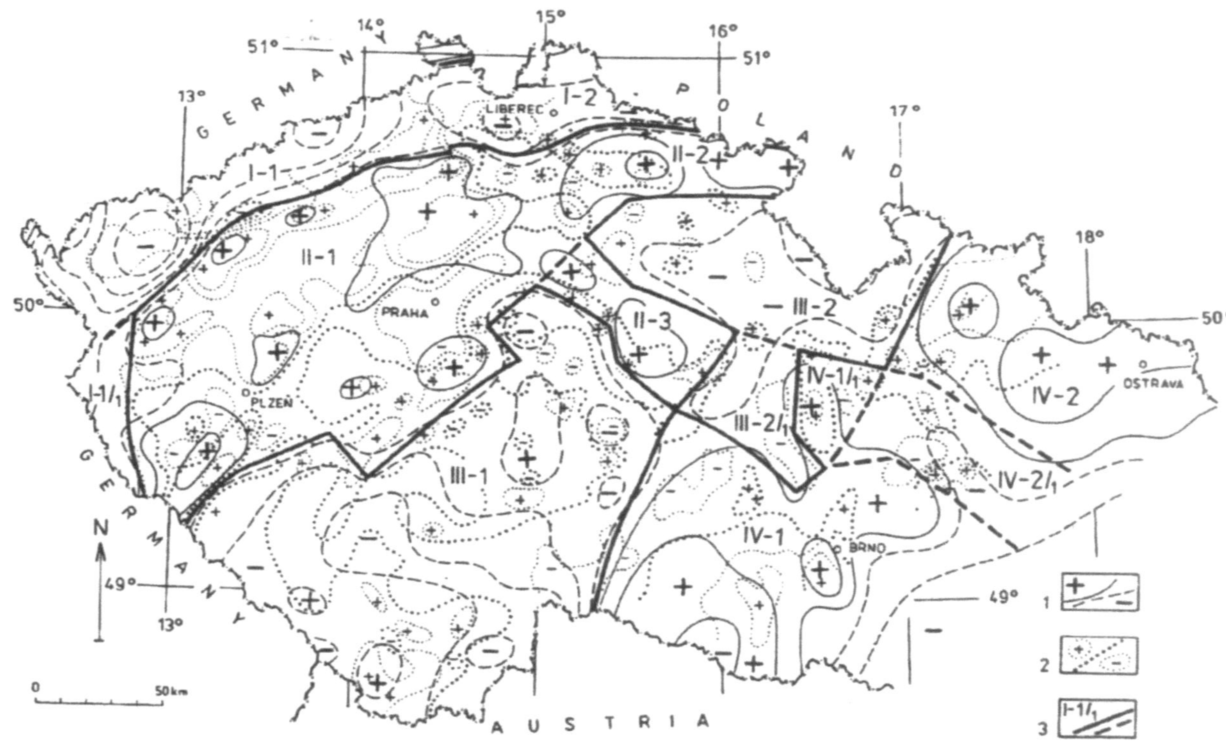

Fig. 9.11 Basic gravity structures of the Bohemian Massif on the territory of Czechoslovakia. 1 – regional positive and negative contours with lows and highs, 2 – schematic regional positive and negative magnetic field contours with highs and lows, 3 – boundaries of areas and subareas based on gravity data, I – Krušné hory – Krkonoše area, I–1 Krušné hory Mts. subarea, I–1/1 transition subarea of the Český les Mts., I–2 Krkonoše Mts. subarea, II – Barrandian – Železné hory Mts. – Broumov area, II–1 Teplá – Barrandian subarea, II–2 Železné hory Mts. subarea, II–3 Broumov subarea, III – Moldanubicum – Klodzko area, III–1 Moldanubicum subarea, III–2 Orlické hory Mts. – Klodzko subarea, III–2/1 transition Svratka subarea, IV – Moravo–Silesian area, IV–1 Moravian subarea, IV–1/1 transition Svitavy subarea, IV–2 Silesian subarea, IV–2/1 transition Upper Moravian Basin subarea.

hory subarea (II–3). In the SW the subarea is delimited by B22 (The Poděbrady deep fault) and by B21, which in the NE borders the Říčany – Kutná hora depression and the Melechov massif. In the basic Ransko massif the border of the subarea turns to the NE along the shear zone A18 and along discontinuity E09 (presumably a continuation of the Přibyslav deep zone). The northern limitation of subarea II–3 (lines B28, D11, F06) is obscured by the effect of sediments of the Bohemian Cretaceous Basin. The Labe – Broumov subarea (II–2) is in the S bordered by discontinuity A13 and NE of Jičín by the striking equatorial D07 at the southern border of the Krkonoše-piedmont Basin.

The inner structure of the Moldanubicum – Klodzko negative gravity area (III) reflects the occurrences of light granitoid massifs. Striking negative structures occur between České Budějovice and Písek, in the area of the presumed Říčany – Kutná Hora massif, in the Central Moldanubian Massif, and in the Orlické hory – Klodzko Dome. The boundary between the Moldanubian subarea and the Moravo–Silesian area (IV) is formed by discontinuity E08 (the Přibyslav deep zone) at the eastern contact of the Central Moldanubian massif. The transition Svratka subarea (III–2/l) contacts the Moldanubian subarea at the intersection of numerous tectonic lines near Staré Ransko. Its boundary with the Moravian subarea (IV–l) runs along line B21 (the southeastern continuation of the Železné hory fault) towards the transverse discontinuity A21 and further along the discontinuity to the NE where its course is complicated by the striking Svitavy gravity high which can be interpreted as the effect of an allochthonous body resting on an older structure, or as a separate subarea (IV–l/l). Its boundary with the Orlické hory – Klodzko subarea (III–2) is line F06, while C23 and B24 form its boundary with the Svratka subarea (III–2/l). The position of the Svratka subarea suggests that it is a secondary separate block of the Earth's crust within the mobile Labe zone along B21 shifted by more than 50 km to the SE in regard to the Moldanubian subarea. A displacement of the Železné hory block along this discontinuity can analogously be assumed.

The Moravo–Silesian area (IV) is characterized by positive gravity field due to a smaller thickness caused by upwarped basic lower crust. The top parts of positive gravity structures occur in the Moravian Moldanubicum, in the Brno Massif, in the Letovice crystalline formation, and in the Jeseníky Mts. Basic and ultrabasic rocks are abundant, while Hercynian granitoids are very rare. The southeastern boundary of the Moravo–Silesian area is often identified with the southeastern boundary of the entire Bohemian Massif or with the boundary of the so called autonomous block (Weiss 1977). The boundary between the Moravian subarea (IV–l) and the transition subarea of the Upper Moravian Basin (IV–2/l) is formed by a significant discontinuity B29 in the Nectava – Konice fault zone. The boundary with the Silesian subarea (IV–2) formed by line B36, representing the northern marginal fault of the Upper Moravian Basin. The striking structure E12, in the N, indicating the deep zone of Červenohorské sedlo (the Kouty fault) separates the Silesian positive gravity subarea from the Orlické hory – Klodzko depression subarea.

The individual system of geophysical indications play different roles in the internal structure of gravity areas and subareas. While in subareas I–1 and II–1 the structures of system A and the transverse dislocations of system B predominate, in subareas (I–2) and (II–2) systems D and F and transverse dislocations of system C play the chief role. In the negative gravity area (III) the frequency and significance of system A gradually decrease, while indications D, C, and in the last also indication E prevail. In the positive gravity area (IV) system A regains importance, but it is often combined (or substituted) with lines of system E. System F has practically no influence in this area. Transcurrent faults of system B are abundant within the entire Bohemian Massif.

Conclusion

The analysis of the characteristics and functions of the systems of discontinuities indicated by geophysical data yields the following conclusions.

Presumably the oldest linear structural elements of the Bohemian Massif are the regularly distributed discontinuities of orthoglobal systems C and D. They have various functions and are often active. System D is clearly manifested in the deeper structure of the Moldanubicum and in metallogenetic zones of various ages within the massif. System C was involved in neovolcanic activity and can be traced in the recent morphology of the terrain.

The Cadomian structure was evidently formed by system A, clearly shown in the Teplá – Barrandian subarea. The lines of this system in their eastern continuation appear as false elements of system D.

System A is evidenced in the Moldanubicum – Klodzko area, and in the Moldanubicum keeps a SW–NE direction. Of the newly interpreted structures the most significant is A10, linking with the tectonic zone of the river Rodl. In the eastern part of the Moldanubian subarea the structures gradually turn to the E and SE, and lines of system O appear. In the Orlické hory – Klodzko subarea system O and F occur. Analogous phenomena can be observed in the Krušné hory – Krkonoše area. The general deformation of the structure of northeastern and eastern parts of the Bohemian Massif are obviously related to the mobile transverse Labe zone.

System B is independent of the environment, it has transverse transcurrent tectonic features. The Jáchymov deep fault (B09, B10) and the Poděbrady deep fault (B21, B22) may be palaeotransform faults. System B also forms the basis of the mobile Labe zone. The horizontal displacements along system B can be associated with the thrust of the Železné hory and Svratka subareas far towards the SE.

The discontinuities of system E indicate the young Hercynian NNE–SSW striking furrows of the Bohemian Massif. Some of the furrows are deep-seated and form boundaries of different gravity areas. They also affected the origin of young Hercynian granitoid intrusions (the Plzeň deep fault – E01).

It has been proved that deposits and indications of uranium mineralization concentrate in negative gravity areas, particularly in their margins and within linear structures near rocks with increased radioactivity.

9.5 Changes in Palaeogeographic Latitudes of the Barrandian Interpreted on the Basis of Palaeomagnetic Data

M. Krs[1]

Palaeomagnetic data provide a possibility of palaeogeographic reconstructions. The presumption of the prevailingly dipolar character of the mean geomagnetic field in the geologic past has partly been confirmed for Phanerozoic rocks by a statistical processing of newly introduced parameters, the so called palaeorotation, the difference of palaeo-inclination and palaeodivergence of palaeomagnetic data (Krs 1982). On the basis of a confirmation of the validity of the dipolar character of the palaeomagnetic field, values of palaeogeographic latitudes (palaeolatitudes) can be calculated for particular blocks (plates) in relation to present geographical latitudes. Interpretations of this kind have usually been undertaken on a global scale, first of all for constructing global-tectonics models, but when the palaeomagnetic data are abundant, they can also be used on a regional scale.

Let us present the data from the Barrandian as an example.

Fig. 9.12 presents the palaeogeographic latitudes as a function of age of rocks, calculated from the position of palaeomagnetic poles for the Barrandian. Open circles designate palaeolatitudes, which were computed for the Barrandian from the mean positions of poles, statistically derived for the North European platform. Black circles (with accompanied numerals 1 to 5) designate the palaeolatitudes computed for the Barrandian from palaeomagnetic poles derived on the rocks of the Bohemian Massif or from the poles derived on the Barrandian rocks. Palaeolatitude 1 was computed from the positions of poles for the Autunian and Saxonian rocks, and palaeolatitude 2 from those for rocks of Westphalian and Stephanian of the Bohemian Massif (Krs 1968). Number 3 marks the Lower Ordovician rocks (Krs et al. 1986), no. 4 marks the Upper Cambrian rocks, and no. 5 those of the Lower Cambrian (Krs et al. 1987), altogether of the Barrandian provenance. The palaeomagnetic data which were derived from measurements on rocks strongly affected by the Variscan folding are based on the multi-component analysis of magnetization. In order to derive them, highly efficient demagnetizer equipped with the MAVACS system was used. Palaeomagnetic investigations of the Barrandian rocks is under way. Rocks of Devonian, Silurian, and Lower Cambrian are paid special attention as the data from them could contribute to clarification of the palaeogeographic position of marginal continental blocks of the proto-Atlantic ocean. The existing results, however, have already suggested a certain affinity to the data from the Armorican Massif (Perroud 1985); the palaeolatitudes computed for the Barrandian are rather lower than those obtained for the Armorican Massif.

The changes of palaeolatitudes for the Barrandian depending on geologic structure indicate a more or less uninterrupted movement of the Barrandian block from the Middle Cambrian from ca 40° of the Southern Hemisphere to the equatorial zone during

1 Geological Institute, Rozvojová 135, 165 00 Praha 6, Czech Republic

the Upper Carboniferous and Lower Permian, from where it had been shifted to the present position already as a part of the North European platform. Figure 9.12 documents the necessity to complete the palaeomagnetic investigations of rocks of the Lower Cambrian, Silurian, and Devonian with the primary aim of clarifying the palaeogeographical position of the Barrandian in regard to other blocks, to the Baltic and Armorican ones.

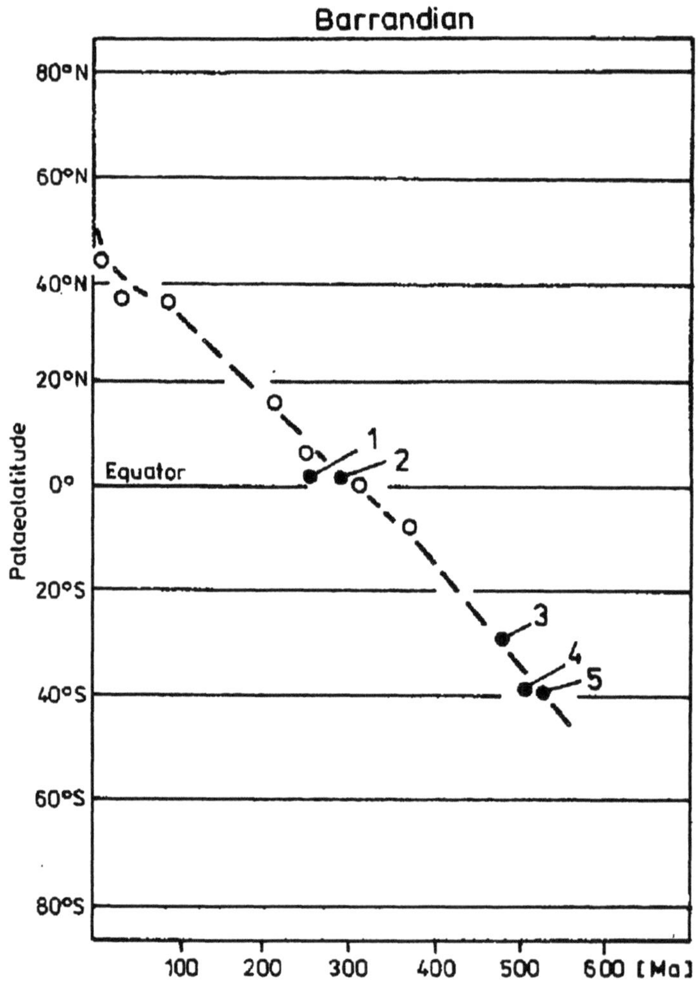

Fig. 9.12 Barrandian area. Change of palaeogeographic latitudes (palaeolatitudes) in relation to geologic time, according to palaeomagnetic results.

References

Ádám A (1976) Connection between geo-electric and geothermal parameters in the Earth. In: Ádám A (ed) Geoelectric and Geothermal Studia, KAPG Geophys. Monograph., Budapest, pp 567 - 571
Ádám A (1978) Geothermal effects in the formation of electrically conducting zones and temperature distribution in the Earth. Phys Earth Planet Inter, Amsterdam 17: 21 - 28
Ádám A, Pospíšil L (1984) Crustal conductivity anomalies in the Carpathian region. Acta Geod Geophys Montan, Budapest 19 (1 - 2): 19 - 34
Ádám A, Verö J, Wallner A (1972) Regional trends of geomagnetic induction arrows in Europe. Acta Geod Geophys Montan, Budapest 7 (3 - 4): 251 - 287
Ádám A, Landy K, Nagy Z (1989) New evidence for the special distribution of the electric conductivity in the Earth's crust and upper mantle in the Pannonian Basin as a "hot spot". Tectonophysics, Amsterdam 164: 361 - 368
Ananin IV (1983) Karta koeficientov zatukhaniya intensivnosti seysmoaktivnykh rayonov SSSR. - In: Geofizicheskie usloviya v ochagovykh zonakh silnykh zemletraseniy. Moskva Nauka, pp 44 - 68
Ando CJ, Czuchra BL, Klemperer SL, Brown LD, Cheadle MJ, Cook FA, Oliver JE, Kaufman S, Walsh T, Thompson JB, Lyons JB, Rosenfeld JL (1984) Crustal profile mountain belt: CCCORP deep seismic reflection profiling in New England Appalachians and implications for architecture of convergent mountain chains. Am Assoc Petrol Geol Bull, Tulsa 68; 7: 819 - 837
Andrusov D (1968) Grundriss der Tektonik der Nördlichen Karpaten. Veda, Bratislava, 188 p
Andrusov D, Bystrický J, Fusán O (1973) Outline of the structure of the West Carpathians. Guide - book for geological excursion X. Congress of Carpathian - Balkan. Geol Assoc. Geol. úst. D. Štúra, Bratislava, pp 32 - 53
Antonini M (1988) Variations in the focal mechanisms during the 1985/1986 Western Bohemia earthquake sequence - correlation with spatial distribution of foci and suggested geometry of faulting. Proc. workshop on "Induced seismicity and associated phenomena", I., Geoph Inst Czechosl Acad Sci, Praha pp 250 - 270
Arnold A, Scharbert HG (1973) Rb - Sr Altersbestimmungen an Granuliten der südlichen Böhmischen Masse in Österreich. Schweiz Mineral Petrogr Mitt Zürich 53: 61 - 78
Babuška V (1966) Velocity of compressional waves and elastic anisotropy of some igneous and metamorphic rocks. Trav Géophys 1965, Academia Praha, pp 223, 275
Babuška V (1984) P - wave velocity anisotropy in crystaline rocks. Geophys J R Astr Soc, Oxford, 76: 113 - 119
Babuška V, Plomerová J, (1989) Seismic anisotropy of the subcrustal lithosphere in Europe: another clue to recognition of accreted terranes. In: Hillhouse J W (ed) Deep structure and path kinematics of accreted terranes. Geoph. Mont. 50. Submitted to AGU Geodyn Ser, Washington pp 209 - 218
Babuška V, Plomerová J (1991) The lithosphere in central Europe - seismological and petrological aspects. Tectonophysics (in press)
Babuška V, Pros Z, Franke W (1977) Effect of fabric and cracks on the elastic anisotropy in granodiorite. Publ Inst Geophys Warszawa A - 6: 117, 179
Babuška V, Plomerová J, Šílený J (1984a) Spatial variations of P residuals and deep structure of the European lithosphere. Geophys J R Astr Soc Oxford 79: 363 - 383
Babuška V, Plomerová J, Šílený J (1984b) Large scale oriented structures in the subcrustal lithosphere of Central Europe. Ann Géophys Paris 2: 649 - 662
Babuška V, Plomerová J, Šílený J (1987) Structural model of the subcrustal lithosphere in Central Europe. In: Fuchs K Froidevaux V (eds) Structure and Dynamics of the Lithosphere - Asthenosphere System, Geodyn Ser, AGU Washington 16: 239 - 251

Babuška V, Plomerová J, Pajdušák P (1988) Lithosphere - asthenosphere in Central Europe model derived from P residuals. In: Nolet G, Dost B (eds) Proc. 4th EGT workshop" "The upper mantle". Strasbourg, pp 37 - 48
Babuška V, Plomerová J, Granet M (1990) The deep lithosphere in the Alps: a model inferred from P residuals. Tectonophysics, Amsterdam 76: 137 - 165
Badham JPN (1982) Strike - slip orogens - an explanation for the Hercynides. J Geol Soc, Oxford 139: 493 - 504
Baer M (1980) Relative travel time residuals for teleseismic events at the new Swiss seismic station network. Ann Géophys, Paris 36: 119 - 126
Bagdasarjan GP, Gukasjan RCh, Cambel B, Veselský J (1982) The age of Malé Karpaty Mts. granitoid rocks determined by Rb - Sr isochrone method. Geol Zb Geol Carpathica, Geol. úst. D. Štúra, Bratislava 33; 2: 131 - 140
Bajuk EI, Volarovich MP, Klíma K, Pros Z, Vaněk J (1967) Velocity of longitudinal waves in eclogite and ultrabasic rocks under pressure to 4 kilobars. Stud Geophys Geod, Academia, Praha 11: 271
Balatka B, Roth Z, Sládek J, Zeman A (1983) Český masív: diskuse o jeho vymezování a tektonickém členění. Věstník ÚÚG, Praha 58; 6: 369 - 376
Balla Z (1984) The Carpathian and the Pannonian Basin: a kinematic analysis. Geophys Trans, Budapest 30; 4: 313 - 353
Bamford D (1977) P_n velocity anisotropy in a continental upper mantle. Geophys J R Astron Soc, Oxford 49, 29
Bard JP et al. (1980) La chaine hercynienne d'Europe occidentale en termes de tectonique des plaques. Mém BRGM, 108, 26 IGC, C, Paris 6: pp 233 - 246
Bartošek J, Jelínek V, Pros Z (1985) Comparison of elastic and magnetic anisotropies of two types of metamorphosed rocks demonstrated on spherical specimens. Physical Properties of the Mineral System of the Earth's Interior. Publ. Union of Czechoslovak Mathematicians and Physicists, Praha, 59 p
Beck R (1907) Untersuchungen über einige südafrikanische Diamantlagerstätten. Z Dtsch Geol Ges, Hannover 59: 275 - 307
Behr HJ (1983) Intracrustal and subcrustal thrust-tectonic at the northern margin of the Bohemian Massif. In: Martin H, Eders W (eds) Intracontinental belts. Springer, Berlin Heidelberg New York, pp 365 - 403
Behr HJ (1987) KTB und kontinentale Krustenforschung. In: KTB-report 87-1, Hannover, pp 1 - 35
Behr HJ, Fritsch E, Mansfeld L (1965) Die Granulite von Zöblitz im Erzgebirge als Beispiel für Granulitbildung in tiefreichenden Scherhorizontes. Krystalinikum, Praha 3: 7 - 29
Behr HJ,Engel W, Franke W, Giese P, Weber K (1984) The Variscan belt in central Europe: main structures, geodynamic implication, or open question. Tectonophysics, Amsterdam 109: 15 - 40
Behr HJ, Stehli FG, Vidal H (1987) Observation of the continental crust through drilling II. Springer, Berlin Heidelberg New York, 230 p
Beránek B (1971) Přínos hlubinných seismických sondáží pro vyhledávání hlubinných zlomů. Geol. úst. D. Štúra, Bratislava, Geol Práce Spr 57: 189 - 196
Beránek B (1978) Výzkum stavby zemské kůry na území ČSSR metodami explozívní seizmologie a transformovaných tíhových polí. (Doktorská disertační práce) Universita Karlova. Fakulta přírodovědecká, Praha
Beránek B (1980) Výzkum hlubinné stavby zemské kůry v ČSSR. Geol Průzk, Praha 22; 4: 98 - 101
Beránek B, Dudek A (1972) The results of deep seismic sounding in Czechoslovakia. Z Geophys, Würzburg 38: 415 - 427
Beránek B, Zátopek A (1977) Rezul'taty izucheniya stroeniya zemnoy kory metodom GSZ v ChSSR. In: Sollogub VB, Čekunov AV (eds) Stroenie zemnoy kory i verkhnej mantii po dannym seismicheskikh issledovanij. Naukova dumka, Kiev pp 35 - 40
Beránek B, Zátopek A (1981) Earth's crust structure in Czechoslovakia and in Central Europe by methods of explosion seismology. In: Zátopek A (ed) Geophysical Syntheses in Czechoslovakia. Veda, Bratislava, pp 243 - 264
Beránek B, Zounková M (1977) Investigations of the earth's crust in Czechoslovakia using industrial blasting. Stud. Geophys. Geod, Academia, Praha 21: 273 - 280
Beránek B, Zounková M (1979) Principal results of deep seismic soundings. In: Vaněk J (ed) Geodynamic Investigations in Czechoslovakia. Veda, Bratislava, pp 105 - 111
Beránek B, Weiss J, Hrdlička A, Zounková M, Fejfar M (1971) Chekhoslovackaya Socialisticheskaya Respublika. Rezul'taty glubinnogo seismicheskogo zondirovaniya. In: Strojenie zemnoy kory centralnoy i jugovostočnoy Evropy (po dannym vzryvnoy seismologii). Naukova dumka. Kijev, pp 182 - 204

Beránek B, Dudek A, Fejfar M, Hrdlička A, Suk M, Zounková M, Weiss J (1972) The crustal structure of Central and South - Eastern Europe based on the results of explosion seismology. Czechoslovakia. Geophys Trans, spec eds, Budapest pp 87 - 98

Beránek B, Dudek A, Zounková M (1975) Rychlostní modely stavby zemské kůry v Českém masívu a Západních Karpatech. Sbor Geol Věd UG, Praha 13: 7 - 20

Beránek B, Konírová L, Mayerová M, Zounková M (1977) Výsledky hlubinných seizmických sondáží na mezinárodním profilu VII. Sbor. Geol. Věd UG, Praha 14: 99 - 117

Beránek B, Leško B, Mayerová M (1979) Interpretation of seismic measurements along the Trans - Carpathian profile K III. In: Geodynamic Investigations in Czechoslovakia, Veda, Bratislava, pp 201 - 205

Beránek B, Suk M, Weiss J (1980) Geological sections through the Variscan Orogene in the Bohemian Massif. Sbor Geol Věd Geol, Praha 34: 7 - 29

Beránek B, Schenk V, Firbas P, Mayerová M, Zátopek A, Schulze A (1981) The Fennolora experiment and crustal structure in the Bohemian Massif. Proc. of the 17th Assembly of the ESC, Budapest 1980, Akadémiai Kiadó, Budapest, pp 479 - 481

Berdichevski M N, Dmitriev V I (1976) Basic principles of interpretation of megnetotelluric sounding curves. In: Ádám A (ed) Geoelectric and Geothermal studies, KAPG Geophysical Monograph. Akadémiai Kiadó, Budapest, 165 p

Bernard JH, Klomínský J (1975) Geochronology of the Variscan plutonism and mineralization in the Bohemian Massif. Věstník ÚÚG, Praha 1; 50: 71 - 81

Bernard JH, Klomínský J, Polanský J (1976) Relationship between the metallogeny and the gravimetric pattern of the Bohemian Massif. Věstník ÚÚG, Praha 2; 51: 65 - 74

Bernard JH, Griffiths M, John B M (1981) REE geochemistry of eclogites and associated rocks from Sauviat sur Vige, Massif Central France. Lithos, Oslo 14: 263 - 274

Bernard JH, Krs M, Morávek P, Polanský J (1983) Endogenous mineralization in the Bohemian Massif in relation to global - tectonics concepts: confrontation of geophysics and metallogeny. Sbor Geol Věd UG, Praha 18: 11 - 51

Berthelmes F, Kautzleben H (1983) A new method of modelling the gravity field of the Earth by point masses. General Assembly of IAG, Symposium Improved gravity field estimation on a global basis, Hamburg

Bezuglaja LS, Kapička A (1981) Piezomagnitnye kharakteristiki basaltov bogemskogo massiva. Sb. II. s'yazda geom. pole i magnetizma gornykh porod. Tbilisi, 94p .

Biela A (1978) Hlboké vrty v zakrytých oblastiach vnútorných Západných Karpát. Region Geol Západ Karpát, Bratislava 10, 11

Bielik M, Burda M, Fusán O, Hübner M, Škorvanek M, Vyskočil V (1986a) Geologicko - hustotný model zemskej kory pozdĺž profilu HSS V. In: Blížkovský M Geofyzikální model litosféry. Záv. zpráva úkolu OFTR ČGÚ č. 82 - 160 za období 1982 - 1985. Praha - Brno MS GFÚ ČSAV - GFB, 127 - 129

Bielik M, Fusán O, Plančár J, Biela A, Túnyi I (1986b) Niektoré nové poznatky o hlbinnej stavbe podunajskej panvy. Geol Práce Spr, Geol. úst. D. Štúra, Bratislava 84: 119 - 134

Bielik M, Majcin D, Fusán O, Burda M, Trešl J, Vyskočil V (1991) Plotnostnye i geotermitcheskie modelirovania Zapadnych Karpat. Litosphera centralnoj i vostočnoj Evropy, T. VIII, Metodika kompleksnoi interpretacii, Naukova dumka, Kiev (in press)

Biely A (1975) Poznámka k termínu "Peripieninský lineament". - Geol. Práce Spr, Geol. úst. D. Štúra, Bratislava 63: 205 - 209

Biely A, Fusán O (1967) Zum Problem der Wurzelzonen der subtatrischen Decken. Správy, Geol. úst. D. Štúra, Bratislava 42: 51 - 64

Birch F (1960) The velocity of compressional waves in rocks to 10 kilobars. Part 1. J Geophys Res, Washington 65; 4: 1083 - 1102

Birch F (1961) The velocity of compressional waves in rocks to 10 kilobars. Part 2. J Geophys Res, Washington 66; 7: 2199 - 2224

Birkenmajer K (1985) Major strike - slip faults of the Pieniny Klippen Belt and the Tertiary rotation of the Carpathians. Publ Inst Geophys, Warszawa A - 16; 175: 101 - 115

Blížkovský M (1973) Isostasy in the eastern part of the Bohemian Massif. Sbor Geol Věd UG, Praha 11: 7 - 18

Blížkovský M (1978) System of programs for calculation of a stripped gravity map of the Bohemian Massif. Geofyz Sbor, Praha XXVI, 504: 103 - 110

Blížkovský M (1982) Comments on the stripped gravity map of the Bohemian Massif. 44th E. A. E. G. Meeting, 1982, Cannes, 312 p
Blížkovský M, Novotný A (1981) Construction of stripped gravity map of the Bohemian Massif. In: Zátopek (ed) Geophysical Syntheses in Czechoslovakia, Veda, Bratislava, pp 149 - 152
Blížkovský M, Pokorný L, Weiss J (1975) Structural scheme of the Bohemian Massif based on geophysical data. Věstník ÚÚG, Praha 50, 1: 1 - 8
Blížkovský M, Dvořák J, Kadlec E, Novotný A (1977) Interpretation of derived gravity maps from the Jesenik Mts. and adjacent areas. Sbor Geol Věd UG, Praha 14: 7 - 21
Blížkovský M, Matolín M, Novotný A, Šalanský K (1984) Regionální geofyzikální mapy pro prognózování rudonosnosti Českého masívu. Geol Průzk, Praha 26; 4: 97 - 101
Blížkovský M, Burda M, Ibrmajer J, Jakubcová I, Pick M, Suk M, Vyskočil V (1985a) Modelling of the density distribution of the Bohemian Massif. Density distribution of the lithosphere, ETH, Zürich, pp 1 - 11
Blížkovský M, Novotný A, Suk M (1985b) Přehled tíhových struktur Českého masívu. Věstník ÚÚG, Praha 60; 3: 143 - 154
Blížkovský M (ed) (1986a): Geofyzikální model litosféry. Záv. zpráva úkolu OFTR ČGÚ č. 82 - 160 za období 1982 - 1985. MS GFÚ ČSAV, Praha - Geofyzika, Brno
Blížkovský M, Dudek A, Ibrmajer I, Mísa_ Z, Suk M, Tomek Č (1986) Nový reflexně seizmický hlubinný profil v moldanubiku. Geol Průzk, Praha 28; 2: 33 - 36
Blížkovský M, Fusán O, Ibrmajer J, Plančár J, Suk M (1986) Geophysical phenomena of the deep structure in Czechoslovakia. J Geodyn, Amsterdam 5: 165 - 178
Bodri L, Bodri B (1985) On the correlation between heat flow and crustal thickness. Tectonophysics, Amsterdam 120: 69 - 81
Boldizsár T (1968) Geothermal data from the Vienna basin. Geophys Res Lett, Washington 73: 613 - 618
Bonhomme M, Fluck P (1981) Nouvelles donés isotopoques Rb/Sr obtenues sur les granulites des vosgeses. Age protérozoique terminal de la série volcanique calcoalcaline et age acadien du metamorphism régional. C R Acad Sci B, Paris 293; 10: 771 - 774
Bořecký V (1988) Petrologické studium terciérních vulkanitů na Teplicku. Acta Univ Carol Geol, Praha, pp 491 - 496
Borissoff BA, Reisner GI, Sholpo VN (1975) Vydeleniye seismoopasnykh zon v alpinskoy skladchatoy oblasti. Nauka, Moskva
Borissoff BA, Reisner GI, Sholpo VN (1976) Tectonics and maximum magnitudes of earthquakes. Tectonophysics, Amsterdam 33: 167 - 185
Borissoff BA, Reisner GI, Sholpo VN (1978) The geological basis for the analysis of seismic risk. Proc. of the Symposium on the Analysis of Seismicity and on Seismic Risk, Liblice, 1977 Academia, Praha, pp 59 - 67
Bortnickaja VM, Burov VS, Vishnyakov IB, Karpenthuk JR, Petrashevich MI, Pilinchuk AS, Prebtechenskaya NS, Sviridenko VG, Sovchik JV (1977) Obosnovenie napravlenij poiskov nefti i gaza v glubokozalegayushchikh gorizontakh Ukrainskykh Karpat. Naukova dumka, Kiev
Bott MHP (1971) The interior of Earth. Arnold, London, 220 p
Bott MHP, Kusznir N J (1984) The origin of tectonic stress in the lithosphere. Tectonophysics, Amsterdam 105: 1 - 13
Bram K (1979) Heat flow measurements in the Federal Republic of Germany. In: Čermák V, Rybach L (eds) Terrestrial Heat Flow in Europe. Springer, Berlin Heidelberg New York, pp 190 - 196
Brown GC (1977) Mantle origin of Cordilleran granites. Nature, London 265; 5589: 21 - 24
Brown GC (1979) The changing pattern of batholith emplacement during Earth history. In: Atherton MP, Tarney J (eds) Origin of granite batholiths: Geochemical evidence. Orpington, Shiva, pp 106 - 115
Brueckner HK (1977) A crustal origin for eclogites and mantle origin for garnet peridotites: strontium isotopic evidence from elinopyroxenes. Contrib Mineral Petrol 60: 1 - 15
Brune J, Dorman J (1963) Seismic waves and Earth structure in the Canadian shield. Bull Seism Soc Am 53: 167 - 205
Bucha V (1957) Normální geomagnetické pole vertikální složky k epoše 1950,0. Trav Géophys 1956, Academia, Praha, pp 577 - 600
Bucha V (1968) Vorläufige Karten der Anomalien der geomagnetischen Vertikalintensität auf dem Gebiet von Mitteleuropa. Věstník ÚÚG, Praha 43, 2

Bucha V (1980) Geomagnetism of the External Flysch Czechoslovakian Carpathians and the possible causes of anomalous geophysical manifestations. Stud Geophys Geod, Academia, Praha 24: 227 - 251

Bucha V (1981) Possible causes of geophysical anomalies in the Czechoslovakian Carpathians. In: Zátopek A (ed) Geophysical Syntheses in Czechoslovakia. Veda, Bratislava, p 289 - 296

Bucha V (1986) Využití geomagnetických a seismických dat pro zjišťování hloubky krystalinika v oblasti JV Moravy. In: Sborník referátů "Výzkum hlubinné stavby Československa", Loučná, pp 149 - 152

Buday T, Kodym O, Maheľ M, Máška M, Matějka A, Svoboda J, Zoubek V (1961) Tektonický vývoj Československa. Sborník prací a tektonická mapa 1:1 000 000, Ústřední ústav geologický, NČSAV, Praha, p 254

Buday T, Dudek A, Ibrmajer J (1969) Některé výsledky interpretace gravimetrické mapy ČSSR v měřítku 1:500 000. Sbor Geol Věd UG, Praha 8: 7 - 30

Bune VI, Gorshkov GP, (1980) Seysmicheskoe rayonirovanie territorii SSSR. Nauka, Moskva

Bune VI, Kárník V, Schenková Z, Schenk V (1983) Seismo - Geological definition of earthquake source regions. Proc. of the Meeting of the W. G. 4.3. KAPG, Bratislava 1979, 79 - 86

Burchart L (1968) Rubidium - strontium isochrone ages of the crystalline core of the Tatra Mountains, Poland. Am J Sci, New Haven 266: 895 - 905

Burda M, Vyskočil V (1991) Some question of density modelling of the Earth's crust and upper mantle. In: Proceedings from seminary "Advances in Gravimetry", Smolenice 1990,, Geophysical Institute, Slov. Acad. Sci., Bratislava, pp 19 - 24

Burda M, Hübner M, Vyskočil V (1988) Density inhomogenity in the upper mantle of central Europe and their gravitational effect. Stud Geophys Geod, Academia, Praha, 32: 54

Burg JP, Matte Ph (1978) A cross section trough the French massif Central and the scope of its Variscan geodynamic evolution. Z Dtsch Geol Ges, Hannover 129: 429 - 460

Burša M (1970) Základy kosmické geodezie, MNO. Praha

Burša M (1971) Effect of removed topography and condensation on deflections of the vertical on the territory of Czechoslovakia. Trav Géophys 1969. Academia, Praha 303: 9 - 30

Buryanov VB, Gordienko VV, Kulik SN (eds) (1978) Komplexnaya geofizicheskaya model' litosfery vostochnych Karpat. Geofiz Sbor, Naukova dumka, Kiev 83: 3 - 17

Buryanov VB, Gordienko VV, Kulik SN, Logvinov IM (1983) Kompleksnoe geofizicheskoe isledovanie tectonosfery kontinentov. Naukova dumka, Kiev, 133 p

Buryanov VB, Gordienko VV, Zavgorodnyaya N, Kulik SN, Logvinov IM, Shuman VN (1987) Ggeofizicheskaya model' tectonosfery Europy. Naukova dumka, Kiev, 38 p

Calcagnile C, Panza GF (1980) The lithosphere - asthenosphere system in the Italian area. In: Udias A, Chanel J (eds) Evolution and Tectonics of the Western Mediterranean and Surrounding Areas. Inst Geogr Nac, Madrid, Spec Publ 201: 225 - 230

Calota F, Caltan I, Ionescu A, Soare A, Steflea V (1971) Inregistrari de variatii geomagnetice pe teritoriul Romaniei (Partea I). Stud Cerc Geol Geofiz Geogr Ser Geofiz, Bucuresti 9: 323 - 331

Cambel B, Vilinovič V (1987) Geochémia a petrológia granitoidn_ch hornínMalých Karpát. Veda, Bratislava, p 248

Cambel B, Bagdasarjan CP, Veselský J, Gukasjan RCh, (1980) The problem of interpretation of nuclear - - geochronological data on the age of crystalline rocks of the West Carpathians. Geol Zbor Geol Carpath, Geol. úst. D. Štúra, Bratislava 31; 1 - 2: 27 - 48

Cambel B, Petrík I, Vilinovič V (1985) Variscan granitoids of the West Carpathians in the light of geochemical - petrochemical study. Geol Zbor Geol Carpath, Geol. úst. D. Štúra, Bratislava 36; 2: 204 - 217

Carswell DA, Dawson JB (1970) Garnet peridotite xenoliths in South African kimberlite pipes and their petrogenesis. Contrib Cad Mineral Petrogr Geol Paleontol, Curitiba 25: 163 - 184

Čech F (1982) Ložiská palív - vzťah k hlbinnej stavbe panonskej panvy a karpatského oblúka. Západ Karpáty Sér Geol, Geol. úst. D.Štúra, Bratislava, 8: 146

Čekunov AV (1987) Litosfera centralnoy i vostochnoy Evropy. Geotraverzy I, II, V. Naukova dumka, Kiev, p 167

Čermák V (1967a) Results of geothermic investigation of heat flow in Czechoslovakia in 1964 - 1966. Stud. Geophys Geod, Academia, Praha 11: 342 - 344

Čermák V (1967b) Heat flow in the Kladno - Rakovník coal basin. Gerlands Beitr Geophys, Leipzig 76: 461 - - 466

Čermák V (1967c) Heat flow near Teplice in North Bohemia. Geophys J R Astron Soc, Oxford 13: 547 - 549
Čermák V (1967d) Coefficient of thermal conductivity of some sediments and its dependence on density and water content of rocks. Chem Erde, Jena 26: 271 - 278
Čermák V (1968a) Terrestrial heat flow in Czechoslovakia and its relation to some geological features. Proc 23rd Int Geol Congr 5: 75 - 85
Čermák V (1968b) Heat flow in the Upper Silesian coal basin. Pure Appl Geophys, Genova 69: 119 - 130
Čermák V (1968c) Heat flow in the Žacléř - Svatoňovice basin. Acta Geophys pol, Warszawa 16: 3 - 9
Čermák V (1968d) Terrestrial heat flow in the Alpine - Carpathian foredeep in South Moravia. J Geophys Res, Washington 73: 820 - 821
Čermák V (1968e) Terrestrial heat flow in eastern Slovakia. Trav Géophys 1967, Academia, Praha 275: 3065 - - 319
Čermák V (1972) Heat flow and gravity in Czechoslovakia. Stud Geophys Geod, Academia, Praha 16: 77 - 87
Čermák V (1974) Deep temperature distribution along the deep seismic sounding profile across the Carpathians (model calculation). Acta geol Acad Sci Hung, Budapest 18: 295 - 303
Čermák V (1975a) Combined heat flow and heat generation measurements in the Bohemian Massif. Geothermics, Pisa 4: 19 - 26
Čermák V (1975b) Terrestrial heat flow in the Neogene foredeep and the Flysch zone of the Czechoslovak Carpathians. Geothermics, Pisa 4: 8 - 13
Čermák V (1975c) Temperature - depth profiles in Czechoslovakia and some adjacent areas derived from heat flow measurements, deep seismic sounding and other geophysical data. Tectonophysics, Amsterdam 26: 103 - 119
Čermák V (1976a) High heat flow measured in the Ostrava - Karviná coal basin. Stud Geophys Geod, Academia, Praha 20: 64 - 71
Čermák V (1976b) Zemský tepelný tok ve vrtu Lidečko v magurském flyši ve vnějších Karpatech. Čas. Mineral Geol, Praha 21: 193 - 198
Čermák V (1976c) Terrestrial heat flow in two deep holes in the Ostrava - Karviná coal basin. Věstník ÚÚG, Praha 51: 75 - 84
Čermák V (1977a) Geothermal measurements in Paleogene, Cretaceous and Permo - Carboniferous sediments in northern Bohemia. Geophys J R Astron Soc, Oxford 48: 437 - 541
Čermák V (1977b) Heat flow in five holes in Eastern and Central Slovakia. Earth Planet Sci Lett, Amsterdam 34: 67 - 70
Čermák V (1977c) Geothermal models of the Bohemian Massif (Variscian) and the Western Carpathians (Alpine) and their mutual relation. Tectonophysics, Amsterdam 41: 127 - 137
Čermák V (1977d) Some comments on the effect of the past climatic changes on the underground field. Publ Inst Geophys, Warszawa A - 3 (103): 45 - 56
Čermák V (1978) First heat flow map of Czechoslovakia. Trav Géophys, 461, Academia, Praha, pp 254 - 261
Čermák V (1979a) Review of heat flow measurements in Czechoslovakia. In: Čermák V, Rybach L (eds) Terrestrial Heat Flow in Europe. Springer, Berlin Heidelberg New York, pp 152 - 160
Čermák V (1979b) Heat flow map of Europe. In: Čermák V, Rybach L (eds) Terrestrial Heat Flow in Europe. Springer, Berlin Heidelberg New York, pp 3 - 40
Čermák V (1979c) Geothermal studies and heat - flow map of Czechoslovakia. In: Geodynamic Investigations in Czechoslovakia. VEDA, Bratislava, pp 129 - 132
Čermák V (1981) Heat flow investigations in Czechoslovakia. In: Zátopek A (ed) Geophysical Syntheses in Czechoslovakia. Veda, Bratislava, pp 427 - 439
Čermák V (1982a) Regional pattern of the lithosphere thickness in Europe. In: Čermák V, Haenel R (eds) Geothermics and Geothermal Energy. Schweizerbart Stuttgart, pp 1 - 10
Čermák V (1982b) Crustal temperature and mantle heat flow in Europe. Tectonophysics, Amsterdam 83: 123 - - 142
Čermák V (1989a) Crustal heat production and mantle heat flow in Central and Eastern Europe. Tectonophysics, Amsterdam 159: 195 - 215
Čermák V (1989b) Heat flow in a sedimentary basin in Czechoslovakia: evaluation of data with special attention to hydrogeology. In: Beck AE, Garven G, Stegena L. (eds), Hydrogeological Regime and Their Subsurface Thermal Effect. Geoph. Monogr., 47, IUGG Vol. 2, pp 75 - 80

Čermák V, Bodri L (1986) Two - dimensional temperature modelling along five East - European geotraverses. J Geodyn, Amsterdam 5: 133 - 163
Čermák V, Hurtig E (1979) Heat flow map of Europe 1:5 000 000. In: Čermák V, Rybach L (eds) Terrestrial Heat Flow in Europe. Springer, Berlin Heidelberg New York
Čermák V, Jetel J (1985) Heat flow and ground water movement in the Bohemian Cretaceous Basin (Czechoslovakia). J Geodyn, Amsterdam 4: 285 - 303
Čermák V, Krčmář B (1967) Tepelný tok ve vrtu NV - 1 (Nová Ves u Chýnova). Věstník ÚÚG, 42, Praha, pp 445 - 448
Čermák V Krčmář B (1968a) Mě_ení tepelného toku ve dvou šachtách v západních a jižních Čechách. Věstník ÚÚG, Praha 43: 415 - 422
Čermák V Krčmář B (1968 b) Der Erdwärmeluss in der südwestlichen Slowakei. Acta Geod Geophys Montan, Budapest 3: 319 - 329
Čermák V, Krešl M (1968) Thermal conductivity of rocks and its measurements on a divided - bar apparatus. Trav Géophys 274, 1967 Academia, Praha , pp 287 - 304
Čermák V, Rybach L (1982) Thermal conductivity and specific heat of minerals and rocks. In: Angenheister G (ed) Physical Properties of Rocks. Landolt - Börnstein New Series, Vol Vla. Springer, Berlin Heidelberg New York, pp 305 - 343
Čermák V, Jetel J, Krčmář B (1968a) Terrestrial heat flow in the Bohemian Massif and its relation to the deep structure. Sbor Geol Věd UG, Praha 7: 25 - 38
Čermák V, Kárník V, Krčmář B (1968b) Měření teploty a tepelný tok v příbramských dolech. Čas Mineral Geol, Praha 13: 215 - 216
Čermák V, Krešl M, Veselý I (1968c) Experimental determination of the coefficient of heat transfer during hole boring and the re - establishment of the temperature field equilibrium. Earth Planet Sci Lett, Amsterdam 5: 153 - 158
Čermák V, Krešl M, Novák V, Venhodová D (1975) Heat flow in Czechoslovakia: Relationship with geological age. In: Recent Research in Geology.. Hindustal Publ. Corp. (India), New Delhi, pp 283 - 288
Čermák V, Pěčová J, Praus O (1976) Heat flow, crustal temperatures and geoelectrical cross - section in Czechoslovakia.In: Ádám A (ed) Geoelectric and Geothermal Studies (east - Central Europe and Soviet Asia), Akadémiai Kiadó, Budapest, pp 538 - 542
Čermák V, Krešl M, Lizoň I, Marušiak I, Veselý I (1977): Geothermal investigation in Czechoslovakia 1974 - - 1975 (progress Report). Publ Inst Geophys, Warszawa A - 3, (103): 13 - 26
Čermák V, Šafanda J, Guterch A (1989) Deep temperature distribution along three profiles crossing the Teisseyere - Tornquist Tectonic zone in Poland. Tectonophysics, Amsterdam 164: 151 - 163
Čermák V, Král M, Krešl M, Kubík J, Šafanda J (1991a) Heat flow, regional geophysics and lithosphere structure in Czechoslovakia and adjacent part of Central Europe. In: Čermák V and Rybach L (eds) Terrestrial Heat Flow and the Lithosphere Structure. Springer Berlin Heidelberg New York, 133 - 165
Čermák V, Král M, Kubík J, Šafanda J, Krešl M, Kučerová L, Jánčí J, Lizoň I, Marušiak I (1991b) Subsurface temperature and heat flow density maps on the territory of Czechoslovakia. In: Hurtig E, Čermák V, Haenel R, Zui V (eds) Geothermal Atlas of Europe, Gotha, (in press)
Červ V, Pek J, Praus O (1984) Models of geoelectrical anomalies in Czechoslovakia. J Geophys, Berlin 55: 161 - - 168
Červ V, Pek J, Praus O (1987) Numerical modelling of geoelectrical structure in Czechoslovakia. Phys Earth Planet Inter, Amsterdam 45: 170 - 178
Chaloupský J (1986) Tectonostratigraphic units of the Bohemian Massif. Čas Mineral Geol, Praha 31: 387 - 393
Chapman DS, Pollack H, Čermák V (1979) Global heat flow with special reference to the region of Europe. In: Čermák V, Rybach L (eds) Terrestrial Heat Flow in Europe. Springer, Berlin Heidelberg New York, 41 - 48
Cidlinský K (1980) Seismic refraction problem in anisotropic media of the cretaceous plateau in Czechoslovakia. Stud Geophys Geod, Academia, Praha 24: 98
Cidlinský K, Ibrmajer I (1986) Příspěvek reflexní seizmiky k poznání geologické stavby plzeňské pánve a jejího hlubšího podloží. Věstník ÚÚG, Praha 61; 1: 11 - 21
Clark SP Jr (1961) Heat flow in the Austrian Alps. Geophys J R Astron Soc, Oxford 6: 54 - 63
Cogné J (1983) The Precambrian in the metamorphic terrains of the Ligerian domain and the problem of Precambrian geodynamic evolution in the mobile Variscan belt in Western France. In: Zoubek V (ed) Precambrian of the younger foldbelts .....(in press)

Csillag J, Földessy J, Zelenka T, Balázs E (1983) The plate tectonic setting of the Eocene colcanic belt in the Carpathian Basin. In: Bisztricsány E, Szeidowitz G (eds) Prof. of the 17th Assembly of the ESC, Budapest pp 589 - 599

Davis PM (1983) Tectomagnetism. Rev Geophys Space Phys 21; 3: 685 - 693

DEKORP Research Group (1985) First result and preliminary interpretation of Decorp 2 - South. J Geophys, Berlin 57: 137 - 163

Dempsey HJ, Scharbert HG (1975) Phase relations in Moldanubian garnet pyroxenites (Bohemian Massif of lower Austria) An experimental study. In: Ford C E (ed) Progress in experimental petrology. N. E. R. C. Publ Ser 18: 70 - 73

Dewey JF (1982) Plate tectonics and the evolution of the British Isles. J Geol Soc, Oxford 139: 371 - 412

Doležal J (1980) Hlubinná stavba autochtonního podloží Karpat na jihovýchodní Moravě podle gravimetrických a geomagnetických dat. Západ Karpaty Sér Geol, Geol. úst. D. Štúra, Bratislava 6: 113 - 140

Dövényi P, Horváth F, Liebe P, Gálfi J, Erki I, (1983) Geothermal conditions of Hungary. Geophys Trans Geof Közlem, Budapest 29: 1 - 114

Drimmel J, Procházková D (1985) Austrian earthquake of April 14th, 1983, with uncommon epicentre in Northern Styria. Ann Geophys, Paris 3: 539 - 543

Drummond BJ (1985) Seismic P - wave anisotropy in subcrustal lithosphere of the north - west Austria. Geophys J R Astron Soc, Oxford 81: 497

Dudek A (1971) Chemical composition of Moldanubian eclogites and their ganets. Krystalinikum, Praha 7: 167 - 181

Dudek A (1980) The crystalline basement block of the Outer Carpathians in Moravia: Bruno-Vistulicum. Rozpr Čs Akad Věd, _ mat přír Věd, Academia, Praha 90; 8: 1 - 85

Dudek A, Fediuková E (1974) Eclogites of Bohemian Moldanubicum. N Jb Miner Abh Stuttgart 121: 127 - 159

Dudek A, Matějovská O, Suk M (1974) Gföhl orthogneiss in the Moldanubicum of Bohemia and Moravia. Krystalinikum, Praha 10: 67 - 78

Eliáš M, Uhmann J (1968) Hustoty hornin ČSSR. ÚÚG, Praha, p 60

Embey - Isztin A (1976) Amphibolite/lherzolite composite xenoliths from Szigliget, North of the Lake Balaton, Hungary. Earth Planet Sci Lett, Amsterdam 31: 297 - 304

Emmermann R, Giese P (eds) KTB-Report 90-4. Beiträge zum 3. KTB-Kolloquium, Giessen

Emmermann R, Rischmüller H (1990) Das kontinentale Tiefbohrprogramm der Bundesrepublik Deutschland (KTB). Geowissenschaften 8

Erst T (1965) Do peridotitic inclusions in basalts represent mantle material. The Upper Mantle symposium, New Delhi 1984, Copenhagen, pp 180 - 186

Farský F (1976) Mineralogische Notizen 1. Mineralien aus den Kosekover Basaltkugeln. Verh K K Geol Reichsanst, Wien, pp 205 - 208

Fediuk F (1973) Ultramafitity krkonošsko - jizerské oblasti. Acta Univ Carol, 1971, Praha 4: 319 - 343

Fediuk F (1981) Ultramafické pecky v bazaltoidech. In: Bernard J H (ed) Mineralogie Československa, Academia, Praha

Fediuk F, Fediuková E (1985) The ultrabasic body with eclogite in the Berek cuarry. Guidebook of 2nd Intern Eclogite Conf, Praha, pp 38 - 47

Fediuk F, Fediuková E (1988) Ultramafické nodule severomoravských basaltoidů. Sbor Geol Věd, Praha 6: 355 - 382. .

Fediuková E (1988) Eclogites in Czechoslovakia. Krystalinikum, Praha 20: 24 - 52

Fiala F (1954) Některé uzavřeniny z andenzitů Slovenského Středohoří. Sbor Ústř Úst Geol, Odd Geol, Praha 21: 309 - 333

Fiala J (1966) The distribution of elements in mineral phases of some garnet peridotites from the Bohemian Massif. Krystalinikum, Praha 4: 31 - 53

Fiala J (1985) Serpentinized garnet lherzolite with intercalations of garnet clinopyroxenitee. Guidebook of 2nd Intern. Eclogite Conf., Praha, pp 48 - 53

Fiala J, Paděra K (1984) Eclogitic rocks from the basement of the Český středohoří Mts. Krystalinikum, Praha 17: 33 - 45

Fiala J, Matějovská O, Vaňková V (1987) Moldanubian granulites and related rocks: petrology, geochemistry and radioactivity. Rozp Čs Akad Věd _ mat přír Věd, Praha 97; 1: 1 - 102

Filo M, Medo S (1977) Výsledky interpretácie magnetických anomálií Slovenska. In: Banícke listy, mimoriadne číslo. Geofyzikálne interpretačné metódy. Veda. Bratislava, pp 122 - 130
Finckh P, Ansorge J, Mueller St (1984) Deep crustal reflections from a Vibroseis survey in northern Switzerland. Tectonophysics, Amsterdam 109; 1/2: 1 - 14
Finlayson DM (1982) Geophysical differences in the lithosphere between Phanerozoic and Precambrian in Australia. Tectonophysics, Amsterdam 84: 287 - 312
Fišera M (1981) Geologické a petrografické poměry okolí Plaňan. Výzk Práce Ústř Úst Geol, Praha 26: 1 - 42
Fisher R (1953) Dispersion on a sphere. Proc R Soc, A, London 217: 295 - 305
Forbes RB, Kuno H, (1967) Peridotite inclusions and basaltic host rocks. In: Wyllie PJ Ultramafic and related rocks. Wiley, New York, pp 328 - 337
Fountain DM, Hurich ChA, Smithson SB (1984) Seismic reflectivity of mylonite zones in the crust. Geology, Boulder, 12; 4: 125 - 198
Franke W (1986) Development of the Central European Variscides - A review. In: Freeman R, Mueller St, Giese P, Munsch B, Fratta M (eds) Proc. of the 3rd Workshop on the EGT Project, ESF Strasbourg, pp 65 - 72
Frey FA (1970) Rare earth abundances in alpine ultramafic rocks. Phys Earth Planet Inter, Amsterdam 3: 323 - - 330
Fuchs G, Matura A (1980) Die Böhmische Masse in Österreich. In: Der geologische Ban Österreichs, Wien, pp 121 - 142
Fusán O (1985) Náčrt hlbinnej stavby Západných Karpát. Miner Slov, Spišská Nová Ves 17, 3, 1985: 204
Fusán O, Kuthan M, Ďuratný S, Plančár J, Zbořil L (1969) Geologická stavba podložia stredoslovenských neovulkanitov. Zbor Geol Vied, Západ. Karpaty, Bratislava 10: 159
Fusán O, Ibrmajer J, Plančár J, Slávik J, Smíšek M (1971) Geologická stavba podložia zakrytých oblastí južnej časti vnútorných Západných Karpát. Západ Karpaty Sér Geol, Geol. úst. D. Štúra, Bratislava 15: 1 - 173
Fusán O, Ibrmajer J, Plančár J (1979) Neotectonic Blocks of the West Carpathians. Geodyn Invest Czech (Final Rep): 187 - 192
Fusán O, Ibrmajer J, Kvitkovič J, Plančár J (1981) Block dynamics of the West Carpathians. In: Zátopek A (ed) Geophysical Syntheses in Czechoslovakia. Veda, Bratislava, pp 153 - 158
Fusán O, Biely A, Ibrmajer J, Plančár J, Rozložník L (1987) Podložie terciéru vnútorných Západných Karpát. Geol. úst. D. Štúra, Bratislava, pp 1 - 124
Garmann LB, Brunfelt AO, Finstad KG, Heier KS (1975) Rare Earth element distribution in basic and ultrabasic rocks from East Norway. Chem Geol, Amsterdam 15: 103 - 116
Gebauer D, Gruenenfelder M (1979) U - Pb zircon and Rb/Sr mineral dating of eclogites and their country rocks. Example: Munchberger gneiss Massif, Bavaria. Earth Planet Sci Lett, Amsterdam 42: 35 - 44
Giese P (1980) Krustenstruktur der Alpen. Berl Geowiss Abh., R. A., Bd 20
Giese P (1884) Struktura zemnoj kory orogennych oblastěj. 27. MGK, C 08, Geofyzika, Brno pp 21 - 27
Giese P (1985) Activities on the central segment of the European geotraverse. Terra Cognita, Paris 5; 3/4: 160
Giese P, Prodehl C, Stein A (eds) (1976) Explosion Seismology in Central Europe. Springer, Berlin Heidelberg New York
Giesel W (1963) Elastische Anisotropie in tektonisch verformten Sedimentgesteinen. Geophys Prospect, The Hague, p 423
Girdler RW (1985) Density Contrast at the Lithosphere - Asthenosphere Boundary (abstract). - In: Bericht Nr. 102, Inst. für Geodäsie and Photogrammetrie ETH, 21. Zürich
Godard G (1982) Lambeaux protolites d'une croute océanique sabductée: les eclogites de Vendée, Massif Armorican, France.Terra Cognita, Paris 2: 212
Godard G, Cabanis B (1985) Eclogites in the "Pays de Leon", North - West Brittany, France. Terra Cognita, Paris 421
Golub G (1965) Numerical methods for solving linear least squares probleme. Numer Math, Heidelberg 7: 206 - - 216
Grecula P (1982) Gemerikum - segment riftogénneho bazénu Paleotetýdy. Miner Slov, Monogr Ser, Spišská Nová Ves p 288
Grecula P, Roth Z (1978) Kinematický model Západních Karpat v souborném řezu. Sbor Geol Věd Geol, Praha 32: 49 - 73

Grecula P, Hovorka D (1987) Early Paleozoic volcanism of the Western Carpathians. In: Pre-Variscan and Variscan events in the Alpine-Mediterranean mountain belt. Flügel, Sassi, Grecula (eds) Miner Slov, Monogr Ser, Alfa, Bratislava, pp 251 - 270

Green DH, Ringwood AE (1972) A comparison of recent experimental data on the gabbro-garnet granulite-eclogite transition. J Geol 80: 277 - 288

Grosser H, Köhler W (1988) Fault plane solutions of the main events during the earthquake swarm 1985/86 in Western Bohemia/Vogtland. Proc. Workshop on "Induced Seismicity and Associated Phenomena", II, Liblice, pp 49 - 64

Grummt S (1988) Die Untersuchung der Struktur der Erdkruste unter den seismischen Stationen KHC und KSP mit langperiodischen Raumwellen. Trav Geophys, Geofyzikální Sborník XXXVI, Academia, Praha

Grünthal G, Bankwitz P, Bankwitz E, Bednarik B, Schenk V, Schenková Z, Zeman A (1985) Seismicity and geological features of the eastern part of the west European platform. Gerlands Beitr Geophys, Leipzig 94: 276 - 289

Grünthal G, Schenk V, Zeman A, Schenková Z (1990) Seismotectonic model for the earthquake swarm of 1985 - 1986 in the Vogtland/West Bohemia focal area. Tectonophysics, Amsterdam 174: 369 - 383

Gutdeutsch R, Aric K (1977) Eine Diskusion geophysikalischer Modelle des Grenzbereiches zwischen Ostalpen und pannonischen Becken. Acta Geol, Zagreb 21: 287 - 296

Haack I (1982) Radioactive isotopes in rocks. In: Angenheister G (ed) Physical Properties of Rocks. Landolt - Börnstein New Ser, Springer, Berlin Heidelberg New York, pp 433 - 560

Haenel R, Zoth C (1973) Heat flow measurements in Austria and heat flow maps of Central Europe. Z Angew Geophys, Berlin 38: 425 - 439

Halousek J, Pešina B, Příhoda K (1986) A new rotating coil magnetometer. In: Internat. Assoc. Geomag. & Aeronomy, 5th General Assembly, Prague. Abstractum 2: 351 (abstr). Praha

Herrmann H (1986) Jednoduchý přepočet tíhové anomálie do čtvercové sítě. Geol Průzk, Praha 28; 2: 52 - 53

Hirajima T, Banno S, Hiroi Y, Ohta Y (1985) Eclogites and related rocks from the Vestgotabreen formation in Spitzbergen. Sec. Inter. Eclogite Conf. Wien, Terra cognita 5: 421

Hohenegger L (1881) Die geognostischen Verhältnisse der Nordkarpaten in Schlesien und den angrenzenden Teilen von Mähren und Galizien, Gotha. p 186

Horváth F, Bodri B, Ottlik P (1979) Geothermics of Hungary and the tectonophysics of the Pannonian Basin "Red Spot". In: Čermák V, Rybach L (eds) Terrestrial Heat Flow in Europe. Springer, Berlin Heidelberg New York, pp 206 - 217

Hovland J, Husebye ES (1982) Upper mantle heterogeneities beneath Eastern Europe. Tectonophysics, Amsterdam 90; 1/2: 137 - 150

Hovorka D (1974) The lithology and the chemical composition of metasediments of the Jarabá Group (West Carpathians). Krystalinikum, Praha 11: 87 - 99

Hovorka D (1979) Origin of the West Carpathian Hercynian Magmatites. Geol Práce Spr, Geol. úst. D. Štúra, Bratislava 72: 131 - 148

Hovorka D (1980) The West Carpathians crust origin and plutonite formations. Geol Zbor Geol Carpath, Geol. úst. D. Štúra, Bratislava 31; 4: 523 - 535

Hovorka D (1986) Staropaleozoický kyslý vulkanizmus centrálnej zóny Západných Karpát. Miner Slov, Spišská Nová Ves 18; 1: 15 - 32

Hovorka D, Fejdi P (1980) Spinel, peridotite, xenoliths in the West Carpathian late Cenozoic alkali basalts and their tectonic significance. Bull Volcanol, Napoli 33; 1: 95 - 106

Hovorka D, Fejdi P (1983) Garnets of peraluminous granites of the Suchý and the Malá Magura Mts. (the Western Carpathians) - their origin and petrological significance. Geol Zbor Geol Carpath, Geol. úst. D. Štúra, Bratislava, 34; 1: 103 - 116

Hovorka D, Lukáčik E (1972) Xenoliths in andesites of the massifs Karanč and Šistor (Southern Slovakia) and their geological interpretation. Geol Zbor Geol Carpath, Geol. úst. D. Štúra, Bratislava 23; 2: 297 - 309

Hovorka D, Méres Š (1989a) Relics of high-grade metamorphic in Tatraveporic crystalline of the West Carpathians (English summary of the Slovak text). Miner Slov, Alfa, Bratislava 21; 3: 193 - 201

Hovorka D, Méres Š (1989b) Polycyclic development of Pre-Carboniferous complexes in the Western Carpathian Mts.: New data and their interpretation. XIVth Congr. Carpatho-Balksn Geol. Assoc., Extended Abstracts, Sofia 294 - 297

Hovorka D, Spišiak J (1983) REE geochemistry and petrological model for the generation of Variscan granitoids in the West Carpathians. Miner Slov, Spišská Nová Ves 15; 2: 97 - 116
Hovorka D, Spišiak J (1987) Vulkanizmus mezozoika Západných Karpát. Veda, Bratislava, pp 1 - 247
Hovorka D, Ivan P, Jaroš J, Kratochvíl M, Reichwalder P, Rojkovič I, Spišiak J, Turanová L (1985) Ultramafic Rocks of the Western Carpathians, Czechoslovakia. Geol. úst. D. Štúra, Bratislava, p 258
Hurtig E, Oelsner C (1979) The heat flow on the territory of the German Democratic Republic. In: Čermák V, Rybach L (eds) Terrestrial Heat Flow in Europe. 1986 - 190. Springer, Berlin Heidelberg New York
Hurtig E, Oesberg P, Ritter E, Grünthal G, Jacobs F (1975) Studies of the crustal structure in the Northern part of Central Europe. Gerlands Beitr Geophys, Leipzig 84; 3/4: 317 - 325
Ibrmajer J (1963) Gravimetric map of Czechoslovakia on 1:200 000 Scale. Stud Geophys Geod, Praha 7, 303 - - 308
Ibrmajer J (1965) Přehledná gravimetrická mapa ČSSR 1:500 000. ÚGÚ, Praha
Ibrmajer J (1981) Geological interpretation of gravity maps of Czechoslovakia. In: Zátopek A (ed) Geophysical Syntheses in Czechoslovakia, Veda, Bratislava, pp 135 - 148
Ibrmajer J, Blížkovský M, Fejfar M, Ibrmajer I, Mayerová M, Novotný M, Tomek Č (1983) Some new results of deep seismic sounding in Czechoslovakia. 28th International geophysical symposium Balatonszémes. Budapest, pp 819 - 835
Ibrmajer J, Suk M et al. (1989) Geofyzikální obraz ČSSR, 354. Academia, Praha
Illies JH, Fuchs K (1983) Plateau - uplift of the Rhenish Massif - Introductory remarks. In: Fuchs K, von Gehlen K, Mälzer H, Morawski H, Semmel A (eds) Plateau Uplift. Springer, Berlin Heidelberg New York, pp 1 - 8
Jakeš P (1969) Retrogressive changes of granulite facies rocks - an example from the Bohemian Massif. Spec Publ Geol Soc Aust, Canberra 2: 367 - 374
Jakeš P (1978) Raná kůra Země. Vesmír, Praha 56: 195 - 196
Jakeš P, Vokurka K (1987) Central Europe. In: Nixon J P (ed) Mantle xenoliths. Wiley, London 149 - 154
Jakeš P, Jelínek E, Mísař Z (1985) Ultramafic rocks within granulites - a type of crust - mantle transition in mountain belt. Terra Cognita, Paris 5; 3/4: 338
Jankowski J, Petr V, Pěčová J, Praus O (1984) Geoelectric anomaly in the Czechoslovak - Polish section of the Carpathians on the basis of geomagnetic and magnetotelluric soundings. Acta Geod Geophys Montan, Budapest 19 (1 - 2): 81 - 91
Jankowski J, Tarlowski Z, Praus O, Pěčová J, Petr V (1985) The results of deep geomagnetic soundings in the West Carpathians. Geophys J R Astron Soc, Oxford 80: 561 - 574
Janků J, Pospíšil L, Vass D (1984) Príspevok diaľkového prieskumu Zeme k poznaniu stavby Západných Karpát (západný úsek). Miner Slov, Spišská Nová Ves 16; 2: 121 - 137
Jaroš J, Mísař Z (1974) Deckenbau der Svratka - Kuppel und seine Bedeutung für das geodynamische Modell der Böhmischen Masse. Sbor Geol Věd Geol, Praha 26: 29 - 79
Jaupart C, Sclater JC, Simmons G (1981) Heat flow studies: Constrains on the distribution of uranium, thorium and potassium in the continental crust. Earth Planet Sci Lett Amsterdam 52: 328 - 344
Jech J, Babuška V, Pros Z (1985) Quantitative correlation of velocity anisotropy with the distribution of microcracks in rocks, Physical Properties of Mineral System of the Earth's Interior. Publ Union of Czechoslovak Mathematicians and Physicists. Praha 175
Jegorkin AV (1966) Analiz tochnosti opredeleniya skorostnykh parametrov razreza zemnoy kory po godografam voln. Izv Akad Nauk SSSR Fiz Zemli, Moskva 9: 72 - 81
Jelínek E (1985) Garnet peridotite with three types of eclogite - Nové Dvory and Mstěnice. 2nd Intern. Eclogite Conf. Excursion guidebook, Praha, pp 11 - 16
Jelínek E, Souček J (1981) Geochemie sobotínského a jesenického amfibolitového masívu. Acta Univ Carol Geol, Praha 4: 379 - 411
Jelínek E, Pačesová M, Mísař Z, Martinec P, Weiss J (1984) Geochemistry of a dismembered metaophiolite complex Letovice, Czechoslovakia. Trans R Soc Earth Sci, Edinburgh 75: 37 - 48
Jelínek E, Mísař Z, Jakeš O (1985) Inclusions of ultrabasic rocks and eclogites in granulite of the Moldanubian complex, Bory Quarry. Guidebook of 2nd Intern. Eclogite Conf, Praha, pp 30 - 37
Jessop AM, Hobart MA, Sclater JC (1976) The world heat flow data collection 1975. Geothermal Service of Canada. Geotherm Ser. Ottawa 5: 125

Jödicke H, Untiedt J, Olgemann W, Schulte L, Wagenitz V (1983) Electrical conductivity structure of the crust and upper mantle beneath the Rhönisch Massif. In: Fuchs K (ed) Plateaux Uplift. Springer, Berlin Heidelberg New York, pp 288 - 302
Joó I, Arabadžijski D, Mladenovski M, Vanko J, Füry M, Thury J, Wyrzykowski T, Mihaila M, Meščerskij IN, Semič VS (1985) The map of vertical crustal movements in the Carpatho - Balkan region. Scale 1:1 000 000. 2 sheets. Kartográfia Budapest
Joó I, Arabadžijski D, Füry M, Meščrskij IN, Mihaila M, Mladenovski M, Németh Z, Steinberg J, Thury J, Vanko J, Wyrzykowski T (1987) New investigations of recent vertical movements in the Carpatho-Balkan region. J Geodyn, Amsterdam 8; Nr. 2-4
Kahle HG, Werner D (1980) A geophysical study of the Rhinegraben. - II. Gravity anomalies and geothermal implications. Geophys J R Astron Soc, Oxford 62: 631
Kapička A (1984) Magnetic susceptibility anisotropy of deformed rocks. Studia Geoph Geod Praha 28: 90 - 100
Kapička A (1988) Anisotropy of magnetic susceptibility in a weak magnetic field induced by stress. Phys Earth Planet Int, Amsterdam 51: 349 - 354
Kapička A (1990) Variations of the mean susceptibility of rocks under hydrostatic and non-hydrostatic pressure. Phys Earth Planet Int, Amsterdam 63: 78 - 84
Kapička A, Savěnko BJ (1979) Srovnitelnaya kharakteristika izmeneniy ostatočnoy namagnichenosti bazaltov pri gidrostaticheskom i odnoosnom davlenijakh. Geof Zhurnal (SSSR), Kiev 1: 57 - 59
Kárník V, Vaněk J (1957) On the investigation of tectonic microtremors in the Komárno Basin. Stud Geoph Geod, Praha 1: 95 - 100
Kárník V, Procházková D, Schenk V, Schenková Z (1981a) Seismicity of Czechoslovakia and of Europe. In: Zátopek A (ed) Geophysical Syntheses in Czechoslovakia. Veda, Bratislava, pp 221 - 242
Kárník V, Procházková D, Schenková Z (eds) (1981b) Atlas of Seismological Maps for Central and Eastern Europe. Geophys. Inst. Acad. Sci. Praha
Kárník V, Schenková Z, Schenk V (1982) Analysis of seismic hazard at a selected site in the Western Carpathians. Trav Géophys 1979, 27, 523. Academia, Praha, pp 149 - 168
Kárník V, Schenková Z, Schenk V (1984a) Earthquake activity in the Bohemian Massif and in the Western Carpathians. Trav Géophys 1981, 29, 547. Academia, Praha, pp 9 - 33
Kárník V, Schenk V, Schenková Z (1984b) Earthquake provinces of Central and Eastern Europe. Proc. of the 27th Inter. Geological Congress, 8. Geophysics VNU Science Press, pp 215 - 256
Kárník V, Schenk V, Schenková Z (1987) Earthquake swarm in West Bohemia 1985-86 - some characteristic. Ann Geoph 5B (No.6): 691 - 700
Kárník V, Procházková D, Schenk V, Schenková Z, Brouček I (1988) Seismic zoning map of Czechoslovakia - version 1987. Stud Geophys Geod, Praha 32: 144 - 150
Kastl E, Tonika J (1984) The Mariánské Lázně metaophiolite complex, West Bohemia. Krystalinikum, Praha 17: 59 - 76
Kay RW, Senechal RG (1976) The rare earth geochemistry of the Troodos ophiolite complex. J Geophys Res, Washington, 81: 964 - 970
Kern H, Wenk HR (1990) Fabric-related velocity anisotropy and shear wave splitting in rocks from the Santa Rosa Mylonite Zone, California. J Geophys Res 95: 11213 - 11223
Kiesl W, Kluger HW, Wieseneder H (1985) Eclogitization in a presumed Variscan zone in the Eastern Alps, Austria. Terra Cognita, Paris 436
Klápová H (1985) Petrology and geochemistry of eclogites from the Krušné hory area - NW Bohemia. Terra Cognita, Paris, pp 422 - 423
Klíma J (1965) Československý vojenský atlas, Naše vojsko, Praha
Klinec A (1975) Geologická mapa Slovenského Rudohoria a Nízkých Tatier: Slovenské Rudohorie stred - Nízké Tatry východ. Geol. úst. D. Štúra, Bratislava
Knaislová D, Procházková D, Tobyáš V (1988) Weak seismic events located on the territory of the ČSR in 1982. Trav Géophys 1984, No 605. Academia, Praha, pp 47 - 52
Knopoff L, Müller St, Pilant W (1966) Structure of the crust and upper mantle in the Alps from phase velocity of Rayleigh waves. Bull Seism Soc Am 56: 1009 - 1044
Knopoff L (1972) Observation and inversion of surface - wave dispersion. In: Ritsema A R (ed) The Upper Mantle. Tectonophysics, Amsterdam 13: 497 - 519

Knothe H, Schroeder E (1972) The crustal structure of Central and Southeastern Europe based on the results of explosion seismology: 23 The German Democratic Republic. Geophys Trans Spec Ed, Budapest, pp 80 - 86
Kodym O st (1963) Geologie II. dfl, SPN Praha, pp 1 - 220
Konečný V, Slávik J (1974) Tectonic background of Miocene vulcanism in the West Carpathians. In: Tectonic of the Carpathian Balkan Regions, Geol. úst. D. Štúra, Bratislava, pp 136 - 145
Kopecký A, Vyskočil P (1972) Recentní pohyby povrchu ČSSR naměřené geodeticky. Věstník ÚÚG 47: 129 - - 141
Kopecký L (1972) Riftové struktury, fenity a mladý alkalický magmatismus jako nositelé rudní mineralizace v Českém masívu. Geol Průzk, Praha 14: 195 - 198
Kornprobst J (1980) Le metamorphism en France. IN: Autrant A, Dercourt (eds) Evolution géologique de la France, 26th IGC, Coll, Orleáns 7: 161 - 189
Kotásek J, Krs M (1965) Paleomagnetic study of tectonic rotation in the Carpathian Mountains of Czechoslovakia. Palaéogeogr Palaeoclimatol Palaeoecol, Amsterdam 1: 39 - 49
Kozlovskij E A (1984) Kolskaya svyerkhglubokaya. Nyedra, Moskva, pp 1 - 490
Kráľ J (1982) Dating of young tectonic movements and distribution of uranium in apatite of granitoid and metamorphosed crystalline rocks of the West Carpathians. Geol Zbor Geol Carpath, Geol. úst. D. Štúra, Bratislava 33; 2: 663 - 664
Krešl M (1976) Dynamic properties of the probe for measuring the temperature gradient. Stud Geophys Geod, Academia, Praha 20: 415 - 422
Krešl M (1981) Technique of heat flow measurements. In: Zátopek A (ed) Geophysical Syntheses in Czechoslovakia, Veda, Bratislava, pp 449 - 454
Krešl M, Veselý I (1973) Instrument for the automatic measurement of the coefficient of heat conductivity of rocks. Trav Géophys, 1971, 361. Academia, Praha, pp 317 - 334
Kröll A, Schimunek K, Wessely G (1981) Ergebnisse und Erfahrungen bei der Exploration in der Kalkalpenzoe in Österreich. Erdöl - Erdgas - Z, Hamburg 97; 4: 134 - 147
Krs M (1966) Palaeomagnetism of some Central European mineral deposits and its geophysical significance. Geol Mijnbow, The Haque 45 Juli 1966: 210 - 230
Krs M (1968) Rheological aspects of palaeomagnetism. Ing Geol Congr, Praha 13; 5: 87 - 96
Krs M (1969) Palaeomagnetismus. Knih Úst_ úst Geol, Academia, Praha 40: 1 - 202
Krs M (1982) Implication of statistical evolution of Phanerozoic palaeomagnetic data (Euroasia, Africa. - Rozpr Čs Akad Věd, - mat. p_ír. Věd, Academia, Praha 92: 3
Krs M, Chvojka R, Valín F (1988) A contribution to methodology of magnetostratigraphic studies of strongly remagnetozed Lower Permian rocks, Bohemian Massif. Z Geol Wiss, Berlin, Band 16, Heft 10, 945-957
Krs M, Šťovíčková N (1966) Paleomagnetic investigation of hydrothermal deposits in the Jáchymov (Joachimsthal) region, Western Bohemia. Trans Inst Min Metallurgy, London 75; 711: 851 - 857
Krs M, Muška P, Pagáč P (1982) Review of Palaeomagnetic investigations in the West Carpathians of Czechoslovakia. Geol Práce Zpr, Geol. úst. D. Štúra, Bratislava 78: 39 - 58
Krs M, Krsová M, Pruner P, Havlíček V (1986) Paleomagnetism, paleogeography and multi - component analysis of magnetization of Ordovician rocks from the Barrandian area of the Bohemian Massif. Sbor Geol Věd UG, Praha 20: 9 - 45
Krs M, Chvojka R, Valín (1987) A contribution to the methodology of magnetostratigraphic studies of strongly remagnetized Lower Permian rocks, Bohemian Massif. Z Geol Wiss, Berlin 10; 10: 945 - 957
Krutichovskaya Z A, Rusakov O M, Paškevič I K, Posgay K (1971) Magnitnoe pole Karpatskogo regiona i nekotorye voprosy ego geologicheskoy interpretacii. Acta Geol Acad Sci Hung, Tomus Budapest 15: 377 - - 382
Kubík J (1990) Subsurface temperature field of the Bohemian Massif. Stud Geoph Geod 34: 110 - 128
Kubík J, Čermák V (1986) Heat flow in the Upper Silesian coal basin, re - evaluation of data with special attention to the lithology. Stud Geophys Geod, Academia, Praha 30: 376 - 393
Kumpera O, Suk M (1985) Some problems of development and structure of the Bohemian Massif. Krystalinikum, Praha 18; 1985: 53 - 74
Kuno H (1965) Petrology and Vulcanism in the Upper Mantle project. - The Upper Mantle symposium, New Delhi, Copenhagen 1964: 109 - 11
Kutas RI (1979) A geothermal model of the earth's crust on the territory of the Ukrainian shield. In: Čermák V, Rybach L (eds) Terrestrial Heat Flow in Europe. Springer, Berlin Heidelberg New York, pp 309 - 315

Kutas RI (1984) Heat flow, radiogenic heat and crustal thickness in southwest USSR. Tectonophysics, Amsterdam 103: 167 - 174
Kuthan M (1967) Young volcanic rocks of the Carpathians in Slovakia. In: Mahel' M, Buday T (eds) (1967) Regional Geology of Czechoslovakia. Part II - The West Carpathians. Academia, Praha, pp 628 - 667
Kvitkovič J, Plančár J, Vyskočil V (1976) The isostatic conditions in relation to the recent vertical movements of the earth's crust in the West Carpathians. Geogr Čas Slov Akad Vied, Bratislava 28: 122
Lachenbruch AH (1968) Preliminary geothermal model of the Sierra Nevada. J. Geophys Res, Washington 73: 6977 - 6989
Lardeaux JM, Santalier DS (1985) Thermal evolution of eclogites in W. Alpine and French paleozoic domain: a comparison. Terra Cognita, Paris, p 423
Lasnier B (1982a) A review of the eclogites in the Hercynides. Terra Cognita, Paris, p 303
Lasnier B (1982b) Recrystalization processes in eclogite. Terra Cognita, Paris, pp 333 - 334
Leško B, Beránek B, Varga Z (1980) Cisaillements horizontaux profonds sous les Karpates occidentales á la lumiére des connaissances geophysiques. Rev Géol Dyn Géogr Phys, Paris 22; 4 - 5: 225 - 226
Leško B, Ibrmajer I (1983) Geologický výskum seizmického výskumu v Liptovskej kotline so zreteľom na výskyt ropy. Miner Slov, Spišská Nová Ves 15; 2: 117 - 128
Lévéque J J, Cara M (1983): Long - period Love - wave overtone data in North America and the Pacific Ocean: new evidence for upper mantle anisotropy. Phys Earth Planet Inter 33: 164 - 179
Lexa J, Konečný V (1974) The Carpathian volcanic arc: a discussion. Acta Geol Acad Sci Hung, Budapest 18; 3 - 4: 275 - 293
Lidster R C (1980) A seismic network study in the Eastern Alps. Ph. D. Thesis, Univ. of Birmingham
Liu HS (1980) Convection generated stress field and intraplate vulcanism. Tectonophysics, Amsterdam 65: 225
Lizoň I (1982) Geotermické pole Západných Karpát. Kand. disertační práce. Geofyzika, Bratislava
Lubimova EA, Polyak BC, Smirnov AyB, Kutas RI, Firsov FV, Sergienko SI, Lyusova LN (1973) Katalog dannykh po teplovomu potoku na teritorii SSSR. Mat Mir Centra Dannykh, Moskva B - 3: 64
Machart J (1984) Ultramafic rocks in the Bohemian part of the Moldanubicum and Central Bohemian Islet zone. Krystalinikum, Praha 17: 13 - 22
Maggetti M (1985) The eclogites from Silvretta, Switzerland. Terra Cognita, Paris, pp 423 - 424
Maheľ M (ed) (1974) Tectonics of the Carpathian - Balkan system. explanation of the tectonic map Carpathian mountain system and adjacent areas 1:1 000 000. Geol. úst. D. Štúra, Bratislava
Maheľ M (ed) (1979) Czechoslovak geology and global tectonics. Veda, Bratislava
Maheľ M (ed) (1980a) Významnejšie problémy geologického vývoja a stavby Československa. Veda. Bratislava
Maheľ M (1980b) Heterogeneity of crust and further fundamental factors of pecularity of development and structure of the West Carpathians. Geol Zbor Geol Carpath, Geol. úst. D. Štúra. Bratislava 37; 4: 397 - 406
Maheľ M (1984a) Nová koncepcia vývoja a stavby Západných Karpát. Miner Slov, Spišská Nová Ves 16; 6: 505 - 600
Maheľ M (ed) (1984) Zemská kora a jej vzťah k nerastným surovinám, 151. Geol. úst. D. Štúra, Bratislava
Maheľ M (1986) Geologická stavba československých Karpát - Paleoalpínské jednotky 1. Veda, Bratislava
Maheľ M, Buday T et al. (1968) Regional geology of Czechoslovakia. Part II, West Carpathians, Academia, Praha, pp 723
Majorowicz J (1984) Problems of tectonic interpretation of geothermal field distribution in the platform areas of Poland. Publ Inst Geophys, Warszawa A - 13 (160): 149 - 166
Majorowicz J, Plewa S (1979) Study of heat flow in Poland with special regard to tectonophysical problems. In: Čermák V, Rybach L (eds) Terrestrial Heat Flow in Europe, Springer, Berlin Heidelberg New York, pp 240 - 252
Man O (1987) The interactive 2 - D interpretation of magnetic/gravity anomalies. - Proceedings 32nd International Geophysical Symposium, Dresden. VEB Geophysik, Leipzig, pp 264 - 273
Man O, Mašín J (1967) Regional aeromagnetic measurements in Czechoslovakia. Stud Geoph Geod, Praha 11: 361 - 363
Marsh B D (1982) On the mechanism of igneous diapirism, stoning and zone melting. Amer J Sci, New Haven 282; 6: 808 - 885
Marušiak I, Lizoň I (1976) Geotermické pole Západných Karpát. Západ Karpaty Sér Geol, Geol. úst. D. Štúra. Bratislava 1: 181 - 206

Máška M (1961) Základní rysy struktury a vývoje Českého masívu. Tektonický vývoj Československa. In: Buday T,...(eds). NČSAV. Praha. pp 15 - 26
Matějka A (1927) Geologická studie z okolí Ružomberka na Slovensku. Sbor St Geol Úst Čs Republ, Praha 7: 529 - 575
Matějka A, Andrusov D (1931) Apercu de la géologie des Carpathes occidentales'. Knih St Geol Úst Čs Republ, Praha 13: 19 - 165
Mathé G (1969): Granat - Klinopyroxen - Paragenese des sachsischen Grundgebirges und ihre petrographisch - mineral fazielle Einordnung. Freiberg Forsch C 226: 49 - 64
Matolín M (1970) Radioaktivita hornin Českého masívu. Academia, Praha, p 98
Matte Ph (1986) La chaine varisque parmi les chaines paléozoiques périatlantiques models d'evolution et position des grands blocks continenteaux au Permo - Carbonifere. Bull Soc Géol Fr, Paris 8: 9 - 24
Mayerová M, Nakládalová Z, Ibrmajer I, Herrmann H (1985) Plošné rozložení M - diskontinuity v ČSSR z výsledků profilových měření HSS a technických odpalů. In: Sbor. referátů 8. celostát. konference geofyziků, Český Budějovice, Sekce S 1, MS Geofyzika. Brno, pp 41 - 53
McGuire RK (1976) FORTRAN computer programme for seismic risk analysis. U S Geol Surv Open File Rep, pp 86 - 87
Medaris LG (1980) Petrogenesis of the Lien peridotite and associated eclogites, Almklovdalen, Western Norway. Lithos, Oslo 13; 339 - 353
Medaris LG, Wang HF (1986) A thermal - tectonic model for high - pressure rocks in the basal gneiss complex of Western Norway. Lithos, Amsterdam 19: 299 - 315
Medaris LG, Carswell DA (1990) The petrognesis of Mg-Cr garnet peridotites in European metamorphic belts. In: Carswell DA (ed) Eclogite Facies Rocks. Blackie Glasgow, pp 261 - 290
Medaris LG, Wang HF, Mísař Z, Jelínek E (1990) Thermobarometry, diffusion modelling and cooling rates of crustal garnet peridotites: Two examples from the Moldanubicum zone of the Bohemian Massif. Lithos, 25: 189 - 202
Meschersky IN, Kusnetsov JG (1986) The map of recent vertical crustal movements on the territories of Bulgaria, Hungary, the GDR, Poland, Rumania, the European part of the USSR, Czechoslovakia. Abstracts of the 7th Interpretation Symposium on recent crustal movements of the Earth. Tallin
Miller H, Mueller St, Perrier G (1984) Structure and dynamics of the Alps. A geophysical inventory. In: Berckhemer H, Hsü K (eds) AGU Geodyn Ser 7: 166 - 175
Mísař Z (1966) Structural history of ultrabasic rocks of the Letovice crystalline unit. Krystalinikum, Praha 4: 109 - 126
Mísař Z (ed) (1974) The Ransko gabbro - peridotite massif and its mineralisation. Univ Carol Geol, Praha, p 207
Mísař Z (1974b) Feeding channels of Pre - Triassic ultrabasic/basic rocks in the Bohemian Massif. Krystalinikum, Praha 10: 133
Mísař Z (1979) The position of ultrabasic rocks in geotectonic cycles and geological units of the Bohemian Massif. Geodynamic Investigation in Czechoslovakia. Veda, Bratislava, pp 167 - 177
Mísař Z (1980) Putevoditel k exkursiam, PK IX Akad. soc. zemí. PřF UK, Praha, 121 p
Mísař Z (1984) Ophiolites and related rocks of Czechoslovakia and their correlation. Krystalinikum, Praha 17: 7 - 11
Mísař Z, Jelínek E (1981) Uzavřeniny peridotitů, pyroxenitů, eklogitů a opálů v leptinitech moldanubika na lokalitě Bory. - Věstník ÚÚG, Praha 56, 13 - 20
Mísař Z, Procházková D (1982) The relation between the seismic activity, basement - and deep fault structures in Central and South - Eastern Europe. Acta Univ Carol, Praha, pp 1 - 18
Mísař Z, Dudek A, Motlová M, Suk M, Weiss J (1972) Interpretace tíhového pole moldanubika a p_ilehlých jednotek. Sbor Geol Věd UG, Praha 10: 7 - 34
Mísař Z, Dudek A, Havlena V, Weiss J (1983) Regionální geologie ČSSR. I., Český masív. SNP, Praha
Mísař Z, Jelínek E, Jakeš P (1984a) Inclusions of peridotite and eclogite in granulite rocks of Pre - Hercynian upper mantle and lower crust in the Eastern Bohemian Massif. Ann Sci Univ Clermont Sér Géol Minéral, Clermont - Ferrand II: 85 - 96
Mísař Z, Jelínek E, Pačesová M (1984b) The Letovice dismembered metaophiolites in the framework of the Saxothuringian zone if the Bohemian Massif. Miner Slov, Spišská Nová Ves 16: 13 - 28
Mísař Z, Jelínek E, Souček J, Tonika J (1984c) The correlation of gabbro - peridotite complexes of the Bohemian Massif. Krystalinikum, Praha 17: 99 - 113

Mísař Z, Dudek A, Paděra K (1985) Contact between garnet peridotite and granulite. Jihlava river valley at Mohelno. Guidebook of 2nd Intern. Eclogite Conf.Praha, pp 17 - 22
Morgan P, Sass JH (1984) Thermal regime of the continental lithosphere. J Geodyn, Amsterdam 1: 142 - 166
Mottlová L, Suk K (1970) K rozšíření granitických hornin v hlubší stavbě moldanubika. Věstník ÚÚG, Praha 15; 4: 383 - 392
Mrázová K, Ruprechtová L (1982) Location of near shocks recorded by the Czechoslovak seismological network. Bull. Seismographic Stations Bratislava, Šrobárová and Skalnaté Pleso for the Year 1976, Veda, Bratislava, pp 56 - 58
Mueller S (1974) The structure of the Earth's crust based on seismic data. Tectonophysics, Amsterdam 20; 1973: 1 - 391
Mueller S (1980) Evolution of the lithosphere - Asthenosphere system in the Earth. Berl Geowiss Abh, R A, Berlin 20: 155
Mueller S (1982) Deep structure and recent dynamics of the Alps. In: Hsú K J (ed) Mountain Building Processes. Academic Press, London, pp 181 - 199
Muška P, Vozár J (1985) Different of Remanent Magnetic Polarization of Crystalline in the West Carpathians. Proceeding reports of the XIII - th Congress of KBGA, Part II, Cracow, pp 230 - 234
Muška P, Vozár J (1987) The paleomagnetic character of the late Paleozoic of the West Carpathians in relation to solution of palinspastic problems. Miner. Slov Monogr, Spišská Nová Ves, pp 361 - 366
Mysen B, Holloway JR (1977) Experimental determination of rare earth fractionation patterns in partial melts from peridotite in the upper mantle. Earth Planet Sci Lett, Amsterdam 34: 231 - 237
Nelson KD, Arnow JA, McBride JH, Wellemin JH, Huang J, Rheng L, Oliver JE, Brown LD, Kaufman S (1985) New COCORP profiling in the southeastern United States. Part I: Late Paleozoic suture and Mesozoic rift basin. Geology, Boulder 13; 11: 714 - 718
Nemčok J, Rudinec P (1979) Geologické profily cez východoslovenské flyšové pásmo. In: Maheľ M (ed) Tektonické profily Západných Karpát. Geol. úst. D. Štúra, Bratislava, pp 65 - 76
Neumann W (1974) Zur erdgeschichtligen Entwicklung des Sächsischen Granulitmassifs. Z Angew Geol, Berlin 30: 183 - 190
Neumann W (1978) Zur Paleogeographie, Tektonik und Metamorphose des Sachsischen Granulitgebirges. In: La chaine varisque d'Europe neogenne et occidentale. CNRS, Paris 243: 97 - 112
Neunhöfer H (1985) Primär - und Sekundäreffekte bei der Ausbreitung seismischer Oberflächenwellen und ihre geophysikalischen Ursachen. Veröff. ZIPE Potsdam 85: 13 - 26
Neunhöfer H, Güth (1975) Dispersion of seismic surface waves in Middle and Northern Europe. Veröff ZIPE 31/2: 399 - 403
Neunhöfer H, Güth D (1983) Bulletin der Mikroerdbeben im Gebiet des Vogtlandes aus der Zeit von August 1962 bis August 1983. ZIPE AW DDR Mitt Berlin 1002: 41 - 58
Neunhöfer H, Güth D (1984) Bulletin der Mikroerdbeben Vogtlands aus der Zeit von August 1962 bis August 1983. ZIPE AW DDR Mitt, Berlin 20: 32 - 65
Neunhöfer H, Güth D (1988) Mikrobeben seit 1962 im Vogtland. Z Geol Wiss, Berlin 16 (2): 135 - 146
Neunhöfer H, Marillier F, Panza G F (1981) Crust and upper mantle structure in the Bohemian Massif from the dispersion of Rayleigh waves. Gerl Beitr Geophys, Berlin 90 (6): 514 - 520
Neunhöfer H, Plešinger A, Kracke D (1983) Crust and upper mantle structure between Moxa and Kašperské Hory from Rayleigh waves. Gerl Beitr Geophys, Berlin 92 (2 - 4): 284 - 290
Neunhöfer H, Güth D, Procházková D, Tobyáš V, Soukup J, Michek R, Knaislová D, Vavryčuk V, Zedník J, Jedlička P, Horálek J, Čermák F, Schmedes E, Antonini M, Bucha V Jr, Klimeš L, Mohapl P, Růžek B, Strauch W (1989a) Joint bulletin of local seismological stations. earthquake swarm 1985 - 86 in Western Bohemia. Geoph. Inst. Czechosl. Acad. Sci. Praha, 270 p
Novák V (1971) Zemský tepelný tok v hlubinných vrtech Žarošice 1 a 2 v oblasti Ždánického lesa. Věstník ÚÚG, Praha 46: 277 - 284
Novotný M (1981) Two methods of solving the linearized two - dimensional inverse seismic kinematic problem. Geophys Beob - Ergebn, Berlin 50: 7 - 15
Novotný M, Nehybka V, Sedlák P (1981) First applications of the method DIME regional refraction profiles in Czechoslovakia. Proceedings of the 26th Geophysical Symposium, Leipzig 1981, pp 163 - 174
Novotný O (1983) Theoretical dispersion curves for the seismic profile Kašperské Hory (Czechoslovakia) - Ksiaz (Poland). Studia Geoph Geod, Praha 27: 157 - 163

Obata M (1980) The Ronda peridotite: garnet-, spinel- and plagioclase- lherzolite facies and the P-T trajectories of a high-temperature mantle instrusion. J Petrol Oxford 21: 533 - 572
Obata M, Dickey JS (1976) Phase relations of massif layers in the Ronda peridotite. Carnegie Inst Wash Geophys Lab Year Book, Washington 75: 562 - 566
Ondra P, Hanák J (1981) Mapa hustot geologicko-petrografických jednotek Českého masívu v mě_ítku 1:200 000. Geol Průzk, Praha 23; 4: 117 - 118
Opletal M (1967) Předběžná zpráva o nálezu xenolitů charnockitických hornin v čedičových horninách Českého středohoří. Čas Mineral Geol, Praha 12; 3: 301 - 303
Pačes T, Čermák V (1976) Subsurface temperature in the Bohemian Massif: Geophysical measurements and geochemical estimates. Proceedings 2nd UN. Symp Dev Use Geotherm Resour, San Francisco 1: 803 - 807
Paděra K. /1972/: Qriquaites in the Moldanubicum of the Bohemian Massif. - Upper Mantle Project in Czechoslovakia, Final rep., 51 - 59. Praha.
Panza GF, Müller St, Calcagnile G (1978a) Lithosphere asthenosphere system in the European Mediterranean Area. Paper presented at the EBS/EGS Symposium on Deep Seismic Sounding and Earthquakes. Strasbourg
Panza GF, Neunhöfer H, Calcagnile G (1978b) Contribution to the phase velocity investigation of Rayleigh waves in Middle Europe. Pure Appl Geophys 116: 1299 - 1306
Panza GF, Mueller S, Calcagnile G (1980) The gross features of the lithosphere - asthenosphere system in Europe from seismic surface waves and body waves. Pure Appl Geophys, Genova 118: 1209 - 1213
Paterson NR, Reeves CV (1985) Applications of gravity and magnetic surveys: The state - of - the - art in 1985. Geophysics, Tulsa 50: 2558 - 2594
Petr V, Pěčová J, Praus O (1985) Anomaly of geomagnetic variations in the southeastern margin of the Bohemian Massif. Stud Geophys Geod, Praha, 29: 169.- 190
Pěč K, Pěčová J, Praus O (1988a) Single- and inter-station transfer functions for non-synchronous geomagnetic arrays - I. Theory of methods used in data analysis. Stud Geophys Geod, Praha 32: 300 - 312
Pěč K, Pěčová J, Praus O (1988b) Single- and inter-station transfer functions for non-synchronous geomagnetic arrays -II. The results of the field data analysis. Stud Geophys Geod, Praha 32: 368 - 391
Pěčová J (1985) Determination of the depth of a conductivity anomaly by separating the geomagnetic variation field into its external and internal part. Trav Géophys, Academia. Praha, p 179
Pěčová J, Praus O, Petr V, Tobyášová M (1974) Results of MT - sounding in the Bohemian Massif. Trav Géophys, Academia, Praha 426: 337 - 353
Pěčová J, Petr V, Praus O (1976) Depth distribution of the electric conductivity in Czechoslovakia from electromagnetic studies. In: Ádám A (ed) Geoelectric and Geothermal Studies. KAPG Geophys Monogr, Budapest, pp 517 - 537
Pequignot P, Caron J M (1985) The transition between blueschists and lawsonit - bearing eclogites on the example of Corsican metabasalts. Terra Cognita, Paris, p 424
Perroud H (1985) Syntése des résultats paléomagnetiques sur le Massif Armoricain. Hercynica, 1: 65-72
Petr V (1980) Fyzikálně-statistická metoda výpočtu hloubky poruchového tělesa na základě magnetovariačních měření. Sborník referátů ze 7. celostátní konference geofyziků; sekce S4, Gottwaldov, pp 35 - 38
Peucat JJ, Vidal P, Postaire B (1982) U - Pb zircon dating of eclogites garnet pyroxenites and associated serpentinites from Hercynian belt of south Brittany, France. Terra Cognita, Paris, pp 325 - 326
Philipotts JA, Schneltzer CC, Thomas HH (1972) Petrogenetic implications of some new geochemical data on eclogitic and ultrabasic inclusions. - Geochim. cosmochim. Acta Oxford l0: 1131 - 1166
Piboule MM, Briand B (1980) Geochemie de quelques eclogites et rocks associtées de l'Est du Massif Central Francais. Terra Cognita, Paris, p 312
Pícha B, Červ V, Pek V (1984) Magnetotelluric inversion along the Osvětimany - Brezová pod Bradlom profile. Stud Geophys Geod, Academia, Praha 28: 1101 - 112
Pick M (ed) (1971) Upper Mantle Project Programme in Czechoslovakia 1962 - 1970. Geophysics, Final Report, Academia. Praha, 186 p
Pick M, Jakubcová I (1986) Trojrozměrné hustotní modely kůry v Českém masívu, metodika řešení. In: Blížkovský M et al. (eds) Geofyzikální model litosféry. Záv. zpráva úkolu OFTR ČGÚ č. 82 - 160 za období 1982 - 1985, 143. GFÚ ČSAV, Praha - Geofyzika Brno, 143 p
Pick M, Charvátová - Jakubcová, I (1988) Modification of the Runcorn's Equations on Convection Flows. Stud Geophys Geod, Academia Praha 32: 47

Pick M, Pícha J, Vyskočil V (1973) Theory of the Earth's gravity field. - Academia - Elsevier, Praha - msterdam 156 p
Pin C, Peucat JJ (1986) Ages des épisodes de métamorphisme paléozoiques dans le Massif Central et Massif Armorician. Bull Soc Géol Fr, Paris 8: 461 - 469
Plančár J, Filo M, Šefara J, Snopko L, Klinec A (1977) Geofyzikálna a geologická interpretácia tiažových a magnetických anomálií v Slovenskom rudohorí. Západ. Karpaty Sér Geol, Bratislava 2: 7 - 144
Plešinger A (1981) Acquisition, processing and interpretation of broadband seismic data. In: Zátopek A (ed) Geophysical Syntheses in Czechoslovakia. Veda, Bratislava, pp 59 - 78
Plešinger A, Horálek J (1976) The seismic broadband recording and data processing system FBV/DPS and its seismological applications. Z Geophys, Berlin 42(3): 201 - 217
Plíva G, Klímková V (1979) Reinterpretace refrakční - seismických dat v oblasti flyšových a vnit_ních Karpat západního Slovenska. Manuscript. Geofyzika Brno, 60 p
Plíva G (1980a) Příspěvek k hlubinné geologické stavbě v oblasti jihovýchodních svahů Českého masivu. Sborník referátů 7. celostátní konference geofyziků, Gottwaldov 89 - 94
Plíva G (1980b) Další poznatky o hlubinné stavbě autonomního bloku. Sborník referátů 4. geologické konference, Bratislava, pp 81-88
Plomerová J, Babuška V (1988) Lithosphere thickness in the contact zone of the Moldanubicum and Saxothuringicum in Central Europe. Phys Earth Planet Inter, Amsterdam 59: 159 - 165
Pokorný L, Polanský J, Šťovíčková N (1970) Evidence of the deep segmented structure of the Bohemian Massif based on geophysical data. Sbor Geol Věd, UG, Praha 9: 7 - 10
Pokorný L, Šťovíčková N, Beneš L (1985) Geofyzikální indikované zlomové struktury Českého masívu a jejich význam pro metalogenezi. Sbor Geol Věd, UG, Praha 19: 127 - 150
Pola I (1967) Náhrada gravitačního pole soustavou hmotných bodů. Kandidátská disertační práce. Charles University, Faculty of Natural Sciences, Praha
Polanský J (1973) Hloubkové řezy Českým masívem. Geol Průzk, Praha 15; 6: 161 - 167
Pollack HN, Chapman DS (1977) Mantle heat flow. Earth Planet Sci Lett, Amsterdam 34: 174 - 184
Popelář J (1968) Gravity field and isostasy in the area of the Czechoslovak West Carpathians. Sbor Geol Věd, UG, Praha 7: 7 - 24
Porstendorfer G, Göthe W, Lengning K, Oelsner Ch, Tanzer R, Ritter E (1976) Nature and possible causes of the anomalous behaviour of electric conductivity in the north of the GDR, Poland and the FRG. In: Ádám A (ed) Geoelectric and Geothermal Studies. KAPG Monograph. Akadémai Kiadó, Budapest, 517 p
Posgay K (1967) A comprehensive survey of geomagnetic masses in Hungary. Geofizikai Közlemények, Budapest XVI; 4: 118
Pospíšil L, Schenk V, Schenková Z (1985) Relation between seismoactive zones and remote sensing data in the West Carpathians. Proc. of the 3rd Symposium on the Analysis of Seismicity and on Seismic Risk. Liblice, Prague, pp 256 - 263
Pospíšil L, Nemčok J, Graniczny M, Doktór S (1986) Contribution of remote sensing to the identification of the strike - slip faults in the West Carpathians. Miner Slov, Spišská Nová Ves 18; 5: 385 - 402
Pouba Z (1970) Precambrian banded magnetite areas of the Desná Dome. Sbor Geol Věd, -. Praha, L 6 12: 7 - 64
Pouba Z (1973) Acid rocks at the contacts of basic and ultrabasic intrusions with biotite gneisses (Vysoký Jeseník Mts, Czechoslovakia). Acta Univ Carol Geol, Praha 1; 2: 123 - 139
Pouba Z, Pertold Z (1979) Relations between subduction and metallogenesis. In: Czechoslovak geology and global tectonics. Veda. Bratislava, pp 215 - 234
Poupinet G (1979) On the relation between P - wave travel time residuals and the age of continental plates. Earth Planet Sci Lett,Amsterdam 43: 149 - 161
Praus O, Tobyášová M (1992) Electrical conductance of sediments overlying the basement of the Czech Republic. Stud Geophys Geod, Praha 36: 161-167
Praus O, Pěčová J (1991) Anomalous geomagnetic fields of internal origin in Czechoslovakia. Stud Geophys Geod, Praha 35: 81 - 89
Praus O, Pěčová J, Petr V, Hvoždara M, Červ V, Pek J, Laštovičková M (1981) Elektromagnetic induction and electrical conductivity of the Earth's body. In: Zátopek (ed) Geophysical Synthesis in Czechoslovakia. Veda, Bratislava, 297 p

Praus O, Pěčová J, Petr V, Červ V, Pek J (1984) Geoelektrické charakteristiky na styku základních geologických bloků v ČSSR. In: Zemská kôra a jej vzfah k nerostným surovinám. Geol. úst. D. Štúra Bratislava, pp 13 - - 146

Praus O, Pěčová J, Petr V, Babuška V, Plomerová J (1990) Magnetoteluric and seismological determination of the lithosphere - asthenosphere transition in Central Europe. Phys Earth Planet Inter, Amsterdam 60: 212 - -228

Příhoda K, Krs M, Pešina B, Bláha J (1989) MAVACS - a new system creating a non-magnetic environment for palaeomagnetic studies. Cuad Geol Iberica, Madrid 12: 223-250.

Příhoda K, Krs M, Pešina B, Bláha J (1991) MAVACS a new system creating a non - magnetic environment for palaeomagnetic studies. Special Issue of "Cuadernos da Geología Ibérica", Barcelona (in press)

Procházková D (1976) The character of seismic activity in Europe. Acta Univ Carol, Math Phys, Praha 17: 49 - - 74

Procházková D (1981) Earthquake parameters determined on the basis of macroseismic material. Gerlands Beitr Geophys, Leipzig 90: 255 - 261

Procházková D (1983) Relationship among the source parameters of earthquakes in Europe. Tectonophysics, Amsterdam 93: 307 - 311

Procházková D (1984a): Earthquake in Central Europe. Contrib Geoph Inst Slov Akad Sci, Bratislava

Procházková D (1984b) Analýza zemět_esení ve St_ední Evropě. Disertační práce. GFÚ ČSAV, Praha

Procházková D (1985a) Attenuation of macroseismic effects in Central Europe. Trav Géophys 1982, Academia. Praha 30; 569: 47 - 93

Procházková D (1985b) Macroseismic fields in the Balkans. Trav Géophys 1983 Academia, Praha, 580: 5 - 44

Procházková D (1988a) The 1985/86 earthquake swarm in Western Bohemia. Seismol Res Lett 59: 71 - 78

Procházková D (1988b) Earthquakes in Central Europe. Contrib Geophys Inst Slov Akad Sci 1985, Veda Bratislava, pp 62 - 101

Procházková D (1988c) Some finding of the 1985/86 earthquake swarm in Western Bohemia. In: Proc Induced Seismicity and Associated Phenomena, Geoph. Inst., Czechosl. Acad. Sci., Prague, pp 211 - 226

Procházková D (1990) Seismicity of Central Europe. Publ. of the Inst. of Geophys., Pan. wyd. naukowe B-14, Warszawa, 231 p

Procházková D, Brouček I (1986) Map of maximum intensities on the territory of Czechoslovakia. Contr Geophys Inst Slov Akad Sci, Veda, Bratislava 16: 25 - 32

Procházková D, Dudek A (1982) Some parameters of earthquakes originated in Central and Eastern Europe. Trav Géophys 1980, Academia. Praha 28; 538: 43 - 81

Procházková D, Kárník V (eds) (1978) Atlas of isoseismal maps, Central and Eastern Europe. Geophys Inst Acad Sci Praha

Procházková D, Zeman J (1982a) Vztah mezi hlubší stavbou kůry a šířením seismických účinků ve střední Evropě. Čas Mineral Geol, Praha 27: 259 - 171

Procházková D, Zeman J (1982b) Vztah mezi šířením seismických účinků a hlubší stavbou kůry v alpinsky mobilní Evropě a jejím předpolí. Čas Mineral Geol,Praha 27: 359 - 370

Procházková D, Brouček I, Guterch A, Lewandowska - Marciniak (1978) Map and list of maximum observed macroseismic intensities in Czechoslovakia and Poland. Publ. Inst. Geophys., Warszawa, pp 1 - 76

Procházková D, Dudek A, Mísař Z, Zeman J (1986) Earthquakes in Europe and their relation to basement structures and fault tectonics. Rozpr. Čs. Akad. Věd, -. mat. přír. věd, Academia, Praha 96: 1 - 80

Procházková D, Udias A, Buforn E (1987a) Regional fault-plane solutions observed during earthquake swarm 1985/86 in western Bohemia. In: Earthquake Swarm 1985-1986 in Western Bohemia. Geoph. Inst. of Czechoslov. Acad. Sci. Praha, pp 277 - 281

Procházková D, Tobyáš V, Soukup J, Michek R, Knaislová D (1987b) List of earthquakes in Western Bohemia, 1985 - 1986. Supplement to Proc. Workshop on Earthquake Swarm in Western Bohemia Dec. 1985 - Feb. 1986, Mariánské Lázně, 1986. Geoph. Inst. of Czechoslov. Acad. Sci. Praha, pp 30 - 238

Pros Z (1977) Izuchenie anizotropii uprugikh svojstv gornykh porod pri vsestoronnikh davleniyach na sharovych obrazcach. High Pressure and Temperature Studia of Physical Properties of Rocks and Minerals. Naukova Dumka, Kiev, p 56

Pros Z, Babuška V (1967) A method for investigating the elastic nisotropy on spherical rock samples. Z Angew Geophys, Berlin 33: 289

Pros Z, Babuška V (1968) An apparatus for investigating the elastic anisotropy on spherical rock samples. Stud Geophys Geod, Praha 12: 192
Pros Z, Podroužková Z (1974) Apparatus for investigating the elastic anisotropy on spherical rock samples at high pressure. Veröff. Zent. - Inst. Phys. Erde Berlin, 22: 42
Rait RW, Shor GG, Francis TJG, Morris GB (1969) Anisotropy of the Pacific upper mantel. J Geophys Res, Washington 74: 3095
Riedel M (1983) Pyroux - počítačový program k výpočtu teploty a tlaku krystalizace z chemického složení pyroxenů. Acta Univ Carol Geol, Praha 4: 289 - 296
Ringwood AE (1975) Composition and petrology of the Earth's mantle. McGray Hill, New York, p 618
Ritter E (1975) Results of geoelectromagnetic deep soundings in Europe. Gerlands Beitr Geophys, Leipzig 84; 3/4: 261 - 273
Röhlich P, Šťovíčková N (1968) Die Tiefenstörungs - Tektonik und der Entwicklung im zentralen Teil der Böhmischen Masse. Geologie, Berlin 17; 6/7: 670 - 694
Roth Z (1980) Západní Karpaty - terciérní struktura střední Evropy. Knih Ústř Úst Geol, Praha 55: 128
Roth Z, Procházková D (1988) Neotectonic movements and earthquakes in Central Europe. Proc. Induced Seismicity and Associated Phenomena, Geoph. Inst., Czechosl. Acad. Sci., Prague, pp 432 - 441
Roy RF, Blackwell DD, Birch F (1968) Heat generation of plutonic rocks and continental heat flow provinces. Earth Planet Sci Lett, Amsterdam 5: 1 - 12
Royden L, Horváth F, Rumpler J (1983) Evolution of the Pannonian basin system, 1. Tectonics. Tectonics, Amsterdam 2: 63 - 90
Runcorn SK (1964) Satellite gravity measurements and a laminar viscous flow model of the Earth's mantle. J. Geophys Res, Washington 69: 4389
Runcorn S K (1967) Flows in the mantle inferred from the low degree harmonics of the geopotential. Geophys J Roy Astron Soc, Oxford 14: 375
Rybach I (1973) Wärmeproduktionsbestimnungen an Gesteinen der Schweizer Alpen. Beitr Geol Schweiz Geotech Ser, 43. Kümmerly and Frey, Zürich.
Rybach L, Buntebarth G (1984) The variation of heat generation, density and seismic velocity with rock type in the continental lithosphere. Tectonophysics, Amsterdam 103: 335 - 344
Sacks IS, Snoke JA (1984) Seismological determinations of the crustal continental lithosphere. In: Pollack H N, Rama Murthy V (eds) Structure and Evolution of Continental Lithosphere, pp 3 - 39
Šafanda J (1985) Calculation of temperature distribution in two - dimensional geothermal profile. Stud Geophys Geod, Praha 29: 191 - 199
Šafanda J (1986) On the estimation of heat flow at the earth's crust lower boundary. Stud Geophys Geod, Praha 30: 192 - 188
Šafanda J, Žáková B, Bunteberth G (1990) Temperature paleogradient estimations in the Central Bohemian Massif. Stud Geoph Geod, Praha 34: 208 - 219
Šalanský K (1983) Regionální magnetické struktury Českého masívu na území ČSR. Věstník ÚÚG, Praha 58; 5: 275 - 286
Sanders IS (1985) Eclogite in transpression zones near the Grenville front. Terra Cognita, Paris, p 436
Schaubert S (1977) Neue ergebnisse radiometrischer Altersdatierungen an gesteinen des Waldviertels. Arb Geol Bundesanst, Wien, pp 11 - 15
Schellhorn O, Gebauer D, Grünenfelder M (1982) Eclogites of the limousin area: A key for the geodynamic interpretation of the western part of French Massif Central. Terra Cognita, Paris 326
Schenk V (1984a) Conversion of observed macroseismic fields up to higher intensity level by generalized circular isoseismals. Engng Geol, Amsterdam 20: 153 - 160
Schenk V (1984b) Výskyt zemětřesení a hlavní strukturní směry Českého masívu. Sborník "Výzkum hlubinné geologické stavby Československa". Loučná n/Des, pp 97 - 101
Schenk V, Mantlík F (1985) Map of maximum expected macroseismic intensity in Czechoslovakia, Proc. of the 3rd Symposium on the Seismicity and on Seismic Risk. Liblice, Prague, pp 17 - 22
Schenk V, Schenková Z (1982) Seismotektonické schema hlavních zlomových linií st_ední a východní Evropy. Sborník "Výzkum hlubinné geologické stavby ČSSR", Brno, pp 133 - 147
Schenk V, Schenková Z (1983) Seismické ohrožení vybrané lokality z hlediska současných poznatků: In: Bezpečnost a provozní spolehlivost reaktoru VVER - seizmická zatížení. ÚISJP, Praha, pp 7 - 12

Schenk V, Schenková Z, Kárník V (1981) Seismic hazard assessment in the Bohemian massif. Proc. of 2nd Symposium on the Analysis of Seismicity and on Seismic Hazard. Liblice, Prague 2: 541 - 544
Schenk V, Kárník V, Schenková Z (1982) Seismotectonic scheme of Central and Eastern Europe. Stud Geoph Geod, Praha 26: 132 - 144
Schenk V, Grünthal G, Schenková Z (1984) Seismic hazard assessment for the GDR. Eng Geol, Amsterdam 20: 73 - 88
Schenk V, Schenková Z, Pospíšil L, Zeman A (1986) Seismotectonic model of the upper mantle part of the Earth's crust of Czechoslovakia. Stud Geoph Geod, Praha 30: 321 - 330
Schenk V, Schenková Z, Mantlík F, Kottnauer P, Grünthal G (1989a) Seismic hazard assessment for Central Europe. Proc. of the 4th Intern. Symp. on the Analysis of Seismicity and on Seismic Risk, Bechyně, 4 - 9 Sept. 1989. Geoph. Inst. Czechosl. Acad. Sci., II, Praha, pp 494 - 503
Schenk V, Schenková Z, Grünthal G (1989b) Problems of seismic hazard assessment for the epicentral area of swarm quakes in Western Bohemia/Vogtland. Tectonophysics, Amsterdam 167: 217 - 222
Schenk V, Schenková Z, Pospíšil L (1989c) Fault system dynamics and seismic activity - examples from Bohemian Massif and Western Carpathians. Geophys Trans Budapest 35; 1 - 2: 101 - 116
Schenková Z, Kárník V (eds) (1985) Catalogue of earthquakes of Central and Eastern Europe. MS, GFÚ ČSAV, Praha
Schenková Z, Schenk V, Kárník V (1981) Seismic hazard estimate for a low seismicity regime - example of Bohemia. Pure Appl Geophys 119: 1077 - 1092
Schenková Z, Schenk V, Zahradník J (1983) Studium zemět_esného ohrožení v Československu. Čs Čas Pro Fyziku, Sekce A, Praha 33: 496 - 504
Schlue JW, Knopoff L (1976) Shear - wave anisotropy in the upper mantle of the Pacific Basin. Geophys Res Lett 3: 359 - 362
Schmidt K (1982) The eclogites of the Münchberg gneiss area. Terra Cognita, Paris, p 313
Schovánek P (1977) Petrologie vybraných ultramafických hornin Českého masívu a chemismus jejich minerálů. MS PřF UK. Praha, p 313
Šefara J (1977) Spojité horizontálne hustotné mapy a ich vplyv na separáciu tiažového pola. Sbor Geol Věd UG, Praha 8: 7 - 36
Šefara J, Pospíšil L, Šutora A, Filo M, Krus S (1985) New Geophysical Maps of the Czechoslovak Part of the West Carpathians. Proc. of the 30th Inter. Geol. Symp., Moscow 1985, C - Panel Discussion, Part 3, Moskva, pp 74 - 80
Shatsky N (ed) (1962): Tectonic map of Europe. International Geological Congress.
Shebalin VN (1971) O predelnoy magnitude i predelnoy ballnosti. Izv Akad Nauk SSSR. Fiz Zemli, Moskva 6: 12 - 20
Shimamura H, Aseda T, Suychiro K, Yamada T, Inatani H (1983) Longshot experiments to study velocity anisotropy in oceanic lithosphere of the northwestern Pacific. Phys Earth Planet Inter, Amsterdam 31: 348
Slavkay M (1979) àalšie výskyty ultramafických efuzív v chočskom a križňanskom príkrove pri Banskej Bystrici. Miner Slov, Spišská Nová Ves 11; 3: 239 - 245
Smíšek M, Kršák J, Plančár J (1972) The gravity effect of bodies of arbitrary shape computed by the horizontal prism method. Contributions of the Geoph Inst Slov Acad Sci, Veda, Bratislava 3: 59 - 67
Smith DC (1980) A tectonic melange of foreing eclogizes and ultramafites in West Norway. Nature, London 287: 366 - 368
Smulikowski K (1968) Differentiation of eclogites and its possible causes. Lithos, Oslo 1: 89 - 101
Smulikowski K (1972) Classification of eclogites and allied rocks. Krystalinikum, Praha 9: 107 - 130
Smulikowski K, Smulikowski W (1982) On the porphyroblastic eclogites of the Sneznik Mts. in the Polish Sudeten. Terra Cognita, Paris, pp 313 - 314
Sokolowski S, Znosko J (1958) Mapa tektonicza Polski. Warszawa
Sollogub VB, Čekunov AV (eds) (1977) Strojenie zemnoj kory i verkhnej mantii po dannym sejsmicheskikh issledovanij. Naukova dumka, Kiev, p 310
Sollogub VB, Čekunov AV (eds) (1988) Litosfera centralnoy i vostochnoy Evropy. Geotraversy I, II, V. Naukova dumka, Kiev, p 168
Sollogub VB, Prosen D, Militzer H (eds) (1971) Stroenie zemnoy kory Central'noy i Jugo - Vostochnoj Evropy (po dannym vzryvnoy seismologii). Naukova dumka, Kiev, p 288

Sollogub VB, Guterch A, Prosen D (eds) (1980) Struktura zemnoy kori i verkhney mantiyi po dannim geofizicheskikh issledovaniyi. Naukova Dumka, Kiev, p 208
Spear FS, Franz G, Thomas S (1985) Evolution of metasediments from the eclogite zone, Tauemwindow, Austria. Terra Cognita, Paris, p 425
Špičák A (1987) Fault plane solution of 1985 Dec. 21 and Jan. 20 events. Proc. Workshop on Earthquake Swarm in Western Bohemia Dec. 1985 - Feb. 1986, Mariánské Lázně, 1986. Geoph. Inst. Czech. Acad. Sci., Praha, pp 268 - 273
Stanica D, Stanica M, Visarion (1986) The structure of the crust and upper mantle in Romania as deduced from magnetotelluric data. Rev Roum Géol Géophys Géograf, Sér Géophys, Bucurest 30: 25 - 35
Stegena L, Géczy B, Horváth F (1975) Late Cenozoic evolution of the Pannonian Basin. Tectonophysics, Amsterdam 26: 71 - 90
Štelcl J, Kocák A, Rozehnal L, Weiss J (1981) Příspěvek k řešení hlubší geologické stavby brněnského masivu s využitím seismického výzkumu s použitím metody Vibroseis. Geofyzika. Brno 63
Štelcl J, Weiss J, Gregerová M, Staněk J, Štelcl J ml (1986) Brněnský masív. Univ. JEP Brno, 255 p
Stromeyer D (1984) Downward continuation of heat flow data by means of the least squares method. Tectonophysics, Amsterdam 103: 55 - 56
Suess FE (1912) Die moravischen Fenster und ihre Beziehung zum Grundgebirge des Hohen Gesenkes. Denkschr (Österr - Akad Wiss) Math - Naturwiss Kl., Wien 88: 541 - 631
Suk M (1983) Petrology of metamorphic rocks. Academia, Praha, 324
Suk M (1986) Das Präkambrium der Böhmischen Masse und des Brunovistulikums in der Tschechoslowakei. Fortschr Mineral, Stuttgart 64: 227 - 234
Suk M (1990) Drilling investigation of the lithosphere in Czechoslovakia. Scientific drilling 1. Springer, Heidelberg New York Berlin, pp 177 - 183
Suk M, Vacek J (1984) Horizontální řezy Českým masívem. Geol Průzk, Praha 26; 8 - 9: 226 - 230
Szénás G (ed-in-chief) (1972) The crustal structure of central southeastern Europe based on the result of explosion seismology, collected papers. Geofizikai Közlemények, Spec Edn, Budapest, 172 p
Takasu A (1985) Prograde eclogites. Terra Cognita, Paris, pp 425 - 426
Talwani M (1973) Computer usage in the computation of gravity anomalies. In: Alder B (ed). Methods in computational physics, 13.. Academic Press, New York, pp 343 - 389
Taylor SR, Gorton MP (1977) Geochemical application of speak source mass spectrometry III. Element sensitivity, precision and accuracy. Geochim Cosmochim Acta, Oxford 41: 1375 - 1380
Texden E (1979) A Pre - Variscan continental rift system in NW Spain. Krystalinikum, Praha 14: 19 - 31
Thill RE, Willard RJ, Bur TR (1969) Correlation of longitudinal velocity variation with rock fabric. J Geophys Res, Washington 74: 4897
Todt WA, Busch WW (1981) U - Pb investigation on zircons from Pre - Variscan gneisses. I. A study from Schwarzwald, West Germany. Geochim Cosmochim Acta, Oxford 45: 1789 - 1801
Tollman J (1985) Das Ausmass der variscischen Deckenbaues im Moldanubikum. Krystalinikum, Praha 18: 117 - 132
Tomek Č (1982) Zdroj západokarpatského tíhového minima a některé nové možnosti pro průzkum Živic. Geol Průzk, Praha 4: 105 - 108
Tomek Č, Herrmann H (1986) Geologická interpretace hustotních profilů v Západních Karpatech. In: Blížkovský M et al. Geofyzikální model litosféry. Záv. zpráva úkolu OFTR ČGÚ č. 82 - 160, 134 - 142. GFÚ ČSAV, Praha - Geofyzika, Brno
Tomek Č, Švancara J, Budík L (1979) The depth and the origin of the West Carpathian gravity low. Earth Planet Sci Lett, Amsterdam 44: 39 - 42
Tomek Č, Dvořáková L, Ibrmajer I, Jiříček R, Koráb T (1987a) Crustal profiles of active continental collisional belt: Czechoslovak deep seismic reflection profiling in the West Carpathians. Geophys J R Astron Soc 89: 383 - 388 Oxford
Tomek Č, Ibrmajer I, Cidlinský K (1987b) Variscan thrust tectonics in the Bohemian Massif - first results revealed reflection seismology - The Rhenish Massif, Berlin, pp 138 - 152
Tomek Č, Ibrmajer I, Koráb T, Biely A, Dvořáková L, Lexa J, Zbořil A (1989) Korové struktury Západních Karpat na hlubinném reflexním seizmickém profilu 2T. Miner Slov, Spišská Nová Ves 21: 3 - 26
Tonika J (1971) Klasifikace metamorfovaných amfibolických hornin mariánsko - lázeňského komplexu. Čas. Mineral Geol, Praha 16: 191 - 194

Trešl J (1989) Stress field in the lithosphere due to topography and density inhomogeneities. Proceedings 6th Int. Symp. "Geodesy and Physics of the Earth", Veroff. Zentralins. Phys Erde, Potsdam 102: 276 - 283

Trešl J, Marvanová V (1988) Stresses and displacements in the lithosphere due to terrain topography. Stud Geoph Geod, Praha 32: 350 - 361

Trešl J, Vyskočil V, Hübner M (1987) Stress in the lithosphere generated by terrain topography and density inhomogeneities in the Earth's crust. Stud Geophys Geod, Praha 31: 145 - 155

Tschermak G (1886) Felsarten von ungewöhnlichen Zusammensetzung in der Umgebung von Teschen und Neutitschein. Sitz Ber Akad Wiss, Wien 12: 126 - 136

Unrung L (1984) Geodynamic Evolution of the Carpathians. Rocz Pol Tow Geol, Kraków 52; 114: 39 - 66

Vacek et al (1983) Prognózní ocenění rudonosnosti Českého masívu. Úst_. úst. geol. Praha, 242 p

Vai GB, Cocozza T (1986) Tentative schematic zonation of the Hercynides in Italy. Bull Geol Soc, Paris 8: 95 - - 114

Van Breemen O, Aftalion M, Bowes D, Dudek A, Mísař Z, Povondra P, Vrána O (1982) Geochronological studies of the Bohemian Massif, Czechoslovakia and their significance in the evolution of Central Europe. Trans Roy Soc, Earth Sci, Edinburgh 73: 89 - 108

Vaněk J (ed) (1979) Geodynamic Investigations in Czechoslovakia, Veda. Bratislava, p 282

Vaňková V, Fiala J, Krešl M, Matějovská O (1981) Radiogenic heat production and distribution of radioactive elements in rocks. In: Zátopek A /ed/ Geophysical Syntheses in Czechoslovakia, Veda. Bratislava, pp 455 - - 466

Veliciu S, Demetrescu C (1979) Heat flow in Romania and some relations to geological and geophysical features. In: Čermák V, Rybach L (eds) Terrestrial Heat Flow in Europe. Springer, Berlin Heidelberg New York, 253 - 260

Vitorello I, Pollack HN (1980) On the variation of continental heat flow with age and the thermal evolution of continents. J Geophys Res, Washington 85: 983 - 995

Vogel DE, Kuipper RP, den Tex E (1982) Eclogiteplagiopyrigarnite relations in the catazonal complexes of NW Spain. Terra Cognita, Paris 314

Vokurka K (1983) Význam vápníku pro geobarometrii lherziolitů. Acta Univ Carol, Praha 4: 344 - 345

Vokurka K (1988) Složení olivínu a izotopy Sr ve vzbraných vulkanitech Českého masivu. MS Úst_ed. ústav geol., Praha

Vokurka K, Povondra P (1983) Geothermometry and geobarometry of lherzalite nodules from Kozakov, NE Bohemia, Czechoslovakia. Acta Univ Carol Geol, Praha 4: 261 - 272

Vyskočil P (1983) Vodorovné gradienty rychlosti svislých pohybů v Českém masivu (research report). VÚGTK, Zdiby

Vyskočil P (1984) Results of recent crustal movement studies: Academia, Praha, 182 p

Vyskočil P (1987) Horizontal recent tectonic deformations in Western Bohemia. Proceedings Earthquake swarm 1985/1986 in Western Bohemia. GFÚ ČSAV, Praha, 143 p

Vyskočil P, Kopecký A (1973) Neotectonics and recent crustal movements in the Bohemian Massif. Edition of VÚGTK, Praha, 82 p

Vyskočil P, Talich M, Zeman A (1987) Rozbory dynamických vlastností v oblasti ostravsko - karvinské pánve a jejím okolí. - Scientific Communication, VÚGTK, Zdiby u Prahy (not published)

Vyskočil V (1971) Correlation analysis of gravity anomalies and isostasy. In: Programme in Czechoslovakia 1962 - 1970. Geophysics, Praha, pp 40 - 46

Vyskočil V (1977) Isostatic properties of the territory of Czechoslovakia and their association with the motions of the Earth's crust. Trav Geophys 1975, Praha 432: 8 - 15

Vyskočil V (1978) On the problem of constructing density models of the lithosphere. Studia Geoph Geod, Praha 22, 402 - 410

Vyskočil V, Burda M (1976) On the computation of the gravitational effect if threedimensional density models of the Earth's crust. Stud Geophys Geod, Praha 20: 213 - 218

Vyskočil V, Burda M, Plančár J (1983) On the manifestation of shallow and deep density inhomogeneities in the gravity field of the Carpathians. Geologica Carpatica, Bratislava 34: 429 - 438

Vyskočil V, Burda M, Kožíšková M (1985) On the gravitational effect of the crust - mantle boundary in Central Europe. Trav Géophys 1982, Academia, Praha, pp 9 - 31

Vyskočil V, Burda M, Bielik M, Fusán O. (1991) Further density models of the Western Carpathians. Contrib Geophys Inst Slov Acad Sci, V.22. Veda, Bratislava (in press)

Waldhauserová J (1984) Proterozoic volcanics and intrusive rocks of the Jílové zone in Central Bohemia. Krystalinikum, Praha 17: 77 - 97
Weber K (1986) Metamorphisme and crustal rheology - implications for the structural development of the continental crust during prograde metamorphism. - In: The Nature of the Lower Continental Crust. Geol Soc Spec Public, New York 24; 1986: 95 - 106
Weber K, Behr HJ (1983) Geodynamic interpretation of the Middle European Variscides. In: Martin H, Eder FW (eds) Intracontinental fold belts, Berlin. pp 427 - 469
Weiss J (1977) Fundament moravského bloku ve stavbě evropské platformy. Folia Univ Purkyn Brun Geol, Praha 18: 13, 1 - 64
Weiss J (1978) Position of the moravian ultrabasites in relation to the intracrustal geological structure. Sbor Geol Věd Geol, Praha 31: 127 - 154
Weiss J (1983) Intracrustal manifestations of the basalt layer in the development of the Moravian block. Miner Slov, Spišská Nová Ves 16; 1: 51 - 60
Werner CD (1981) Sächsische Granulitgebirge - Saxonian granulite massif. In: Ophiolites and initialites of northern border of the Bohemian Massif, Freiberg 1: 129 - 161
Wielandt E, Knopoff L (1982) Dispersion of very long - period Rayleigh waves along the east Pacific Rise: evidence for S - wave velocity anomalies to 450 km depth. J Geophys Res 87: 8631 - 8641
Wielandt E, Sigg A, Plešinger A, Horálek J (1987) Deep structure of the Bohemian Massif from phase velocities of Rayleigh and Love waves. Studia Geoph Geod Praha 31: 1 - 7
Wielandt E, Plešinger A, Sigg A, Horálek J (1988) Anisotropy in the uppermost mantle under Bohemia? In: Nolet G. Dost B Proc. of the 4th Workshop on the European Geotraverse Project. ISF, Strasbourg, pp 145 - 147
Wieladek P, Ernst T (1977) Application of the method of least squares to determining impulse response and transfer functions. Publ Inst Geophys, Warszawa 110: 3 - 12
Wimmenauer W, Ádám A (1985) High - pressure fenomena in acid rocks accompanying eclogites in the Schwarzwald, West Germany. Terra Cognita, Paris 426
Yegorov EL, Chrnyak EL, Palshin NA, Demidova TA, Kaikkonen P (1983) Numerical thin - sheet modelling of the telluric field distortion by the Hybrid technique. In: Hjelt SE, Vanyan LL (eds) The development of the deep geo-electrical model of the Baltic Shield. Part I, Numerical methods. Rep. No 7, Univ. Oulu 39 p
Yoder HS, Tilley C (1962) Origin of basalt magmas, and experimental studies of natural and synthetic rock systems. J Petrol Technol, Dallas 3: 342 - 352
Zahradník J, Čermák V (1979) Two - dimensional correlation of geophysical data applied to the heat flow and earth's crust thickness patterns in Czechoslovakia. Stud Geophys Geod, Praha 23: 251 - 262
Zahradník J, Vavryčuk V, Jánský J, Zedník J (1988) Focal mechanisms of selected events of West Bohemia earthquake swarm 1985/86 constrained by P - wave amplitudes. Proc. Workshop on Induced Seismicity and Associated Phenomena, Liblice, 1988, I, Geoph. Inst., Czech. Acad. Sci. Praha, pp 271 - 289
Zátopek A (ed) (1981) Geophysical Syntheses in Czechoslovakia. Veda. Bratislava, 497 p
Zátopek A (1984) Stavba zemské kůry na území ČSSR a geofyzikální syntéza na podkladě explozívní seizmologie. - Zemská kora a jej vzťah k nerostným surovinám. Geol. úst. D. Štúra, Bratislava, pp 33 - 40.
Zátopek A, Beránek B (1974) Correlation of geophysical data with structural phenomena of Central Europe. Sbor Geol Věd Geol, Praha 26: 113 - 124
Zeman J (1981a) Kritéria pro prognózování skrytých rudních ložisek založené na výzkumu hlubinné stavby a vývoje kůry. Geol Průzk, Praha 3: 65 - 68
Zeman J (1981b) Hlubinné stavby, vývoj kůry Českého masívu a možnosti prognózování ložisek. Geol Průzk, Praha 7: 199 - 200
Zoubek V (1967) Některé kritické problémy krystalinika Českého masívu. Čas Mineral Geol, Praha 12; 2: 151 - - 155
Zoubek V (ed) (1971) Upper Mantle project programme in Czechoslovakia 1962-1970. Geology-Final Report. 162. Academia, Praha
Zoubek V (1972) Quelques problémes et résultats de la corrélation géologique du socle précambrien de l'Europe varisque et alpine. Not Mém Serv Géol, Rabat 236, 177 - 192
Zoubek V (ed) (1988) Precambrian in younger fold belts. J. Wiley Chichester, 885 p

Zoubek V, Vyskočil V (1977) La champ gravimétrique anomal en relation aux structures géologiques de l'Europe centrale in la chaine Variscique d'Europe moyenne et occidentale; colloque international du CNRS, 243: 358 - 369

Zverev A, Tulina I (1974) Peculiarities in the deep structure of the Sakhalin - Hokkaido - Primorye. Tectonophysics Amsterdam 20: 111 - 128

MIX
Papier aus verantwortungsvollen Quellen
Paper from responsible sources
FSC® C105338

If you have any concerns about our products,
you can contact us on
ProductSafety@springernature.com

In case Publisher is established outside the EU,
the EU authorized representative is:
Springer Nature Customer Service Center GmbH
Europaplatz 3, 69115 Heidelberg, Germany

Printed by Libri Plureos GmbH
in Hamburg, Germany